# Advances in Hydro-Geological Research in Arid and Semi-Arid Areas

# Advances in Hydro-Geological Research in Arid and Semi-Arid Areas

Topic Editors

**Ahmed Elbeltagi**
**Quanhua Hou**
**Bin He**

MDPI

Basel • Beijing • Wuhan • Barcelona • Belgrade • Novi Sad • Cluj • Manchester

*Topic Editors*
Ahmed Elbeltagi
Agricultural Engineering
Mansoura University
Mansoura
Egypt

Quanhua Hou
School of Architecture
Chang'an University
Xi'an
China

Bin He
Institute of
Eco-Environmental and Soil
Sciences
Guangdong Academy of
Sciences
Guangzhou
China

*Editorial Office*
MDPI AG
Grosspeteranlage 5
4052 Basel, Switzerland

This is a reprint of the Topic, published open access by the journals *Hydrology* (ISSN 2306-5338), *Water* (ISSN 2073-4441), *Atmosphere* (ISSN 2073-4433), *Agriculture* (ISSN 2077-0472) and *Geosciences* (ISSN 2076-3263), freely accessible at: https://www.mdpi.com/topics/RWO6EU12C6.

For citation purposes, cite each article independently as indicated on the article page online and as indicated below:

Lastname, A.A.; Lastname, B.B. Article Title. *Journal Name* **Year**, *Volume Number*, Page Range.

**ISBN 978-3-7258-5947-4 (Hbk)**
**ISBN 978-3-7258-5948-1 (PDF)**
**https://doi.org/10.3390/books978-3-7258-5948-1**

# Contents

**About the Editors** . . . . . . . . . . . . . . . . . . . . . . . . . . . . . . . . . . . . . . . . . . . . . . . **vii**

**Sha Sha, Lijuan Wang, Die Hu, Yulong Ren, Xiaoping Wang and Liang Zhang**
Agricultural Drought Model Based on Machine Learning Cubist Algorithm and Its Evaluation
Reprinted from: *Hydrology* **2024**, *11*, 100, https://doi.org/10.3390/hydrology11070100 . . . . . . **1**

**Shiang-Jen Wu and Han-Yuan Yang**
Modeling Reliability Analysis for the Branch-Based Irrigation Water Demands Due to Uncertainties in the Measured Surface Runoff
Reprinted from: *Agriculture* **2024**, *14*, 1107, https://doi.org/10.3390/agriculture14071107 . . . . **20**

**Yuxiang Shao, Buqing Yan, Lubaiyang Liu, Xiao Yu, Gang Feng, Kun Zhang and Kang Gong**
Hydrochemical Characteristics and Quality Evaluation of Irrigation and Drinking Water in Bangong Co Lake Watershed in Northwest Tibetan Plateau
Reprinted from: *Water* **2023**, *15*, 2655, https://doi.org/10.3390/w15142655 . . . . . . . . . . . . . **47**

**Man Li, Shanshan Li, Huancai Liu and Junjie Zhang**
Balancing Water Ecosystem Services: Assessing Water Yield and Purification in Shanxi
Reprinted from: *Water* **2023**, *15*, 3261, https://doi.org/10.3390/w15183261 . . . . . . . . . . . . . **69**

**Jiamin Huang, Kebin Shi, Xingpeng Shi, Guocheng Hao and Yimin Yang**
Arid AREAS Water-Piled Photovoltaic Prevents Evaporation Effects Research
Reprinted from: *Water* **2023**, *15*, 3716, https://doi.org/10.3390/w15213716 . . . . . . . . . . . . . **85**

**Sisay S. Mekonen, Scott E. Boyce, Abdella K. Mohammed and Markus Disse**
Using an Open-Source Tool to Develop a Three-Dimensional Hydrogeologic Framework of the Kobo Valley, Ethiopia
Reprinted from: *Geosciences* **2024**, *14*, 3, https://doi.org/10.3390/geosciences14010003 . . . . . . **100**

**Zebin Luo, Mingbo Yang, Xiaocheng Zhou, Guiping Liu, Jinlong Liang, Zhe Liu, et al.**
Evaluation of Various Forms of Geothermal Energy Release in the Beijing Region, China
Reprinted from: *Water* **2024**, *16*, 622, https://doi.org/10.3390/w16040622 . . . . . . . . . . . . . **127**

**Yaling Chen, Jun Wen, Xianhong Meng, Qiang Zhang, Xiaoyue Li, Ge Zhang and Run Chen**
An Assessment of the Coupled Weather Research and Forecasting Hydrological Model on Streamflow Simulations over the Source Region of the Yellow River
Reprinted from: *Atmosphere* **2024**, *15*, 468, https://doi.org/10.3390/atmos15040468 . . . . . . . . **141**

**Akida Salam, Qing He, Alim Abbas, Tongwen Wu, Jie Zhang, Weihua Jie and Junjie Liu**
Boundary Layer Height and Trends over the Tarim Basin
Reprinted from: *Atmosphere* **2024**, *15*, 541, https://doi.org/10.3390/atmos15050541 . . . . . . . . **159**

**Lucas Bouba, Pierre-Alain Ayral and Sophie Sauvagnargues**
Landscape Drivers of Floods Genesis (Case Study: Mayo Mizao Peri-Urban Watershed in Far North Cameroon)
Reprinted from: *Water* **2024**, *16*, 1672, https://doi.org/10.3390/w16121672 . . . . . . . . . . . . . **173**

**John A. Webb, Jaafar Jotheri and Rod J. Fensham**
Springs of the Arabian Desert: Hydrogeology and Hydrochemistry of Abu Jir Springs, Central Iraq
Reprinted from: *Water* **2024**, *16*, 2491, https://doi.org/10.3390/w16172491 . . . . . . . . . . . . . **192**

**Lijuan Fan, Lanxin Liu, Jing Hu, Fen Zhao, Chunhui Li and Yujun Yi**
A Long-Term Evaluation of the Ecohydrological Regime in a Semiarid Basin: A Case Study of the Huangshui River in the Yellow River Basin, China
Reprinted from: *Hydrology* **2024**, *11*, 168, https://doi.org/10.3390/hydrology11100168 . . . . . . **210**

**Abdelfattah Elsayed Elsheikh, Mahmoud A. El Ammawy, Nessrien M. Hamadallah, Sedky H. A. Hassan, Sang-Eun Oh, Kotb A. Attia and Mahmoud H. Darwish**
The Surface Water Potentiality in Arid and Semi-Arid Basins Using GIS and HEC-HMS Modeling, Case Study: Gebel El Sibai Watershed, Red Sea
Reprinted from: *Water* **2024**, *16*, 3111, https://doi.org/10.3390/w16213111 . . . . . . . . . . . . . **227**

**Xiangqian Wei, Yi Liu, Cong Cheng, Xinyu Chang and Jun Guo**
Topographic Precipitation Diagnosis: Model Design and Validation in a Two-Dimensional Context
Reprinted from: *Atmosphere* **2025**, *16*, 593, https://doi.org/10.3390/atmos16050593 . . . . . . . . **251**

**Yuanhong You, Yanyu Lu, Yu Wang, Houfu Zhou, Ying Hao, Weijing Chen and Zuo Wang**
Sensitivity of Soil Moisture Simulations to Noah-MP Parameterization Schemes in a Semi-Arid Inland River Basin, China
Reprinted from: *Agriculture* **2025**, *15*, 2286, https://doi.org/10.3390/agriculture15212286 . . . . **266**

# About the Editors

**Ahmed Elbeltagi**

Ahmed Elbeltagi is an Assistant Professor in the Agricultural Engineering Dept., Faculty of Agriculture, Mansoura University, Egypt, with a Ph.D. in Environmental and Resource Sciences from Zhejiang University (top 50 in QS World University Rankings). He has an interdisciplinary background spanning environmental science, engineering, and advanced technologies, focusing on agricultural water management, soil and water conservation, hydrological risk assessment, water footprint modeling, and sustainable water resource planning. He applies remote sensing, GIS, hydrological modeling, simulation tools, and AI and machine learning in areas such as irrigation efficiency and watershed management, integrating climate change impact analysis and environmental sustainability metrics to promote system-based resource conservation. His achievements include authoring over 200 SCI-indexed papers; inclusion in Stanford University's Top 2% Scientists list (2024); receiving two Egyptian National State Prizes in Science (2022/2023); participation in international collaborative projects across Egypt, China, and Hungary (on climate-resilient agriculture, urban flooding, and wastewater reuse); and holding editorial/reviewer roles in journals like *Remote Sensing*, *Journal of Hydrology*, and *Science of the Total Environment*. He has also co-authored two international books and several book chapters with Springer and Taylor & Francis, maintaining a strong academic network via collaborations and editorial work.

**Hou Quanhua**

Hou Quanhua (b. 1972, Guang'an, Sichuan), a professor, doctor, and national-registered planner, serves as Dean of the School of Architecture at Chang'an University. Engaged in urban–rural planning teaching, research, and practice, he focuses on urban planning compilation techniques, land and transportation, and rural planning (recently emphasizing rural planning and urban land-transport collaboration). He has published over 50 papers (SCI/EI/CSSCI/CSCD indexed), 1 monograph, and 1 edited textbook; presided over 10 international, national, and provincial-ministerial research projects; participated in 5 such projects; and led/joined over 100 planning practice projects, with total funding exceeding CNY 10 million. His honors include the Shaanxi Provincial Science and Technology Progress Second Class Award and over 10 industry awards. He teaches undergraduates, postgraduates, and doctoral students, and has been named Excellent Teacher and Advanced Scientific and Technological Individual multiple times. His social roles include memberships in the Academic Committee of Urban Transportation Planning (Urban Planning Society of China) and Shaanxi Urban–Rural Planning Expert Committee, deputy director of Shaanxi Rural and Town Construction Expert Committee, and expert roles in Guangdong and Shaanxi Science and Technology Departments' consulting/evaluation databases.

**Bin He**

Bin He has more than 20 years of experience in hydrological engineering and water pollution control. He holds a PhD in "Hydrology for Environmental Engineering" from Ehime University in Japan and subsequently obtained the Research Fellow Award from Japan Society for the Promotion of Science (JSPS). Following that, he has worked as an Assistant Professor at the University of Tokyo (Japan), an Associate Professor at Kyoto University (Japan), and a Professor at the Chinese Academy of Sciences. His main interests include GIS-based hydrological modeling for nonpoint source pollution (NPS), developing engineering techniques for water pollution control. He has published more than 100 papers in journals, 3 books, and 20 patents in environmental engineering.

He is the leader of more than 30 domestic and international projects, an Associate Editor-in-Chief of the Journal *Geoenvironmental Disasters,* and Editor of journals *Landslides* and *Hydrological Research Letter.* He also works as a reviewer for more than 50 international journals including *Water Research, Hydrology, Earth System Sciences,* etc.

hydrology MDPI

Article

# Agricultural Drought Model Based on Machine Learning Cubist Algorithm and Its Evaluation

Sha Sha, Lijuan Wang *, Die Hu, Yulong Ren, Xiaoping Wang and Liang Zhang

Key Laboratory of Arid Climatic Change and Reducing Disaster of Gansu Province, Key Open Laboratory of Arid Climatic Change and Disaster Reduction of CMA, Institute of Arid Meteorology, CMA, Lanzhou 730020, China; shas@iamcma.cn (S.S.); hud@iamcma.cn (D.H.); renyl@iamcma.cn (Y.R.); wangxp@iamcma.cn (X.W.); zhangl@iamcma.cn (L.Z.)
* Correspondence: wanglj@iamcma.cn

**Abstract:** Soil moisture is the most direct evaluation index for agricultural drought. It is not only directly affected by meteorological conditions such as precipitation and temperature but is also indirectly influenced by environmental factors such as climate zone, surface vegetation type, soil type, elevation, and irrigation conditions. These influencing factors have a complex, nonlinear relationship with soil moisture. It is difficult to accurately describe this non-linear relationship using a single indicator constructed from meteorological data, remote sensing data, and other data. It is also difficult to fully consider environmental factors using a single drought index on a large scale. Machine learning (ML) models provide new technology for nonlinear problems such as soil moisture retrieval. Based on the multi-source drought indexes calculated by meteorological, remote sensing, and land surface model data, and environmental factors, and using the Cubist algorithm based on a classification decision tree (CART), a comprehensive agricultural drought monitoring model at 10 cm, 20 cm, and 50 cm depth in Gansu Province is established. The influence of environmental factors and meteorological factors on the accuracy of the comprehensive model is discussed, and the accuracy of the comprehensive model is evaluated. The results show that the comprehensive model has a significant improvement in accuracy compared to the single variable model, which is a decrease of about 26% and 28% in RMSE and MAPE, respectively, compared to the best MCI model. Environmental factors such as season, DEM, and climate zone, especially the DEM, play a crucial role in improving the accuracy of the integrated model. These three environmental factors can comprehensively reduce the average RMSE of the comprehensive model by about 25%. Compared to environmental factors, meteorological factors have a slightly weaker effect on improving the accuracy of comprehensive models, which is a decrease of about 6.5% in RMSE. The fitting accuracy of the comprehensive model in humid and semi-humid areas, as well as semi-arid and semi-humid areas, is significantly higher than that in arid and semi-arid areas. These research results have important guiding significance for improving the accuracy of agricultural drought monitoring in Gansu Province.

**Keywords:** comprehensive agricultural drought index (CADI); CART; relative soil moisture; DEM; climate zone

## 1. Introduction

Soil moisture is an important indicator for monitoring agricultural drought, and methods for obtaining soil moisture include soil drilling, remote sensing inversion, numerical model simulation, and data assimilation. The most accurate method for obtaining soil moisture is the soil drilling and weighing method, but this method is based on single-point observation, with limited spatial representativeness and very low efficiency in obtaining data. With the development of remote sensing science and technology, remote sensing technology is widely used for soil moisture surface layer monitoring due to its wide spatial coverage and high spatial resolution [1]. It overcomes the shortcomings of site observation. Land surface models or assimilation systems can also obtain continuous soil moisture from surface layer to root zone layer based on numerical simulation or data assimilation techniques [2,3]. However, the spatial resolution of soil moisture simulated by this method is low, and it is difficult to verify due to large spatial scale differences; furthermore, it is also difficult to meet the spatial resolution requirements of provincial-level agricultural drought monitoring. Gansu Province is located in the inland region of China, and has a severe water scarcity [4]. Most of the area is located in the climate transition zone of China [5], and it is also a sensitive area for global climate change [6]. The east–west elevation drop of Gansu Province is nearly 5 km, and the geographical environment is very complex, which poses great difficulties for soil moisture estimation based on remote sensing and land surface models or data assimilation technology, making the accuracy of soil moisture estimation in Gansu Province not ideal. Further improving the accuracy of soil moisture estimation has extremely important guiding significance for agricultural production in Gansu Province.

Soil moisture is an important factor in the water cycle mechanism [7], which is influenced by the atmospheric system, but it also affects the atmospheric system through feed-back from vegetation, soil, and other factors [8]. The physical mechanisms that affect soil moisture are very complex. The amount of precipitation and the temperature directly affect the soil moisture, and it is also indirectly affected by environmental factors, such as climate zone, surface vegetation type, soil type, altitude, irrigation conditions, etc. [9]. It is often difficult to describe soil moisture comprehensively and accurately with a single index. The application of multi-source data has been a research hotspot in recent years for agricultural drought monitoring based on soil moisture. Multivariate statistical modeling is an important method for establishing a comprehensive drought monitoring model [10–16]. However, these methods often insufficiently consider environmental factors, and the determination of weight coefficients is somewhat artificial, resulting in limited spatiotemporal applicability of this method. In recent years, the rapidly developing machine learning (ML) methods [17] have had strong nonlinear mapping capabilities, providing new methods for solving nonlinear problems such as soil moisture. They are currently the main method for drought monitoring based on multi-source data. Because they have fewer parameters, faster modeling speed, and higher accuracy, ML algorithms based on decision tree (DT) classification are most widely used to estimate soil moisture [18–23]. Previous studies have focused more on using ML algorithms to estimate soil moisture, and no matter which drought index is considered, environmental factors such as digital elevation models (DEMs), climate zone, land cover types, irrigation, etc. are important factors that DT models have considered. However, there have been few reports on the evaluation of environmental factors of ML algorithms in previous studies. In this paper, based on multi-source data, such as meteorological data, remote sensing data, numerical model data, and environmental factors, an appropriate number of indexes are selected according to the results of cross-verification, and a comprehensive agricultural drought monitoring model is established by using the Cubist algorithm based on classification and regression trees (CARTs). On the basis of evaluating the influence of environmental and meteorological factors on the accuracy of the comprehensive model, the accuracy of the comprehensive model is evaluated. The results aim to provide a new technology for agricultural drought monitoring in Gansu Province, so as to improve the accuracy of agricultural drought monitoring.

## 2. Materials and Methods

### 2.1. Study Area

The natural environment in Gansu Province is very complex. Gansu Province is located on the Loess Plateau in western China, in the upper reaches of the Yellow River. It is the intersection of the Loess Plateau, Qinghai Tibet Plateau, and Mongolian Plateau, with a sloping terrain from southwest to northeast and a narrow terrain from east to west (as shown in Figure 1a). However, due to the proximity of Gansu Province to the Qinghai Tibet Plateau to the west, the elevation drops from Gannan, in the west, to Pingliang and Qingyang, in the east, is nearly 5 km, which leads to a significant difference in temperature, precipitation, climate types, and vegetation types between the west and east. Gansu

Province has a rich variety of vegetation types, including cultivated vegetation, forest land, grassland, desert, and a total of 12 vegetation types (as shown in Figure 1b). The climate of Gansu Province is complex, mainly including arid, semi-arid, semi-arid and humid, humid, and semi-humid regions (as shown in Figure 1c) [24]. Gansu Province is dry, with little rainfall and high evapotranspiration, resulting in significant spatiotemporal differences in precipitation [25]. In terms of agricultural cultivation, the east area of the Yellow River in Gansu Province belongs to a rain-fed agricultural area called Hedong area, and crop growth relies entirely on natural precipitation. The western area of the Yellow River, which is called the Hexi area, is an irrigated agricultural area due to very little natural precipitation, and crop growth mainly relies on irrigation.

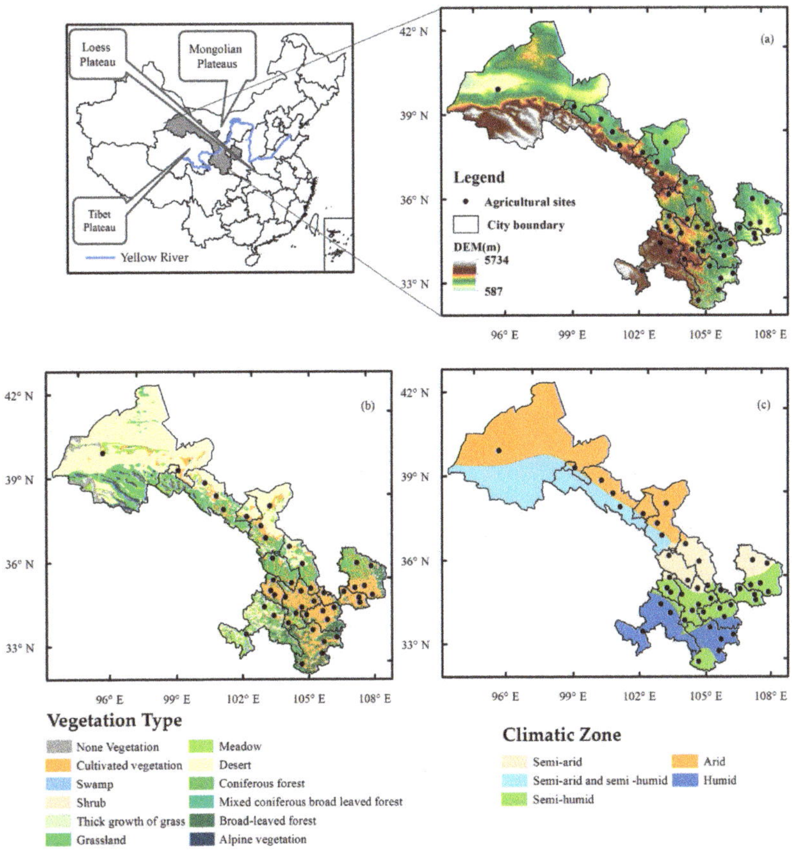


**Figure 1.** Geographical location (**a**), vegetation type (**b**), and climate zone diagram of Gansu Province (**c**).

*2.2. Dataset Descriptions*

2.2.1. Relative Soil Moisture

This study uses relative soil moisture (RSM) as a monitoring indicator for agricultural drought and as the dependent variable of the comprehensive agricultural drought monitoring model based on the Cubist algorithm. The RSM data used is the ten-day soil relative humidity data of 44 stations in Gansu Province from February 2003 to November 2016, with a depth of 10, 20, and 50 cm. The majority of the station data is up to 2012, and the average value is calculated from the three values to the monthly RSM data. The monthly soil relative humidity datasets at depths of 10, 20, and 50 cm are respectively referred to as RSM_10, RSM_20, and RSM_50. Drought classification based on RSM is shown in Table 1 [26].

**Table 1.** Drought classification based on RSM.

| Agricultural Drought Grade | Classification |
|---|---|
| None drought D1 | 60% < RSM |
| Light drought D2 | 50% < RSM ≤ 60% |
| Moderate drought D3 | 40% < RSM ≤ 50% |
| Severe drought D4 | 30% < RSM ≤ 40% |
| Extreme drought D5 | RSM ≤ 30% |

#### 2.2.2. Meteorological Data

From a physical mechanism perspective, precipitation and temperature are the most important factors affecting soil moisture, while radiation, wind speed, etc. are also non-ignorable factors affecting soil moisture. Sunlight hours are the main factor in calculating radiation, so the meteorological data used in this article include the multiple time scales Standard Precipitation Index (SPI_1, SPI_3, SPI_6, SPI_9), meteorological drought composite index (MCI) [25], days of no rain, max days of no rain (DNR_max), temperature anomaly (TA), wind speed anomaly (WSA), relative humidity anomaly (RHA), sun hour anomaly (SHA), etc., with a time range of 2000 to 2018.

#### 2.2.3. Remote Sensing Data

The remote sensing drought index used in the study includes two types: optical and microwave, with a time range of 2000–2018.The vegetation condition index (VCI) [27], temperature condition index (TCI) [28], and temperature vegetation drought index (TVDI) [29] were calculated using MODIS data (MOD09A1 and MO-D11A2) (https://search.earthdata.nasa.gov/search/granules?p=C2343111356-LPCLOUD&pg[0][v]=f&pg[0][gsk]=-start_date&q=MOD09A1&tl=1523328714!4!!, https://search.earthdata.nasa.gov/search/granules?p=C2269056084-LPCLOUD&pg[0][v]=f&pg[0][gsk]=-start_date&q=MOD11A2&tl=1523328714!4!!, (accessed on 11 May 2024)) as optical remote sensing soil moisture indicators, with a spatial resolution of 500 m and 1 km. Using the active–passive microwave fusion soil moisture dataset of version 04.5 from the Climate Change Initiative (CCI) project of the European space agency (ESA) [30,31] (https://catalogue.ceda.ac.uk/uuid/38b8e5e524e1449ab4b4994970752644 (accessed on 11 May 2024)), extract its volumetric moisture content (v%) and calculate its percentile [32] based on historical data, as the soil moisture index based on microwave remote sensing, which is denoted as SM_CCI, with a spatial resolution of 25 km.

#### 2.2.4. Land Surface Model Data

This article uses soil moisture products of several land surface modes, including community atmosphere biosphere land exchange (CABLE) and Noah and variable infiltration capacity (VIC) of the global land data association system (GLDAS) over the period of 2000–2018. The CABLE model is a land surface model developed by the commonwealth scientific and industrial research organization (CSIRO) in Australia, which has been well applied to drought monitoring in China [33,34]. The percentile of soil moisture in the model was used as a soil moisture indicator based on the land surface model, denoted as SM_ CABLE. The GLDAS is a joint project between the National Aeronautics and Space Administration (NASA), the National Centers for Environmental Prediction (NCEP), and the National Oceanic and Atmospheric Administration (NOAA) of America [35,36]. GLDAS adopts advanced data assimilation technology to integrate satellite observation data and ground-based observant ion data into a unified model. Currently, GLDAS includes four land surface models, including Noah, Mosaic, community land model (CLM), and VIC land surface models. This study uses the latest version 2.1 of the Noah model soil moisture (in kg/m$^2$) at depths of 0–10, 10–40, and 40–100 cm (https://disc.gsfc.nasa.gov/datasets/GLDAS_NOAH025_M_2.1/summary?keywords=GLDAS, (accessed on 11 May 2024)) and VIC model soil moisture (in kg/m$^2$), at depths of 0–30 cm, 18–27 cm, 50–400 cm, and root zone (https://disc.gsfc.nasa.gov/datasets/GLDAS_VIC10_M_2.1/summary?keywords=GLDAS, (access on 11 May 2024)), with a monthly time scale from 2000 to 2018. The spatial resolutions of Noah and VIC models are 0.25° and 1°, respectively. After extracting the data, calculate the percentage of anomaly (PA) separately. The PA of soil moisture at depths of 0–10, 10–40, and 40–100 cm in Noah mode is denoted as NOAH_PA_10, NOAH_PA_40, NOAH_PA_100, respectively. The PA at depths of 0–30 cm, 18–27 cm, 50–400 cm, and root zone in the VIC mode are recorded as VIC_PA_30, VIC_PA_d3, VIC_PA_d2, and VIC_PA_root, respectively.

#### 2.2.5. Environmental Data

In such a complex environment as Gansu Province, the spatiotemporal applicability of various drought monitoring indexes is very different. Therefore, when establishing a comprehensive agricultural drought monitoring model, this article mainly considers environmental factors such as season, DEM, climate zone, vegetation type, and the presence or absence of irrigation. They are denoted as Season, Envi_DEM, Envi_ClimateZone, Envi_VegeType, and Envi_Irrigation, respectively.

### 2.3. Method

#### 2.3.1. CART Algorithm

The CART algorithm [21] is based on the "recursive binary segmentation" method, which constructs a decision tree by gradually decomposing the dataset into two subsets. For the regression problem, the algorithm uses the criterion of minimizing the square difference to find a boundary point in dataset D, which is used to divide D into two parts: D1 and D2, and to minimize the square difference of each part in datasets D1 and D2. Then, find similar boundary points in D1 and D2, respectively, and continue cycling until the termination condition is met. To prevent overfitting, it is also necessary to prune the generated tree model to obtain the optimal decision tree. The leaf nodes of the decision tree correspond to a predicted value.

#### 2.3.2. Random Forest Algorithm

In order to address the issue of overfitting in CARTs, Breiman [37] proposed random forest (RF) in 2001, which is an ensemble learning method for CARTs. RF conducts n random samplings with replacement of the dataset, and each sampling establishes a CART. RF establishes n CARTs, and finally uses the average of n CARTs prediction results as the final prediction result of RF. Therefore, like the CART algorithm, each leaf node in the RF tree corresponds to a predicted value.

#### 2.3.3. Cubist Algorithm

Cubist is also a CART-based algorithm [38,39]. Unlike CART and RF, the leaf nodes of Cubist are not predicted values, but rather a regression equation. Cubist generates a series of rules at leaf nodes of trees for each rule, such as "if condition x is met, then use the associated regression model", for example:

Rule 1:
If
Season in {Winter, Spring}
Envi_DEM <= 1382
Envi_ClimateZone in {semi-arid zone, humid zone, semi humid zone}
MCI <= −0.049
Then
Output = −66.5 + 0.0788 × Envi_DEM + 10.7 × MCI + 7.8 × SPI1 + 10.6 × SPI9 + 53 × TVDI

#### 2.3.4. Evaluation Method

All of the analysis of the algorithm in this article was completed in the Python environment. There are many indexes to be considered in the comprehensive model, but not all indexes may improve the accuracy of the comprehensive model. In order to find the optimal combination, this study uses the 5-fold cross-test method to evaluate various combinations and parameter settings before establishing the comprehensive model. Based on the results of cross validation, the indexes and parameter settings of the comprehensive model were determined.

This article evaluates the accuracy of the constructed comprehensive model using indicators such as correlation coefficient R, determination coefficient $R^2$, root mean square error (RMSE), and mean absolute percentage error (MAPE).

## 3. Results

### *3.1. Construction of Comprehensive Model*

#### 3.1.1. Correlation between Multi-Source Drought Indexes

Figure 2 shows a heat map of the correlation coefficients between various drought indexes and RSM at a depth of 10 cm. It shows that there is strong collinearity between MCI, SPI3, and SPI6, and there is also strong collinearity between soil moisture anomalies at different depths in the GLDAS NOAH and VIC modes. Indexes, except for WSA and VIC_PA_30, are significantly correlated with RSM. The meteorological index, especially the MCI index, has a better correlation with RSM, but its correlation coefficient R is 0.45, and the determination coefficient $R^2$ is about 0.2. In the meteorological index, RHA and SHA also show significant positive and negative correlation with RSM, followed by the correlation between optical remote sensing indexes and RSM. The correlation between the land surface model drought index and RSM is relatively the worst. It is obvious that it is difficult for a single index to accurately describe the temporal and spatial distribution characteristics of RSM. To simplify the model, collinear variables with poor correlation with RSM at different depths were removed.

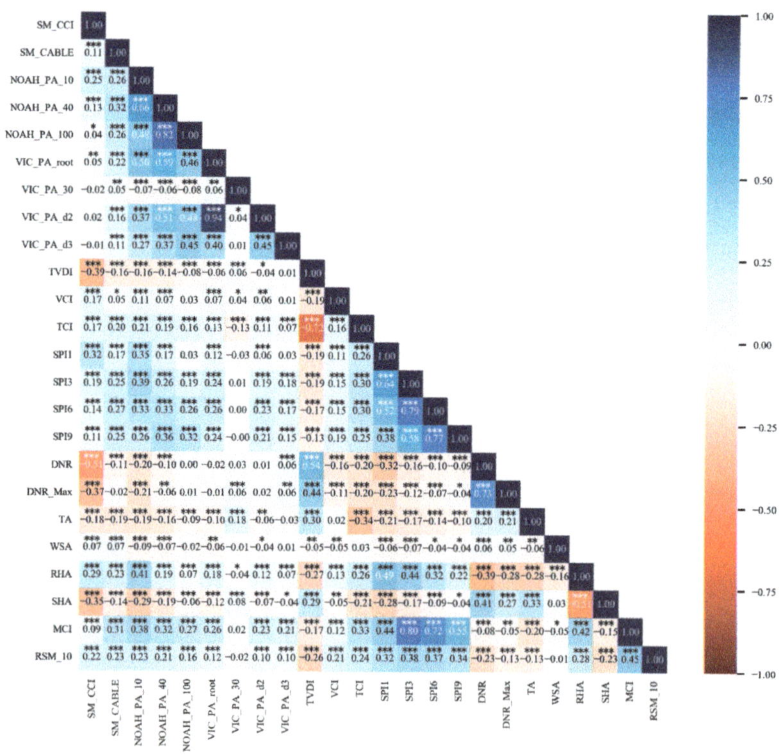

**Figure 2.** Heat map of the correlation between drought indexes and RSM at a depth of 10 cm. In the figure, * represents $p < 0.05$, ** represents $p < 0.01$, *** represents $p < 0.001$.

#### 3.1.2. Selection of Algorithm

In the study, CART, Cubist and random forest (RF) algorithms, which were based on CARTs, were compared by using different numbers of input variables. $R^2$ of these three algorithms on the training and test datasets is shown in Figure 3. It shows that although the $R^2$ of the CART model in the training set is close to 1, the $R^2$ on the test set is negative, indicating that the CART model performs very poorly on the test set and is severely overfitting. Although RF has significantly improved on the test set compared to CARTs, but RF is also significantly overfitting. Whereas the Cubist algorithm has similar $R^2$ on the training and testing sets, it still performs well on the testing set and is better than RF. Therefore, the Cubist algorithm was chosen for estimating the soil moisture.

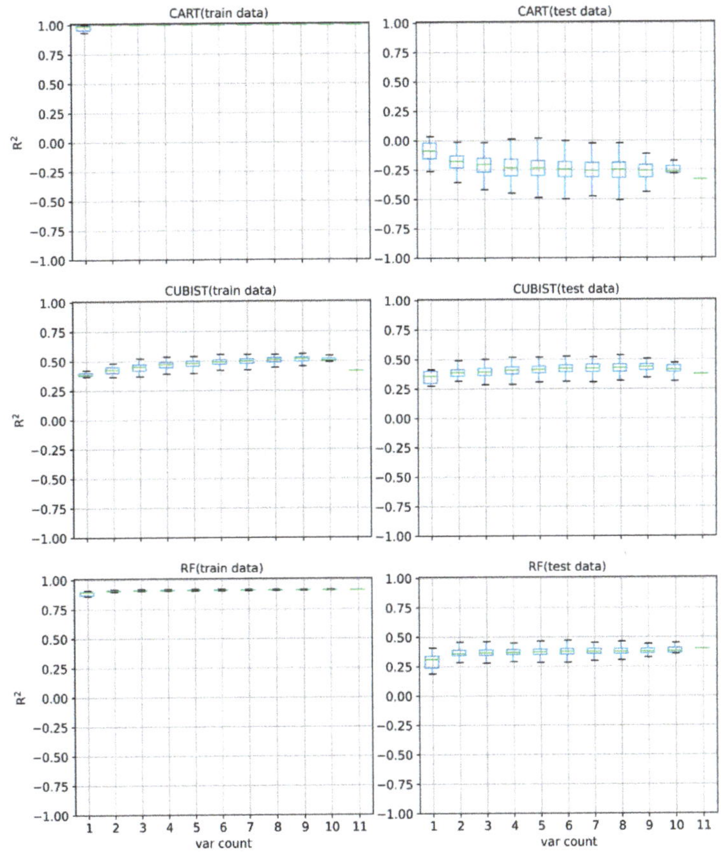


**Figure 3.** The box diagram of R2 of CART, Cubist, and RF in training and test data sets by using different numbers of input variables.

#### 3.1.3. Selection of Environmental Factors

In the study, environmental factors such as season, DEM, climate zone, and irrigation type were considered. Taking a depth of 10 cm as an example, Figure 4 shows the $R^2$ for cross validation of models considering different environmental factors. It shows that without considering any environmental factors, the $R^2$ of the model on the test set increased with the number of variables, but the highest did not exceed 0.3, and the model performance was low. After adding season factors, the model performance was improved, but the improvement was not significant. After adding the DEM factor to the model, the performance of the model was significantly improved, and the impact of the number of variables on the model performance decreased. After adding the climate zoning factor to the model, the performance slightly improved. After adding the irrigation zoning factor, the model performance no longer increased. In addition, the accuracy of the comprehensive model increased slowly with an increase in the number of variables. After the number of variables exceeded 6, the accuracy of the model no longer increased. Overall, season, DEM, and climate zone environmental factors can improve model performance, especially the DEM factor. The results for depths of 20 cm and 50 cm are similar. Therefore, when constructing a comprehensive model, environmental factors such as season, DEM, and climate zone will be considered.

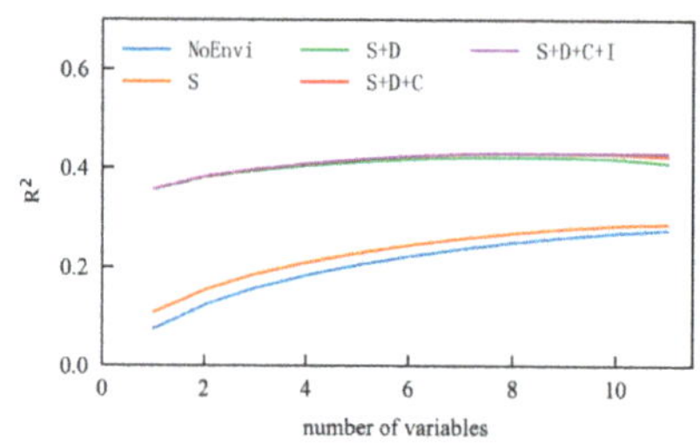


**Figure 4.** Cross-validation results of model accuracy under different environmental factors at a depth of 10 cm (In the figure, "NoEnvi" represents no environmental factors, while "S", "D", "C", "I" represent season, DEM, climate zone, and irrigation, respectively).

3.1.4. Selection of Variables

After adding variables one by one from the set of variables that removed collinear variables, a 5-fold cross validation was performed. The results are shown in Figure 5. It shows that TVDI and MCI can significantly improve the accuracy of the comprehensive model. SM_CABLE, SPI1, SPI9, WSA, and SHA can also improve the accuracy of the comprehensive model to a certain extent, so variables that are conducive to improving the accuracy of the model are selected to construct a comprehensive drought monitoring model (shown in Table 2).

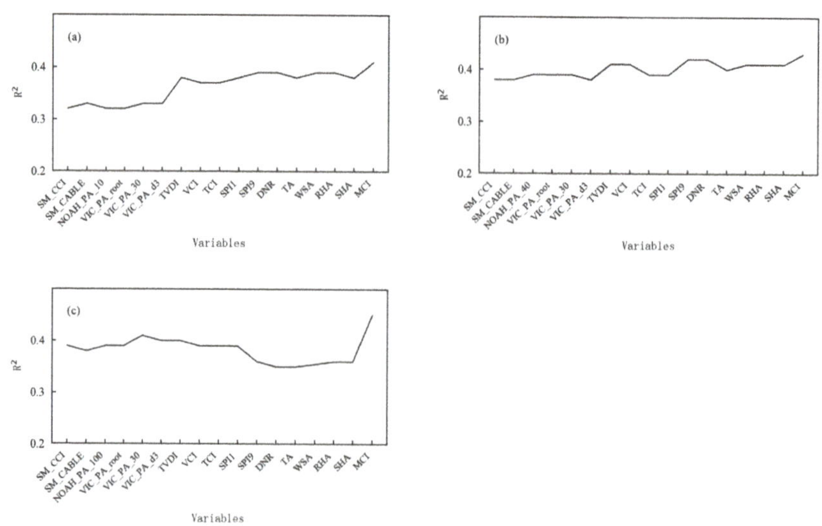


**Figure 5.** Cross-validation results after adding different variables to the variable sets at 10 cm depth (**a**), 20 cm depth (**b**), and 50 cm depth (**c**).

**Table 2.** Variables used to build the comprehensive drought monitoring model.

| Depth | Variables |
|---|---|
| 10 cm | MCI, TVDI, SPI1, SPI9, SM_CABLE, WSA |
| 20 cm | NOAH_10_40 cm_PA, TVDI, SPI9, WSA, SHA, MCI |
| 50 cm | NOAH_40_100 cm_PA, VIC_PA_30, MCI |

### *3.2. Evaluation of the Comprehensive Model*

#### 3.2.1. Comprehensive Evaluation

Based on the previous analysis, considering three environmental factors: season, DEM, and climate zone, a comprehensive agricultural drought monitoring model (hereinafter referred to as the comprehensive model) based on the Cubist algorithm was constructed using the variables shown in Table 1. Eighty percent of the data was used for modeling, 20% for validation and evaluation, and a single-variable model was also constructed. The modeling and validation accuracy of comprehensive models at different depths are shown in Table 3. Figure 6 shows the scatter plots of observed and fitted values of the validation data for comprehensive models and the MCI model with the best correlation at the 10 cm, 20 cm, and 50 cm depths. It shows that the three comprehensive models all have a certain degree of overfitting. There is a phenomenon that high values are underestimated and low values are overestimated. The accuracy of the comprehensive models is significantly improved compared with the single MCI model. RMSE and MAPE of the 10 cm, 20 cm, and 50 cm depth comprehensive models have an average decrease of about 26% and 28%, respectively, compared with the MCI model. Comparing the $R^2$, RMSE, and MAPE of comprehensive models in different climate zones, it was found that the fitting results were significantly better in semi-humid areas, humid areas, and semi-arid areas. The shallow comprehensive model at a depth of 20 cm has the best fitting results in different climate zones, with a MAPE of about 11%~25% for each climate zone, followed by the comprehensive model at a depth of 10 cm, whose MAPE of each climate zone is about 12%~38%, and the comprehensive model at a depth of 50 cm, whose MAPE of each climate zone is about 13%~32%. At the same time, it is noted that the RSM in arid areas is mostly above 60%. Due to the low natural precipitation in arid areas, agricultural production uses irrigation, and the fitting of MCI models in arid areas is also poor. Irrigation may be an important reason for the relatively poor fitting of the comprehensive model and MCI model in this area.

Figure 7 shows the time series diagrams of actual and fitted values for several typical stations in different climate zones at a depth of 10 cm. It shows that the comprehensive model can simulate the variation characteristics of typical stations over time. The MAEP of actual and fitted values for stations in Gaotai, Minle, Yuzhong, Lintao, and Lixian are 10.8%, 8.8%, 19.4, 15.8%, and 12.6%, respectively. The actual RSM of Lixian, Lintao, and Yuzhong, located in humid, semi-humid, and semi-arid areas, fluctuates around 60%. Among them, Lixian and Lintao have more accurate simulations of high values, while Yuzhong has a lower simulation of high values and a heavier simulation of drought conditions. For example, after 2010, the actual RSM of Yuzhong was mostly above 60%, while the simulated RSM was lower than the actual one, reaching the level of light drought. The three stations have higher simulations of low values and a lighter simulation of drought conditions, such as in Lintao from 2009 to 2012, where the actual RSM was mostly below 60%; even below 40%, it reached the level of severe drought, while the simulated RSM was relatively high, only reaching the level of light to moderate drought. The actual RSM of Minle, located in semi-arid and semi-humid areas, and the Gaotai, located in arid areas, fluctuates around 80%, and the actual RSM also fluctuates around 80%, which is consistent with the actual situation without drought. The actual RSM of Minle has shown a downward trend since 2011, and the comprehensive model has simulated this change feature. The Gaotai fluctuates significantly around 80%, while the fluctuation simulated by the comprehensive model is slightly smaller.

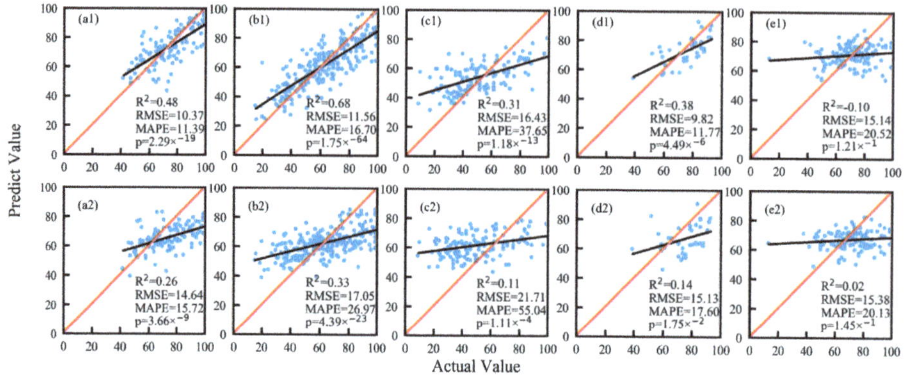


(**a**)

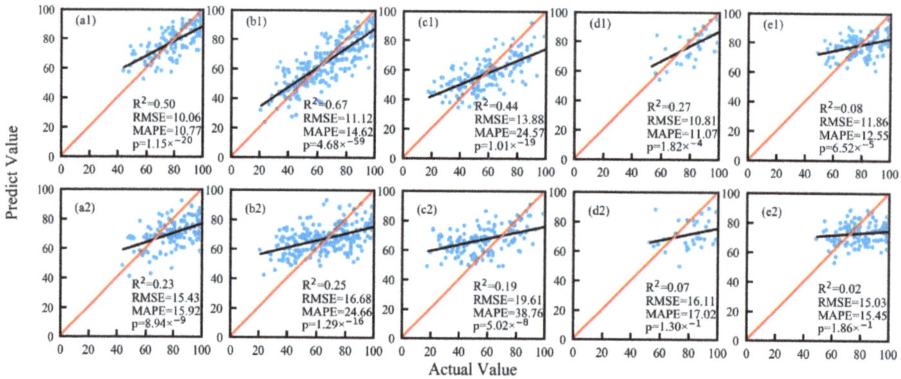


(**b**)

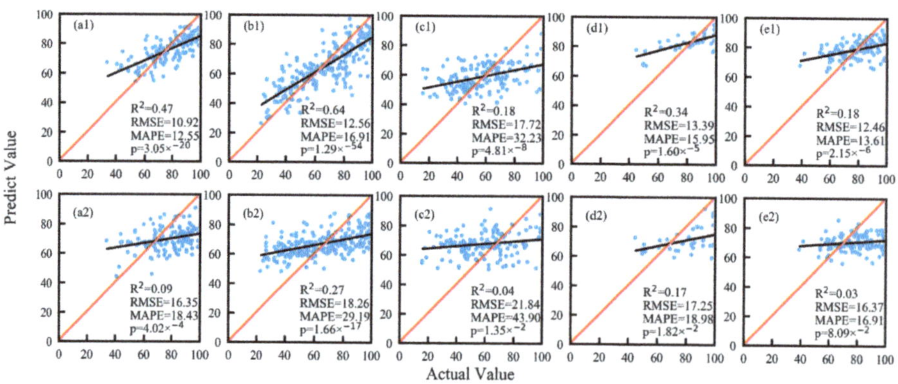


(**c**)

**Figure 6.** Scatter plots of observation and fitting values for the comprehensive model and MCI model at depths of 10 cm (**a**), 20 cm (**b**), and 50 cm (**c**) on the test dataset (In the figure, (a)~(e) represent humid areas, semi-humid areas, semi-arid areas, semi-arid and semi-humid areas, and arid areas, respectively; 1 and 2 represent the comprehensive model and MCI model, respectively).

**Table 3.** Comprehensive model accuracy evaluation.

| Depth | Training Data Set | | | | | Test Data Set | | | | |
|---|---|---|---|---|---|---|---|---|---|---|
| | Sample Size | $R^2$ | *p*-Value | RMSE | MAPE | Sample Size | $R^2$ | *p*-Value | RMSE | MAPE |
| 10 cm | 2636 | 0.71 | 0 | 10.19 | 15.24 | 659 | 0.56 | $2.19 \times 10^{-118}$ | 13.10 | 20.33 |
| 20 cm | 2621 | 0.75 | 0 | 9.44 | 12.72 | 656 | 0.57 | $1.88 \times 10^{-124}$ | 12.77 | 18.57 |
| 50 cm | 2687 | 0.67 | 0 | 11.31 | 15.81 | 672 | 0.54 | $1.07 \times 10^{-116}$ | 13.23 | 18.62 |

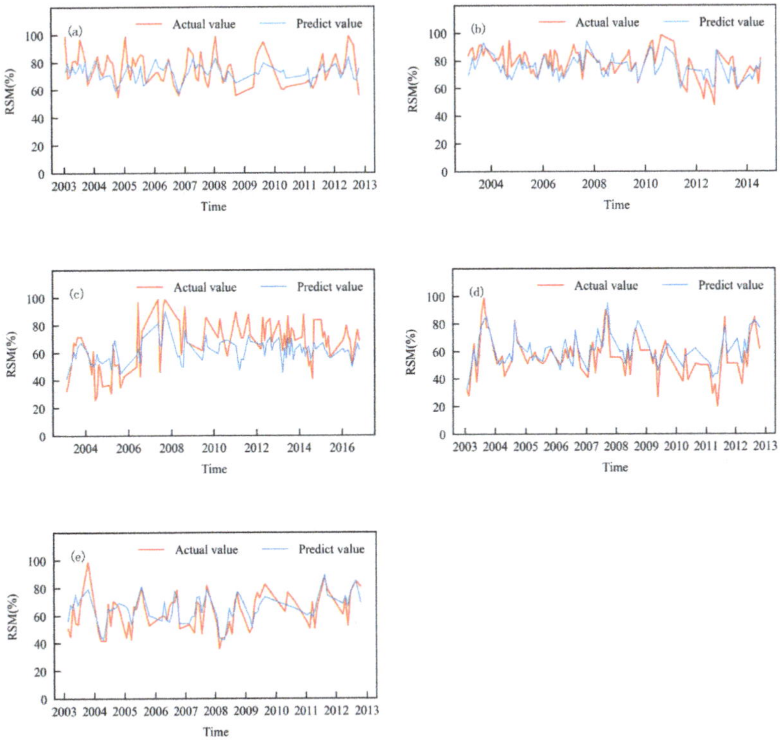


**Figure 7.** Time series diagrams of the true and predicted RSM at a depth of 10 cm depth at Gaotai (**a**), Minle (**b**), Yuzhong (**c**), Lintao (**d**), and Lixian (**e**) in different climate zones (Gaotai, Minle, Yuzhong, Lintao, and Lixian are located in arid areas, semi-arid and semi-humid areas, semi-arid areas, semi-humid areas, and humid areas, respectively).

### 3.2.2. Evaluation of the Impact of Environmental Factors on the Model

The RSME difference between the model without considering environmental factors and the model considering environmental factors for each station at 10 cm depth was calculated, and the spatial distribution is shown in Figure 8. After considering season, DEM, and climate zone, the errors of most stations decreased, and the model accuracy was greatly improved. Among them, the model's accuracy was significantly improved after considering DEM factors. The station-averaged error difference at three depths is shown in Table 4. It shows that DEM, climate zone, and season factor alone can averagely reduce the error of the comprehensive model at three depths by about 20%, 7.5%, and 2.5%, respectively, and all three factors can averagely reduce the error by about 25%. This indicates that environmental factors, especially the DEM factor, are very crucial for establishing a comprehensive model based on the Cubist algorithm, even much more than increasing the number of input variables.

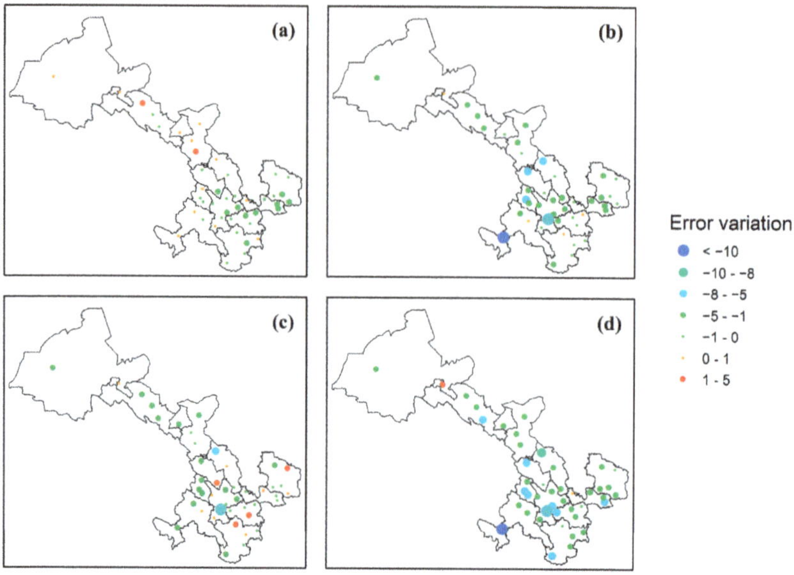


**Figure 8.** Spatial distribution diagram of the RSME difference between the model without considering environmental factors and the model considering environmental factors ((**a**) is only considering season factors, (**b**) is only considering DEM factors, (**c**) is only considering climate zone factors, and (**d**) is considering all three factors simultaneously).

**Table 4.** Statistical table for the station-averaged RMSE difference, considering different environmental factors.

| Depth | Error Variation of Season | | Error Variation of DEM | | Error Variation of Climate Zone | | Error Variation of Three Factors | |
|---|---|---|---|---|---|---|---|---|
| | Absolute Value | Relative Value | Absolute Value | Relative Value | Absolute Value | Relative Value | Absolute Value | Relative Value |
| 10 cm | −0.4 | −3% | −2.4 | −14.3 | −1 | −5.7% | −3.2 | −22% |
| 20 cm | −0.4 | −3% | −2.8 | −18% | −1.2 | −7.6% | −3.7 | −26% |
| 50 cm | −0.2 | −1.3% | −3.2 | −20% | −1.5 | 9% | −4.2 | −26% |

In order to understand the reasons why DEM can drastically improve accuracy in the comprehensive model, the following experiments were conducted. Using DEM and the variables in Table 2, the comprehensive model was constructed using the multiple regression algorithm and Cubist algorithm, respectively. The results of the training and testing sets are shown in Table 5. It shows that the fitting result of the multiple regression algorithm is significantly inferior to that of the Cubist algorithm. The core idea of the Cubist algorithm is to generate several multiple regression equations under the classification of rules. The multiple regression algorithm uses a single multiple regression equation throughout the entire research area. From the fitting results of both, the generation of rules in the Cubist algorithm is an important reason for the improvement of the comprehensive model accuracy of the Cubist algorithm. Table 6 shows the importance of each variable in the Cubist algorithm-based comprehensive model at 10 cm depth in both rules and models. It shows that DEM ranks first in terms of importance in rules, indicating that DEM is a very important classification indicator. It is the impact of DEM on rule generation that significantly improves the accuracy of the comprehensive model.

**Table 5.** The fitting results of multiple regression algorithm and Cubist algorithm.

| Algorithm | Depth | Training Data Set | | | Test Data Set | | |
|---|---|---|---|---|---|---|---|
| | | $R^2$ | RMSE | MAPE | $R^2$ | RMSE | MAPE |
| multiple regression | 10 cm | 0.27 | 16.24 | 25.73 | 0.28 | 16.74 | 27.64 |
| | 20 cm | 0.26 | 16.15 | 23.04 | 0.25 | 16.93 | 26.16 |
| | 50 cm | 0.17 | 17.86 | 27.22 | 0.15 | 18.06 | 26.89 |
| Cubist | 10 cm | 0.65 | 11.28 | 17.24 | 0.47 | 14.29 | 22.94 |
| | 20 cm | 0.68 | 10.61 | 14.25 | 0.50 | 13.85 | 20.15 |
| | 50 cm | 0.57 | 12.79 | 18.18 | 0.45 | 14.55 | 21 |

**Table 6.** The importance of various variables in the 10 cm Cubist algorithm integrated model in rules and models.

| Variables | Rules | Models |
|---|---|---|
| DEM | 86 | 87 |
| MCI | 41 | 93 |
| TVDI | 12 | 82 |
| SPI1 | 6 | 68 |
| SPI9 | 4 | 59 |
| SM_CABLE | 2 | 45 |
| WSA | 1 | 29 |

### 3.2.3. Evaluation of the Impact of Meteorological Factors on Comprehensive Models

The RSME difference between the model without considering the meteorological factors and the model considering the meteorological factors for each station was also calculated, and spatial distribution at 10 cm depth is shown in Figure 9. It shows that after considering MCI, SPI1, SPI9, and WSA, the error of most stations had decreased, while the error of some stations had increased. The stations with increased error are mostly distributed in the Hexi area, which is an arid area. Comparing Figure 8a–c, it is found that after considering SPI1, the errors of almost all stations had decreased to varying degrees, indicating that the precipitation in the past month was more conducive to reducing the error of the comprehensive model. The station-averaged error difference at three depths is shown in Table 7. Meteorological factors mainly reduce the error of shallow, comprehensive models. MCI, SPI1, SPI9, WSA, and SHA alone can averagely reduce the error of the comprehensive model at three depths by about 6.4%, 5.6%, 3.5%, 0.9%, and 1%, respectively. The comprehensive consideration of all five meteorological indexes can reduce the error at three depths by about 10.6%, 2.9%, and 5.9%, respectively. Overall, precipitation is still the most important meteorological factor affecting the accuracy of the comprehensive models, and WSA and SHA can slightly improve the accuracy of the comprehensive models at 10 cm and 20 cm.

**Table 7.** Statistics of mean station error for models considering different meteorological factors.

| Depth | MCI | | SPI1 | | SPI9 | | WS | | SH | | Considering All | |
|---|---|---|---|---|---|---|---|---|---|---|---|---|
| | a | r (%) | a | r(%) | a | r (%) | a | r (%) | a | r(%) | a | r (%) |
| 10 cm | −0.8 | −6.7 | −0.7 | −5.6 | −0.5 | −4.1 | −0.2 | −1.48 | - | - | −1.27 | −10.6 |
| 20 cm | −0.7 | −6.5 | - | - | −0.3 | −2.8 | −0.01 | −0.25 | −0.13 | −1 | −0.51 | −2.9 |
| 50 cm | −0.7 | −5.9 | - | - | - | - | - | - | - | - | −0.7 | −5.9 |

Note: In the table, a and r represent absolute and relative values, respectively, and "-" indicates that this factor has not been considered.

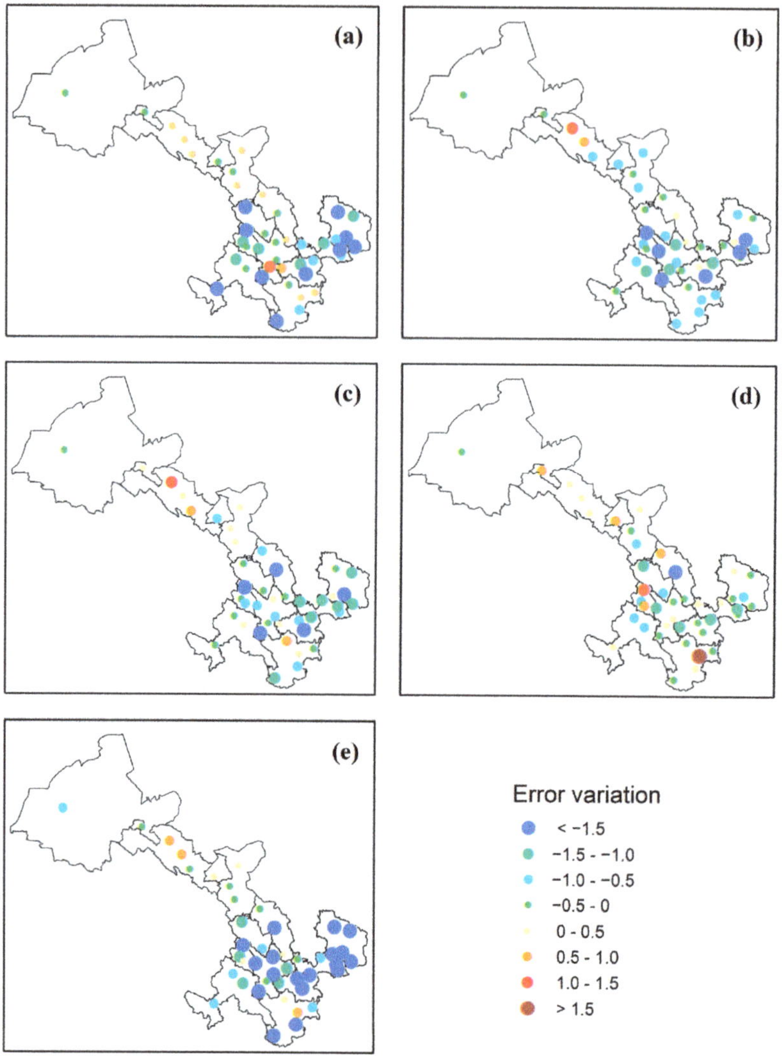


**Figure 9.** Spatial distribution diagram of the RSME difference between the model without considering meteorological factors and the model considering meteorological factors at 10 cm depth ((**a**) is only considering MCI, (**b**) is only considering SPI1, (**c**) is only considering SPI9, (**d**) is only considering WSA, (**e**) is considering MCI, SPI1, SPI9, and WSA simultaneously).

### *3.3. Application of the Comprehensive Model*

Since late July 2016, the average temperature in the central and eastern parts of Gansu Province has been around 2 °C higher, with the highest average temperature in nearly 56 years and the lowest precipitation in nearly 44 years. The high temperature and lack of precipitation have led to drought in the central and eastern parts of Gansu, with most of Dingxi city and the northwest of Tianshui city experiencing severe drought. Due to the lack of effective precipitation during the critical period of crops, local crops have suffered serious disasters, and potatoes and corn even have experienced widespread yield reductions or crop failures. Figure 10 shows the spatial distribution of the comprehensive drought index at 10 cm depth, percentage of precipitation anomaly, and actual RSM from June to August of 2016. It shows that after effective precipitation appeared in Lanzhou, Baiyin, northern Dingxi, and northern Qingyang in July, the drought situation in these areas was somewhat alleviated. However, by August, with the decrease in precipitation, the drought situation in these areas further intensified. Although the time scales, depths, and thresholds for drought classification are different, there is a certain degree of spatial consistency between drought classification classified by the comprehensive model and site RSM, which can indicate the spatial–temporal variation characteristics of this drought event in central Gansu province. Among them, CADI accurately reflects drought in northern Dingxi, Qingyang, Tianshui, and Longnan. However, CADI has a weak indication of drought in central Dingxi. There is a clear boundary phenomenon between semi-arid and semi-humid areas. The analysis for Figure 6 shows that the lower-fitting RSM of the typical semi-arid station and the higher-fitting RSM of the typical semi-humid station, especially the higher fitting in the semi-humid areas, are the reasons for the "boundary phenomenon" in this case.

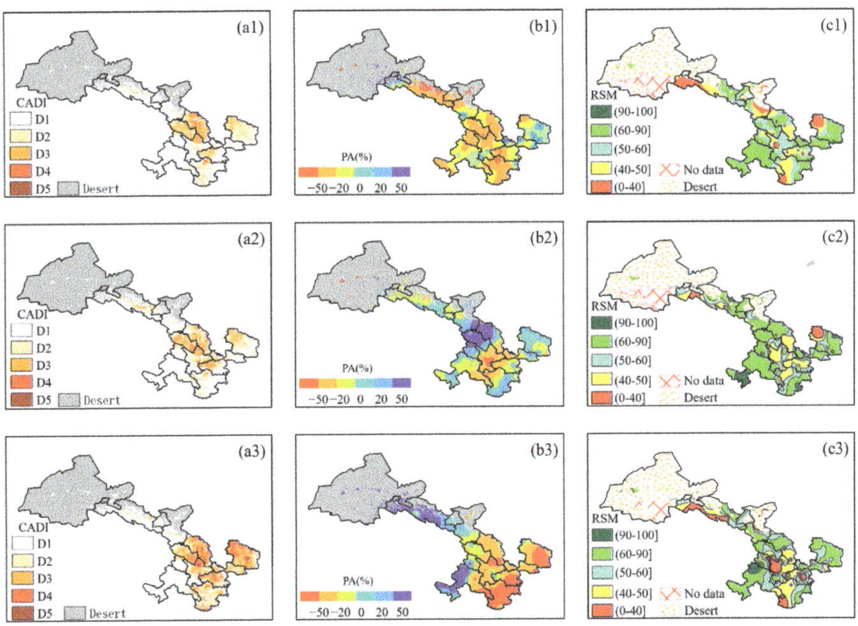


**Figure 10.** The drought classification map of the comprehensive agricultural drought index at 10 cm depth in June (**a1**), July (**a2**), and August (**a3**) of 2016, and the percentage of precipitation anomalies in June (**b1**), July (**b2**), and August (**b3**), as well as the actual RSM at 0–30 cm depth on June 16 (**c1**), July 16 (**c2**), and August 16 (**c3**).

## 4. Discussion

In the early study of soil moisture estimation models based on tree models, environmental factors were considered important factors regardless of the drought indexes considered. However, the importance of environmental factors in machine learning models was rarely discussed in previous research. This article attempts to reveal the importance of environmental factors in improving the accuracy of the Cubist algorithm, which provides a reference for a better understanding of the Cubist algorithm.

The core idea of the Cubist algorithm is to generate a series of rule-based regression equations, and environmental factors are important references for establishing rules. Establishing rules is actually an automatic division of regions in space, thereby achieving the automatic selection of indexes in regression equations. The large east–west span, large DEM span, and complex terrain in Gansu Province are important reasons for the underlying surface and complex climate. DEM is also the most important environmental factor affecting the accuracy of the comprehensive model. This indicates that selecting appropriate environmental factors for the study area using the Cubist algorithm is crucial for improving the model's fitting ability.

However, the fitting accuracy of the integrated model is poor in the arid areas of Hexi and the semi-arid areas in the central region. On the one hand, although the average annual precipitation in the Hexi region is relatively low, the water resources provided by the Qilian Mountains can meet the irrigation needs of agriculture in the Hexi region. The lack of irrigation data may be an important reason for the low fitting accuracy of the comprehensive model of the Hexi arid area. In addition, SPI1 is the only meteorological factor that can improve the accuracy of the model in the Hexi region (Figure 9b), indicating that short-term precipitation has a more important impact on SM in the region. SPI1, which represents short-term precipitation, and MCI and SPI9, which represent long-term precipitation, have a positive effect on improving model accuracy in the Hedong region. Due to its arid climate characteristics, the soil sand content in the Hexi arid region is relatively high, while the soil clay content in humid and semi-humid areas is relatively high. The difference in soil properties may be an important factor in the different patterns of precipitation's impact on SM in different regions. However, due to the high sand content, large rock and soil voids, and high groundwater level in the Hexi region, the $ET_0$ (1049.3–1269.9 mm) is much higher than the precipitation (42–200 mm) [40]. Moreover, Zhang et al.'s [41] research shows that there is a strongly coupled nonlinear relationship between typical stations' SM and evapotranspiration (EF) located in semi-arid areas. It can be seen that soil properties and evaporation are important factors affecting water cycling. The lack of soil properties and evaporation data may be another important reason for the low fitting accuracy in these regions.

At the same time, it is noted that the comprehensive model based on the Cubist algorithm exhibits obvious "boundary phenomena" in different climate zones, especially in semi-arid and semi-humid areas (Figure 10). The study uses the Cubist algorithm to achieve regression estimation of continuous soil moisture values, and divides soil moisture into drought levels with intervals of 10. Therefore, small errors may cause differences in drought levels. Low fitting accuracy in semi-arid areas is the important reason for the "boundary phenomena". Apparently, the "boundary" is consistent with the climate zone. Although the consideration of climate zones can help improve the accuracy of comprehensive models, "boundary phenomena" also exist due to climate zones.

## 5. Conclusions

(1) Among the comprehensive models constructed by the Cubist algorithm, the model at 20 cm depth has the highest accuracy, followed by the models at 10 cm and 50 cm depths. The validation $R^2$ of the comprehensive model at 10 cm, 20 cm, and 50 cm depth is 0.56, 0.57, and 0.54, and the RMSE is 13.1, 12.8, and 13.2, respectively. The MAPEs are about 20.3%, 18.6%, and 18.6%, respectively. The accuracy of the comprehensive model has been significantly improved compared to the single-variable model. The RMSE and MAPE of the comprehensive model has decreased by about 26% and 28% compared to the best MCI model, on average, at the 10 cm, 20 cm, and 50 cm depths.

(2) The fitting accuracy of the comprehensive model in humid areas and semi-humid areas, as well as semi-arid and semi-humid areas, is significantly higher than that in arid and semi-arid areas. In humid areas, semi-humid areas, semi-arid and semi-humid areas, semi-arid areas, and arid areas, the average validation $R^2$ of the comprehensive model at 10 cm, 20 cm, and 50 cm depth is 0.48, 0.66, 0.31, 0.33, and 0.16, respectively. The average RMSE is 10.5, 11.7, 16, 11.3, and 13.2, and MAPEs are about 11.6%, 16.1%, 31.5%, 12.9%, and 16.5%, respectively.

(3) Environmental factors play a crucial role in improving the accuracy of comprehensive models, with a greater impact than increasing the number of drought indicators.

Considering that the DEM, climate zone, season alone can averagely reduce the error by about 20%, 7.5%, and 2.5%, respectively, and considering that all three factors can averagely reduce the error by about 25%. Compared to environmental factors, meteorological factors have a slightly weaker effect on improving the accuracy of comprehensive models. The consideration of meteorological factors, such as precipitation, WS, and SH, averagely reduce the error by about 6.5%.

(4) The lack of irrigation, soil property, and evapotranspiration data, especially the lack of evapotranspiration data, may be an important reason for the low fitting accuracy of the comprehensive model in the arid and semi-arid areas of Hexi. In the future, efforts will be made to introduce water content information, such as irrigation, soil properties, and evapotranspiration data, into the comprehensive model, in order to improve the soil moisture-monitoring ability in the Hexi region.

(5) Classification is another major task of machine learning algorithms, which can directly obtain drought levels and may improve the accuracy of drought levels. In the future, it will be necessary to compare the results of machine learning regression algorithms and classification algorithms to improve the accuracy of drought levels, especially in the division of drought levels in semi-arid and arid areas. In addition, the "boundary phenomenon" exists due to climate zone, and climate zone is based on multiple indicators, such as precipitation, dryness, temperature, and accumulated temperature [24]. Therefore, in addition to improving the fitting accuracy of regression models or the accuracy of classification models, using zoning indicators instead of climate zone may improve this situation. However, further research is needed to discover the impact of these factors on the comprehensive model.

**Author Contributions:** Conceptualization, S.S.; methodology, S.S. and L.W; software, L.W.; validation, D.H.; formal analysis, S.S.; investigation, X.W.; resources, L.Z.; writing—original draft preparation, S.S.; writing—review and editing, S.S and Y.R.; visualization, S.S. All authors have read and agreed to the published version of the manuscript.

**Funding:** This study was supported by the National Natural Science Foundation of China (Grant Nos. 42105131, 42075120, and 41875020), the Meteorological Science and Technology Research Project of Gansu Provincial Meteorological Bureau (ZcZd2022-26), and the Innovation Team of Lanzhou Institute of Arid Meteorology, CMA (GHACXTD-2020-4).

**Data Availability Statement:** MOD09A1 and MOD11A2 of MODIS data are available at https://search.earthdata.nasa.gov/search/granules?p=C2343111356-LPCLOUD&pg[0][v]=f&pg[0][gsk]=-start_date&q=MOD09A1&tl=1523328714!4!! and https://search.earthdata.nasa.gov/search/granules?p=C2269056084-LPCLOUD&pg[0][v]=f&pg[0][gsk]=-start_date&q=MOD11A2&tl=1523328714!4!! (access on 11 May 2024); CCI soil moisture data of version 04.5 are available at https://catalogue.ceda.ac.uk/uuid/38b8e5e524e1449ab4b4994970752644 (accessed on 11 May 2024); soil moisture data of GLDAS Noah model and VIC model of version 2.1 are available at https://disc.gsfc.nasa.gov/datasets/GLDAS_NOAH025_M_2.1/summary?keywords=GLDAS (accessed on 11 May 2024), https://disc.gsfc.nasa.gov/datasets/GLDAS_VIC10_M_2.1/summary?keywords=GLDAS (accessed on 11 May 2024), respectively.

**Conflicts of Interest:** The authors declare no conflict of interest.

## References

1. Guo, N.; Wang, X. Advances and Developing Opportunities in Remote Sensing of Drought. *J. Arid. Meteorol.* **2015**, *33*, 1–18.
2. Liu, H.; Wang, F.; Zhang, T. Evaluation applicability of CLDAS and GLDAS soil moisture for the Loess Plateau. *Agric. Res. Arid. Areas* **2018**, *36*, 270–276.
3. Kowalczyk, E.A.; Wang, Y.P.; Law, R.M.; Davies, H.L.; McGregor, J.L.; Abramowitz, G. *CSIRO Atmosphere Biosphere Land Exchange Model for Use in Climate Models and as an Offline Model*; CSIRO Marine and Atmospheric Technical Report; Commonwealth Scientific and Industrial Research Organisation: Canberra, Australia, 2006; pp. 1–37.
4. Yang, X.; Yang, Q. Research on meteorological drought severity model for Loess Plateau in Gansu. *J. Nat. Disasters* **2007**, *16*, 30–36.
5. Zhang, L.; Zhang, Q.; Zhang, H.; Yue, P.; Li, H.; Wang, J.; Zhao, F.; Wang, Y.; Wang, J. Environmental factors driving evapotranspiration over a grassland in a transitional climate zone in China. *Meteorol. Appl.* **2022**, *29*, e2066. [CrossRef]
6. Zhang, Q.; Hu, Y.; Cao, X.; Liu, W. On Some Problems of Arid Climate System of Northwest China. *J. Desert Res.* **2000**, *20*, 357–362.

7. Yang, S.; Liu, C. Remote sensing calculation of soil moisture and analysis of water cycle process in the Yellow River Basin. *Sc China Ser. E Technol. Sci.* **2004**, *34*, 1–12. [CrossRef]
8. Zhou, S.; Williams, A.; Lintner, B.; Berg, A.; Zhang, Y.; Keenan, T.; Cook, B.; Hagemann, S.; Seneviratne, S.; Gentine, P. Soi moisture–atmosphere feedbacks mitigate declining water availability in drylands. *Nat. Clim. Chang.* **2021**, *11*, 38–44. [CrossRef
9. Brown, J.; Wardlow, B.; Tadesse, T.; Hayes, M.; Reed, B. The Vegetation Drought Response Index (VegDRI): A New Integrate Approach for Monitoring Drought Stress in Vegetation. *GISci. Remote Sens.* **2008**, *45*, 16–46. [CrossRef]
10. Sánchez, N.; González-Zamora, Á.; Piles, M.; Martínez-Fernández, J. A New Soil Moisture Agricultural Drought Index (SMAD Integrating MODIS and SMOS Products: A Case of Study over the Iberian Peninsula. *Remote Sens.* **2016**, *8*, 287. [CrossRef]
11. Sun, P.; Zhang, Q.; Wen, Q.; Singh, V.; Shi, P. Multisource data based integrated agricultural drought monitoring in the Huai Rive basin, China. *J. Geophys. Res. Atmos.* **2017**, *122*, 10751–10772. [CrossRef]
12. Ji, T.; Li, G.; Yang, H.; Liu, R.; He, T. Comprehensive drought index as an indicator for use in drought monitoring integratin multi-source remote sensing data: A case study covering the Sichuan-Chongqing region. *Int. J. Remote Sens.* **2018**, *39*, 786–80 [CrossRef]
13. Wang, J.; Zhu, X.; Liu, X.; Pan, Y. Research on agriculture drought monitoring method of Henan Province with multi-sources dat *Remote Sens. Land Resour.* **2018**, *30*, 180–186.
14. Bijaber, N.; El Hadani, D.; Saidi, M.; Svoboda, M.D.; Wardlow, B.D.; Hain, C.R.; Poulsen, C.C.; Yessef, M.; Rochdi, A. Developin a Remotely Sensed Drought Monitoring Indicator for Morocco. *Geosciences* **2018**, *8*, 55. [CrossRef] [PubMed]
15. Meng, L.; Dong, T.; Zhang, W. Drought monitoring using an Integrated Drought Condition Index (IDCI) derived from multi-senso remote sensing data. *Nat. Hazards* **2016**, *80*, 1135–1152. [CrossRef]
16. Zhang, X.; Chen, N.; Li, J.; Chen, Z. Multi-sensor integrated framework and index for agricultural drought monitoring. *Remot Sens. Environ.* **2017**, *188*, 141–163. [CrossRef]
17. Peng, X.; Wang, Q.; Yuan, C.; Lin, K. Review of research on data mining in application of meteorological forecasting. *J. Ari Meteorol.* **2015**, *33*, 19–27.
18. Han, J.; Mao, K.; Xu, T.; Guo, J.; Zuo, Z.; Gao, C. A Soil Moisture Estimation Framework Based on the CART Algorithm and It Application in China. *J. Hydrol.* **2018**, *563*, 65–75. [CrossRef]
19. Yang, J.; Zhang, S.; Bai, Y.; Huang, A.; Zhang, J. SPEI Simulation for Monitoring Drought Based Machine Learning Integratin Multi-Source Remote Sensing Data in Shandong. *Chin. J. Agrometeorol.* **2021**, *42*, 230–242.
20. Shen, R.; Huang, A.; Li, B.; Guo, J. Construction of a drought monitoring model using deep learning based on multi-sourc remote sensing data. *Int. J. Appl. Earth Obs. Geoinf.* **2019**, *79*, 48–57. [CrossRef]
21. Breiman, L.; Friedman, J.; Olshen, R.; Stone, C.; Olshen, R. *Classification and Regression Trees*; Chapman and Hall: New York, N USA, 1984.
22. Demisse, G.; Tadesse, T.; Bayissa, Y.; Atnatu, S.; Argaw, M.; Nedaw, D. Vegetation condition prediction for drought monitoring i pastoralist areas: A case study in Ethiopia. *Int. J. Remote Sens.* **2018**, *39*, 4599–4615. [CrossRef]
23. Nam, H.; Tadesse, T.; Wardlow, B.; Hayes, M.; Svoboda, M.; Hong, E.; Pachepsky, Y.; Jang, M. Developing the vegetation drough response index for South Korea (VegDRI-SKorea) to assess the vegetation condition during drought events. *Int. J. Remote Sen* **2018**, *39*, 1548–1574. [CrossRef]
24. Bao, W. *Gansu Climate*; China Meteorological Press: Beijing, China, 2018; 283p.
25. Deng, Z.; Xie, J.; Liu, X.; Yin, D. The Characteristics and Development and Utilization of Climate Resources in Gansu Provinc *J. Arid. Meteorol.* **1998**, *16*, 16–19.
26. *GB/T20481-2006*; Classification of Meteorological Drought. National Climate Center: Beijing, China, 2006; pp. 1–17.
27. Kogan, F. Remote sensing of weather impacts on vegetation in non-homogeneous areas. *Int. J. Remote Sens.* **1990**, *11*, 1405–141 [CrossRef]
28. Kogan, F. Application of vegetation index and brightness temperature for drought detection. *Adv. Space Res.* **1995**, *15*, 91–10 [CrossRef]
29. Sandholt, I.; Rasmussen, K.; Andersen, J. A simple interpretation of the surface temperature—Vegetation index space fo assessment of surface moisture status. *Remote Sens. Environ.* **2002**, *79*, 213–224. [CrossRef]
30. Mcnally, A.; Shukla, S.; Arsenault, K.; Wang, S.; Peters-Lidard, C.D.; Verdin, J.P. Evaluating ESA CCI soil moisture in East Afric *Int. J. Appl. Earth Obs. Geoinf.* **2016**, *48*, 96–109. [CrossRef] [PubMed]
31. Shen, X.; An, R.; Quaye-Ballard, J.; Zhang, L.; Wang, Z. Evaluation of the European Space Agency Climate Change Initiative Soi Moisture Product over China Using Variance Reduction Factor. *J. Am. Water Resour. Assoc.* **2016**, *52*, 1524–1535. [CrossRef]
32. Xu, W. *Educational Statistics*, 2nd ed.; Nanjing Normal University Press: Nanjing, China, 2007.
33. Li, Y.; Zhang, L.; Zhang, H.; Pu, X. Drought Monitoring Based on CABLE Land Surface Model and Its Effect Examination o Typical Drought Events. *Plateau Meteorol.* **2015**, *34*, 1005–1018. (In Chinese) [CrossRef]
34. Zhang, L. Drought Monitoring Technique Based on Land Surface Model and the Study of its Application Effects in China. Ph.D Thesis, Lanzhou University, Lanzhou, China, 2016.
35. Deng, H.; Lu, Y.; Wang, Y.; Chen, X.; Liu, Q. Assessment of Actual Evapotranspiration in the Minjiang River Basin Based on th GLDAS-Noah Model. *Sci. Geogr. Sin.* **2022**, *42*, 548–556.
36. Liu, P.; Song, H.; Bao, W.; Li, J. Applicability Evaluation of CLDAS and GLDAS Soil Temperature Data in Shaanxi Provinc *Meteorol. Sci. Technol.* **2021**, *49*, 604–611.

. Breiman, L. Random Forests. *Machine Learning.* **2001**, *45*, 5–32. [CrossRef]
. Quinlan, J. *C4.5: Programs for Machine Learning*; Morgan Kaufmann Publishers Inc.: San Francisco, CA, USA, 1993.
. Data Mining with Cubist. Available online: http://www.rulequest.com/cubist-info.html (accessed on 4 April 2022).
. Ma, Y.; Sun, D.; Zhang, R.; Xu, J.; Wang, X. Analysis of driving factors for spatiotemporal variation of reference crop evapotranspiration in Gansu Province. *Chin. Agric. Meteorol.* **2022**, *43*, 881–892.
. Zhang, L.; Sha, S.; Zhang, Q.; Zhao, F.; Zhao, J.; Li, H.; Wang, S.; Wang, J.; Hu, Y.; Han, H. Investigating the Coupling Relationship between Soil Moisture and Evaporative Fraction over China's Transitional Climate Zone. *Hydrology* **2023**, *10*, 221. [CrossRef]

agriculture

MDPI

*Article*

# Modeling Reliability Analysis for the Branch-Based Irrigation Water Demands Due to Uncertainties in the Measured Surface Runoff

**Shiang-Jen Wu * and Han-Yuan Yang**

Department of Civil and Disaster Prevention Engineering, National United University, Miaoli City 360302, Taiwan
* Correspondence: sjwu@nuu.edu.tw

**Abstract:** This study aims to model the uncertainty and reliability quantification of estimating the planning irrigation water demands in the multi-canal irrigation zone, named the RA_IWD_Canal model. The proposed RA_IWD_Canal could estimate the zone-based and branch-based water demands and quantify their uncertainties and reliabilities via the weighted frequency quantile curves. The historical planning irrigation water demands and measured surface runoff from 2019 to 2024 in the Zhudong irrigation zone are utilized in the model development and application. Using the proposed RA_IWD_Canal model, the estimated branch-based irrigation water demands exhibit a significant variation (on average, from 0.02 $m^3/s$ to 1.7 $m^3/s$) in time and space attributed to uncertainties in the historical gauged surface runoff. Also, the Zhudong Canal zone is demonstrated to be sufficiently supplied irrigation water subject to existing introduced water demands with a high reliability of 0.85; instead, the associated branches have considerable difficulty achieving the expected irrigation efficiency based on the desired water requirements with low reliability (nearly 0.25). To keep all branches in the irrigation zone consistent in irrigation efficiency, the probabilistic-based water demands could be introduced via the proposed RA_IWD_Canal model with the desired reliability.

**Keywords:** irrigation water demands; optimal water allocation; L-moment; reliability quantification

## 1. Introduction

Recently, irrigation water allocation systems could be grouped into schedule-based and object-based water supply [1]. According to the locations of the irrigation branches, the allocated irrigation water could be achieved from upstream to downstream in schedule-based systems [1,2]; on the contrary, within object-based systems, the irrigation water should be optimally allocated subject to the priorities given in advance [1]. The minimum water requirements are commonly given in the growing seasons to achieve the expected crop production among the above irrigation systems with various water allocation strategies [2–4]. Also, irrigation water demands frequently vary with the crop types [5]. However, if the upstream inflow is less than the desired irrigation water demands, the shortage risk of irrigation water might be induced. On the contrary, sufficient irrigation water over water demands could be supplied with high reliability. Therefore, introducing available water demands and requirements is vital for modern irrigation systems.

However, climate change significantly triggers the water-related shortage risk [6–10]; additionally, increasing industrial and domestic water frequently raises the complexity of irrigation water use due to the irrigation zone utilization [1,3]; it is more likely to lead to shortage risk to the irrigation regions lacking sufficient water supply subject to the existing water demands for the crop growing. To facilitate irrigating efficiency, the reliability of existing irrigation water demands should be necessarily quantified and evaluated to introduce reasonable ones, considering the effect of uncertainties in climatic and hydrological features and land use. In general, the irrigation water demands in the irrigation zones could be introduced by the deterministic models (e.g., CROPWAT model, Penman–Monteith method, SWAT model, and WBM model) subject to the crop types and climatic as well as the hydrological features in the irrigation zones [1,11–17]. For instance, Hussain et al. [16] employed an experiment to estimate the water requirement for crops during the growth seasons based on the difference in the water depths calculated via the deterministic equations with the change in evapotranspiration and soil moisture. Dang et al. [17] estimated the water requirements based on the difference in the historical surface runoff and evapotranspiration during crop growth. In addition, the irrigation water demands could be set based on the sources of the surface runoff and cultivation extents [2,18–20]. For example, Zhang et al. [9] simulated the surface runoff via the hydrological model with the observed hydrological data, such as rainfall, evapotranspiration, and soil moisture, as the boundary condition of the crop growth model to estimate the water demands. In addition to the crop type, hydrological data, and cultivated area, irrigation water demands could be introduced subject to the crop economic values [21,22].

Overall, as mentioned above, irrigation water demands were mainly introduced via the hydrological models with the crop-related factors (crop type and price as well as the cultivation area) and hydrological data (rainfall, evapotranspiration, and soil moisture). However, the resulting irrigation water demands were estimated without considering the temporal and space changes and induced uncertainties in the measurements of irrigation-related surface runoff. Also, in contrast with the zone-based irrigation water demands comprehensively provided, the branch-based ones within multi-canal irrigation zones are rarely introduced, so as to be approximated based on the corresponding cultivation extents to the branches [1,6]; namely, the irrigation water demands probably have uncertainty in space. Accordingly, the above uncertainties are highly likely to impact the availability of irrigation water demand, further reducing irrigation reliability. Nevertheless, a number of investigations have assessed irrigation reliability [13,14]; they mainly focused on evaluating the effect of rainfall variation on irrigation efficiency and reliability. Therefore, this study aims to develop an optimization and uncertainty analysis-derived model for estimating branch-based irrigation water demands and quantifying their quantiles with gauged historical surface runoffs within multi-canal irrigation zones, named the RA_IWD_Canal model; it is expected that the resulting water demand quantiles could not only describe the uncertainties in the estimated branch-based irrigation water demands due to surface-runoff variation but also apply to the reliability quantification of the zone-based and branch-based irrigation water demands of interest.

## 2. Methodology

The proposed RA_IWD_Canal model is mainly developed to estimate branch-based irrigation water demands, quantify uncertainties, and induce reliability in the multi-branch irrigation zone. Thus, within the proposed RA_IWD_Canal model, the optimization and uncertainty analysis should be configured with the observed surface runoff at the discharge gauges within the study area. Accordingly, the detailed model concepts and methods regarding the optimization and uncertainty analysis for the irrigation water could be addressed below.

### *2.1. Model Concept*

The proposed RA_IWD_canal model could consist of three components: data collection, estimation of branch-based irrigation water demands, and uncertainty and reliability quantification of irrigation water demands. At the data collection step, in addition to the measured surface runoff at all discharge gauges and upstream inflow, the historical irrigation water demands introduced should be required in the model development and validation. After that, to estimate the branch-based water demands in the multi-canal irrigation zone via the proposed RA_IWD_Canal model, an optimization-based model for allocating the branch-based irrigation water supplies (OPA_IWS_Canal) [2] is adopted under consideration of the historical runoff data recorded at the discharge gauges. In detail, if the discharge gauge is installed at the branch, the corresponding measured surface runoff could be regarded as the branch-based irrigation water demand; alternatively, at the branch in the middle of two discharge gauges, its irrigation water demands could be estimated with the difference of surface runoff at two discharge gauges. In the case of several branches in the middle of two discharge gauges, their irrigation water demands could be achieved via the OPA_IWS_Canal model with the difference of gauged surface runoff.

Eventually, to quantify and assess the reliabilities of the specific irrigation water demands, the proposed RA_IWD_Canal model proceeds with the uncertainty analysis via weighted frequency curves [23], consisting of quantiles under the desired probabilities, established using the L-Moment approach, estimated via the L-Moment method; thus, the dispersion of estimated branch-based water demands could be quantified in terms of the L-mean and L-CV; accordingly, the exceedance probability of the specific water demands (e.g., introduced planning magnitudes) could be extrapolated from the resulting quantile curves, called overestimated risk as a reference to the reliability.

### 2.2. *Optimization Estimation of Irrigation Water Demands*

Within the optimization water allocation model (OPA_IWA_Canal model) proposed by Wu et al. [1], schedule-based water allocation is adopted to distribute the upstream inflow in the main canal to the target branches located from upstream to downstream; thus, the irrigation water supplied at the target branch (named the branch-based irrigation water supply) could be estimated by comparing the remaining canal-based water supply with the corresponding maximum delivered water volume:

$$Q_{S,IBL} = min\{Q_{S,Canal}, Q_{MDF,IBL}\} \tag{1}$$

where $Q_{S,IBL}$ denotes the irrigation water supply obtained at the ith branch (called branch IBL); $Q_{S,IBL\ canal}$ stands for the water supply that could be provided from the main canal for the branch IBL; and $Q_{MDF,IBL}$ serves as the maximum delivered water supply at the target branch IBL. The above water supply provided from the main canal to the branch IBL $Q_{s,IBL,canal}$ could be calculated by the following equation under consideration of the water supplies received by the branches and water-intake hydraulic structures located both upstream of the canal-based irrigation zone:

$$Q_{S,IBL,canal} = Q_{IN,Canal} - \sum_{I=1=1}^{IBL} Q_{S,IBL,canal} - Q_{S,HYST} \tag{2}$$

in which $Q_{IN,Canal}$ stands for the resulting upstream inflow from the upstream boundary in the canal; $Q_{s,IBL,canal}$ serves as the branch-based irrigation water supply obtained at the upstream IBL branch; and $Q_{S,HYST}$ is the total water intake of the hydraulic structures upstream calculated using the following equation:

$$Q_{S,HYST} = Q_{D,HYST} \times \beta_{IRR,HYST} \tag{3}$$

where $Q_{D,HYST}$ is the expected water demands of the hydraulic structures and $\beta_{IRR,HYST}$ stands for the corresponding water-intake reduction ratios of the hydraulic structures. Also, the maximum derived water volume at the target IBL branch ($Q_{MDF,IBL}$) could be computed as [20]

$$Q_{MDF,IBL} = Q_{D,IBL} \times \alpha_{SR,IBL} \tag{4}$$

where $Q_{D,IBL}$ and $\alpha_{MDR,IBL}$ denote the corresponding irrigation water demand at the branch IBL (named the branch-based irrigation water demand) and the maximum derived supply to the target branch (IBL), respectively. In reference to the branch-based irrigation water demand, it is more likely to be determined based on the type of the crop and expected cultivation extent; it is commonly introduced for all of the canal-based irrigation zones, comprising a group of trenches [20]. Accordingly, in this study, the branch-based irrigation water demand ($Q_{D,IBL}$) could be approximated subject to the canal-based irrigation water demand and the cultivation area of the branches as

$$Q_{D,IBL} = Q_{D,Canal} \times \frac{CA_{IBL}}{CA_{Canal}} \quad (5)$$

where $Q_{D,Canal}$ is the planning irrigation water demand for the canal-based zone; and $CA_{IBL}$ and $CA_{Canal}$ account for the cultivation extents of the target i$^{th}$ branch (IBL) and canal-based zone, respectively.

Within the OPA_IWD_Canal model, achieving the optimal branch-based irrigation water supply ($Q_{S,IBL}$) and demand ($Q_{D,IBL}$), the branch-based supplying satisfaction index ($SI_{IBL}$) is required to evaluate the irrigation efficiency at the branch IBL as [2,20,24]

$$SI_{IBL} = \frac{Q_{S,IBL}}{Q_{D,IBL}} \quad (6)$$

in which $Q_{S,IBL}$ and $Q_{D,IBL}$ account for the branch-based irrigation water supply and demand, respectively; in the case of $SI_{IBL}$ approaching the expected value (e.g., 1.0), the optimal water supply at the target branch is achieved with an acceptable satisfaction index $SI_{IBL}$ given a desired water demand. The detailed concepts of the OPA_IWS_Canal model can be referred to Wu's investigation [2].

### 2.3. *Uncertainty and Reliability Quantification of Estimated Irrigation Water Demands*

Uncertainties in the hydrological data and relevant deterministic models are frequently caused by a lack of information on hydrologic-related phenomena, which might induce the failure risk of the expected performance regarding the hydraulic structures. In general, the uncertainties could be described in terms of the statistical moments of various orders, including the mean, variance, and coefficients of skewness as well as kurtosis; also, the detailed information on the data and model inputs as well as the outputs could refer to the quantiles under consideration of the various occurrence probabilities. However, the resulting quantiles should be calculated via the identified best-of-fit probability density functions (PDF), which significantly change with the goodness-of-fit criteria adopted [17]. To reduce the uncertainties attributed to the selection of the goodness-of-fit criteria, Wu et al. [23] presented a weighted frequency curve method to produce quantities of the desired occurrence probabilities.

Apart from the uncertainty due to PDF selection, the PDF parameters are commonly calibrated by the conventional produce-moment approach; however, the worse bias of the resulting statistical properties and quantiles might be induced with the moment orders; thus, Hosking [25] proposed the L-moments to significantly reduce the above bias attributed to the orders of statistical moments as to provide more accurate quantiles. Therefore, within the proposed RA_IWD_Canal model, the weighted frequency curve approach with the L-moment could be employed in the uncertainty and reliability quantification of the estimated branch-based irrigation water demands. Thus, within the proposed RA_IWD_Canal model, the reliability of the estimated irrigation water demand could be represented in terms of the exceedance probability concerning the specific demand, which could be defined as [2,14]

$$\text{Reliability} = P_r(W_D \geq w_d) \quad (7)$$

where $W_D$ accounts for the quantiles in the weighted frequency curve; and $w_d$ serves as the specific demand, such as the estimated branch-based irrigation water demands or the introduced planning irrigation water demand. The concept regarding the weighted frequency curves is briefly introduced as follows.

To reduce uncertainty in the quantiles attributed to the inappropriate probability distributions selected, Tung [26] proposed a weighted frequency curve (WFC) method by collaborating the multiple probability distributions; when deriving the WFC model, the commonly used probability distributions are considered, including the normal and log-normal distributions, Pearson and log-Pearson distributions, Gamma distribution, generalized extreme value distribution, generalized Pareto distribution, and generalized logistic distribution. In detail, the above-considered probability distributions are assumed to be suitable for estimating the quantiles of given probabilities, which are calculated via the plotting position formula. After that, the weighted quantiles can be obtained by the following equation:

$$X_{p,w} = \sum_{i=1}^{M} (w_i \times X_{p,i}) \tag{8}$$

where $X_{p,w}$ accounts for the weighted quantiles of a given probability (p); $X_{p,i}$ and $w_i$ stand for the quantiles of a given probability (p) and the corresponding weighted factors under the ith probability distribution concerned, respectively. Note that the weighted factors of the candidate probability distributions are calculated based on their fitness performance in terms of the mean square error ($MSE_i$) as

$$MSE_i = \frac{1}{n}\sum_{j=1}^{n} \left(X_{(j)} - Y_{(i,j)}\right)^2 \tag{9}$$

in which n denotes as the sample size; $X_{(j)}$ and $Y_{(i,j)}$ and serve as the ascending sample data and corresponding quantile coming from the ith candidate probability distribution. Eventually, the resulting weighted factors of the candidate probability distributions are computed via the following equation:

$$w_i = \frac{{}^{1}/{MSE_i}}{\sum_{i=1}^{n} \left({}^{1}/{MSE_i}\right)} \tag{10}$$

where $w_i$ is treated as the weighted factor of the ith candidate probability distirouion.

In total, the quantile curves of the estimated branch-based irrigation water demands via the OPA_IWD_Canal model can be established through the weighted frequency curve methods to proceed with the reliability quantification of the specific irrigation water demands within the proposed RA_IWD_Canal model. In particular, the parameter calibration of the candidate probability distributions should be carried out via the L-moment method with a group of estimated branch-based irrigation water demands. Also, the uncertainties of the estimated branch-based irrigation water demands could be quantified using L-moment ratios of the first four orders as

$$L\text{-}CV: \ \tau_2 = \frac{\lambda_2}{\lambda_1} \tag{11}$$

$$L\text{-}Skewness: \ \tau_2 = \frac{\lambda_3}{\lambda_2} \tag{12}$$

$$L\text{-}Kurtosis: \ \tau_2 = \frac{\lambda_4}{\lambda_2} \tag{13}$$

where $\lambda_i$ stands for the ith order L-moment, which could be obtained via the following equation proposed by Hosking [18]:

$$\lambda_r = \frac{1}{r}\sum_{k=1}^{r-1}\left[(-1)\binom{r-1}{k}E(X_{r-k:r})\right], r = 1,2,\ldots \quad (14)$$

where $E(X_{r-k:r})$ denotes the expected value of the kth ascending sample datum among the r sample data extracted from the population.

### 2.4. Model Framework

To summarize the relevant concepts, the proposed RA_IWD_Canal model can be configured using optimal irrigation water allocation and uncertainty quantification methods. Therefore, developing the proposed RA_IWD_Canal model could refer to the following steps:

Step [1] Collecting information on the irrigation system, including the system structure, cultivation extents, and the number and location of the irrigation branches, discharge gauges, and water-intake hydraulic structures; geometric and hydrologic data are also necessary to apply to model development, upstream inflow, gauged surface runoff, and irrigation water demand planning.

Step [2] Carrying out the uncertainty analysis to quantify the stochastic properties of the gauged surface runoff and planning irrigation water demands.

Step [3] Calculating the differences in the measured surface runoffs among the discharge gauges.

Step [4] Grouping the irrigation branches into various clusters based on their locations compared with the spots of the discharge gauges.

Step [5] The observed surface runoff is treated as the estimated water demand at the branch with the single discharge gauge.

Step [6] The difference in the gauged surface runoffs is the estimated water demands in the cluster with a single irrigation branch.

Step [7] Estimating the optimal water supplies based on the water demand at more than one irrigation branch within the cluster via the OPA_IWS model with the difference in the corresponding gauged surface runoffs.

Step [8] Quantifying the uncertainties in the estimations of the branch-based water demands to calculate their corresponding quantiles under the desired probabilities.

Step [9] Quantifying the corresponding reliabilities to the existing water zone-based and branch-based water demands and serve the probabilistic-based water demand estimates under a desired reliability as the introduced ones. The above model development framework could refer to Figure 1.

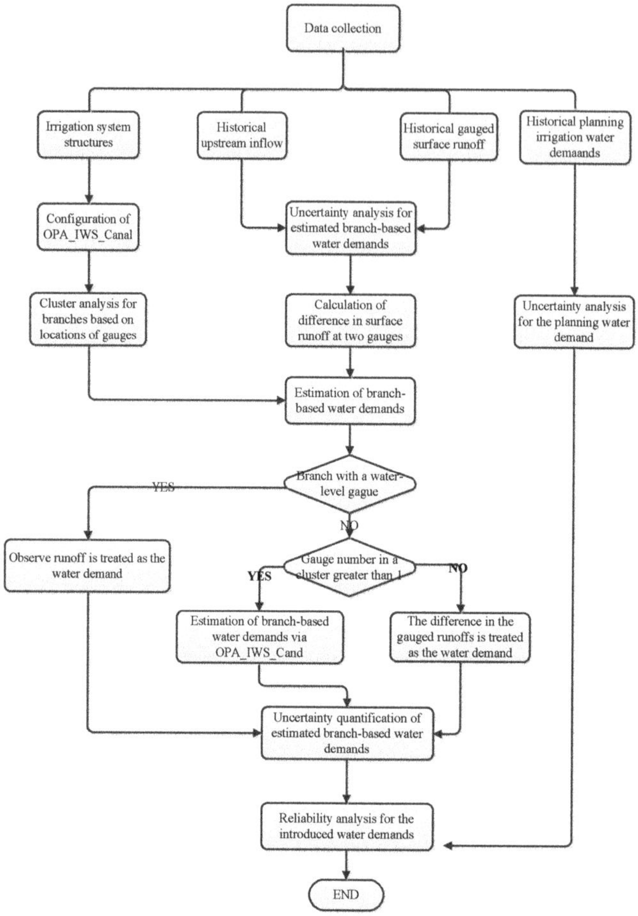


**Figure 1.** The schematic framework of estimating branch-based irrigation water demands and quantifying the corresponding uncertainties and reliabilities via the proposed RA_IWD_Canal model.

## 3. Study Area and Data

To express the development and application of the proposed RA_IWD_Canal model for estimating branch-based irrigation water demands, a multi-branch irrigation zone, Zhudong Canal irrigation zone, is selected in this study. The Zhudong Canal irrigation zone, whose cultivation extent is nearly 800 ha, is located in Northern Taiwan (see Figure 2). The main crop is rice, which grows from the middle of February to the end of November. Within the study area, 15 branches (BL1-BL14 and Bazhuang branch), whose irrigation extents vary from 10 ha to 50 ha (see Figure 3), as well as the two water intake structures, including Baoshan Reservoir located between the $7^{th}$ seventh and $8^{th}$ eighth (i.e., BL7 and BL8) branches and the Yuandon treatment plant situated between the $2^{nd}$ nd and $3^{rd}$ branches.; their detailed physical characteristics and operational rules could be referred to in Wu's investigation [2]. Also, to effectively capture the spatial and temporal changes in the irrigation water in the Zhudong Canal zone, 15 discharge gauges (G1–G15) were set up.

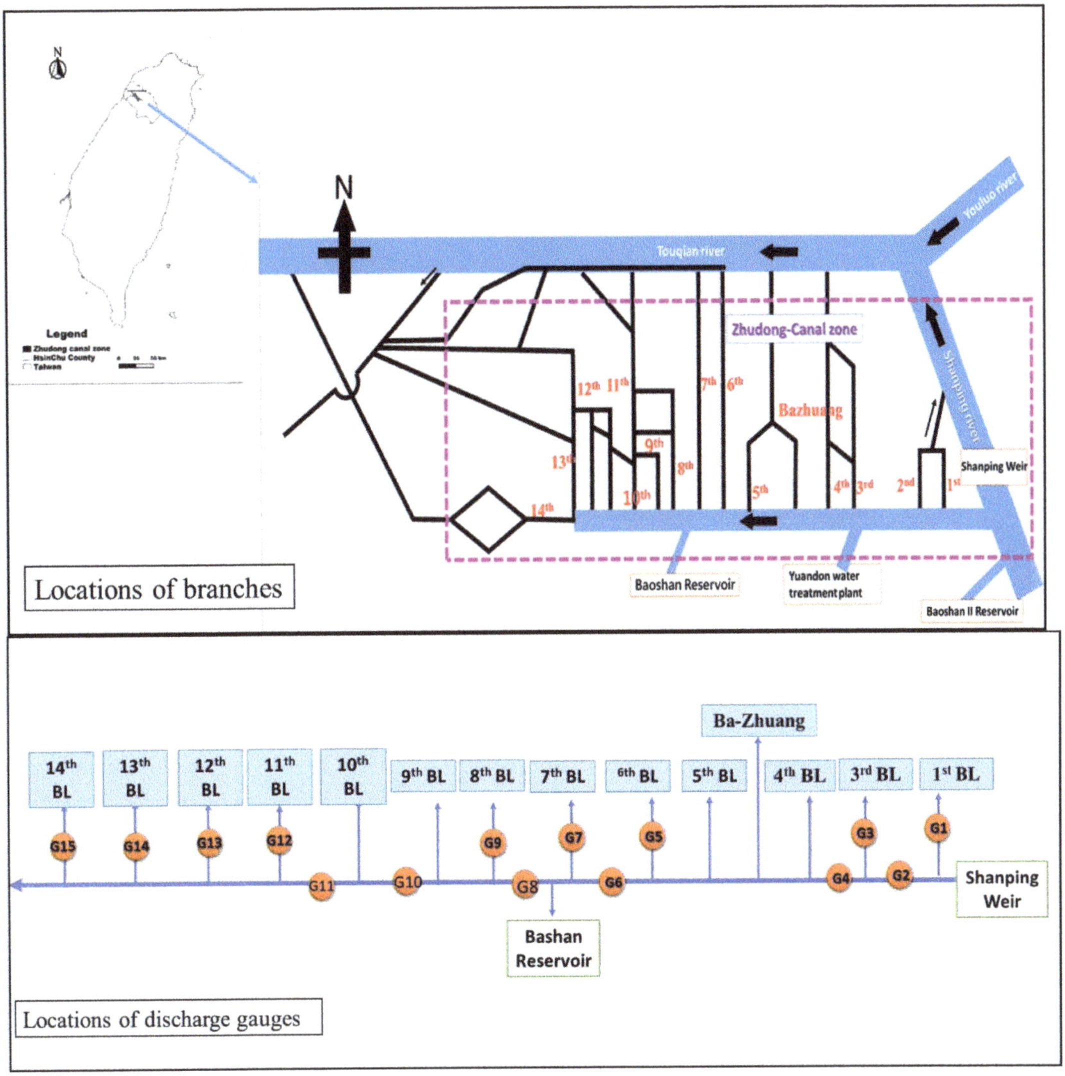


**Figure 2.** Location of the study area Zhudong Canal zone with 15 irrigation channel-based branches and 15 discharge gauges [2,20].

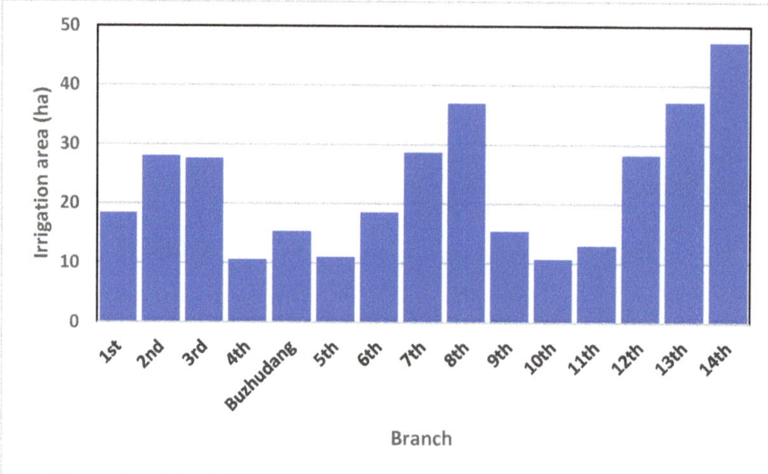


**Figure 3.** Cultivation extents at the fifteen channel-based branches within the Zhudong Canal irrigation zone [2,20].

The Zhudong canal zone supplied the irrigation water converted from upstream inflow at the Shanping Weir. Namely, the upstream inflow mainly comes from the Shanping River, whose historical 10-day surface runoff at the Shaning water-level gauge in 1959–2022 changes from 1 $m^3/s$ to 400 $m^3/s$ (see Figure 4). In addition, to effectively measure the spatial change of the surface runoff in the Zhudong Canal zone, 15 discharge gauges were installed from upstream to downstream in the Zhudong Canal zone to measure the surface runoff, as shown in Figure 2. Among 15 discharge gauges, the gauges G2, G3, G6, G8, G10, and G11 are located in the main channel, and the remaining gauges are set up in the branches, which could directly measure water intake to the branches. Thus, given Figure 5, the 10-day surface runoffs at 15 discharge gauges were measured from 2019 to 2022, indicating that the observed surface runoff at the $4^{th}$ to $33^{rd}$ 10-day period (on average 0.5 $m^3/s$) markedly exceeds the observations at the remaining 10-day periods (nearly 0.14 $m^3/s$); in particular, at the $3^{rd}$, $5^{th}$, $7^{th}$ and $9^{th}$ discharge gauges, their observed surface runoffs are considerably over observations at the remaining gauges by 63%; this indicates that the observations of surface runoff within the study area (Zhudong Canal zone) are associated with a significant variation in time and space. Moreover, the $4^{th}$ to $33^{rd}$ 10-day periods could be regarded as the rainy seasons, with the remaining 10-day period being treated as the drought season. Therefore, this study focuses on the uncertainty quantification and reliability assessment of the irrigation water demands in the rainy season (i.e., $4^{th}$–$33^{rd}$ 10-day period).

In general, ahead of supplying irrigation water, the planning irrigation water demands should be given as the criterion for evaluating the irrigation performance [1,15]. Subsequently, the irrigation water supplies and requirements are represented in 10-day periods (Wu et al., 2023). Therefore, the planning irrigation water demands at various 10-day periods in the Zhudong Canal zone from 2015 to 2024 were introduced for evaluating irrigation performance as shown in Figure 6 (roughly from 0.85 to 1.5 $m^3/s$); this unveils the intro irrigation water demands in the Zhudong Canal zone reach the maximum (approximately 1.8 $m^3/s$) at the $23^{rd}$ 10-day periods. Note that within the Zhudong Canal zone, the third branch (Su-Qi-Lin) was also given the planning irrigation water demand as shown in Figure 6b with the significantly less introduced ones (about 0.11 $m^3/s$).

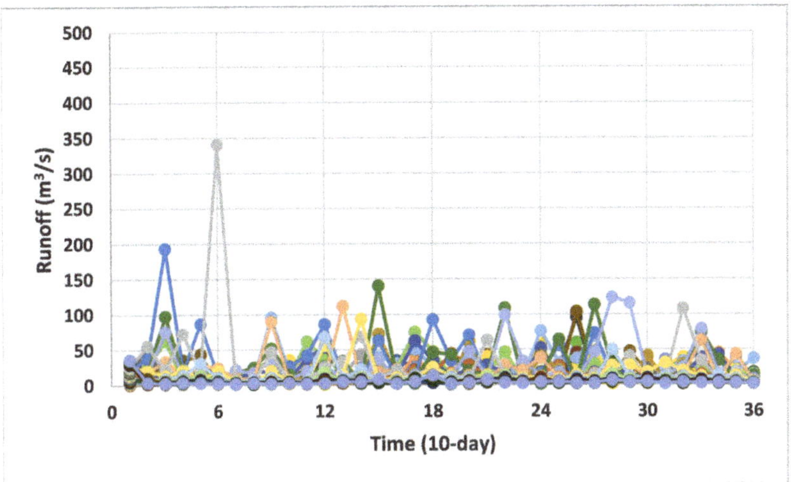


**Figure 4.** Historical 10-day surface runoff at Shanging water-level gauge Weir from 1959–2022.

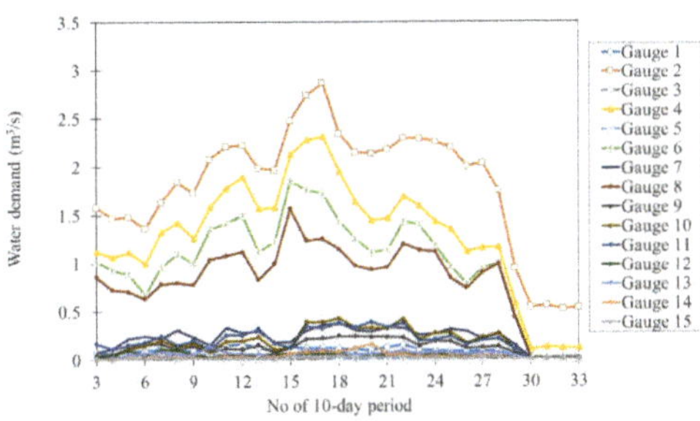


(**a**) 2020

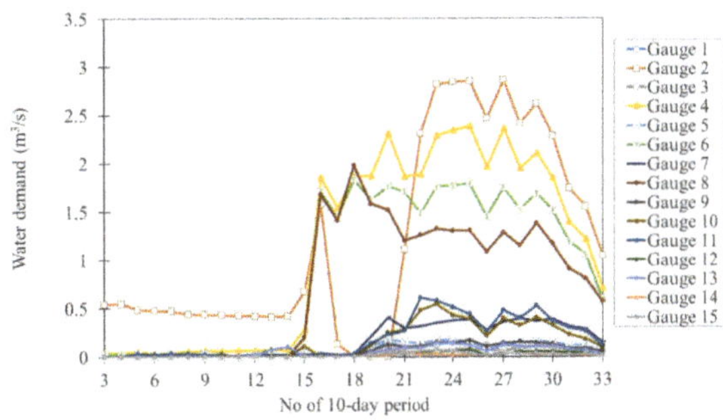


(**b**) 2021

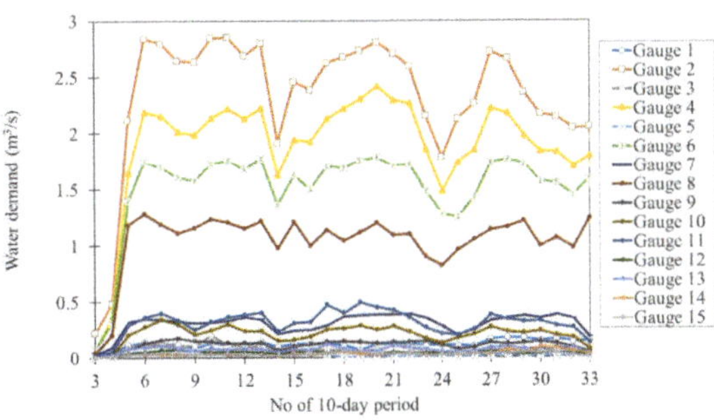


(**c**) 2022

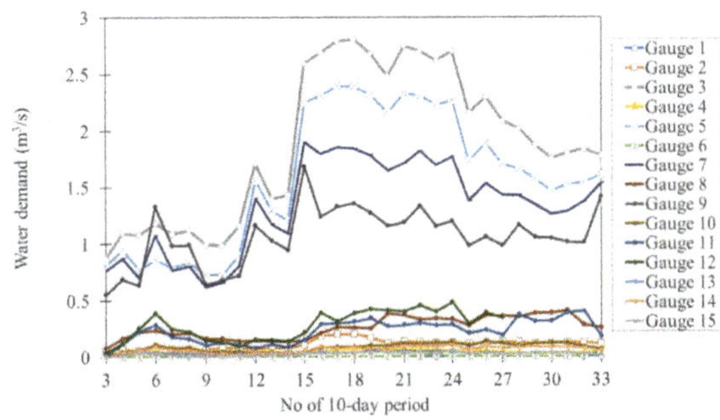


(**d**) 2023

**Figure 5.** Historical surface runoff at 15 discharge gauges within the Zhudong Canal zone.

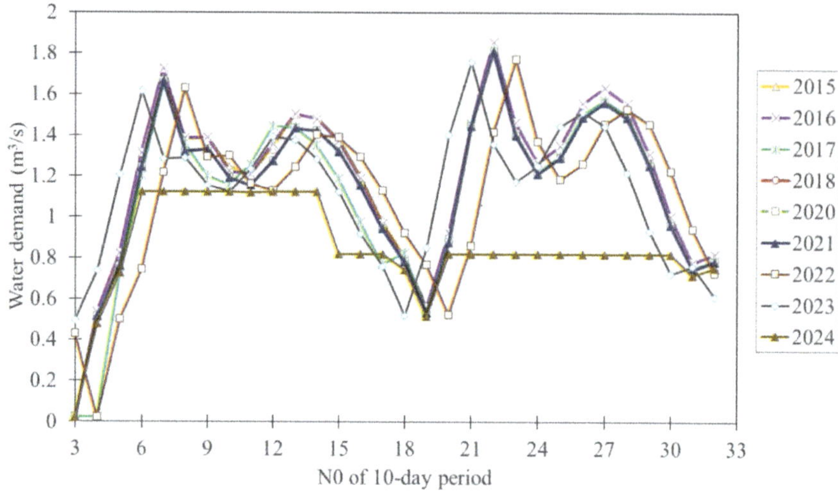


(**a**) Zhudong Canal zone

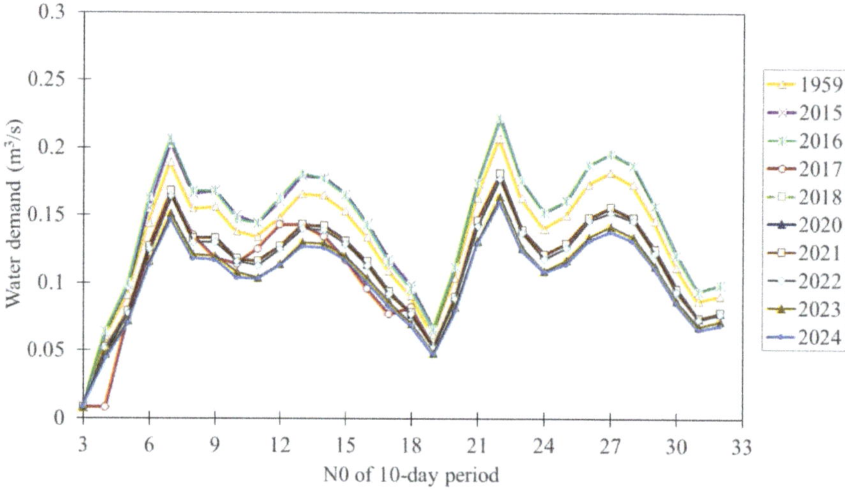


(**b**) The 3rd branch (Su-Qi-Lin branch)

**Figure 6.** Historical 10-day planning irrigation water demands within the study area.

## 4. Results and Discussion

Regarding the development framework of the RA_IWD_Canal model (see Figure 1), the irrigation-related data in the study area (Zhudong Canal zone) should be collected in advance. Then, the proposed RA_IWD_Canal model could estimate the branch-based irrigation water demands, quantify their uncertainties, and include reliabilities. The detailed model development and evaluation of application results can be found below.

### 4.1. Establishment of the Relationship between the Branch-Based Water Demand and Gauged Runoff

Relying on Figure 1, before developing the proposed RA_IWD_Canal model, the irrigation branches should be grouped into the desired cluster based on the locations of the irrigation branches and discharge gauges. As shown in Figure 2, a group of discharge gauges are installed at the 1st, 6th, and 8th branches and from the 11th to 14th branches; their irrigation water demands could be given the measurements of gauged surface runoff. In addition, the irrigation water demands at the 9th and 10th branches could be achieved by calculating the difference in the observed surface runoffs at the two discharge gauges as

$$Q_{D,9th} = Q_{G,8th} - Q_{D,8th} \tag{15}$$

$$Q_{D,10th} = Q_{G,10th} - Q_{G,11th} \tag{16}$$

where $Q_{D,9th}$ and $Q_{D,10th}$ account for the estimated irrigation water demands at the 9th and 10th branches, respectively; and $Q_{G,8th}$, $Q_{G,10th}$ and $Q_{G,11th}$ serve as the observed surface runoffs at the 8th, 10th, and 11th water-level gauges. Thus, estimating the irrigation water demands directly with the measurements of gauged surface runoff is called a data-derived approach. Regarding the remaining branches (i.e., the 4th branch, fifth branch, and the Ba-Zhuang branches), their irrigation water demands should be obtained via the OPA_IWS_Canal model with the difference in the gauged surface runoffs at the 4th and 5th discharge gauges (i.e., $Q_{G,4th}$ and $Q_{G,5th}$), named the model-derived approach. In summary, Table 1 lists the formulas to estimate the irrigation water demands at all branches in the study area. Flowing the estimations of the branch-based irrigation water demands, the zone-based water demand ($Q_{D,zone}$) could be estimated with the water demands at all branches, excluding the 3rd branch (i.e., Su-Qi-Lin), which is given separately, as

$$Q_{D,zone} = \sum_{i=1,i\neq 3}^{14} Q_{D,i} \tag{17}$$

**Table 1.** Formulae for estimating the branch-based irrigation water demands via the proposed RA_IWD_Canal model.

| No of Branch | Gauged Surface Runoff Is Used | Formula |
|---|---|---|
| The 1st branch $Q_{D,1st}$ | Gauged discharge $Q_{G,1st}$ | $Q_{D,1st} = Q_{G,1st}$ |
| The 3rd branch (Shu-Qi-Lin) $Q_{D,3rd}$ | Gauged discharge $Q_{G,3rd}$ | $Q_{D,3rd} = Q_{G,3rd}$ |
| The 4th branch $Q_{D,4th}$ | Difference in gauged discharges $Q_{G,4th}$ and $Q_{G,5th}$ | OPA_IWS_Canal model |
| The 5th branch $Q_{D,5th}$ | | |
| The branch Ba-Zhuang $Q_{D,BZ}$ | | |
| The 6th branch $Q_{D,6th}$ | Gauged discharge $Q_{G,3rd}$ | $Q_{D,3rd} = Q_{G,3rd}$ |
| The 7th branch $Q_{D,7th}$ | Gauged discharge $Q_{G,7th}$ | $Q_{D,7th} = Q_{G,7th}$ |
| The 8th branch $Q_{D,8th}$ | Gauged discharge $Q_{G,9th}$ | $Q_{D,8th} = Q_{G,9th}$ |
| The 9th branch $Q_{D,9th}$ | Difference in gauged discharges $Q_{G,8th}$ and $Q_{G,9th}$ | $Q_{D,7th} = Q_{G,8th} - Q_{G,9th}$ |
| The 10th branch $Q_{D,10th}$ | Difference in gauged discharges $Q_{G,10th}$ and $Q_{G,11th}$ | $Q_{D,10th} = Q_{G,10th} - Q_{G,11h}$ |
| The 11th branch $Q_{D,11th}$ | Gauged discharge $Q_{G,12th}$ | $Q_{D,11th} = Q_{G,12th}$ |
| The 12$^{th}$ branch $Q_{D,12th}$ | Gauged discharge $Q_{G,13th}$ | $Q_{D,12th} = Q_{G,13th}$ |
| The 13th branch $Q_{D,13th}$ | Gauged discharge $Q_{G,14th}$ | $Q_{D,13th} = Q_{G,14th}$ |
| The 14th branch $Q_{D,14th}$ | Gauged discharge $Q_{G,15th}$ | $Q_{D,14th} = Q_{G,7th}$ |

### *4.2. Uncertainty Quantification End Assessment of Introduced Planning Irrigation Water Demands*

Before carrying out the reliability analysis for the branch-based irrigation water demands in the study area, the uncertainties in the officially-introduced water demands for the study area (Zhudong irrigation zone) and the 3rd branch (Su-Qi-Lin) (see Figure 5) should be quantified; Figure 7 shows that their L-moment ratios at 30 10-day periods are calculated via the proposed RA_IWD_Canal model, as shown in Figures 7 and 8, in which the corresponding 95% confidence intervals could also be found. Observed from Figure 7, the mean values of introduced planning irrigation water demands for the Zhuang Canal zone vary from 0.02 $m^3/s$ to 1.7 $m^3/s$, with a noticeable L-CV value (on average, 0.13) and a large confidence interval of around 0.8 $m^3/s$; in detail, the $7^{th}$–$33^{rd}$ 10-day periods were given high irrigation water demands, on average from 0.8 $m^3/s$ to 1.5 $m^3/s$. In summary, the introduced planning irrigation water demands are more likely to have temporal variations in the various 10-day periods; this indicates that the current planning water demands possibly hardly reflect the variations attributed to climate change and extreme events. A similar varying trend could be referred to in Figure 8, showing that the planning water demands at the $3^{rd}$ branch (Su-Qi-Lin) have a low average of 1.2 $m^3/s$ with a considerable L-CV of around 0.1 $m^3/s$.

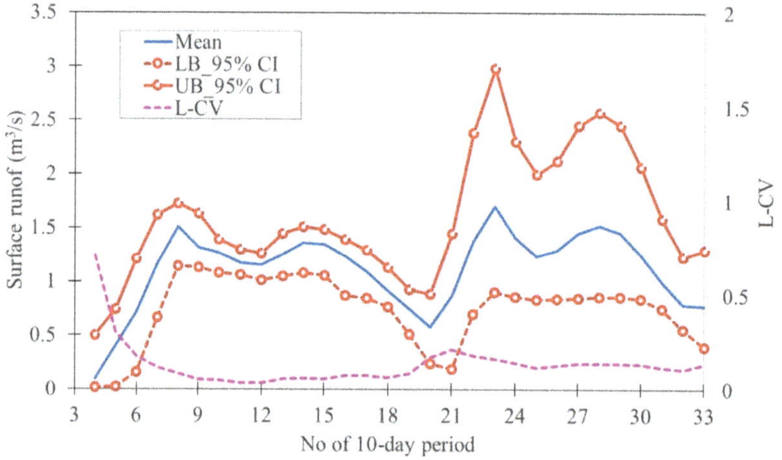


**Figure 7.** Uncertainties of introduced planning irrigation water demands for the Zhudong Canal zone.

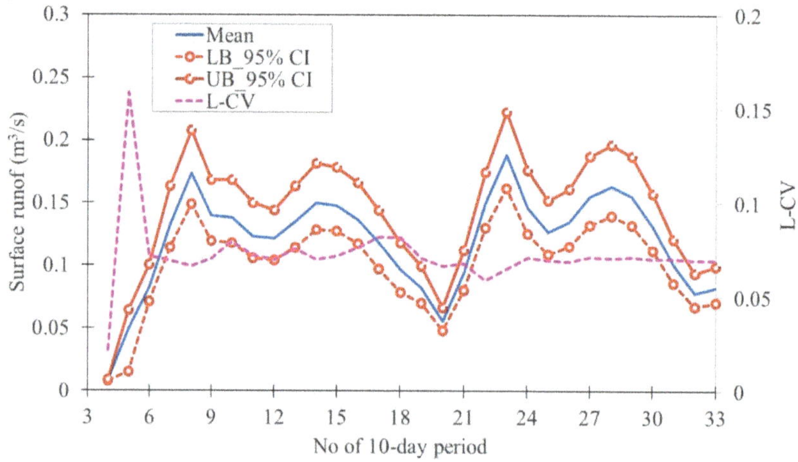


**Figure 8.** Uncertainties of introduced planning irrigation water demands for the $3^{rd}$ branch (Su-Qi-Lin branch).

Therefore, by proceeding with the proposed RA_IWD_Canal model, it could be known that finalizing the irrigation water demands associated with the desired uncertainties is necessary in response to the variations due to the temporal and spatial change in hydrological features. Accordingly, the proposed RA_IWD_Canal model is supposed to proceed with quantifying the reliabilities of introduced irrigation water demands subject to the uncertainties in the surface runoff-related irrigation water supplies.

### 4.3. Uncertainty Quantification and Assessment of Branch-Based Irrigation Water Demands

Concerning Table 1, for a group of the branches in the study area, their corresponding irrigation water demands could be obtained via the data-derived approach with the gauged surface runoffs as shown in Figure 9; alternatively, the branch-based irrigation water demands (i.e., the four and fifth as well as the Ba-Zhang branches) are estimated using the OPA_IWS_Canal model, with a given supplying satisfaction index (see Equation (6)) (i.e., $SI_{IBL}$ = 1.0) set up in the proposed RA_IWD_Canal model (see Figure 10). Comparing the results from Figures 9 and 10, the data-derived branch-based water demands exhibit a considerable change in time; for example, concerning the 3$^{rd}$ branch (i.e., Su-Qi-Lin), the resulting irrigation water demands have a significant change, roughly from 0.4 $m^3/s$ to 0.8 $m^3/s$ in 2020–2023. Nevertheless, the above branches are nearly consistent with the spatial distributions of the data-derived irrigation water demands markedly; for instance, in 2023, the more significant estimated water demands (around from 0.03 $m^3/s$ to 0.3 $m^3/s$) could be found at the 18$^{th}$–30$^{th}$ 10-day periods. Instead, the 4$^{th}$, 5$^{th}$, and Ba-Zhuang branches include a similar spatial and temporal varying trend in the model-derived irrigation water demands; for example, in 2000, their estimated irrigation water demands significantly increase from 0.4 to 0.7, which reach the maximum at the 14$^{th}$–17$^{th}$ 10-day periods and then gradually decline to 0.03 at the 30$^{th}$–33$^{rd}$ 10-day periods; as comparable to Figure 5, the above varying trend in time resembles the change in the difference in the gauged discharges at the 4$^{th}$ and 5$^{th}$ discharge gauges.

Using the proposed RA_IWD_Canal model, the uncertainties of estimated branch-based irrigation water demands could be represented in terms of the L-moment ratio calculated, as shown in Figure 11. It can be seen that the estimated branch-based irrigation water demands during the rainy seasons (4$^{th}$–33$^{rd}$ 10-day periods) display a considerable variation in space due to a large L-CV (on average, from 0.2 to 0.8). It concludes that branch-based irrigation water demands noticeably rely on the measurements of the gauged surface runoff. By doing so, the variations in the observed surface runoff significantly lead to uncertainties in the estimations of the branch-based branches.

Altogether, the regulation of the branch-based irrigation water demands should be considered based not only on irrigation features (e.g., cultivation extents and crop types), but also on the change of measured surface runoff supplied in time and space. That is to say, the discharge gauges comprehensively installed within the irrigation zones play a vital role in accurately and reliably estimating the irrigation water requirements with high spatial resolution.

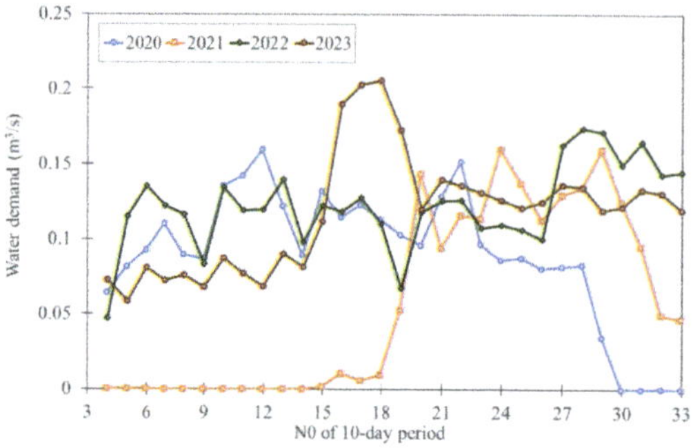


(**a**) The 1st branch

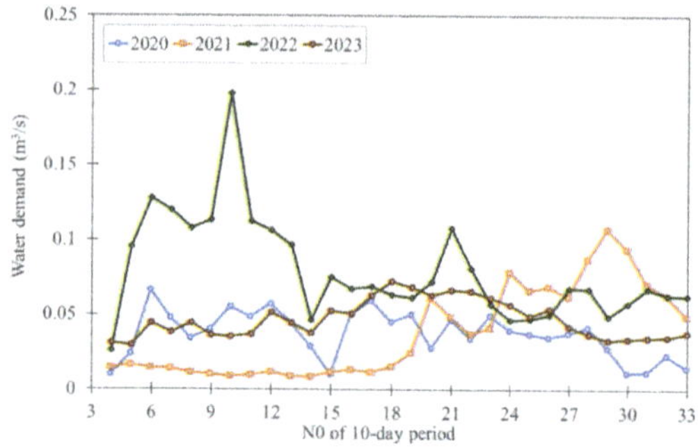


(**b**) The 3rd branch (Su-Qi-Lin)

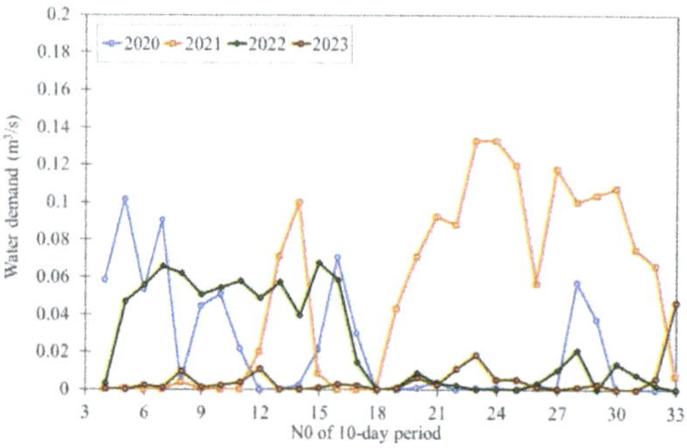


(**c**) The 6th branch

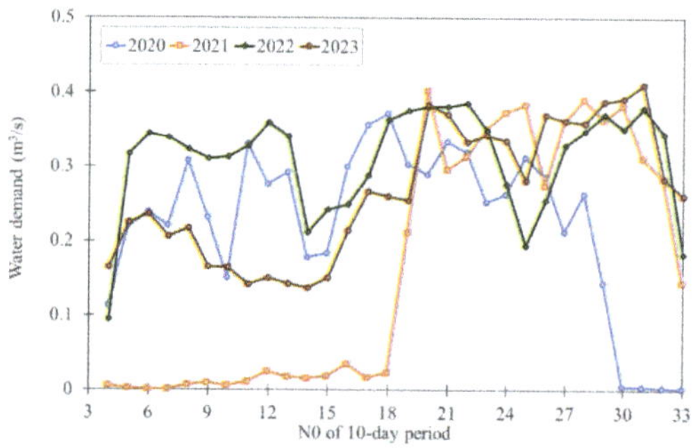


(**d**) The 7th branch

**Figure 9.** *Cont.*

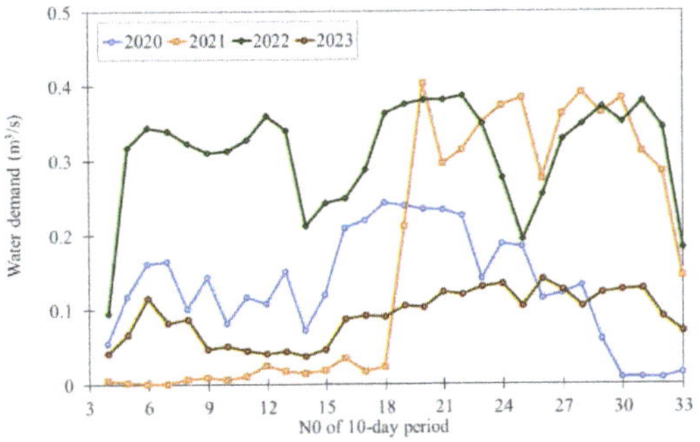


(**e**) The 8th branch

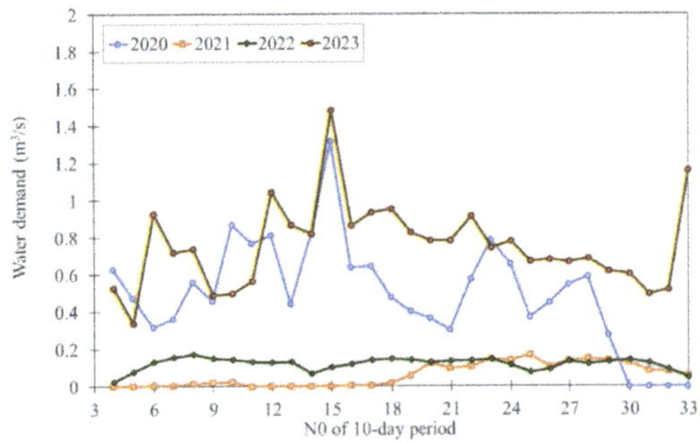


(**f**) The 9th branch

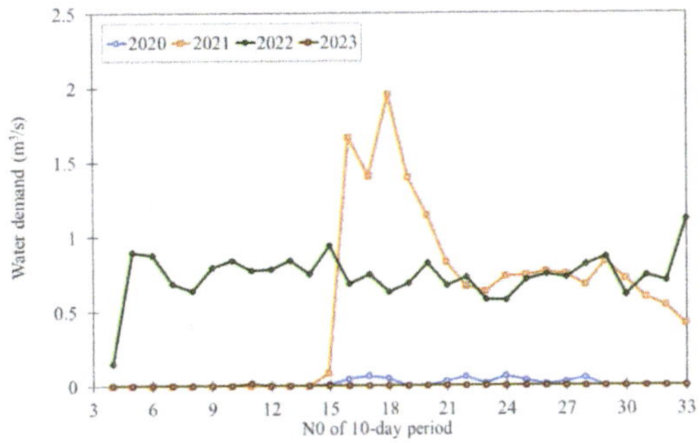


(**g**) The 10th branch

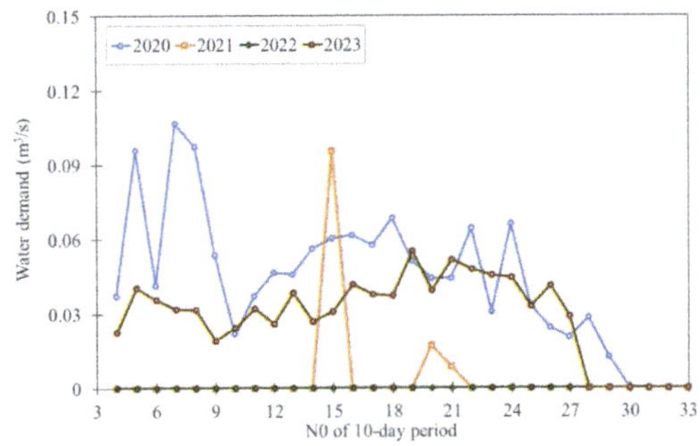


(**h**) The 11th branch

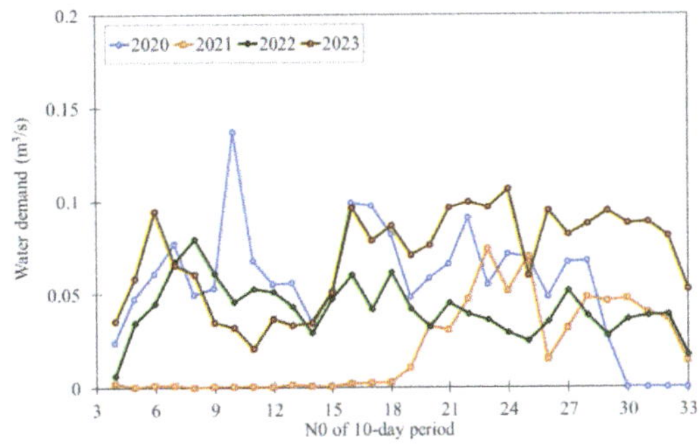


(**i**) The 12th branch

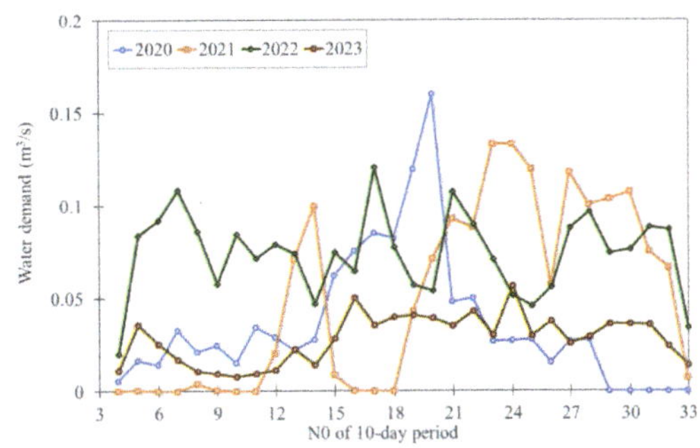


(**j**) The 13th branch

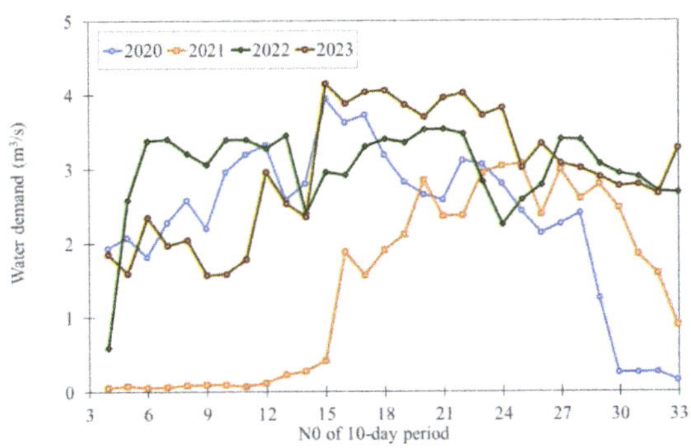


(**k**) The 14th branch

**Figure 9.** Data-derived irrigation water demands at the branches.

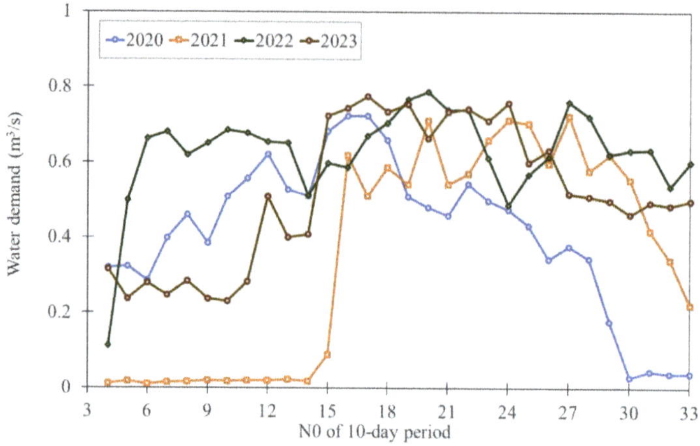


(**a**) The 4th branch

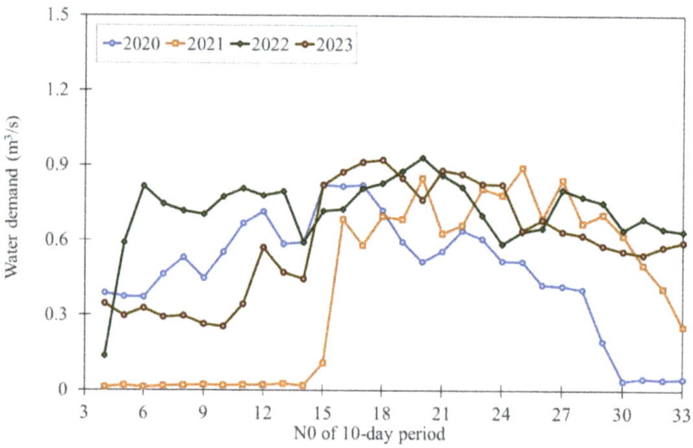


(**b**) Ba-Zhuang branch

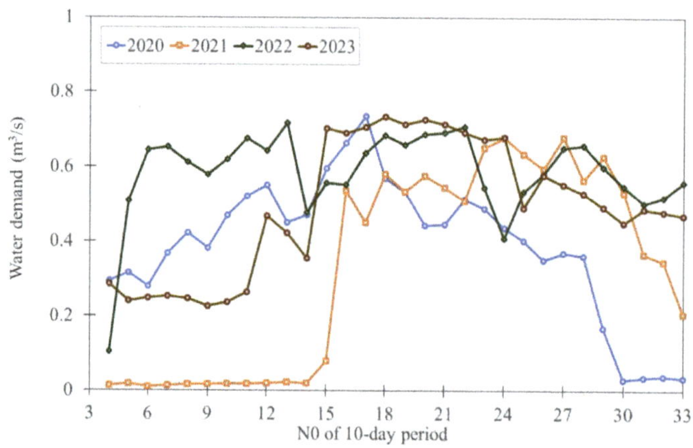


(**c**) The 5th branch

**Figure 10.** Model-derived irrigation water demands at the specific branches.

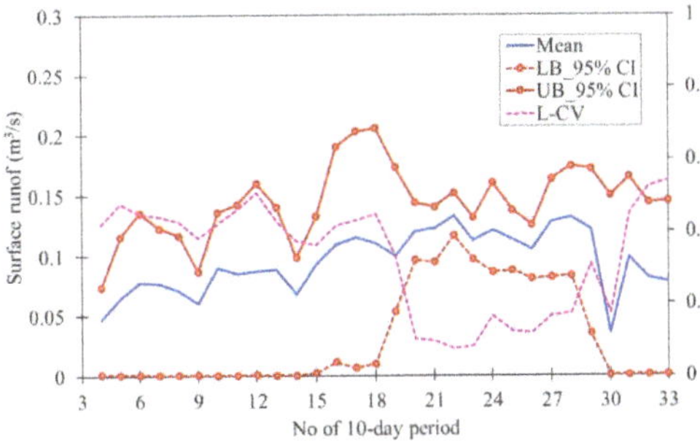


(a) The 1st branch

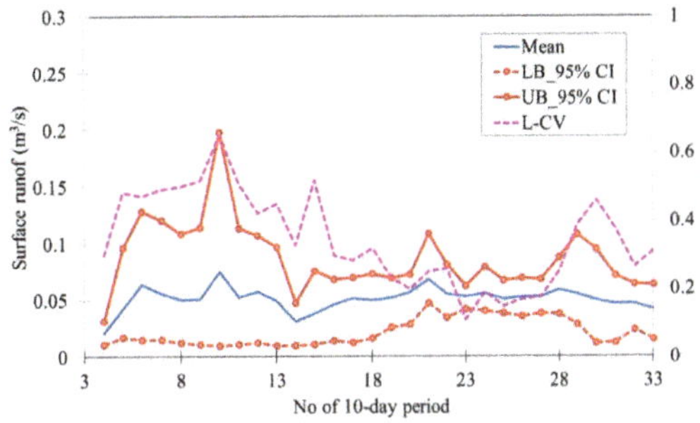


(b) The 3rd branch (Su-Qi-Lin)

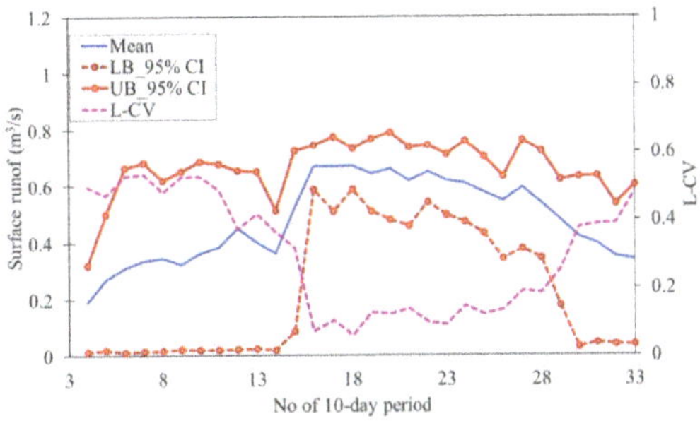


(c) The 4th branch

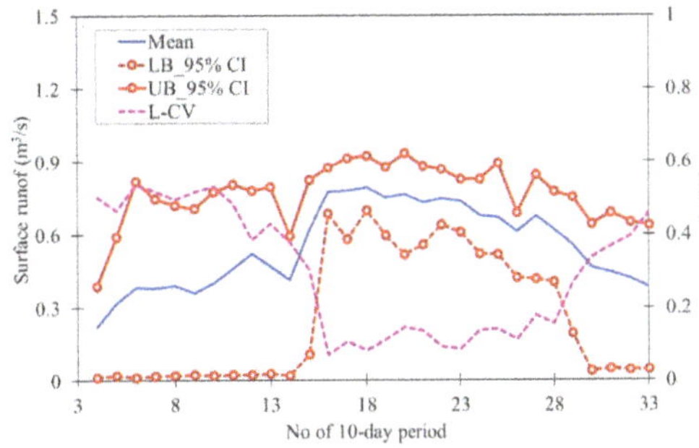


(d) Ba-Zhuang branch

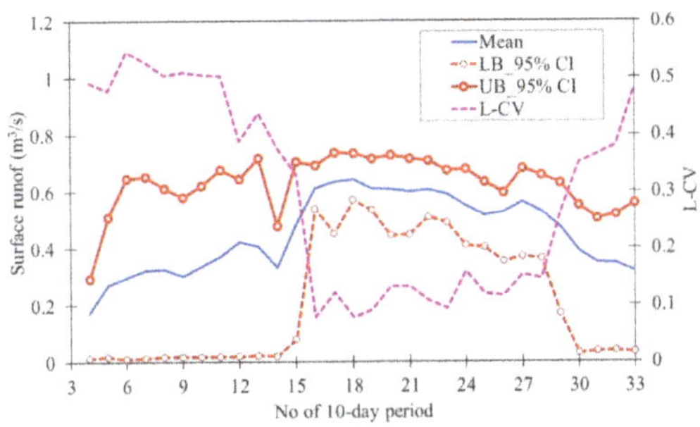


(e) The 5th branch

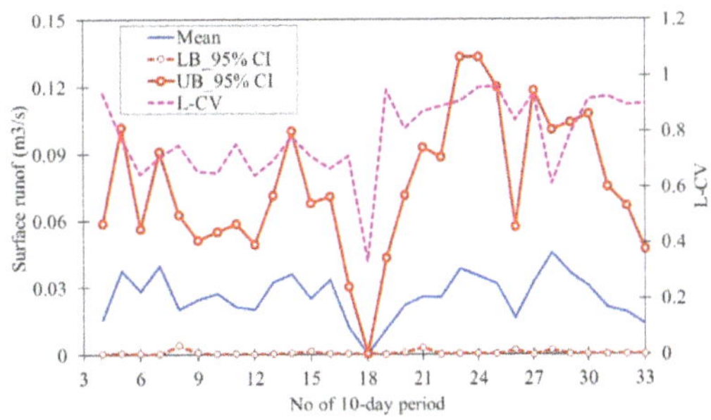


(f) The 6th branch

**Figure 11.** *Cont.*

**Figure 11.** Uncertainties in the estimated branch-based irrigation water demands.

### 4.4. Reliability Quantification of Irrigation Water Demands

As well as estimating the branch-based irrigation water demands, the proposed RA_IWD_Canal model could be applied in the reliability analysis for the introduced planning irrigation water demand by establishing the weighted quantile curves, consisting of the quantiles under the desired probability. Since the two planning irrigation water demands were issued for the study area, including the Zhudong Canal zone and the 3rd branch (Su-Qi-Lin), the irrigation water demands for the Zhudong Canal zone should be achieved by summing up all the resulting branch-based ones via Equation (17). Figure 12 shows the estimated irrigation water demands for the Zhudong Canal zone at the 30 10-day periods in 2020–2024, indicating that the spatial varying trend of the one-based irrigation water demands significantly change with time (year); in detail, the more significant water demands (about from 2 $m^3/s$ to 4 $m^3/s$) could be observed at the 4th–20th 10-day period in 2020 in contrast with results at the 20th and 30th periods, varying from 1.8 $m^3/s$ to 2 $m^3/s$. This unveils that considerable spatial and temporal variation exists in the zone-based irrigation water demands, indicating that the reliabilities are supposed to be quantified to evaluate their irrigation efficiency. Alternatively, the estimated irrigation water demands at the Su-Qi-Lin branch could be referred to in Figure 11b.

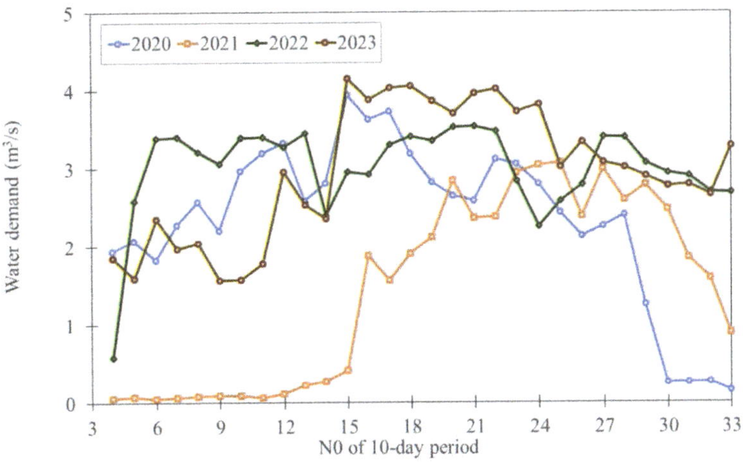


**Figure 12.** Estimations of the irrigation water demands for the Zhudong Canal zone.

Accordingly, to proceed with the reliability assessment of the desired irrigation water demands, the corresponding quantile curves should be established via the weighted frequency curve methods within the proposed RA_IWD_Canal model, as shown in Figure 13. Given Figure 13, the resulting quantiles from the estimated water demands for the Zhudong Canal zone commonly exceed those at the 3rd branch, especially for the lower cumulative probabilities of less than 0.1. By doing so, the corresponding exceedance probabilities to the introduced irrigation water demands in 2019–2024 (i.e., reliability) could be calculated with the above-resulting quantile curves as shown in Figures 14 and 15; it can be seen that the exceedance probabilities of the introduced water demands for the Zhudong irrigation zone (see Figure 14), on average, approximate 0.85; specifically, in the 16th–28th 10-day periods, the corresponding reliability (exceedance probability) reaches 0.97. Moreover, in the 7th–10th 10-day periods, the corresponding introduced water demands could be given with an exceedance probability (i.e., reliability) of less than 0.7. It concludes that the study area (Zhudong Canal zone) could be sufficiently supplied irrigation water higher than the introduced ones in 2019–2024 with a high likelihood. In contrast, the Zhudong Canal zone, the exceedance probabilities of the introduced irrigation water demands at the Su-Qi-Lin (the 3rd branch) (see Figure 15), on average, are noticeably less than 0.25 with a minimum approaching 0.00001 one at the 14th 10-day period, namely treated as the fallow period; this is because surface runoff has been probably converted to the upstream 1st branch to hardly supply the expected water amount to the 3rd branch (Su-Qi-Lin). Namely, the current introduced irrigation water demands at the Su-Qi-Lin branch are markedly overestimated, making it difficult to achieve the desired irrigation requirements. That is to say, the introduced water demands for the Su-Qi-Lin branch should be necessarily declined to boost the irrigation efficiency.

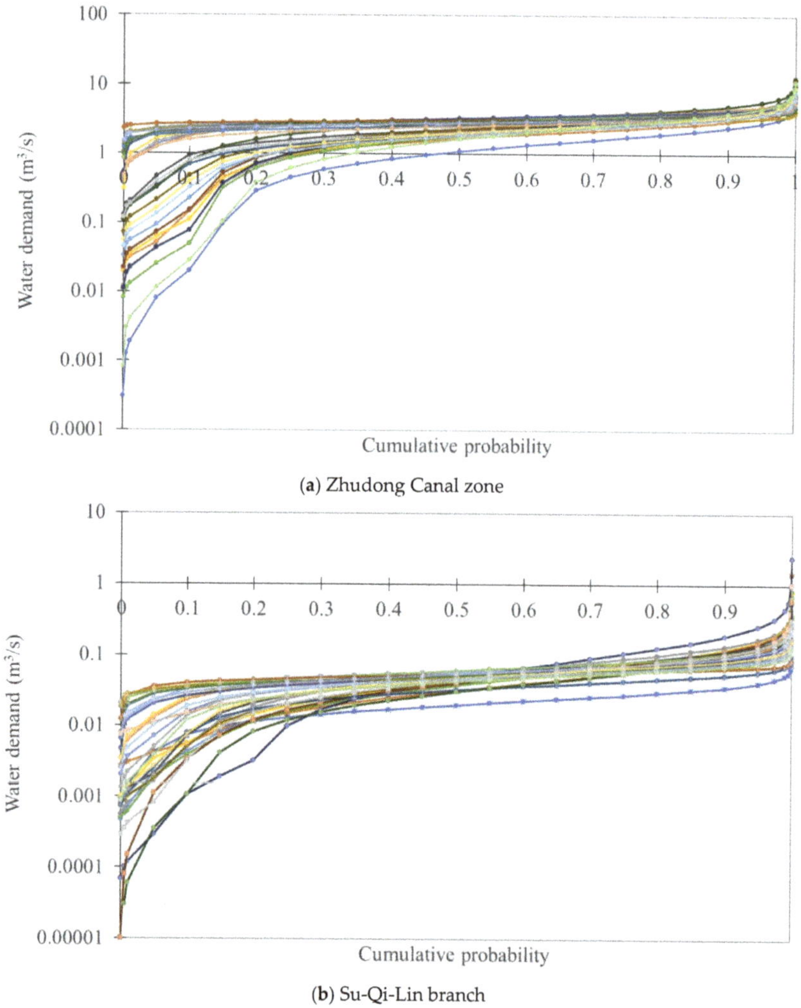


**Figure 13.** Quantiles of estimated irrigation water demands for the Zhudong Canal zone and the 3rd branch (Su-Qi-Lin) at 30 10-day periods.

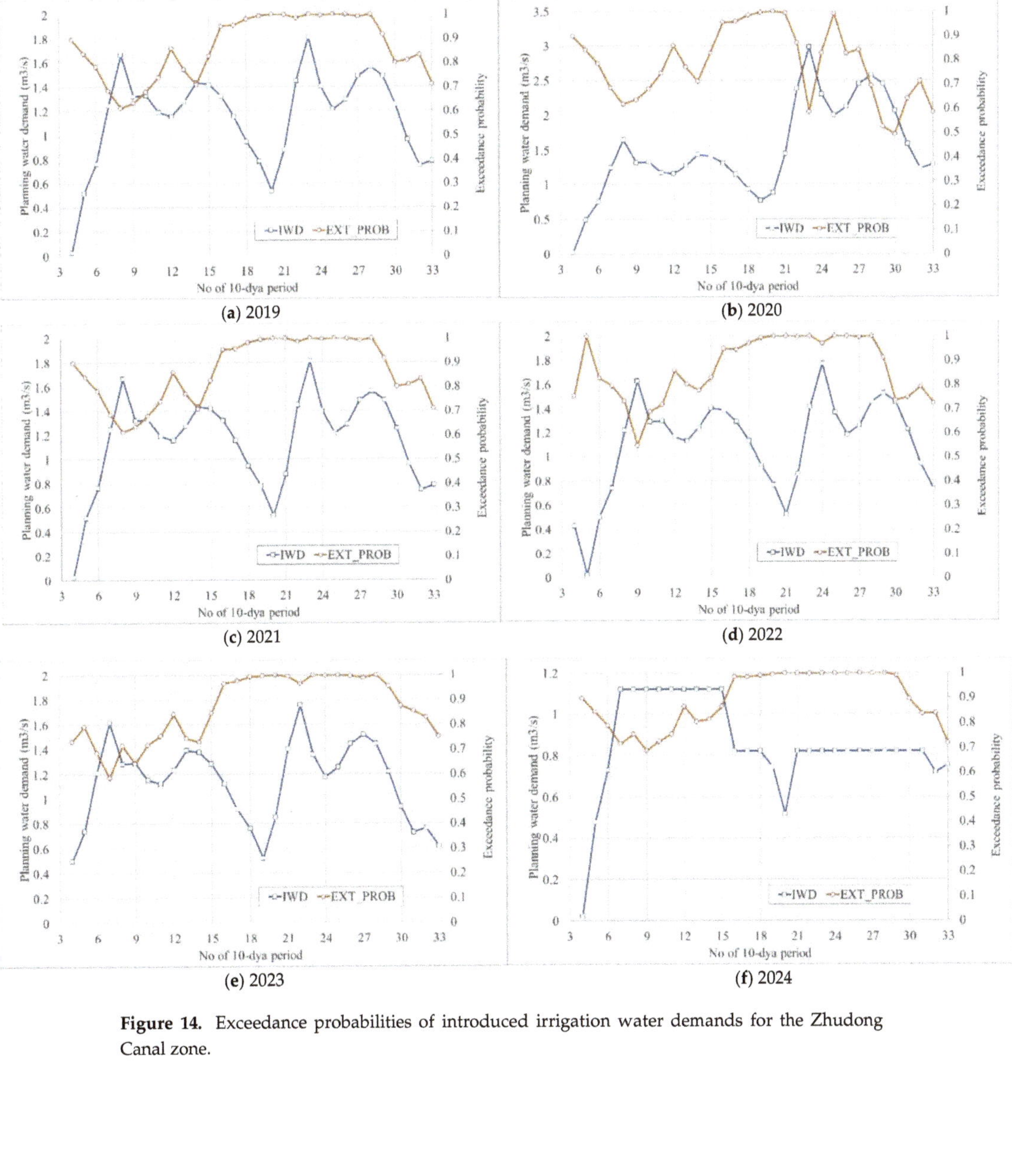


**Figure 14.** Exceedance probabilities of introduced irrigation water demands for the Zhudong Canal zone.

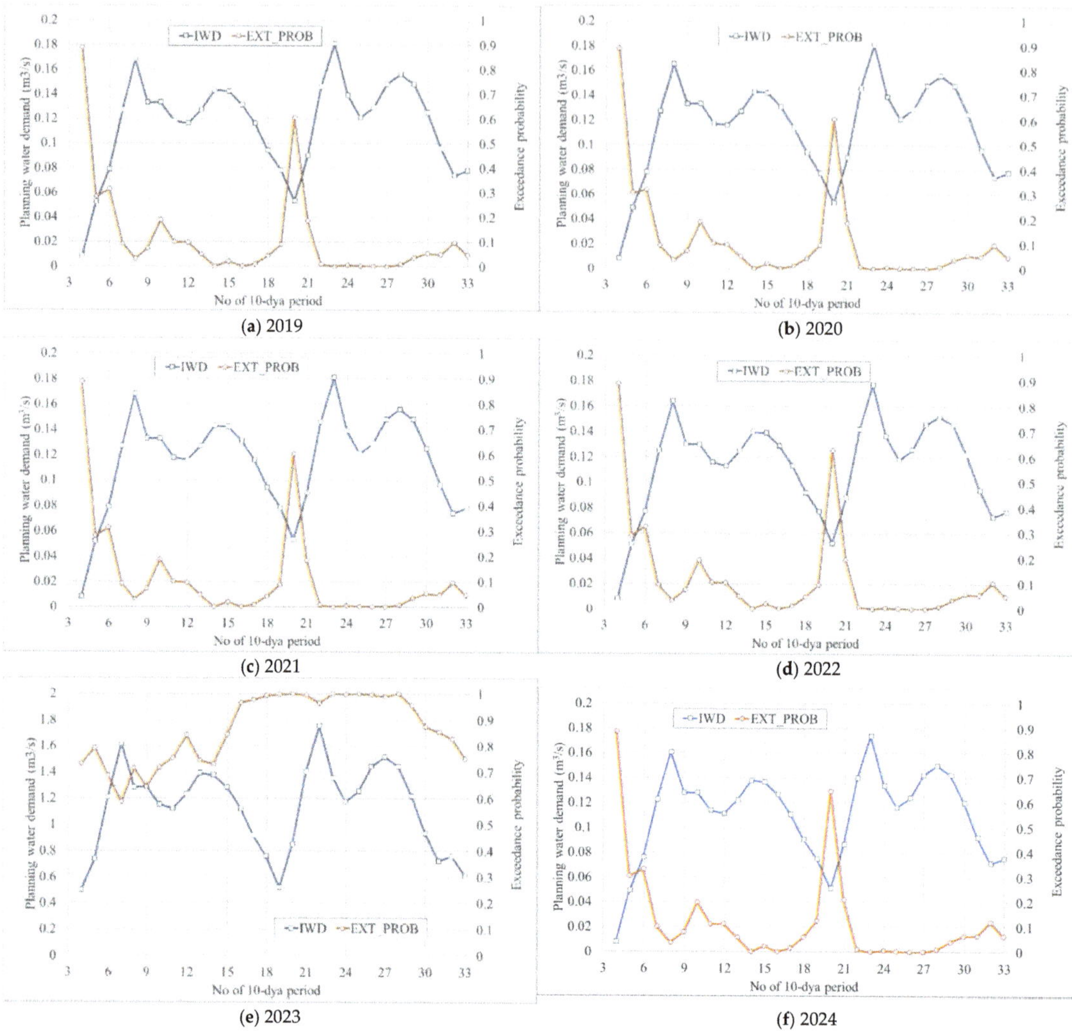


**Figure 15.** Exceedance probabilities of introduced irrigation water demands for the 3rd branch (Su-Qi-Lin).

Altogether, despite the recently introduced irrigation water demands for the Zhudong Canal zone with an average reliability of higher than 0.9, the irrigation water is hardly supplied subject to the introduced water demands at the specific 10-day periods (7$^{th}$–10$^{th}$ periods) merely with a reliability of nearly 0.7. In addition, the introduced ones for the Su-Qi-Lin branch exhibit a low irrigation efficiency with a reliability of less than 0.3, indicating that the introduced water demands at the Su-Qi-Line branch area are considerably overestimated to lead to a shortage risk. To enhance the reliability of the introduced irrigation water demands for the entire branches within the study area, the probabilistic-based irrigation water could be provided by calculating the quantiles of the irrigation water demands under a desired exceedance probability (i.e., reliability) as the introduced magnitudes. Figure 16 illustrates the probabilistic-based irrigation water demands at 30 10-day periods for the Zhudong Canal zone as well as the third branch under a desired reliability of 0.8; also, it could be seen that the above probabilistic-based irrigation water demands are mutually correlated in time with a high correlation coefficient of roughly 0.8, indicating that it is necessary to take the spatiotemporal correlation of irrigation water as setting the introduced water demands into account. As well as the zone-based irrigation water demands, the proposed RA_IWD_Canal could provide the probabilistic-based irrigation water demands with high spatial resolution, revealing that the irrigation water demands at all branches could be quantified under an accepted reliability of 0.8 as shown in Figure 17. According to Figure 17, the 6$^{th}$ and 11$^{th}$ branches are associated with significantly low probabilistic-based water demands (approximately 0.000001 $m^3/s$), revealing that they could be treated as non-irrigation branches in case of insufficient irrigation water supplies.

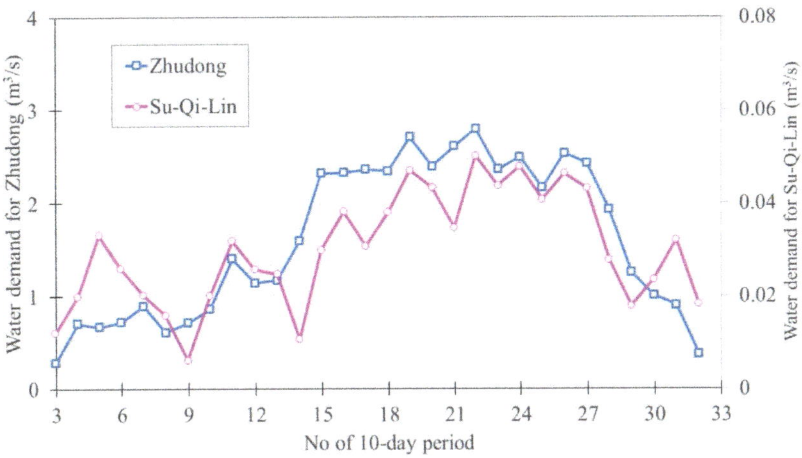


**Figure 16.** Probabilistic-based irrigation water demands introduced at the Zhudong Canal zone and Su-Qi-Lin branch, with a reliability of 0.8.

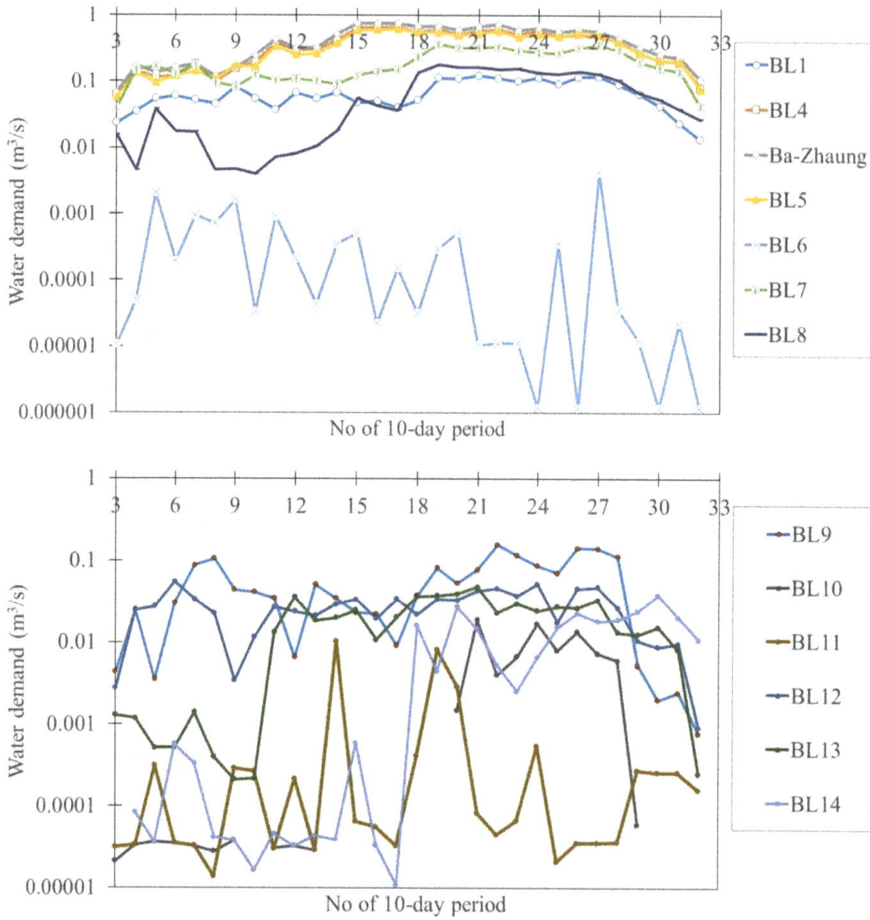


**Figure 17.** Probabilistic-based irrigation water demands at all branches, with a reliability of 0.8.

## 5. Conclusions

This study aims to model reliability analysis for estimating the branch-based irrigation water demands in a multi-canal irrigation zone subject to the uncertainties in the observations of gauged surface runoff, named the RA_IWD_Canal model. Within the proposed RA_IWD_Canal model, the corresponding uncertainties to the estimated irrigation water demands are quantified in terms of L-moment ratios, and induced reliabilities of the introduced irrigation water demands could be achieved by calculating the exceedance probabilities via the weighted quantile-curve method. Additionally, the probabilistic-based branch-based irrigation water demands could be accordingly provided with the desired reliability to boost irrigation performance.

The Zhudong Canal irrigation zone located in Northern Taiwan, with fifteen branches, is selected as the study area, and the associated observed surface runoff at the fifteen discharge gauges from 2019 to 2023 is adopted in the model development and application. As well as the historical gauged surface runoffs, the introduced 10-day irrigation water demands set for the Zhudong irrigation zone and the 3rd branch (Su-Qi-Lin) set from 2019 to 2024 are utilized in the reliability quantification. The application results indicate that the introduced irrigation water demands at the 30 10-day periods exhibit significant variation (on average, from 0.02 $m^3/s$ to 1.7 $m^3/s$) with a noticeable 95% confidence interval of nearly 0.8 $m^3/s$. Also, the average of their corresponding reliabilities approaches 0.85 but with a significant change in time (from 0.63 $m^3/s$ to 0.999 $m^d/s$). In contrast with the Zhudong Canal zone, the introduced irrigation water demands at the Su-Qi-Lin branch, on average, are approximately less than 0.12 $m^3/s$ with a significantly low reliability (around 0.25). This implies that the introduced irrigation water demands within the study area were set without considering the variations of the climatic and hydrological features that impact irrigation reliability and efficiency. Subsequently, quantified via the proposed RA_IWD_Canal model with a desired and acceptable reliability, the resulting probabilism-based irrigation water demand could be treated as the introduced water demands to enable all branches in the study area to achieve high irrigation efficiency consistently.

Although the proposed RA_IWD_Canal model could effectively be applied in estimating the branch-based irrigation water demands, the measured surface runoffs for the four years are adopted in the model development and application. Hence, to reduce the effect of the record length of gauge surface off on the model applicability, more observations of the gauged runoff data are desirably required. In addition, the estimation of the branch-based irrigation water demands is demonstrated via the proposed RA_IWD_Canal model only with the surface runoff observations to have a significant correlation in time. However, the other hydrological features (e.g., rainfall, evapotranspiration, and soil moisture) should impact the irrigation water supplies and requirements [6,10,27,28]. Also, AI-created models have recently been comprehensively applied in irrigation water allocation and management [29,30]. Thus, using the resulting probabilistic-based irrigation water demands under different conditions of gauged surface runoffs, future work could be performed by training AI-derived models to estimate probabilistic-based irrigation water demands with the surface runoff and hydroclimatic features of interest given.

**Author Contributions:** Conceptualization, methodology, writing, and editing, S.-J.W.; data curation and simulation, H.-Y.Y. All authors have read and agreed to the published version of the manuscript.

**Funding:** This research was funded by the National United University Project: Development and Application of Intelligent Disaster Prevention System for Civil and Hydraulic Engineering (Grant SM113004).

**Institutional Review Board Statement:** Not applicable.

**Data Availability Statement:** The data presented in this study are available on request from the corresponding author.

**Conflicts of Interest:** The authors declare no conflicts of interest.

## References

1. Solangi, G.S.; Shah, S.A.; Alharbi, R.S.; Panhwar, S.; Keerio, H.A.; Kim, T.-W.; Memon, J.A.; Bughio, A.D. Investigation of irrigation water requirement for major crops in Pakistan using CROPWAT model based on climate data. *Water* **2022**, *14*, 2578. [CrossRef]
2. Wu, S.J.; Yang, H.Y.; Chang, C.H.; Hsu, C.T. Modeling GA-derived optimization analysis for canal-based irrigation water allocation under variations in runoff-related and irrigation-related factors. *Agric. Water. Manag.* **2023**, *290*, 1085588. [CrossRef]
3. Soomor, S.; Solangi, G.S.; Siyal, A.A.; Golo, A.; Bhatti, N.B.; Soomro, A.G.; Memon, A.H.; Panhwar, S.; Keerio, H.A. Estimation of irrigation water requirement and irrigation scheduling for major crops using the CROPWAT model and climatic data. *Water Pract. Technol.* **2023**, *18*, 685–700. [CrossRef]
4. Zhang, Y.; Yang, P. Agriculture water optimal allocation using minimum cross-entropy and entropy weight based on TOPSIS method in Hetao irrigation district, Northwest China. *Agriculture* **2022**, *12*, 853. [CrossRef]
5. Jamal, Q.K.; Shanthasheela, M.; Sureshverma, R.; Vasanthapriya, S. Factors influencing the knowledge and adoption of sustainable sugarcane initiative (SSI) by the sugarcane farmers of Villupuram District. *Int. J. Res. Appl. Res. Technol.* **2017**, *2*, 106–112. [CrossRef]
6. Dalias, P.; Christou, A.; Neocleous, D. Adjustment of irrigation schedules as strategy to mitigate climate change impact on Agriculture. *Agriculture* **2019**, *9*, 4. [CrossRef]
7. Nie, W.; Zaitchik, B.F.; Rodell, M.; Kumar, S.V.; Arsenault, K.R.; Badr, H.S. Irrigation water demand sensitivity to climate variability across the contiguous United States. *Water Resour. Res.* **2021**, *57*, e2020WR027738. [CrossRef]
8. Page, M.L.; Fakir, Y.; Jarlan, L.; Boone, A.; Berjamy, B.; Khabba, S.; Zribi, M. Projection of irrigation water demand based on the simulation of synthetic crop coefficients and climate change. *Hydrol. Earth Syst. Sci.* **2020**, *25*, 637–651. [CrossRef]

9. Wu, S.J.; Kuo, C.Y.; Yeh, K.C.; Wang, C.D.; Wang, W.J. Reliability analysis for reservoir water supply due to uncertainties in hydrological factors, rainfall-runoff routing and operating rule curves. *J. Hydro-Environ. Res.* **2021**, *34*, 24–45. [CrossRef]
10. Wisser, D.; Frolking, S.; Douglas, E.M.; Fekete, B.M.; Vorosmarty, C.J.; Schumann, H. Global irrigation water demand: Variability and uncertainties arising from agricultural and climate data sets. *Geophys. Res. Lett.* **2008**, *35*, L24408. [CrossRef]
11. Wriedt, G.; Van Der Velde, M.; Aloe, A.; Bouraoui, F. Estimating irrigation water requirements in Europe. *J. Hydrol.* **2009**, *37*3, 527–544. [CrossRef]
12. Wang, F.; Chen, Y.; Li, Z.; Fang, G.; Li, Y.; Xia, Z. Assessment of the irrigation water requirement and water supply risk in the Tarim River Basin, Northwest China. *Sustainability* **2019**, *11*, 4941. [CrossRef]
13. Woznicki, S.A.; Nejadhasemi, A.P.; Parsinejad, M. Climate change and irrigation demand: Uncertainty and adaptation. *J. Hydro. Reg. Stud.* **2015**, *3*, 247–264. [CrossRef]
14. Renault, D.; Vehmeyer, P.W. On reliability in irrigation service preliminary concepts and application. *Irrig. Drain. Syst.* **1999**, *13*, 77–105. [CrossRef]
15. Zhang, Y.; Wu, Z.; Singh, V.P.; Su, Q.; He, H.; Yin, H.; Zhang, X.; Wang, F. Simulation of Crop Water Demand and Consumption Considering Irrigation Effects Based on Coupled Hydrology-Crop Growth Model. *J. Adv. Model. Earth Syst.* **2021**, *13*, e2020MS002360. [CrossRef]
16. Hussain, F.; Shahid, M.A.; Majeed, M.D.; Ali, S.; Ibni Zamir, M.S. Estimation of the Crop Water Requirements and Crop Coefficients of Multiple Crops in a Semi-Arid Region by Using Lysimeters. *Environ. Sci. Proc.* **2023**, *25*, 101. [CrossRef]
17. Dang, C.; Zhang, H.; Yao, C.; Mu, D.; Lyu, F.; Zhang, Y.; Zhang, S. IWRAM: A hybrid model for irrigation water demand forecasting to quantify the impacts of climate change. *Agric. Water. Manag.* **2024**, *291*, 108643. [CrossRef]
18. Wen, T.H.; Lin, C.H.; Chen, C.T.; Su, M.D. Analysis of Spatial Scenarios Aiding Decision Making for Regional Irrigation Water-Demand Planning. *J. Irrig. Drain. Eng.* **2007**, *133*, 445–467. [CrossRef]
19. Shen, Y.; Li, S.; Chen, Y.; Qi, Y.; Zhang, S. Estimation of regional irrigation water requirement and water supply risk in the arid region of Northwestern China 1989–2010. *Agric. Water. Manag.* **2013**, *128*, 55–64. [CrossRef]
20. Wu, S.J.; Mai, J.S.; Lin, Y.H.; Yeh, K.C. Modeling Probabilistic-Based Reliability Analysis for Irrigation Water Supply Due to Uncertainties in Hydrological and Irrigation Factors. *Sustainability* **2022**, *14*, 12747. [CrossRef]
21. Sun, T.; Huang, Q.; Wang, J. Estimation of irrigation water demand and economic returns of water in Zhangye Basin. *Water* **2018**, *10*, 19. [CrossRef]
22. Mohseni, S.; Mehrjerdi, M.R.Z.; Ezzatabadi, M.A.; Boshrabadi, M. Irrigation water demand management with emphasis on pricing policy. *Water Policy* **2022**, *24*, 1095. [CrossRef]
23. Wu, S.J.; Tung, Y.K.; Yang, J.C. Incorporating Daily Rainfall to Derive At-Site Hourly Depth-Duration-Frequency Relationships. *J. Hydrol. Eng.* **2009**, *14*, 992–1001. [CrossRef]
24. Benavides, J.; Hernandez-Plaza, E.; Mateos, L.; Fereres, E. A global analysis of irrigation scheme water supplies in relation to requirements. *Agric. Water Manag.* **2021**, *243*, 106457. [CrossRef]
25. Hosking, J.R.M. L-moments: Analysis and estimation of distribution using linear combination of order statistics. *J. R. Stat. Soc. Ser. B* **1990**, *52*, 105–124. [CrossRef]
26. Tung, Y.K. Weighted model for hydrologic frequency analysis. In Proceedings of the Symposium on Managing the Extremes—Floods and Droughts, Water Resources Planning and Management Conference, Roanoke, VA, USA, 19–22 May 2002; ASCE: Roanoke, VA, USA, 2002.
27. Mahan, J.; Lascano, R.J. Irrigation analysis based on long-term weather data. *Agriculture* **2016**, *6*, 42. [CrossRef]
28. Yan, Z.; Li, M. A stochastic optimization model for agricultural irrigation water allocation based on the field water cycle. *Water* **2018**, *10*, 1031. [CrossRef]
29. Ali, A.; Hussain, T.; Tantashutikun, N.; Hussain, N.; Cocetta, G. Application of Smart Techniques, Internet of Things and Data Mining for Resource Use Efficient and Sustainable Crop Production. *Agriculture* **2023**, *13*, 397. [CrossRef]
30. Sattar, J.; Oamara, M.U.; Kisekkab, I.; Arshada, M.; Anwaroul-Haq, M. Artificial intelligence (AI) inspired an innovative approach for real-time flow measurements—Water distributions at outlets, in Punjab province Pakistan. *Water Policy* **2024**, *26*, 558. [CrossRef]

water

MDPI

rticle

# Hydrochemical Characteristics and Quality Evaluation of Irrigation and Drinking Water in Bangong Co Lake Watershed in Northwest Tibetan Plateau

uxiang Shao [1], Buqing Yan [1,*], Lubaiyang Liu [2], Xiao Yu [1], Gang Feng [1], Kun Zhang [1] and Kang Gong [1]

1 Research Center of Applied Geology of China Geological Survey, Chengdu 610036, China; shaoyuxiangcgs@outlook.com (Y.S.); jiazhuo@126.com (X.Y.); fg17716159968@163.com (G.F.); 18683651975@163.com (K.Z.); 17608058483@163.com (K.G.)

2 Xi'an Institute for Innovative Earth Environment Research, Xi'an 710061, China; liuyang@ieecas.cn

* Correspondence: 127961917@139.com

**Abstract:** Bangong Lake is a narrow and long lake in the arid region of the plateau in northern Tibet. The salinity of the east of the lake is different from that the west, resulting in differences in the natural environment and human living conditions on each side. Watershed hydrochemical analysis and spatial statistical analysis can help to understand regional hydrochemical evolution and water quality evaluation. In this study, the hydrochemical characteristics of surface water (glacier, river, and lake) and groundwater in the Bangong Co Lake Watershed were investigated to reveal the relationships between various water bodies. The drinking water quality index (DWQI) and USSL classification were applied to assess groundwater quality suitability for agricultural and drinking purposes. The hydrochemical characteristics show the differences among water bodies and their spatial distribution. The analyzed groundwater and surface water samples, such as river water and glaciers, were mainly Ca-$HCO_3$-type and the lake water was mainly categorized as Na-Cl-type with some Na-$HCO_3$-Cl type. The lake water's chemical components are mainly affected by evaporative karst decomposition. The main mineralization process of groundwater and river water was related to the dissolution of reservoir minerals, such as dolomite and calcite, as well as halite. The drinking water quality index (DWQI) indicates that 79% of the groundwater samples in the study area were of good enough quality for drinking. In terms of irrigation water quality, the electrical conductivity (EC), calculated sodium adsorption ratio (SAR), and magnesium hazardous ratio (MHR) showed that more than 13% of the total samples were not suitable for irrigation. However, the USSL classification indicated that glacier and river water are relatively suitable for irrigation. Additionally, some groundwater and lake water has very high alkalinity or salinity, which is alarming when considering them for irrigation.

**Keywords:** Bangong Co Lake Watershed; hydrochemical characteristics; water quality; surface water; groundwater

## 1. Introduction

The Tibetan Plateau, known as the roof of the world and the water tower of Asia, is the birthplace of the Yellow River, the Yangtze River, the Ganges River, the Mekong River, the Indus River, and other rivers, and has the largest, highest, and most densely distributed alpine lake cluster in the world [1,2]. In the Tibetan Plateau, lakes play an important role in regulating the ecosystem and hydrochemistry and are a vital indicator of water–rock interaction and environmental change in the basins [3,4]. In the last few years, many academics have conducted investigations on lakes in the Tibetan Plateau, which mainly focused on hydrochemical characteristics and its spatial distribution [5,6], typical water pollutants [7], water quality assessment [8,9], and hydrochemistry evolution and sources [10,11]. Due to the vulnerability of the water environment to climate change and the scarcity of water resources in the alpine desert, water quality assessment and hydrochemical evolution in these areas have attracted increasing attention [12,13].

Natural factors, such as the geological structure, hydrogeological conditions, vegetation, climate, rock weathering, terrain, and human activities, greatly impact the chemical composition of water [14]. Sun and Jin pointed out the differences in chemical characteristics and evolution between tectonic and glacial lakes on the plateau based on the analysis of physicochemical parameters and ion concentrations of lakes in the Tibetan Plateau [6]. Li et al. discussed the relationship between lake change and climate in recent decades and clarified that lakes are important supporters of the water cycle and environmental change [15]. Based on spatial analysis of the heavy metal contents in ores, air pollutants, and lake water samples, Bazova and Moiseenko discussed the sources of heavy metals in lakes in northwestern Russia, and the results showed that the concentration of Cu, Ni, Mn, Zn, Pb, and As in lake water was caused by anthropogenic pollution, such as flue gas emissions of metallurgical enterprises [16]. Baranov et al. studied the chemical composition, natural geochemistry, and human factors of dissolved effluents in the northern Valdai Mountains using statistical methods [17]. The results showed that the soil mineral composition, rainfall intensity, and biogeochemical processes had a great influence on the chemical composition of water bodies in the aeration zone. Moreover, many scholars have gained a significant understanding of regional soil salinization, regional geological background, element migration and enrichment law, petrophysical weathering process, and other aspects by studying ion characteristics of watershed water bodies [18]. It can be seen that the hydrochemical characteristics of natural water bodies are one of the important indicators when studying regional water cycles and evolution.

The reserves and quality of water resources are important for the stable and sustainable development of human society and natural ecology. Good water quality is especially valuable in water shortage areas. High evaporation and low rainfall lead to salinization or heavy metal pollution in shallow groundwater or surface water in lake basins in the arid region of the plateau [19]. Previously, water quality assessments were carried out on many lake basins in the Tibetan Plateau, such as Qinghai Lake [20], the Ebinur Lake Watershed [21], Dal Lake in Kashmir Valley [22], Mapam Yumco [23], and Nam Co [24]. The results showed that the water quality of most lakes was relatively uniform and fluctuated in a small range. The results also indicated that lake water with high salinity had a great influence on the shallow surface water of the lake shore, and the salinity of groundwater had a great influence on plant growth, human life, animal husbandry development, and industrial activities in the region.

Bangong Co Lake is a long and narrow tectonic lake with a nearly east–west distribution. The western part of the Tibetan Plateau is characterized by a dry and cold climate with strong evaporation [25]. As the main source of replenishment for the Lake Bangong Co Watershed, glacial meltwater is insufficient to balance evaporation in the area. Likewise, affected by evaporation across the whole watershed and river water supply in the east of Bangong Co Lake, the salinity of the lake gradually increases from east to west [26]. The water circulation system of the plateau arid watershed is very fragile and sensitive, which makes the water resources in the study area vulnerable to the impacts of human activities, animal husbandry activities, and climate change.

Some scholars investigating lake groups on the Tibetan Plateau have pointed out that the lake water type on the east side of Bangong Co Lake is Cl-Na or $HCO_3$·Cl-Mg·Na [27,28]. Lin et al. researched information on toxic organic pollutants and metals of Bangong Co Lake, and their results indicate that polycyclic aromatic hydrocarbons (ΣPAH) and phthalic acid esters (ΣPAE) concentrations in lake water have no relationship with hydrochemical parameters, and organic pollutants have been the main source of domestic waste related to increasing regional human activities in recent years [29]. The domestic water and irrigation water in the study area are mainly shallow aquifers in the lakeshore zone and river valley or supplemented by snowmelt water and river water. However, there are few relevant studies on the various water bodies of Bangong Co Lake Watershed. Therefore, the hydrochemical characteristics and controlling factors of the river water, lake water, groundwater, and other water bodies in the watershed of Bangong Co Lake are still unknown. Additionally, there is no available information on water quality assessment in this region.

The present investigation was performed to explore the glaciers, rivers, lakes, and groundwater along the lakeshore in the Bangong Co Lake Watershed. Statistical analysis, a Gibbs diagram, a Piper diagram, and an ion ratio analysis were used to analyze the hydrochemical characteristics and formation mechanisms of different water bodies. The drinking water quality index (DWQI) and irrigation water quality parameters were used to assess the water quality and its suitability in the study area. The specific objectives of this work are as follows: (i) To explain the hydrochemistry and evolution of different water bodies in Bangong Lake Basin, (ii) to identify the various factors controlling the water chemistry in this area, and (iii) to assess the suitability of the water quality for drinking and irrigation.

## 2. Materials and Methods

### 2.1. Study Area

Bangong Co Lake Watershed is located in the northwest of Tibet, China. Based on ArcGIS hydrological analysis module and DEM data, the watershed area of Bangong Co Lake is about 33,000 km$^2$, with the location between 32°40′–34°30′ N and 78°10′–81°15′ E, with an elevation range of 3736–6771 m. The watershed reaches the Kanakoram Mountains in the north and the Gangdises Mountains in the south. The highest point is located in the eastern mountains of Zecuo Lake, and the lowest is located in Bangong Co Lake, in the middle of the area (Figures 1 and 2). The geomorphology of the watershed is that of plateau lake basin geomorphic type, which has been fully eroded by snow and water for a long time. The watershed covers the Bangong Lake replenishment river basin and several small intermountain lake basins, such as Spangour Co, Zecuo, Rebangcuo, Aiyongcuo, and Sharda Co [30]. This region has a cold and dry climate, with an average annual temperature of 0.1–2 °C. The annual rainfall is about 70–80 mm, with high variability, and is mostly concentrated in July or August [31]. The main source of water in the study area is glacial meltwater. Due to gravity, the water mainly flows along the slope or gully to feed rivers or veins of rock and eventually drains into the lake. Because the sediments in river valleys in the middle or lower reaches and the sediments in lake shores are thick and loose, the river water is likely to penetrate, forming relatively stable shallow groundwater in the alluvial fan or valley area.

Geologically, the study area is located in the nearly east–west Bangong Co-Nujiang suture zone, with Qiangtang block to the north and Lhasa block to the south (Figure 1) [32,33]. Qiangtang block in the region is mainly composed of carbonate rocks, sandstone, and slate rocks from Yanshiping formation, Tunlonggongba formation, and Nuoco formation, and Lhasa Terrane in the study area is mainly composed of ophiolite mélange, Cretaceous gabbros and granites, and sandwiched limestone and flysch sediments from Mugagangri Group and Langshan formation (Figure 1) [31,34]. Quaternary sediments are mainly distributed on the sides of lake and river valleys. NWW-trending and NE-trending faults are mainly developed in the Bangong Co Lake area. Most faults align with the boundary of the lake basin in geomorphology, and the fault surface is shovel-type dip into the lake. Meanwhile, the NWW-trending faults are the leading faults in the region, belonging to the Bangong Co-Nujiang suture system, which has undergone multiple periods of subduction, collision, and other tectonic processes, and are nearly consistent with the overall tectonic line and the ophiolitic hybrid zone. The NE-trending faults are mostly developed in some north–south gullies and valleys and have the characteristics of water conduction after differentiation and denudation [35].

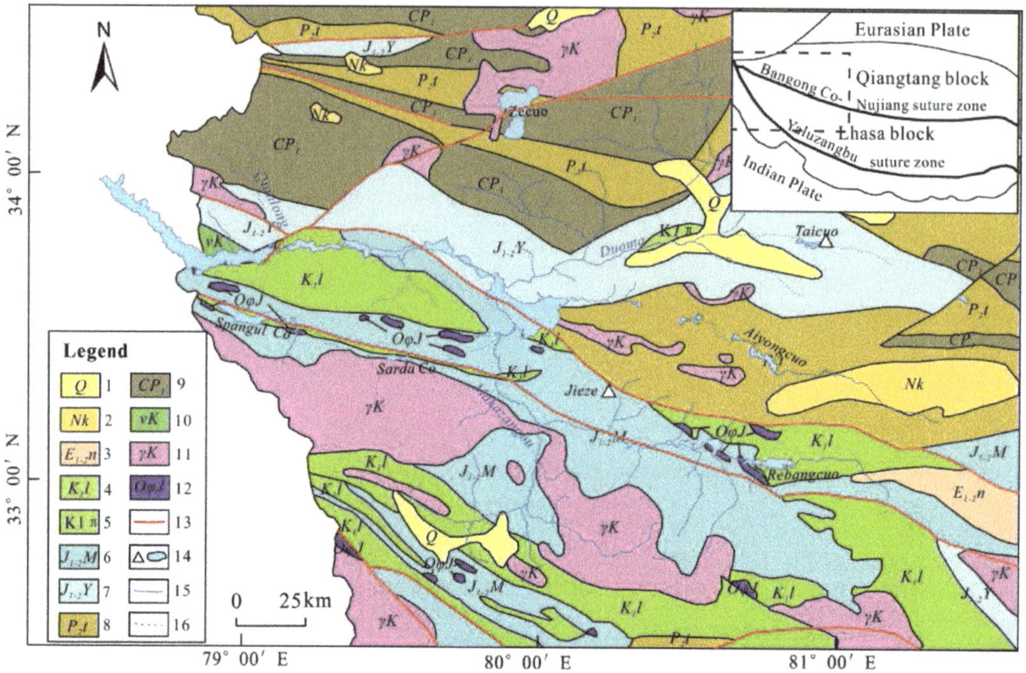


**Figure 1.** 1: Quaternary. 2: Sandstone and conglomerate of Cantuo Fm. 3: Mudstone and conglomerate of Niubao Fm. 4: Limestone of Langshan Fm. 5: Limestone and sandstone of Tielongtan Gr. 6: Argillaceous slate and metamorphic sandstone of Mugagangri Gr. 7: Limestone and clastic rocks of Yanshiping Gr. 8: Sand shale and limestone of Tunlonggongba Fm. 9: Silty slate of Nuoco Fm. 10: Cretaceous gabbro. 11: Cretaceous granite. 12: Ophiolite. 13: Faults. 14: Salt lake (salt mine) and lake. 15: Perennial river. 16: Seasonal river. (this figure is modified from the regional geological map released by the GeoCloud Platform of the China Geological Survey).

There are many aquatic plants and fishes in Bangong Co Lake. Fishes mainly include Schizothoracinae and Triplophysa stenura, and the aquatic plants encompass Pyrrophyta, Bacillariophyta, and Chlorophyta [28]. Moreover, a small field of wetlands lies east and southeast of the lake, mainly growing plateau meadows and reeds. Plateau desert meadow is the main vegetation type in the Bangong Co Lake catchment area. The dry period is a maximum of nine months in this area, so the grass yield of the regional meadow is relatively low [36]. Animal husbandry activities are mainly concentrated in the rainy season, and animal husbandry activities are basically practiced in areas with abundant groundwater, such as alluvial fan front, lakeshore, and valley land.

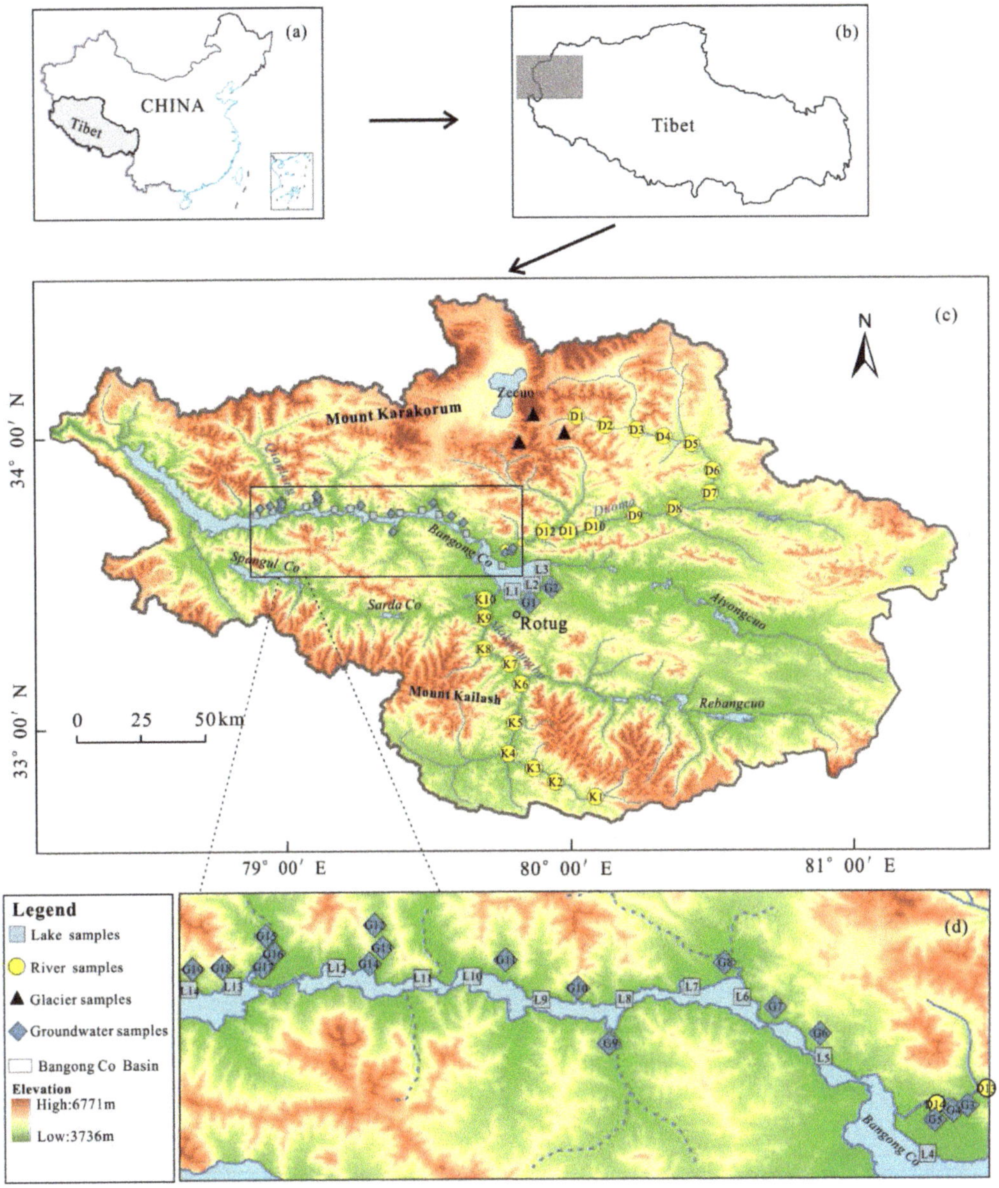


**Figure 2.** (**a**) The location of the Tibet Autonomous Region in China; (**b**) a map of the study area; (**c**,**d**) geomorphology of sampling sites and study areas.

## 2.2. Sampling and Measurement

In early June 2022, field investigation was carried out in Bangong Lake Basin, and a total of 60 water samples were collected, including 14 lake water samples, 19 groundwater samples, 24 river water samples, and 3 glacier water samples (Figure 2). Among them, river samples were mainly distributed in Makazangbu and Doma Rivers, and groundwater samples were mainly collected from wells on alluvial fan bodies in lakeshore zone. The water depth of the collected wells was less than 15 m, and most of the glacier samples were new snow samples. Water samples were mostly collected at a depth of 0.1–0.5 m, filtered through 0.45 μm cellulose acetate membrane, and stored in 1500 mL cleaned polyethylene bottles. Before collection, the collection containers were rinsed with sampling water 2 to 3 times. The lake water samples were taken from a location about 1–2 m away from the lake. In addition, water electrical conductivity (EC) and pH were measured in the field using a multi-parameter portable water analyzer (Manta, Eureka Inc., Austin, TX, USA), with a pH accuracy of ±0.1 pH and conductivity accuracy of ±1 μS/cm.

All collected samples were stored in 0–4 °C incubators until they were analyzed in the laboratory. Parallel samples were used to control the test's accuracy during the experiment. For any water sample, a part of it was acidified with 1:1 nitric acid and stored in a plastic bottle for cation analysis. The remaining part was placed in glass bottles and sealed with parafilm sealing film to analyze anions such as $SO_4^{2-}$ and $Cl^-$. The content of $HCO_3^-$ was measured by acid–base titration, with an accuracy of ±0.05%. Water salinity (TDS) was determined using the dry weight method at 105 °C, with an accuracy of ±0.2%. Total hardness (TH) was determined using the EDTA method with an accuracy of ±0.05%. The main cation ($Na^+$, $K^+$, $Ca^{2+}$, and $Mg^{2+}$) concentrations and trace heavy metal (Fe, Mn, Zn, Cu, and Cr) levels were determined using inductively coupled plasma emission spectrometer (Agilent5110, Agilent Technologies Inc., Santa Clara, CA, USA), and the testing error was less than 2%. The total As content was detected via atomic fluorescence (AFS-933, Beijing Jitian Instrument Co., LTD, Beijing, China) with an accuracy of ±0.5%. Anion concentration ($SO_4^{2-}$, $Cl^-$, $F^-$, and $NO_3^-$) was determined via chromatography (ECO IC, Vantone China Co., LTD, Beijing, China), and the testing error was less than 0.5%. The ion balance of the major soluble cationic equivalent ($TZ^+ = Na^+ + K^+ + 2Ca^{2+} + 2Mg^{2+}$) and soluble anionic equivalent ($TZ^- = Cl^- + NO_3^- + HCO_3^- + F^- + 2SO_4^{2-}$) of all samples was compared for ion balance test, and it was within ±5%. All water samples were finished by Xizang Shengyuan Environmental Engineering Co., LTD, Tibet, China.

### 2.3. Data Processing

SPSS v22.0 (SPSS Inc., Chicago, IL, USA) was used to determine the average (Mean), maximum (Max), minimum (Min), and coefficient of variation (CV). The Piper diagrams [37], principal component analysis, Gibbs diagrams [38], and ion ratio analysis [39] were implemented in Origin v2021 (OriginLab, Northampton, MA, USA). and Microsoft Excel v2013 (Microsoft Corp., Redmond, WA, USA) to explore the classification and main controlling factors of the nature water in Lake Bangong Co Watershed. International standards [40], drinking water quality index [41,42], USSL classification [43], and other irrigation water indicators were used to evaluate groundwater suitability for drinking and irrigation. ArcGIS v10.4 (Esri Inc., Redlands, CA, USA) and CorelDRAW Graphics Suite v18 (Corel Corp., Ottawa, ON, Canada) were used for mapping and modification.

## 3. Results and Discussion

### 3.1. Chemical Composition of Different Water Bodies in Bangong Lake Basin

#### 3.1.1. Statistical Characteristics of Parameters

The parameters of each water body in Bangong Co Watershed are shown in Table 1. The pH of glaciers, lake water, river water, and shallow groundwater along the lakeshore in the study area ranged from 6.6 to 8.5. The pH of glaciers was relatively low (pH = 6.6~7.6), with an average value of 7.2, which is neutral. Other water bodies were weakly alkaline as a whole, among which the pH of Bangong Co Lake water was relatively high (pH = 8.1~8.5), with an average value of 8.4. The coefficient of variation (CV) for the water pH of each water body was <10%, which showed weak variability. Salinity (TDS) and electrical conductivity (EC) are parameters that can be used to determine the salinity of water [44]. The salinity (TDS) of glaciers, river water, groundwater, and lake samples ranged from 13 to 25, 81 to 438, 124 to 972, and 206 to 4091 mg/L, respectively. The salinity (TDS) values of 10 lake samples, mainly distributed in the western or middle part of the lake, exceeded the WHO's standard limit (1 g/L) [40]. The average salinity (TDS) of river water was 282 mg/L, which was also higher than the global average and values for other large rivers in China [45]. Similarly, the electrical conductivity (EC) of glacier water, river water, groundwater, and lake samples ranged from 23 to 43, 145 to 779, 220 to 1706, and 385 to 7861 μS/cm, respectively. Water hardness is a very important parameter in the growth and reproduction of aquatic organisms [46]. The total hardness (TH) of the glaciers, groundwater, river, and lake were in the range of 23–43 mg/L, 27–750 mg/L, 58–467 mg/L, and 258–2854 mg/L, respectively. With the decrease in altitude, the values regarding the total hardness (TH), electrical conductivity (EC), and salinity (TDS) increased gradually, which were the highest around Bangong Co Lake. The coefficient of variation (CV%) for total hardness (TH), electrical conductivity (EC), and salinity (TDS) of different water bodies were between 10% and 66%, which demonstrates medium variability. Based on the total hardness (TH) and salinity (TDS) [47], lake water is hard-brackish water, glacier water is soft-fresh water, and river water changes from soft-fresh water to hard-fresh water along the stream (Figure 3).

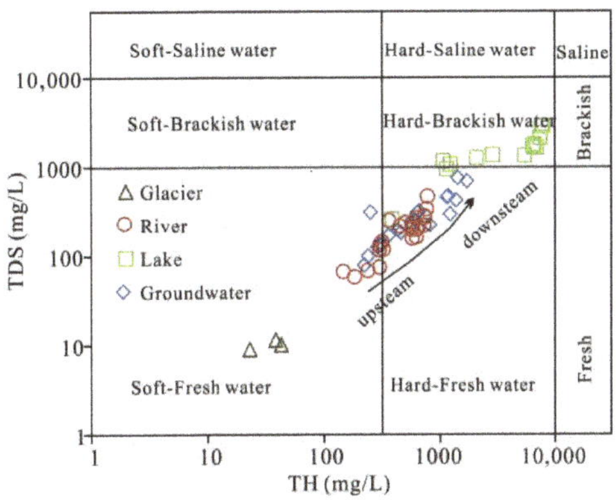


**Figure 3.** Classification of water in Bangong Co Lake Watershed based on total hardness (TH) and salinity (TDS).

3.1.2. Ion Spatial Distribution

The ion diagrams (Figure 4) depict the milligram equivalent and spatial variation in anions and cations in the different water bodies across the basin. In the water samples of groundwater, river water, and glaciers, the cations were dominated by $Ca^{2+}$, and the order of abundance was $Ca^{2+} > Na^{+} > K^{+} > Mg^{2+}$. However, $Na^{+}$ was the main cation of lake water, and the cation content followed the order $Na^{+} > Mg^{2+} > K^{+} > Ca^{2+}$. The major anion in groundwater, river water, and glacier samples was $HCO_3^-$, and the order of the abundance in these samples was $HCO_3^- > SO_4^{2-} > Cl^- > NO_3^-$, whereas the primary anion in lake samples was $Cl^-$, and the anion content followed the order $Cl^- > HCO_3^- > SO_4^{2-} > NO_3^-$. A strong coefficient of variation was found in $Na^+$, $Cl^-$, and $SO_4^{2-}$ of groundwater, and $NO_3^-$ of river and lake water, which was between 1.15 and 2.85, indicating that the spatial distribution of these ionic components was unstable. The coefficient of variation of most chemical parameters of different water bodies in the area was less than 1 (Table 1), which is a weak or medium variation. Spatially, the Doma River and Maka Zangbo River showed a gradually increasing trend of ions from upstream to downstream. The main ions concentration in the lake increased gradually from east to west (Figure 4c). However, for lake samples L7 to L5 (Figures 2 and 4c), the ions and salinity (TDS) increased slowly compared to the other sections, which may be due to a water exchange barrier caused by the narrow body of the lake [50]. According to Figure 4d, there is no obvious spatial rule of groundwater in the lakeshore zone.

**Table 1.** Major ions and heavy metal compositions of water samples from the Bangong Co Lake Watershed.

| Sample Type | | pH | TDS | TH | EC | $K^+$ | $Na^+$ | $Ca^{2+}$ | $Mg^{2+}$ | $Cl^-$ | $SO_4^{2-}$ | $HCO_3^-$ | $NO_3^-$ | $F^-$ | Fe | Mn | As |
|---|---|---|---|---|---|---|---|---|---|---|---|---|---|---|---|---|---|
| Glacier ($n$ = 3) | min | 6.6 | 13 | 9 | 23 | 0.15 | 0.25 | 3.08 | 0.22 | 1.04 | 1.96 | 6.67 | 0.18 | ND | ND | ND | ND |
| | max | 7.6 | 25 | 12 | 43 | 0.26 | 0.66 | 3.66 | 0.58 | 1.25 | 2.83 | 8.00 | 0.64 | ND | ND | ND | ND |
| | Ave | 7.2 | 20 | 10 | 35 | 0.20 | 0.49 | 3.44 | 0.36 | 1.12 | 2.38 | 7.56 | 0.35 | - | - | - | - |
| | CV | 6 | 27 | 10 | 24 | 24 | 36 | 7 | 43 | 8 | 15 | 8 | 60 | - | - | - | - |
| River water ($n$ = 24) | min | 7.1 | 81 | 58 | 145 | 0.41 | 1.41 | 10.40 | 3.23 | 1.76 | 7.63 | 52.20 | ND | ND | ND | ND | ND |
| | max | 8.5 | 438 | 467 | 779 | 6.31 | 65.60 | 88.50 | 52.06 | 50.60 | 132.62 | 427.50 | 63.83 | 0.7 | 0.138 | 0.398 | 0.011 |
| | Ave | 7.8 | 282 | 196 | 498 | 2.28 | 26.98 | 39.93 | 19.89 | 27.65 | 31.89 | 211.56 | 7.68 | 0.2 | 0.044 | 0.115 | 0.005 |
| | CV | 5 | 39 | 47 | 39 | 52 | 61 | 50 | 54 | 54 | 90 | 47 | 213 | 87 | 82 | 143 | 68 |
| Lake water ($n$ = 14) | min | 8.1 | 406 | 258 | 385 | 1.88 | 46.30 | 6.91 | 14.60 | 3.35 | 38.50 | 168.0 | ND | ND | ND | ND | 0.002 |
| | max | 8.5 | 4491 | 2854 | 7861 | 83.10 | 1089.0 | 39.80 | 266.40 | 1163.60 | 1004.00 | 1092.0 | 1.02 | ND | 0.091 | 0.010 | 0.044 |
| | Ave | 8.4 | 2606 | 1573 | 4546 | 37.85 | 638.31 | 15.84 | 134.06 | 618.50 | 605.25 | 700.64 | 0.10 | - | 0.043 | 0.003 | 0.020 |
| | CV | 1 | 62 | 43 | 61 | 73 | 68 | 56 | 58 | 69 | 65 | 52 | 278 | - | 60 | 76 | 63 |
| Groundwater ($n$ = 19) | min | 6.8 | 124 | 79 | 220 | 0.94 | 3.10 | 12.33 | 7.43 | 3.40 | 0.33 | 66.50 | 1.09 | ND | ND | ND | ND |
| | max | 8.4 | 972 | 754 | 1706 | 23.44 | 184.0 | 180.00 | 93.50 | 256.00 | 528.00 | 581.70 | 68.57 | 1.5 | 0.224 | 0.441 | 0.003 |
| | Ave | 7.7 | 414 | 302 | 729 | 5.79 | 45.75 | 66.79 | 32.98 | 49.52 | 92.43 | 287.19 | 28.83 | 0.5 | 0.060 | 0.127 | 0.002 |
| | CV | 5 | 63 | 61 | 62 | 94 | 96 | 75 | 70 | 132 | 134 | 53 | 68 | 90 | 124 | 117 | 16 |
| Jieze salt lake [48] | | - | - | - | - | 18,149 | 89,232 | 115 | 28,961 | 142,324 | 127,351 | 4054 | - | - | - | - | - |
| Taicuo salt lake [49] | | - | - | - | - | 95 ± 37 | 1563 ± 887 | 1.6 ± 1.3 | 133 ± 12 | 1773 ± 653 | 628 ± 227 | 805 ± 413 | - | - | - | - | - |
| WHO limit [40] | | - | 1000 | 500 | - | - | 200 | - | - | 250 | 250 | - | 50 | 1.5 | 0.3 | 0.1 | 0.01 |

The units for ion concentration are mg/L, for EC, μs/cm, and for TDS, mg/L. CV coefficient of variation, %. ND not detected.

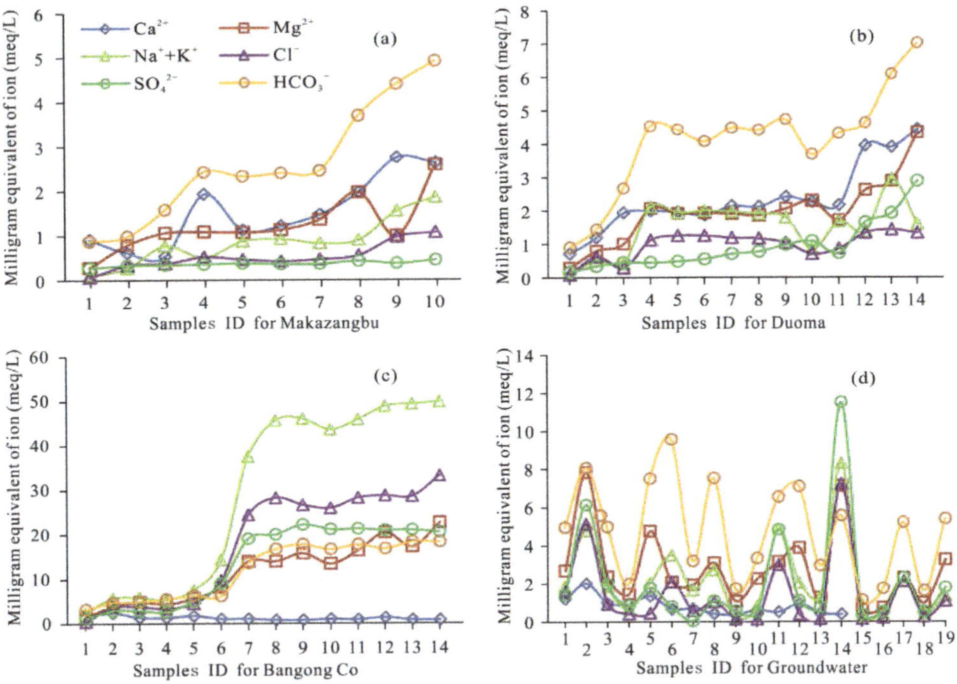


**Figure 4.** The spatial distribution of anions and cations in different water samples.

Al, Cr, Pb, Cd, Hg, Cu, and Ni were below the limit of quantification. Moreover, Fe, Mn, As, and $F^-$ were found at very insignificant concentrations. Table 1 shows the summary of the datasets of trace metals. Concentrations of Fe in various water bodies varied from ND to 0.224 mg/L, which was within the WHO's tolerable limit (0.3 mg/L). Meanwhile, concentrations of As in lake samples varied from 0.002 to 0.035 mg/L, with a mean concentration of 0.021 $\pm$ 0.012 mg/L, most of which exceeded the WHO's tolerable limits (0.01 mg/L). The results were similar to those of previous studies, which suggested that the As in Bangong Co Lake was mainly derived from natural sources such as rock weathering [29]. Mn concentrations in the river and groundwater ranged from ND to 0.398 mg/L and ND to 0.441 mg/L. 4.2% ($n$ = 1) in river water samples and 15.8% ($n$ = 3) in shallow groundwater, exceeding the WHO's limit for Mn content (0.1 mg/L). However, few lake water samples had Mn content exceeding the WHO's standard. The Mn exceedance points of groundwater are located near the shoreline of the lake, with the pH ranging from 6.83 to 7.72. The river water exceedance point is located at a spring recharge in the lower Doma River, and the pH is 7.19. Low water pH provides favorable conditions for Mn enrichment in groundwater. On the contrary, Mn and Fe ions easily form hydroxide precipitation in lake water with relatively high pH (8.06–8.52). The flu-lacustrine purplish neogene conglomerates, rich in Fe, Mn, and other elements, are scattered in some river valleys in the study area, which may provide Mn sources for moderate and acidic groundwater. $F^-$ concentrations of various water bodies were all below the acceptable limit according to WHO standards.

### 3.1.3. Piper Diagram and Hydrochemical Classification

Water hydrochemical characteristics can be classified using the Piper diagram, which shows scatter plots of cations ($Na^+$, $K^+$, $Ca^{2+}$, and $Mg^{2+}$) and anions ($HCO_3^-$, $Cl^-$, and $SO_4^{2-}$) [37,42]. According to Figure 5, water samples in the Bangong Co Lake Watershed were classified into three water types. River samples, glaciers, and most of the groundwater corresponded to zone 2 of the Piper diagram, characterized as Ca-$HCO_3$-type samples. The lake samples were distributed in zone 5, classified as saline water with high $Cl^-$ or $SO_4^{2-}$ and $Na^+$. Due to $Cl^-$ being the main cation of lake samples, lake water was classified as a Na-Cl-type sample. Only a few groundwater samples corresponded to zone 3, representing contaminated water with high $Cl^-$ and $Ca^{2+}$ concentrations. Eastern water samples were in the transition between the river point and the lake point, indicating that these water samples exhibit ion exchange and simple dissolution or mixing.

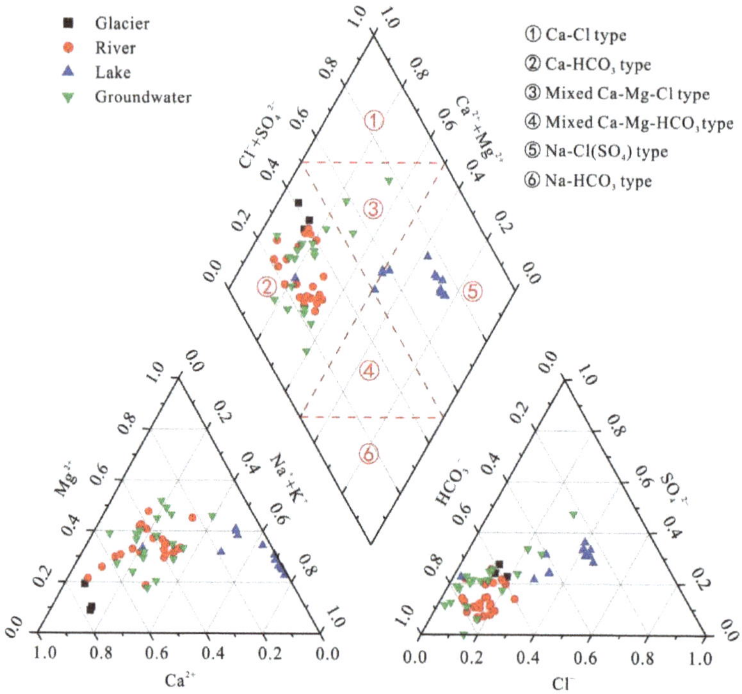


**Figure 5.** Piper plot of water samples in the study area.

## 3.2. Cause Analysis of Hydrochemical Characteristics

### 3.2.1. Gibbs Diagrams

Based on an analysis of many global hydrochemical compositions of surface waters, such as large lakes, rivers, and precipitation water samples, diagrams were proposed by Gibbs to determine the features of ionic distribution in different natural water bodies [38]. Rock weathering, atmospheric precipitation, and evaporation–crystallization process were identified as three major factors controlling the surface water chemistry in Gibbs plots [11,51]. In this paper, a TDS-$Na^+/(Na^+ + Ca^{2+})$ diagram and TDS-$Cl^-/(Cl^- + HCO_3^-)$ diagram were used to distinguish the ionic characteristics of water bodies in the Bangong Co Lake Watershed (Figure 6).

With regard to the river and groundwater samples, the cation ratios $Na^+/(Na^+ + Ca^{2+})$ varied from 0.07 to 0.66, and the anion ratios $Cl^-/(Cl^- + HCO_3^-)$ ranged from 0.02 to 0.43. Additionally, all the samples for both diagrams fell in the rock weathering domain. For the lake water samples, the ranges of the cation ratios $Na^+/(Na^+ + Ca^{2+})$ and the anion ratios $Cl^-/(Cl^- + HCO_3^-)$ were 0.66–0.99 and 0.02–0.52, respectively. The western lake samples mainly fell into the evaporation domain (Figure 6c) or near the evaporation dominance zone (Figure 6b), and the eastern lake samples fell in the transition zone between evaporation and rock weathering (Figure 6b,c). This indicates that the dominant ions of various water bodies were significantly affected by rock weathering and evaporation–crystallization may affect the western lake water and some groundwater. Duoma and Makazangbu rivers which were the main sources of replenishment, joined Bangong Co Lake on the east side, resulting in a characteristic transition between evaporation and rock weathering in the eastern lake. Glacier water is considered to be a precipitation sample in the Bangong Co Lake area. The glacier water samples are part of the end element of rock weathering and are slightly closer to the rain zone in Gibbs diagrams. Therefore, the dissolved ions in the rainwater of the Bangong Co Lake Watershed are slightly affected by ocean evaporation and are mainly controlled by the dissolution of atmospheric $CaCO_3$ particles.

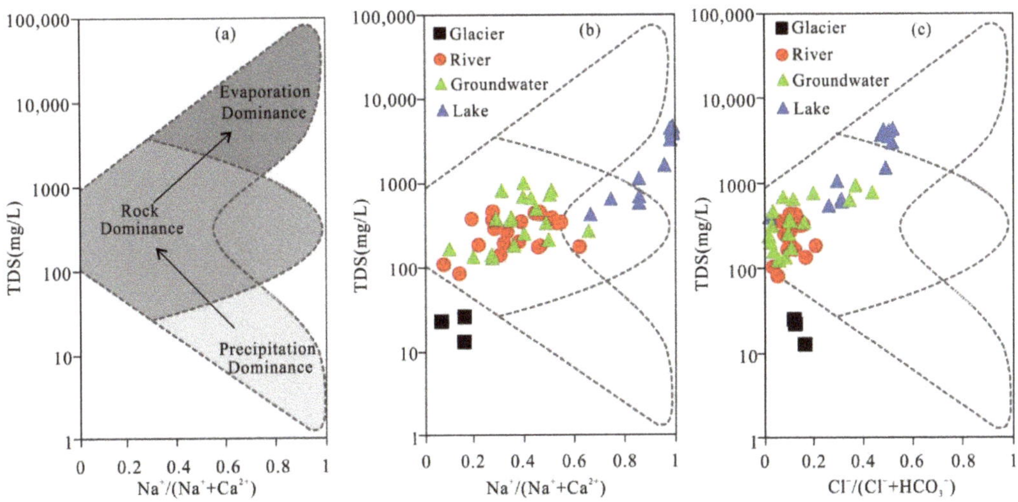


**Figure 6.** Gibbs plot: (**a**) Comparison of natural processes that define the water chemistry of water on the Gibbs (1970) diagram; (**b**) TDS versus $Na^+/(Na^+ + Ca^{2+})$; (**c**) TDS versus $Cl^-/(Cl^- + HCO_3^-)$.

3.2.2. Ion Ratio Analysis

The cations and anions dissolved by the chemical weathering of different rocks contribute to the combination of water in nature [52]. Therefore, the ion ratio characteristics can reflect the water–rock reaction and evolution process between natural water and rock strata [39]. As can be seen in the regional geological map (Figure 1), the main runoff areas of rivers are mainly developed strata, such as the Mugagangri Group, Langshan formation, Yanshiping formation, and Nuoco formation. The lithology of these strata is principally carbonate rock, sandstone, and conglomerate.

$Na^+$ and $K^+$ are mainly derived from evaporite or silicate, and $Ca^{2+}$ and $Mg^{2+}$ are mainly derived from carbonate and silicate weathering and evaporative dissolution [53]. Thus, when the ratio of $\gamma(Na^+ + K^+)$ and $\gamma(Cl^-)$ is close to or above the 1:1 line, dissolution of evaporate is the leading role for $Na^+$ and $K^+$ in water chemical evolution. Figure 7a shows that the sample points of $\gamma(Na^+ + K^+)$ and $\gamma(Cl^-)$ are distributed around a straight line, with a slope of 1.63 and $R^2$ of 0.99. Most of the water bodies (except glaciers) are almost on the upper side of the 1:1 line, indicating that halite (NaCl) and potassium salt (KCl) are the main sources of $Na^+$, $K^+$, and $Cl^-$, and the $Cl^-$ content is not enough to balance the content of $Na^+$ and $K^+$ in water. The excess $Na^+$ and $K^+$ may be derived from the weathering and dissolution of silicate rock.

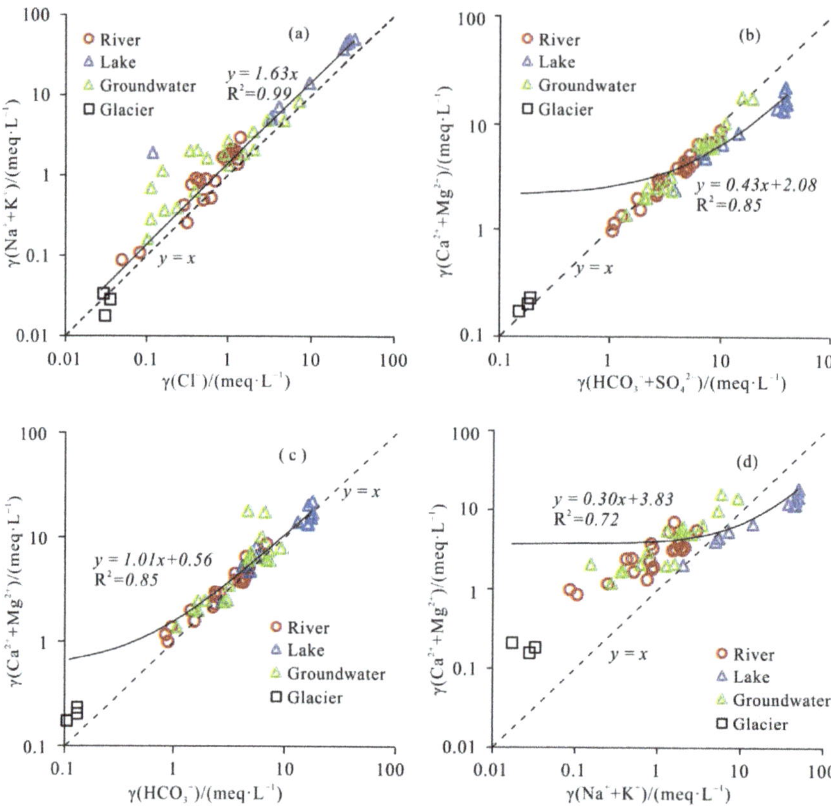


**Figure 7.** Distributions of ionic ratios in water samples: (**a**) $\gamma(Na^+ + K^+)$ versus $\gamma(Cl^-)$; (**b**) $\gamma(Ca^{2+} + Mg^{2+})$ versus $\gamma(HCO_3^- + SO_4^{2-})$; (**c**) $\gamma(Ca^{2+} + Mg^{2+})$ versus $\gamma(HCO_3^-)$; (**d**) $\gamma(Ca^{2+} + Mg^{2+})$ versus $\gamma(Na^+ + K)$.

Carbonates, evaporates, or silicates containing calcium and magnesium are the main sources of $Ca^{2+}$ and $Mg^{2+}$ ions in natural water [24]. Figure 7b shows the ratio of $(Ca^{2+} + Mg^{2+})$ to $(HCO_3^- + SO_4^{2-})$ in the water sample. The slope of the scattered fitting line of the water samples is 0.43, and the $R^2$ is 0.85, indicating that many sample points are below the 1:1 line, especially the lake points. In these water bodies, $Ca^{2+}$ and $Mg^{2+}$ contents are not high enough to reach equilibrium with $HCO_3^-$ and $SO_4^{2-}$, and it is generally believed that the dissolution of evaporites that contain mirabilite ($NaSO_4$), polyhalite ($K_2Ca_2Mg[SO_4]_4{\cdot}2H_2O$), etc., may provide abundant $SO_4^{2-}$ or $HCO_3^-$ for the water bodies.

The ratio of $\gamma(Ca^{2+} + Mg^{2+})$ and $\gamma(HCO_3^-)$ can further be used to investigate the source of $Ca^{2+}$ and $Mg^{2+}$ [6]. As can be seen in Figure 7c, these sample points fit a line with a slope of 1.01, and the $R^2$ is 0.85, which is close to 1:1 line. This indicates that $Ca^{2+}$, $Mg^{2+}$, and $HCO_3^-$ in river water, groundwater, and other water bodies are mainly derived from the dissolution of dolomite ($CaMg(CO_3)_2$) and calcite ($CaCO_3$) rather than gypsum ($CaSO_4$).

Generally, silicates are more difficult to weather than carbonates. Thus, the ratio of $(Ca^{2+} + Mg^{2+})/(Na^+ + K^+)$ or $\gamma(HCO_3^-)/\gamma(Na^+ + K^+)$ in the water can be used as a symbol for judging the main types of weathered rocks in natural water [24]. As shown in Figure 7d, there are obvious differences between lake water and other water points. The high ratios of $\gamma(Ca^{2+} + Mg^{2+})/\gamma(Na^+ + K^+)$ in the river water and groundwater indicate flow through the carbonate weathering regions (Figure 7d). Additionally, the lake samples were located on the lower side of the 1:1 line, which confirmed that silicate dissolution or evaporate dissolution contributes to the main ion characteristics (Figure 7d).

3.2.3. Hydrochemical Modeling of Mineral Saturation Index (*SI*)

The saturation index (*SI*) of the main minerals in natural water indicates the equilibrium state between minerals and water and distinguishes the dissolution and precipitation reaction of minerals to simulate the trend of water chemistry in the environment. Phreeqc software is a hydrochemical simulation software developed by the United States District Survey, which can calculate the saturation index of minerals in groundwater. The saturation index (*SI*) is calculated as

$$SI = \lg\frac{IAP}{K}$$

where *IAP* is the ionic activity product, and *K* is the equilibrium constant. $SI > 0$ indicates that the mineral is saturated, and precipitate occurs. $SI = 0$ indicates that the mineral reached the equilibrium state, relatively stable. $SI < 0$ indicates that the mineral is not saturated, and a dissolution reaction occurs [51].

The ion ratio analysis of water in the study area showed that halite and carbonate minerals were the main minerals of the water–rock reaction. The solubility relationship between dolomite, calcite, and halite is shown in Figure 8. There are some differences in the saturation index (*SI*) of minerals in various water bodies. The salt mines have the largest halite *SI*, varying from −3.96 to 0.27. The ranges of halite *SI* in the lake, groundwater, river water, and glaciers are −8.36–−4.59, −9.48–−5.96, −10.13–−7.06, and −11.07–−10.65, respectively. The saturation index (*SI*) of calcite and dolomite in the river and groundwater samples has a smaller variation and is generally within ±1 or ±2, indicating that these minerals tend to be saturated continuously during the hydrochemical evolution. The ranges of calcite and dolomite *SI* in lake water are 0.29–0.87 and 0.71–3.17, respectively. On the contrary, the calcite and dolomite *SI* values in glaciers are less than 0 and have a range of −3.3–−2.18 and −7.41–−4.82, respectively.

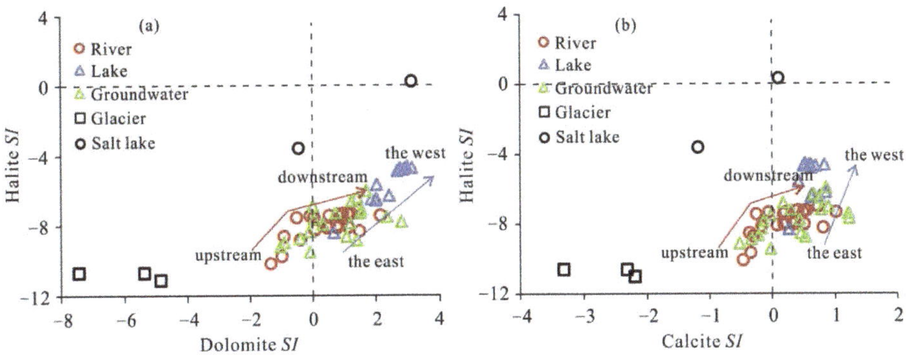


**Figure 8.** *SI* relationship between halite/dolomite (**a**) and halite/calcite (**b**) (the salt lake samples are data collected from the literature, as shown in Table 1).

The results showed that the saturation state of calcite, dolomite, and halite was not reached in glacier meltwater and upstream river water bodies. With the contact between the river and the carbonate rock strata in the flowing area, calcite and dolomite in the river bodies in the middle and lower and Bangong Co Lake water reached saturation. Overall, the halite saturation index of Bangong Co Lake showed an increasing trend from east to west, but it did not reach the saturation state like the river water. This indicates that the calcite and dolomite gradually become saturated from unsaturated to saturated, while halite dissolves during water flow from upstream to downstream and into the lake. The data collected from the salt lakes in this area show that the saturation index of calcite, dolomite, and halite in some salt lakes is greater than 0, indicating the precipitation of calcite, dolomite, and halite.

Previous studies [35,48] pointed out that the Bangong Co Lake basin experienced a period of basin shrinkage during the plateau's uplift. During this time, some independent intermountain lakes entered the evaporite-deposition period due to abrupt regional drought climate, forming the characteristics of dispersed salt lake salt mines in the area, such as the Jieze salt mine [48] and the Taicuo salt mine [49] (Figure 1). Through the collection of chemical composition data on the salt lake water body and salt mine soil layer (Table 1), it can be seen that there are rich minerals such as halite (NaCl), dolomite ($CaMg(CO_3)_2$), calcite ($CaCO_3$), and mirabilite ($NaSO_4$) in these lacustrine chemical sedimentary profiles. Under rain erosion or lateral river water intrusion, the interlacustrine group continuously replenishes rivers and lakes, providing abundant $Na^+$, $K^+$, $Cl^-$, and $SO_4^{2-}$ ions. Therefore, the rich carbonate strata, the rock salt minerals produced in the historical environment, and the arid climate environment in the study area lead to the diverse hydrochemical characteristics in the basin.

#### 3.2.4. Principal Component Analysis (PCA)

Principal component analysis (PCA) can reduce the dimension of the data set by converting the original variables into new and uncorrelated variables generated by retaining original information [54]. Many studies have used the principal component analysis (PCA) technique to identify important water quality parameters [55,56]. Before component analysis, all data passed the Kaiser–Meyer–Olkin (KMO) value (0.704) and Bartlett's test of sphericity statistics ($p < 0.05$) by using Origin2022 to evaluate the feasibility of principal component analysis (PCA) for source apportionment [57].

By filtering principal components (PCs) with an eigenvalue greater than 1 [29], only two PCs were extracted from the scree plot and explained 89.0% of the total variance (Table 2 and Figure 9). $K^+$, $Na^+$, $Mg^{2+}$, $HCO_3^-$, $Cl^-$, $SO_4^{2-}$, TDS, As, and EC have a relatively high loading on the principal component PC1, accounting for 75.4% of the total variance. PC2 explained 13.6% of the total variance and has a strong positive loading for $Ca^{2+}$ and $NO_3^-$ and weak negative loading for pH. Given the water sample loading distribution (Figure 9a), the loading of lake water points in PC1 is higher than that of other water bodies. The high salinity (TDS) and other ions in lake water are mainly caused by the dry environment and the lithology of the regional strata. Therefore, PC1 mainly indicates the characteristics of natural sources. Lin, Dong, and Wang pointed out that the concentration of heavy metals, such as As, Mn, and Ni, in the water body of Bangong Co Lake was lower than the background value of the water body in Tibetan areas, indicating that those heavy metals in the water were mainly derived from natural rock weathering [29]. This also verified that PC1 containing high loadings of As and TDS in this study mainly represented natural sources. PC2 comprised $Ca^{2+}$, $NO_3^-$, and pH. According to the loading distribution of the sample points (Figure 9a), the points with high PC2 loadings are mainly groundwater and a few river water points, which are mainly located in the human living, farming area, and the valley area where the meadow grows in the alluvial fan of the Doma River. This indicates that PC2 mainly illustrates animal husbandry activities and human activities. The middle and lower reaches of the Doma River and Makazangbu River are important animal husbandry areas, agricultural irrigation areas, and human living areas, indicating that agricultural fertilization and human activities have a certain impact on the water environment.

**Table 2.** Loadings of each variable on principal component (PC) comparison of Bangong Co Lake Watershed.

| **Elements** | **Component** | |
|---|---|---|
| | **PC1** | **PC2** |
| $K^+$ | 0.305 | 0.041 |
| $Na^+$ | 0.317 | 0.012 |
| $Ca^{2+}$ | −0.081 | 0.676 |
| $Mg^{2+}$ | 0.312 | 0.115 |
| $Cl^-$ | 0.317 | 0.034 |
| $SO_4{}^{2-}$ | 0.311 | 0.102 |
| $HCO_3{}^-$ | 0.300 | 0.172 |
| $NO_3{}^-$ | −0.079 | 0.636 |
| TDS | 0.317 | 0.041 |
| EC | 0.318 | 0.041 |
| TH | 0.302 | 0.047 |
| pH | 0.182 | −0.248 |
| As | 0.289 | −0.126 |
| Eigenvalue | 9.801 | 1.767 |
| Variance (%) | 75.393 | 13.593 |

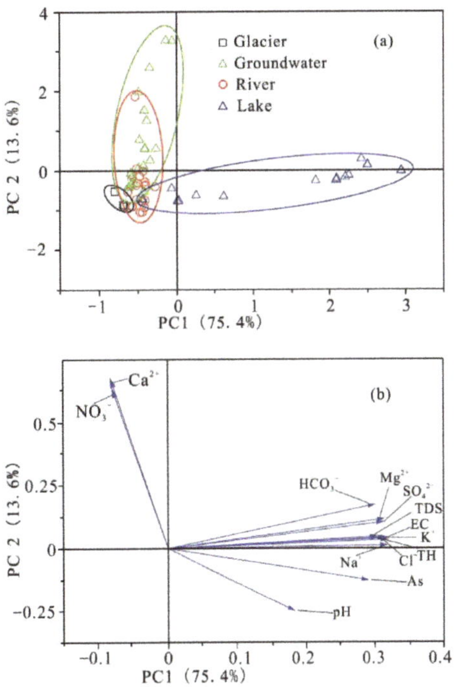


**Figure 9.** (**a**) Score plot of principal component analysis of major ions in water samples from the study area; (**b**) loading plot of principal component analysis of major ions in water samples from the study area.

### *3.3. Water Quality Evaluation*

#### 3.3.1. Assessment of Groundwater Suitability for Drinking Purposes

A drinking water quality index (DWQI) method was used to assess the suitability of groundwater for drinking [41,42]. The drinking water quality index (DWQI) is a water quality assessment method proposed by Horton RK and widely used [58]. This technique used the weighted index method to change many water quality parameters into a single index that can be compared, effectively providing a comprehensive groundwater quality evaluation model [59]. The calculation of the drinking water quality index (DWQI) can be divided into several steps: (a) Assignment of each parameter's weight according to its relative importance (values from 2 to 5), (b) calculation of the relative weight for each index ($W_i$), (c) the rate calculation of the quality parameter ($Q_i$), and (d) the water quality index computation of each sample. These data formulas are calculated via the following equations:

$$W_i = w_i / \sum_{i=1}^{n} w_i$$

$$Q_i = (C_i - C_{ip}) / (S_i - C_{ip}) \times 100$$

$$DWQI = \sum (Q_i \times W_i)$$

where $W_i$ is the relative weight, $w_i$ is the weight of each parameter, $n$ is the total number of parameters considered for the drinking water quality index (DWQI) calculation, and $C_i$ is the concentration of each parameter (mg/L). $C_{ip}$ is the ideal value of the parameter in pure water, and $S_i$ is the standard value of each parameter.

In this study, an analysis of eleven water quality parameters, namely, pH, TH, TDS, $Cl^-$, $SO_4^{2-}$, $NO_3^-$, $F^-$, $Na^+$, As, and Mn, was used to evaluate the suitability of groundwater for drinking. The relative weight of each parameter and its weight used in these calculations are presented in Table 3. Based on drinking water quality index (DWQI) values [41,51,60], the groundwater quality status can be categorized into five types: Excellent water (<20), good water (≥20 and <40), poor water (≥40 and <80), very poor water (≥80 and <120) and water unsuitable for drinking (≥120). In the Bangong Co Lake Watershed, the computed DWQI values vary from 12 to 86, with an average value of 32 ± 20. According to the DWQI classification, 26% ($n$ = 5) of the total groundwater samples fall under the excellent category, and 53% ($n$ = 10) are classified as good water, respectively. About 21% ($n$ = 4) of groundwater samples fell under moderate water quality (poor and very poor), and no sample was labeled as unsuitable for drinking (Figure 10). Figure 10 shows the spatial distribution of the drinking water quality index (DWQI), which illustrates that the chemical characteristics of groundwater in the lakeshore zone do not have a regular distribution in the east–west direction. Meanwhile, wells located near the lakeshore have higher drinking water quality index (DWQI) values than those away from the lakeshore, indicating that groundwater near the lakeshore may be affected by lake water (Figure 10).

**Table 3.** The relative weight for each index and weight assigned for drinking water quality index (DWQI).

| Parameters | Weight | Relative Weight |
|---|---|---|
| $Na^+$ | 2 | 0.047 |
| $Cl^-$ | 4 | 0.093 |
| $SO_4^{2-}$ | 3 | 0.070 |
| $NO_3^-$ | 5 | 0.116 |
| TDS | 5 | 0.116 |
| pH | 3 | 0.070 |
| As | 5 | 0.116 |
| TH | 3 | 0.070 |
| Fe | 4 | 0.093 |
| Mn | 4 | 0.093 |
| $F^-$ | 5 | 0.116 |

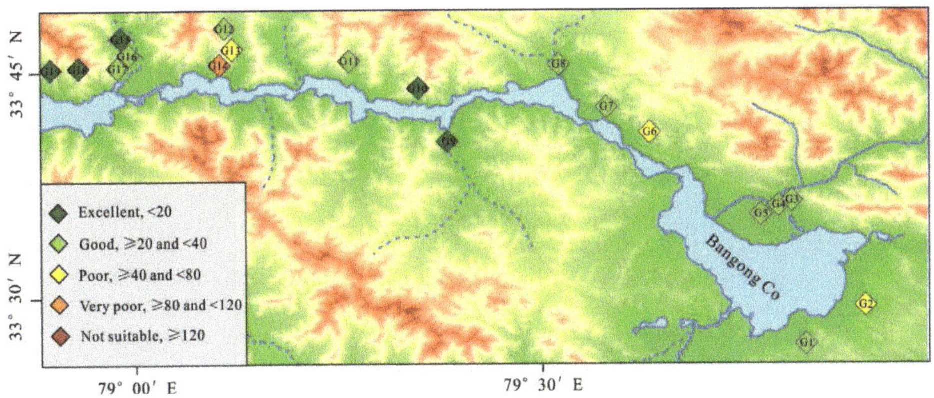


**Figure 10.** Assessment map of groundwater suitability in Bangong Lake Watershed.

### 3.3.2. Water Evaluation for Irrigation

In the plateau and semi-arid region of Bangong Co Lake Watershed, agriculture and livestock husbandry are the main occupations, and the main crops in the region are barley, rape, wheat, and peas [61,62]. Agriculture is the basic department of regional economic and social development. In addition, afforestation is an initiative advocated by the local government to improve the ecological environment [61]. Due to the fact that water quality plays an important role in crops, the survival of tree planting, and soil characteristics in the study area, groundwater irrigation suitability evaluation is necessary. To better evaluate the suitability of regional surface water and groundwater, this study uses the USSL classification proposed by Richards [43] and several other methods, such as electrical conductivity (EC), adsorption ratio (SAR), permeability index (PI), and magnesium hazardous ratio (MHR) [43,63,64]. Table 4 shows the calculation methods and sources of these parameters.

**Table 4.** Equations used for irrigation and irrigation water quality assessment.

| IWQ Parameters and Equations | Irrigation Problem | Degree of Restriction on Use | | | Number of Samples (%) | | | Reference |
|---|---|---|---|---|---|---|---|---|
| | | None | Slight to Moderate | Severe | None | Slight to Moderate | Severe | |
| EC (μS/cm) | Salinty | <700 | 700~3000 | >3000 | 35 (58%) | 17 (28%) | 8 (13%) | [43] |
| $SAR = \frac{Na^+}{\sqrt{0.5 \times (Ca^{2+} + Mg^{2+})}}$ | Permeability | >700 | 700~200 | <200 | 51 (85%) | 1 (2%) | 8 (13%) | [65] [43] |
| $PI = \left(\frac{Na^+ + \sqrt{HCO_3^-}}{Ca^{2+} + Mg^{2+} + Na^+}\right) \times 100$ | | <25 | 25~75 | >75 | 19 (32%) | 41 (68%) | 0 | [63] |
| $MHR = \left(\frac{Mg^{2+}}{Ca^{2+} + Mg^{2+}}\right) \times 100$ | | <50 | | >50 | 33 (55%) | - | 27 (45%) | [64] |

High salinity and high sodium concentration in irrigation water are the main causes of soil salinization, which affects the growth of plants and crops [66]. The sodium adsorption ratio (SAR), recommended by Richards [43], is one of the important indices used to calculate the harm of sodium in irrigation water. In the Bangong Co Lake Watershed, the sodium adsorption ratio (SAR) value ranged from 0.03 to 16.72, with a mean value of 3.05 ± 4.92. Based on the classification of sodium adsorption ratio (SAR) values [43,67], water samples in the Bangong Co Lake Watershed are classified into good ($n = 51$, 85%), poor ($n = 1$, 2%), and unsuitable ($n = 8$, 13%). The water points with sodium adsorption ratio (SAR) values exceeding the limits are located in the west and middle parts of Bangong Co Lake. Electrical conductivity (EC), the salinity parameter of irrigation water, ranges from 23 to 7861 μS/cm with an average value of 1493 ± 2179 μS/cm. According to electrical conductivity (EC) value classification, water samples in the Bangong Co Lake Watershed can be classified as good ($n$ = 35, 58%), poor ($n$ = 17, 28%), and unsuitable ($n$ = 8, 13%). Similarly, the water points with severe EC values are located in the west and middle parts of Bangong Co Lake, and points with poor electrical conductivity (EC) values are mainly located in the eastern part of Bangong Co Lake and groundwater near the lake's shoreline and lower sections of streams.

Doneen pointed out that poor soil permeability is usually caused by irrigation water with high $Na^+$ and $HCO_3^-$ concentrations [63]. Thus, the permeability index (PI) is often used to evaluate the harm of a high Na ion concentration in soil in irrigation water. In this study, the permeability index (PI) ranged from 36 to 181, averaging 71 $\pm$ 26. According to Wilcox's classification, watershed water samples were classified as good ($n$ = 49, 32%) or suitable ($n$ = 41, 68%) [67,68].

Ayers and Westcot noticed that irrigation water with high $Mg^{2+}$ could lead to soil $Ca^{2+}$ deficiency and crop yield reduction [64]. In this study, the magnesium hazardous ratio (MHR) ranged from 10 to 98, with a mean of 54 $\pm$ 23. The results show that 55% of water samples are suitable for irrigation (MHR < 50), and the rest of the water samples ($n$ = 27, 45%) are unsuitable (MHR $\geq$ 50). The unsuitable points for irrigation of MHR are mainly the lake's water points, some river water samples, and some groundwater samples.

The USSL classification suggested by US Salinity Laboratory Staff (1954) is a practical way to assess the suitability of irrigation water [43]. The USSL diagram best explains the combined effect of the salinity hazard and sodium hazard. In this paper, the USSL diagram (Figure 11) shows that 13% of the total samples are in the C1-S1 field, including three glacier samples, two groundwater samples, and three river samples (distributed upstream) which indicates relatively low alkalinity and salinity hazards. In the C2-S1 field, there are 10 groundwater samples, 19 river samples, and 1 lake sample with a medium of alkalinity hazard and low salinity. A total of seven groundwater samples and four lake samples (mainly in the eastern lake) are classified as part of the C3-S1 region. These samples with low sodium (S1) and low, medium, or high alkalinity (C1, C2, C3) apply to all soil types of irrigation. Only one lake sample was plotted on the C4-S2 region, and western lake samples were almost all plotted on the C5-S4 region, which is unsuitable for crops and soil irrigation. In conclusion, glacial meltwater and river water are more suitable for irrigation than lake water and some groundwater.

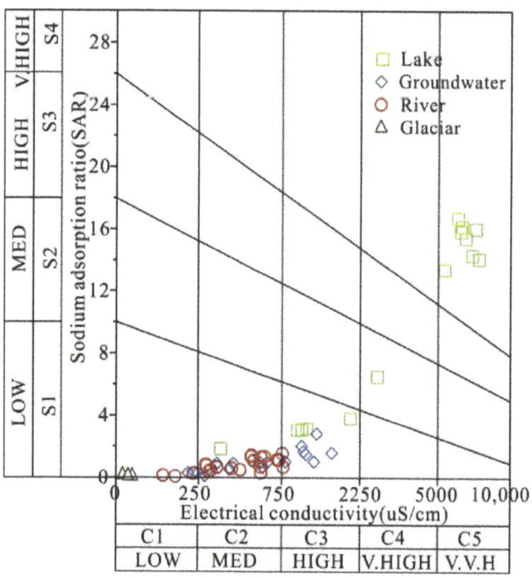


**Figure 11.** USSL diagram showing the suitability of groundwater for irrigation purposes.

## 4. Conclusions

The hydrochemical characteristics of surface water and lakeshore groundwater in Bangong Lake Watershed, Northwest Tibet, China, were analyzed using a multivariate statistical method, Piper diagrams, Gibbs diagrams, the ion ratio method, and principal component analysis (PCA). At the same time, the drinking water quality index (DWQI), irrigation water indexes, and spatial analysis were used to evaluate the quality of irrigation water and drinking water. The main conclusions are summarized below.

Most of the water samples, such as lake and river samples in the study area, were found to be slightly alkaline. Glacier water was categorized as soft-fresh, and lake water was classified as hard-brackish. Groundwater and river water were categorized as soft-fresh and hard-fresh. Na-Cl was the primary water type for lake water, while glacier, river, and groundwater were dominated by the Ca-$HCO_3$ type. Only 78% of groundwater samples were Ca-Cl types. The ionic concentrations of river waters showed an increased trend from upstream to downstream in terms of their spatial distribution, and high concentrations were found in the lake's coastal region. Regarding the spatial patterns of water in the Bangong Co Lake, ionic concentrations increased from east to west. There was no obvious spatial distribution of groundwater ion concentrations in the lakeshore zone in the east–west direction.

Precipitation dominance was observed in rivers and groundwater, and evaporation and crystallization dominance was observed in lakes, according to the Gibbs diagram. The ion ratio showed that the dissolution of carbonates and evaporites plays a leading role in the regional hydrochemical characteristics, corresponding to the limestone and dolomite formations widely exposed in the region. The calculation of the saturation index showed that calcite and dolomite were supersaturated in the water body of the basin, except for glaciers and some upstream rivers. Although all the lake samples from the study area were undersaturated regarding halite, their saturation index showed an increasing trend from east to west, indicating strong evaporation and continuous halite concentration characteristics. The collected data revealed salt lakes supersaturated with halite in this area. Thus, the main mineralization is related to the dissolution of calcite, dolomite, and halite in the strata. PCA showed that the regional hydrochemistry was mainly affected by natural factors, and only a few river sections and groundwater points had an anthropogenic input of $NO_3^-$ and $Ca^{2+}$ plasma, which may be related to animal husbandry activities in the valley and human activities in the river sector.

Regarding drinking water quality, the drinking water quality index (DWQI) value classified the groundwater as excellent (67%), good (5%), and poor (19%). The map of the suitability assessment for drinking water demonstrated that the closer the shallow groundwater was to the lake shore, the worse the water quality. With regard to the irrigation water quality, the irrigation water indexes indicated that glaciers, rivers, most groundwater, and a small part of the east lake are suitable for irrigation, while these indicators also indicated that the west and middle lake ($n$ = 8, 13%) are not suitable for irrigation. The permeability index (PI), electrical conductivity (EC), and USSL classification showed that some shallow groundwater near the lake shoreline and the lake water on the east side have moderate hazards in terms of salinity and permeability, and it is necessary to consider rational use for different types of land irrigation. Overall, the primary contribution of this research will help the Bangong Co Lake area sustainably utilize water resources.

**Author Contributions:** Writing—Review and Editing, Y.S. and B.Y.; Writing—Original Draft Preparation, Y.S. and L.L.; Investigation, K.G. and K.Z.; Project Administration, X.Y.; Software, G.F. All authors have read and agreed to the published version of the manuscript.

**Funding:** The research is sponsored by the China Geological Survey Project ([2023]10-02-01), and supported by State Key Laboratory of Loess and Quaternary Geology Open Fund (SKLLQG2241).

**Data Availability Statement:** The data that support the findings of this study are available from the corresponding author B.Y. upon reasonable request.

**Acknowledgments:** The authors gratefully acknowledge J.W. at laboratory of Xizang Shengyuan Environmental Engineering Co., Ltd. for the constant help. We also thank the contributions of H.Y., H.H., and X.Z. at Research Center of Applied Geology of China Geological Survey.

**Conflicts of Interest:** The authors declare no conflict of interest.

## References


1. Wang, X.P.; Gong, P.; Wang, C.F.; Ren, J.; Yao, T.D. A review of current knowledge and future prospects regarding persisten organic pollutants over the Tibetan Plateau. *Sci. Total Environ.* **2016**, *573*, 139–154. [CrossRef] [PubMed]
2. Yafeng, S.; Jiawen, R. Glacier recession and lake shrinkage indicating the climatic warming and drying trend in central Asia. *Ann Glaciol.* **1990**, *14*, 261–265. [CrossRef]
3. Yan, L.J.; Zheng, M.B.; Wei, L.J. Change of the lakes in Tibetan Plateau and its response to climate in the past forty years. *Front Earth Sci.* **2016**, *23*, 310–323.
4. Zhang, R.; Zhu, L.; Ma, Q.; Chen, H.; Liu, C.; Zubaida, M. The consecutive lake group water storage variations and their dynami response to climate change in the central Tibetan Plateau. *J. Hydrol.* **2021**, *601*, 126615. [CrossRef]
5. Jin, Z.D.; You, C.F.; Wang, Y.; Shi, Y. Hydrological and solute budgets of Lake Qinghai, the largest lake on the Tibetan Plateau *Quat. Int.* **2010**, *218*, 151–156. [CrossRef]
6. Sun, M.P.; Jin, H.A. Hydrochemistry differences and causes of tectonic lakes and glacial lakes in Tibetan Plateau. *Water* **2020** *12*, 3165. [CrossRef]
7. Ren, J.; Wang, X.; Wang, C.; Gong, P.; Yao, T. Atmospheric processes of persistent organic pollutants over a remote lake of the central Tibetan Plateau: Implications for regional cycling. *Atmos. Chem. Phys.* **2016**, *14*, 1401–1415.
8. Wang, C.; Zhou, H.; Kuang, X.; Hao, Y.; Shan, J.; Chen, J.; Li, L.; Feng, Y.; Zheng, Y. Water quality and health risk assessment o the water bodies in the Yamdrok-tso basin, southern Tibetan Plateau. *J. Environ. Manag.* **2021**, *300*, 113740. [CrossRef]
9. Zhang, Y.; Zhang, Y.; Shi, K.; Zhou, Y.; Li, N. Remote sensing estimation of water clarity for various lakes in China. *Water Res* **2021**, *192*, 116–128. [CrossRef]
10. Li, H.; Li, J.; Liu, X.L.; Yang, X.; Zhang, W.; Wang, J.; Niu, Y.Q. Composition characteristics and source analysis of major ions in four small lake-watersheds on the Tibetan Plateau, China. *Huan Jing Ke Xue Huanjing Kexue* **2015**, *36*, 430–437.
11. Wang, P.; Shang, Y.N.; Shen, L.C. Characteristics and evolution of hydrochemical compositions of freshwater lake in Tibetan plateau. *Huan Jing Ke Xue Huanjing Kexue* **2013**, *34*, 874–881. [PubMed]
12. Gunkel, A.; Shadeed, S.; Hartmann, A.; Wagener, T.; Lange, J. Model signatures and aridity indices enhance the accuracy of wate balance estimations in a data-scarce Eastern Mediterranean catchment. *J. Hydrol. Reg. Stud.* **2015**, *4*, 487–501. [CrossRef]
13. Medici, G.; Lorenzi, V.; Sbarbati, C.; Manetta, M.; Petitta, M. Structural classification, discharge statistics, and recession analysis from the springs of the Gran Sasso (Italy) carbonate aquifer; comparison with selected analogues worldwide. *Sustainability* **2023** *15*, 10125. [CrossRef]
14. He, W.M.; Song, Y.G.; Zhang, C.S.; Zhang, Q.F. Mineral elements exchange patterns in biogeochemical cycle of Jinshuhe Rive Basin. *Ecol. Environ. Sci.* **2011**, *20*, 217–225.
15. Li, X.; Gao, S.; Li, W.; Qiu, G.; Tian, Z. Temporal and spatial evolution of lakes in the Bashang Plateau for nearly recent 30 years. *Water Resour. Prot.* **2022**, *2*, 698–708. [CrossRef]
16. Bazova, M.; Moiseenko, T. Migration activity of elements in the water of lakes of northwestern Russia. *Geochem. Int.* **2021** *59*, 970–982. [CrossRef]
17. Baranov, D.Y.; Moiseenko, T.I.; Dinu, M.I. Geochemical trends in the formation of atmospheric precipitation in the conditionally background area of the Valdai National Park. *Geochem. Int.* **2020**, *58*, 1159–1173. [CrossRef]
18. Xiao, C.L.; Liang, X.J.; Wang, B. *Hydrogeology*, 2nd ed.; Tsinghua University Press: Beijing, China, 2010.
19. Jiang, H.C.; Lv, Q.Y.; Yang, J.; Wang, B.C.; Dong, H.L.; Gonsior, M.; Schmitt-Kopplin, P. Molecular composition of dissolved organic matter in saline lakes of the Qing-Tibetan Plateau. *Org. Geochem.* **2022**, *167*, 104400. [CrossRef]
20. Xu, H.; Hou, Z.; An, Z.; Liu, X.; Dong, J. Major ion chemistry of waters in Lake Qinghai catchments, NE Qinghai-Tibet plateau China. *Quat. Int.* **2010**, *212*, 35–43. [CrossRef]
21. Zhu, S.; Zhang, F.; Zhang, Z.; Kung, H.-T.; Yushanjiang, A. Hydrogen and oxygen isotope composition and water quality evaluation for different water bodies in the Ebinur Lake Watershed, Northwestern China. *Water* **2019**, *11*, 2067. [CrossRef]
22. Kumar, R.; Parvaze, S.; Huda, M.; Allaie, S. The changing water quality of lakes-a case study of Dal Lake, Kashmir Valley. *Environ Monit. Assess.* **2022**, *194*, 228. [CrossRef] [PubMed]
23. Sun, R.; Zhang, X.Q.; Wu, Y.H. Major ion chemistry of water and its controlling factors in the Yamzhog Yumco Basin, South Tibe *J. Lake Sci.* **2012**, *24*, 600–608.
24. Guo, J.M.; Kang, S.C. Temporal and spatial variations of major ions in Nam Co Lake water, Tibetan Plateau. *Huan Jing Ke Xue Huanjing Kexue* **2012**, *33*, 2295–2302. [PubMed]
25. Fontes, J.C.; Gasse, F.; Gibert, E. Holocene environmental changes in Lake Bangong basin (Western Tibet). Part 1: Chronology and stable isotopes of carbonates of a Holocene lacustrine core. *Palaeogeogr. Palaeoclimatol. Palaeoecol.* **1996**, *120*, 25–47. [CrossRef]
26. Xiao, Y.; Xie, S.Y.; Wang, M.D.; He, Y.; Hou, J.Z. Characteristics of water temperature based on fractal and R/S Method in Bangong Co and Dagze Co. *Geol. Sci. Technol. Inf.* **2015**, *34*, 200–206.

7. Hu, M.; Wang, Y.; Du, P.; Shui, Y.; Cai, A.; Lv, C.; Bao, Y.; Li, Y.; Li, S.; Zhang, P. Tracing the sources of nitrate in the rivers and lakes of the southern areas of the Tibetan Plateau using dual nitrate isotopes. *Sci. Total Environ.* **2019**, *658*, 132–140. [CrossRef] [PubMed]
8. Wen, Z.; Zheng, M.; Xu, X.; Liu, X.; Guo, G.; He, Z. Biological and ecological features of saline lakes in northern Tibet, China. *Hydrobiologia* **2006**, *541*, 189–203. [CrossRef]
9. Lin, L.; Dong, L.; Wang, Z. Hydrochemical composition, distribution, and sources of typical organic pollutants and metals in Lake Bangong Co, Tibet. *Environ. Sci. Pollut. Res.* **2021**, *28*, 9877–9888. [CrossRef]
0. Yang, H.A.; Li, Z.Q.; Ye, B.S.; Jiao, K.Q. New result of glacier inventory in the drainage basins of the Bangong lake in china. *J. Glaciol.* **2003**, *25*, 685–691.
1. Guo, G.; Zhi, J.; Ye, B. *Regional Geological Survey of 1:250,000 Ritu County, Tibet Autonomous Region (I44C003002)*; Jiangxi Institute of Geological Survey: Nanchang, China, 2004; pp. 20–120.
2. Yin, A.; Harrison, T.M. Geologic Evolution of the Himalayan-Tibetan Orogen. *Annu. Rev. Earth Planet. Sci.* **2000**, *28*, 211–280. [CrossRef]
3. Zhao, B.; Shi, R.; Zou, H.; Chen, S.; Huang, Q.; Sun, Y.L.; Yang, J. Intra-continental boninite-series volcanic rocks from the Bangong-Nujiang Suture Zone, Central Tibet. *Lithos* **2021**, *12*, 386–387. [CrossRef]
4. Li, H.L.; Gao, C.; Li, Z.H.; Zhang, Z.; Peng, Z.M.; Guan, J.L. Age and tectonic significance of Jingzhushan Formation in Bangong Lake Area, Tibet. *Geotecton. Et Metallog.* **2016**, *40*, 663–673.
5. Liu, F.; Liu, D.; Zhou, T. Causes and Tectonic Evolution of Bangong Lake Basin. *Earth Sci. J. China Univ. Geosci.* **2013**, *38*, 745–754.
6. Li, Y.; Li, M.; Cao, Y. Cashmere Diameter Deviation Of Tibetan Cashmere Goats. *Acta Vet. Et Zootech. Sinica* **1999**, *30*, 432–437.
7. Piper, M. A graphic procedure in the geochemical interpretation of water-analyses. *Trans. Am. Geophys. Union* **1944**, *25*, 914–923. [CrossRef]
8. Gibbs, R.J. Mechanisms controlling world water chemistry. *Science* **1970**, *170*, 1088–1090. [CrossRef]
9. Meybeck, M. Global chemical weathering of surficial rocks estimated from river dissolved loads. *Am. J. Sci.* **1987**, *287*, 401–428. [CrossRef]
0. WHO. *Guidelines for Drinking-Water Quality*; Fourth Edition Incorporating the First Addendum; WHO: Geneva, Switzerland, 2017; pp. 308–475.
1. Adimalla, N.; Li, P.; Venkatayogi, S. Hydrogeochemical evaluation of groundwater quality for drinking and irrigation purposes and integrated interpretation with water quality index studies. *Environ. Process.* **2018**, *5*, 363–383. [CrossRef]
2. Sarkar, M.; Pal, S.C.; Islam, A.R.M.T. Groundwater quality assessment for safe drinking water and irrigation purposes in Malda district, Eastern India. *Environ. Earth Sci.* **2022**, *81*, 11–20. [CrossRef]
3. Richards, L.A. Diagnosis and improvement of saline and alkali soils. *Soil Sci.* **1954**, *78*, 290. [CrossRef]
4. Wu, J.; Xue, C.; Tian, R.; Wang, S. Lake water quality assessment: A case study of Shahu Lake in the semiarid loess area of northwest China. *Environ. Earth Sci.* **2017**, *76*, 232. [CrossRef]
5. Xiao, J.; Zhang, F.; Jin, Z. Spatial characteristics and controlling factors of chemical weathering of loess in the dry season in the middle Loess Plateau, China. *Hydrol. Process.* **2016**, *30*, 4855–4869. [CrossRef]
6. Kim, Y.; Mo, H.; Son, J.; Lee, Y.S.; Lee, S.E.; Cho, K. Interactive effects of water pH and hardness levels on the growth and reproduction of Heterocypris incongruens (Crustacea: Ostracoda). *Hydrobiologia* **2015**, *753*, 97–109. [CrossRef]
7. Ganguli, S.; Rifat, M.a.H.; Howlader, S.; Hasan, M.A.; Islam, S.; Alam, M.N.E.; Islam, M.N. Assessment of Bhatiari Lake water quality: Pollution indices, hydrochemical signatures and hydro-statistical analysis. *J. Indian Chem. Soc.* **2022**, *99*, 105–115. [CrossRef]
8. Hu, W. *Geological Characteristics and Uplifting Environment Evolution of Salt Lake in the Eastern Section of Pangong Lake in Xizang Province*; China University of Geosciences (Wuhan): Wuhan, China, 2014; pp. 1–61.
9. Liu, J.; Wei, L.; Zheng, M. Distribution and Characteristics of Carbonate and Elementary Geochemistry of Profile TT-1 in Dahyab Tso (Tai Cuo) Tibet and Its Palaeoenvironment Significance. *Acta Geol. Sin.* **2007**, *81*, 1289–1298.
0. Wang, J.Y.; Long, A.H.; Deng, M.J.; Xie, L. Water balances of east and west lakes Balkhash and their ptimization management. *J. Glaciol.* **2011**, *33*, 1353–1362.
1. Kammoun, A.; Abidi, M.; Zairi, M. Hydrochemical characteristics and groundwater quality assessment for irrigation and drinking purposes: A case of Enfidha aquifer system, Tunisia. *Environ. Earth Sci.* **2022**, *81*, 41. [CrossRef]
2. Fadili, A.; Najib, S.; Mehdi, K.; Riss, J.; Makan, A.; Boutayeb, K.; Guessir, H. Hydrochemical features and mineralization processes in coastal groundwater of Oualidia, Morocco. *J. Afr. Earth Sci.* **2016**, *116*, 233–247. [CrossRef]
3. Zhang, Y.; Zhu, G.F.; Ma, H.Y.; Yang, J.X.; Pan, H.X.; Guo, H.W.; Wan, Q.Z.; Yong, L.L. Effects of Ecological Water Conveyance on the Hydrochemistry of a Terminal Lake in an Inland River: A Case Study of Qingtu Lake in the Shiyang River Basin. *Water* **2019**, *11*, 1673. [CrossRef]
4. Gu, Y.G.; Li, Q.S.; Fang, J.H.; He, B.Y.; Fu, H.B.; Tong, Z.J. Identification of heavy metal sources in the reclaimed farmland soils of the pearl river estuary in China using a multivariate geostatistical approach. *Ecotoxicol. Environ. Saf.* **2014**, *105*, 7–12. [CrossRef]
5. Ahmad, S.; Singh, N.; Mazhar, S.N. Hydrochemical characteristics of the groundwater in Trans-Yamuna Alluvial aquifer, Palwal District, Haryana, India. *Appl. Water Sci.* **2020**, *10*, 75. [CrossRef]
6. Şehnaz, Ş.; Şener, E.; Davraz, A.; Varol, S. Hydrogeological and hydrochemical investigation in the Burdur Saline Lake Basin, southwest Turkey. *Geochemistry* **2019**, *80*, 125–135. [CrossRef]

57. Wang, D.; Shi, L. Source identification of mine water inrush: A discussion on the application of hydrochemical method. *Arab. Geosci.* **2019**, *12*, 121–134. [CrossRef]
58. Horton, R.K. An index number system for rating water quality. *J. Water Pollut. Control Fed.* **1965**, *37*, 300–306.
59. Ali, Z.I.; Gharbi, A.; Zairi, M. Evaluation of groundwater quality in intensive irrigated zone of Northeastern Tunisia. *Groundw Sustain. Dev.* **2020**, *11*, 100482.
60. Kumari, M.; Rai, S.C. Hydrogeochemical evaluation of groundwater quality for drinking and irrigation purposes using wate quality index in semi arid region of India. *J. Geol. Soc. India* **2020**, *95*, 159–168. [CrossRef]
61. Yan, J.H. *2022 Government Work Report of Ali Prefecture, Tibet Autonomous Region*; Tibet Daily: Lhasa, China, 2022; pp. 1–2.
62. Huntington, E. Pangong: A glacial lake in the Tibetan Plateau. *J. Geol.* **1906**, *14*, 599–617. [CrossRef]
63. Doneen, L.D. *Water Quality for Agriculture, Department of Irrigation*; University of California: Davis, CA, USA, 1964; pp. 5–10.
64. Ayers, R.S.; Westcot, D.W. *Water Quality for Agriculture*; Food and Agriculture Organization of The United Nations: Rome Italy, 1976.
65. Bouwer, H. *Groundwater hydrology (McGraw-Hill Series in Water Resources and Environmental Engineering Series*; McGraw-Hil College: New York, NY, USA, 1978; pp. 1–10.
66. Rajmohan, N.; Masoud, M.H.Z.; Niyazi, B.a.M. Assessment of groundwater quality and associated health risk in the aric environment, Western Saudi Arabia. *Environ. Sci. Pollut. Res.* **2021**, *28*, 9628–9646. [CrossRef]
67. Wilcox, L. *Classification and Use of Irrigation Waters*; United States Department of Agriculture: Washington, DC, USA, 1955; pp 1–5.
68. Saleh, A.; Al-Ruwaih, F.; Shehata, M. Hydrogeochemical processes operating within the main aquifers of Kuwait. *J. Arid Environ* **1999**, *42*, 195–209. [CrossRef]

water

Article

# Balancing Water Ecosystem Services: Assessing Water Yield and Purification in Shanxi

Man Li [1], Shanshan Li [1], Huancai Liu [1,2,*] and Junjie Zhang [1]

1 School of Geographical Science, Shanxi Normal University, Taiyuan 030031, China
2 School of Geography and Tourism, Shaanxi Normal University, Shannxi 710062, China
* Correspondence: liuhc@sxnu.edu.cn

**Abstract:** Water yield and purification are important aspects of water ecosystem services, and achieving a balanced development of the two is necessary for the development of aquatic ecosystems. Using the InVEST model, the spatiotemporal variations of regional water yield and purification services in Shanxi, China, from 2000 to 2020 were analyzed. Three future scenarios (natural development, urban development, and ecological protection) were assessed for 2030 using the PLUS model. The results showed that in 2000–2020, the water yield of Shanxi Province in terms of space was generally low in the middle and northwest and high in the southeast, and it was affected by land-use change and climatic change. From 2000 to 2020, the water yield of Shanxi Province changed by 78.8 mm. In 2030, water yield will be highest under the urban development scenario (380.53 mm) and lowest in the ecological protection scenario (368.22 mm). Moreover, the water quality purification capacity improved, with nitrogen loading high in the center and low in the east and west. Due to the implementation of environmental protection policies and the improvement of the technical level, the nitrogen load was the highest in 2000 (0.97 kg/hm$^2$) and lowest in 2015 (0.94 kg/hm$^2$). By 2030, because of the high nitrogen loadings of cultivation and construction land and low nitrogen loadings of forests and grasslands, the nitrogen load was lowest under the scenario of urban development (0.94 kg/hm$^2$) and highest under ecological protection (0.85 kg/hm$^2$).

**Keywords:** water yield; water purification services; InVEST model; PLUS model; Shanxi Province

## 1. Introduction

Water ecosystem services reflect the ability of ecosystems to process and regulate water and can provide guarantees for water resources and environmental security for human society [1]. With the increase in socioeconomic development, human demand for water resources has become increasingly intense, leading to frequent water shortages, water pollution, and severe impacts on the sustainable development of the region [2–4]. Since the 18th National Congress of the Communist Party of China (CPC), as the construction of ecological civilization has been integrated into social and economic construction, the evaluation of ecosystem service capacity has become increasingly important [5]. The balanced development between water yield and purification is of great significance for maintaining the stability, biodiversity, sustainable use, and healthy development of human society [6–8].

With the development of Geographic Information Systems and remote sensing technologies in ecology and hydrology, many models and methods can be used to simulate regional water yield and purification services [9–11]. For example, models such as the DHI MIKE, Topography-based Hydrological Model, Soil and Water Assessment Tool (SWAT), Terrain Lab, and Integrated Valuation of Ecosystem Services and Tradeoffs (InVEST) have been used to simulate and analyze water yield services in different watersheds; models or methods such as the single-factor method, Nemero index method, grey evaluation method, SWAT, DO concentrations, the support vector regression (SVR) model, and InVEST have been used to evaluate water quality conditions in different regions [12,13].

The InVEST model is an open-source ecosystem service function assessment model based on the ecological production process, assessment, and trade-off of ecosystem services. It calculates the physical and value quantities of ecosystem services, incorporates socioeconomic factors into ecosystem conservation decisions, and enables the spatial presentation of ecosystem services.

Owing to fewer parameter requirements [14–16], flexible parameter adjustments, a wide range of applications [17], and other advantages, the InVEST model is widely used by scholars and has achieved good results. However, the model has limitations, such as high data dependence, parameterization and calibration challenges, and the complexity of economic valuation. Internationally, Redhead et al. [18], Daneshi et al. [19], and Bejagam et al. [20], respectively, used the InVEST model to simulate the water yield of the British rivers, the basin flowing into the Caspian Sea in northern Iran, and the Tungabhadra basin in the Indian peninsula. In China, Li et al. [21], Chen et al. [22], and Li et al. [23], respectively, used the InVEST model to analyze the spatial and temporal changes in water resources and water ecosystem services in Shaanxi Province, Hanjiang City, and Taihu Lake Basin. In Shanxi, Pan et al. [24], Wang et al. [25], and Yang et al. [26] used the InVEST model to analyze the spatiotemporal evolution of coalfield ecosystem services in Shanxi Province, the pattern of multiple ecosystem services and ecological security in Shanxi Province, and the coupling and coordination of the sustainable development of the ecosystem in Shanxi Province. However, there are fewer assessments of water ecosystem services in Shanxi Province. The geographical location and ecological environment of Shanxi Province are extremely important. Therefore, it is very important to supplement the research on the water ecological environment in Shanxi Province.

To more accurately predict future changes in water ecosystem services, scholars have performed a large number of studies on the possibility of future land-use change [27,28], among which the Patch-generating land-use simulation (PLUS) model is widely used because of its high simulation accuracy, fast data processing, and effective simulation of complex land-use evolution. The PLUS model has been used to predict the land-use and ecosystem services of water ecosystems. Currently, combining the prediction of multi-scenario land-use and ecosystem services has become a popular topic in research. Ferreira et al. [29] provided suggestions for future ecological restoration in the southwestern region of Portugal by predicting the land-use status of the region. Gao et al. [30] analyzed the future ecological risk status of the local ecosystems by predicting the land-use types of Nanjing for 2025 under multiple future scenarios. Li et al. [31] predicted the land-use types and assessed the carbon stock in Kunming City based on the PLUS and InVEST models to provide a reference for mitigating regional carbon loss.

Located in the eastern part of the Loess Plateau and the middle reaches of the Yellow River, Shanxi Province is an important economic and ecological region in China and a vital part of the Yellow River Basin [32]. It is situated at the northern boundary of the East Asian summer wind system. This geographic location exposes the province to the dynamic interplay of mid-latitude westerly winds and Asian monsoon winds [33]. The territory is crisscrossed by mountains and has a complex topography; its unique geographic location and elevation make it a sensitive area to climate change [34]. In Shanxi Province, an important energy and chemical base in China, industrial water consumption is high and causes serious water pollution. The degradation of aquatic ecosystem services has long been a prominent concern [35]. The quality of the water ecological environment in Shanxi Province can directly affect the fragile ecological environment and country of Shanxi Province in terms of the development of the people's economy. As an important water yield area in the Yellow River Basin, it will also be related to the water resources situation of the entire Yellow River Basin.

Quantitative research on water yield and purification in Shanxi Province is essential for government departments to understand the production, distribution, and utilization of water resources, which is conducive to the rational planning for the allocation and utilization of such resources. Such research can help better understand the health status of water ecosystems in Shanxi Province and provide a scientific basis for ecological protection and restoration, thereby contributing to improving the quality of water bodies and reducing the impact of water pollution on ecology and human health, helping the government formulate reasonable development strategies, balancing the relationship between economic growth and environmental protection, and achieving sustainable social, economic, and environmental development [36].

In this study, we conducted a quantitative assessment using the InVEST model to analyze the spatiotemporal variations in water yield and purification services from 2000 to 2020. Additionally, three future development scenarios (natural development, urban development, and ecological protection) were established to estimate the potential water yield and purification services in 2030. This research seeks to provide scientific support for the sustainable development of regional aquatic ecosystems, with a particular emphasis on the importance of maintaining the co-development between water quantity and quality to ensure ecological preservation.

## 2. Methodology

### *2.1. Study Area*

Shanxi Province (34°34′–40°44′ N, 110°14′–114°33′ E) in North China has a total area of $15.67 \times 10^4$ km$^2$ (Figure 1). This area comprises various landforms, such as mountains, hills, plateaus, and basins. Mountains and hills account for 80% of the total area of the province; the terrain is high in the northeast and low in the southwest. It has a temperate continental monsoon climate and an uneven seasonal distribution of precipitation, with more precipitation in summer and fall and less precipitation in winter and spring. Shanxi Province spans two major water systems, the Yellow and Hai Rivers, with a total annual average water resource of $180.59 \times 10^8$ m$^3$, of which the Yellow River Basin includes the basins of the Fen, Yellow Tributary, Sushui, and Qin Rivers; and the Hai River Basin includes the basins of the Yongding, Hutuo, Zhangwei, and Daqing Rivers.

### *2.2. Data Sources*

This study mainly utilized spatial data, including meteorological, natural environment, soil, socioeconomic, land-use, and road information data (Table 1). Owing to the different resolutions of these data sources, the necessary data were resampled and projected by raster processing. The spatial resolution of all data was standardized to 30 m by ArcMap10.7, and the coordinate system used was Krasovsky_1940_Albers.

**Table 1.** Data sources and descriptions.

| Categorization | Data Type | Data Sources and Processing |
|---|---|---|
| Meteorological elements | Measured quantity of rain | Data Center for Resource and Environmental Sciences, Chinese Academy of Sciences (http://www.resdc.cn) |
| | Temperatures | |
| | Potential evapotranspiration | National Meteorological Information Center-China Meteorological Data Network (http://www.cma.cn) |
| Natural environment | DEM | Geospatial Data Cloud (http://www.gscloud.cn) |
| | Slope | Calculated from DEM data in ArcGIS to get |
| Soil features | Soil data | Chinese soil information in the World Soil Database (HWSD) |
| Socioeconomic | GDP | Data Center for Resource and Environmental Sciences, Chinese Academy of Sciences (http://www.resdc.cn) |
| | Demographic | |
| Land use | Land-use type | |

Table 1. *Cont.*

| Categorization | Data Type | Data Sources and Processing |
| --- | --- | --- |
| Road information | Distance to the first level of road | National Geographic Information Resources Catalog Service System (http://www.servicetianditu.gov.cn) |
| | Distance to secondary roads | |
| | Distance to tertiary roads | |
| | Distance to highway | |
| | Distance to railroad | |
| | Distance to Government | |

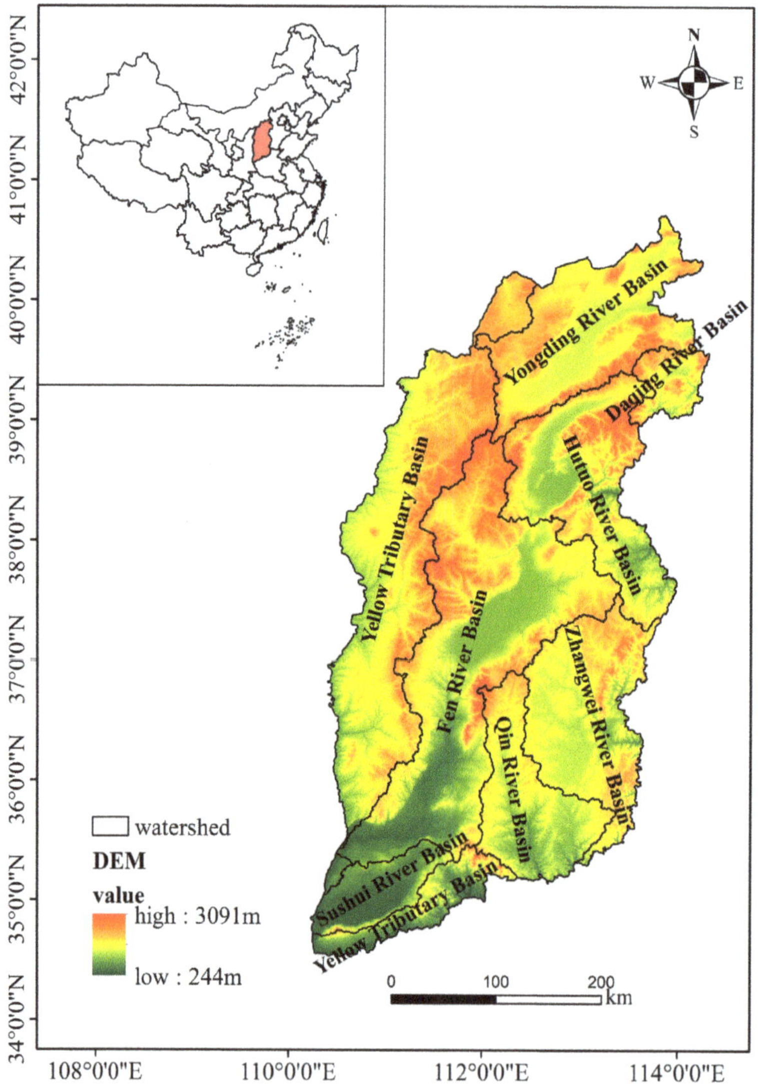


**Figure 1.** Overview of the geographic location of Shanxi Province.

### 2.3. Research Methodology

#### 2.3.1. InVEST Model

The InVEST model is an ecosystem assessment model based on different surface land covers that can be used to simulate the material quality and quantity of ecosystem services and spatialize ecosystem services to improve natural resource management. The model includes multiple submodules, such as water yield, water purification, soil conservation, and habitat quality. For the evaluation of aquatic ecosystem services in Shanxi Province in this study, water yield and purification modules were mainly utilized to calculate the amount of water yield and nitrogen loading. The details of the modules used are as follows:

(1) Water yield module

The water yield module (water yield) is an estimation method based on the water balance method and Budyko's water–heat balance assumption [37], which combines meteorological, topographic, vegetation, soil, and other data, using the difference between precipitation and evapotranspiration to obtain the water yield. The greater the water yield, the greater the water supply. This study compared the impact of land-use changes on water yield under different scenarios for 2020 and 2030, where the water yield in 2030 was based on precipitation in 2020. The calculation formula is as follows:

$$Y(x) = \left(1 - \frac{AET(x)}{p(x)}\right) \cdot p(x) \tag{1}$$

$$\frac{AET(x)}{p(x)} = \frac{1 + w(x) + R(x)}{1 + w(x) \cdot R(x) + 1/R(x)} \tag{2}$$

$$w(x) = Z \cdot \frac{PAWC(x)}{p(x)} \tag{3}$$

$$R(x) = \frac{k(x) \cdot ET_0}{P(x)} \tag{4}$$

where Y(x) is the average annual water yield (mm), AET(x) is the average annual actual evapotranspiration (mm), P(x) is the average annual precipitation (mm), R(x) is the Budyko drying coefficient, and PAWC(x) is the available water content of vegetation (mm), which is the coefficient of evapotranspiration of vegetation and the average annual reference evapotranspiration (mm). Z denotes the Zhang coefficient [38], which was obtained by improving the calculations based on the Budyko curve. Through continuous simulation improvement, when Z was 4.03, the difference between the simulated average annual water yield and actual natural runoff in Shanxi Province was the smallest, and the simulation results were the best.

(2) Water purification module

The nutrient retention module estimates the effect of vegetation and soil on water purification based on the ability of nitrogen nutrients in runoff to remove pollutants and determines the capacity of water purification. The higher the total nitrogen content, the more severe the pollution level of the watershed and the weaker the water quality purification capacity. The calculation formula is as follows:

$$ALV_x = HSS_x * pol_x \tag{5}$$

$$HSS_x = \frac{\lambda_x}{\overline{\lambda_w}} \tag{6}$$

$$\lambda_x = \log(\Sigma_u Y_u) \tag{7}$$

where $ALV_x$ is the adjusted load value, $pol_x$ is the output coefficient, $HSS_x$ is the hydrological sensitivity score, $\lambda_x$ is the runoff index, $\overline{\lambda_w}$ is the average runoff index in the watershed, and $\Sigma_u Y_u$ is the total water yield (mm) within the runoff path.

As different land-use types exhibit different water purification capacities, their biophysical coefficients are different. This study combined the results of relevant existing research to determine the local biophysical coefficients (Table 2) [39], which can be used to reflect the attributes of the soil cover and land-use types. This specifically includes the actual evapotranspiration assignment, root depth, plant evapotranspiration coefficients, nitrogen loading coefficients, nitrogen sequestration efficiency, and other relevant parameters. The actual evapotranspiration assignment was assigned as 1 and 0 based on whether there was vegetation cover or not, respectively.

**Table 2.** Details of biophysical coefficients.

| Land-Use Type | Actual Evapotranspiration Assignment | Depth of Root System (mm) | Plant Evapotranspiration Coefficient | Nitrogen Load Factor | Nitrogen Interception Efficiency |
|---|---|---|---|---|---|
| Arable land | 1 | 350 | 0.75 | 18.23 | 0.4 |
| Woodland | 1 | 2500 | 0.93 | 3.45 | 0.75 |
| Grassland | 1 | 750 | 0.63 | 8.02 | 0.5 |
| Waters | 0 | 1 | 1 | 0.01 | 0.05 |
| Construction land | 0 | 1 | 0.25 | 11.03 | 0.05 |
| Unused land | 1 | 20 | 0.4 | 9.83 | 0.05 |

2.3.2. Bivariate Spatial Correlation Model

Global spatial autocorrelation can describe the overall distribution of a phenomenon and determine whether the phenomenon has agglomeration characteristics in space; however, it cannot precisely identify the areas in which local spatial autocorrelation can project the scope of the agglomeration [40]. Among these, the Geoda tool can analyze raster spatial data, which can effectively reveal the correlation between different elements of spatial units, and the calculation formulas are as follows:

$$\text{Moran's I} = \frac{N\sum_{i}^{N}\sum_{j\neq 1}^{N} W_{ij} z_i^a z_j^b}{(N-1)\sum_{i}^{N}\sum_{j\neq 1}^{N} W_{ij}} \tag{8}$$

$$\text{LISA}_i = \frac{1}{N}\frac{(x_i-\overline{x})}{\sum_{i}(x_i-\overline{x})^2}\sum_{j} W_{ij}(x_i-\overline{x}) \tag{9}$$

where Moran's I denotes the bivariate global Moran's index, $z_i^a$ is the deviation of the attribute of the ith cell from the mean, $W_{ij}$ is the spatial weight matrix, and N is the number of units in the study area. The value of Moran's I is in the range of −1 to 1; Moran's I > 0, < 0, and = 0 indicate positive, negative, and no correlations, respectively. Local indicators of spatial association (LISA) is the bivariate local autocorrelation index, where $x_i$ is the attribute value of unit i and represents the average of all attribute values. When LISA > 0, the spatial cell is a high-high- or low-low-value spatial agglomeration, and when LISA < 0, the spatial cell is a high-low- or low-high-value spatial agglomeration.

2.3.3. PLUS Model

The PLUS model is a recently developed model that is based on traditional land-use simulation and random forest models such as system dynamics [41], Future Land-Use Simulation [42], CLUE-Scanner [43], and Spatial-temporal Markov chain [44] models. It integrates a Create-a-Research-Space model of a multi-type random seed mechanism that can be used to analyze the drivers of land expansion and predict the evolution of land-use patches.

To evaluate the simulation accuracy, the simulated land-use types in 2020 were compared with real land-use data in 2020. The Kappa coefficient was found to be 0.84, which was greater than 0.7, indicating its high simulation accuracy. Therefore, it can be used to predict the land-use types in Shanxi Province by 2030.

Land use is affected by policies, economic development, natural environment, and other aspects. Due to the uncertainty of future development, it is based on the land use changes in Shanxi Province from 2000 to 2020 and the possibility of future development under three future development scenarios (i.e., natural development, urban development, and ecological protection). Simulations and projections of changes in land use in Shanxi Province by 2030 were then developed. These three development scenarios are designed to explore more comprehensively the impacts of land use on ecosystems. These different scenarios are set up for several reasons: the natural development scenario takes into account natural evolution and natural processes and helps to understand the impact of land use on aquatic ecosystems in the absence of human policy interventions; the urban development scenario takes into account that Shanxi Province is in the rising stage of economic transformation, and urban expansion may still exist in the future; the ecological protection scenario takes into account that the environmental awareness of the government and the public is gradually rising, and in order to achieve the sustainable and high-quality development of Shanxi Province, ecological protection measures may be strengthened in the future. Under the natural development scenario, the change in each land-use type continues the current development trend; under the urban development scenario, the transfer of construction land to other land-use types is restricted, but other land can be transferred to construction land; and under the ecological protection scenario, the transfer of ecological land, such as woodland, grassland, and water, to other land is restricted, but other land can be transferred to ecological land types (Table 3; where "+" indicates transferable and "−" indicates non-transferable land).

**Table 3.** Multi-scenario land-use transfer matrix.

| Scenario Setting | Land-Use Type | Arable Land | Woodland | Grassland | Waters | Construction Land | Unused Land |
|---|---|---|---|---|---|---|---|
| Natural Development cenario | Arable land | + | − | + | + | − | + |
| | Woodland | − | + | + | − | − | + |
| | Grassland | + | + | + | + | − | + |
| | Waters | − | − | + | + | − | + |
| | Construction Land | + | − | + | − | + | + |
| | Unused land | + | + | + | + | − | + |
| Urban Development Scenario | Arable land | + | + | + | − | − | + |
| | Woodland | − | + | − | + | + | + |
| | Grassland | − | − | + | + | + | + |
| | Waters | − | − | − | + | − | + |
| | Construction Land | + | + | + | + | + | + |
| | Unused land | + | + | − | − | − | + |
| Ecological Protection Scenario | Arable land | + | − | − | − | − | + |
| | Woodland | + | + | + | − | − | + |
| | Grassland | + | − | + | − | − | + |
| | Waters | + | − | + | + | − | + |
| | Construction Land | + | − | − | − | + | + |
| | Unused land | + | − | − | − | − | + |

## 3. Results and Discussion

### *3.1. Changes in Water Yield in Shanxi Province from 2000 to 2020*

#### 3.1.1. Temporal Variations

Generally, water yield is mainly affected by rainfall and actual evapotranspiration, whereas the main factors affecting actual evapotranspiration include air temperature and land use [45]. In Shanxi Province, the differences in the average water yield of different land-use types were as follows: unutilized land > constructed land > grassland > cultivated land > woodland > waters. Because most unutilized land is not shaded by vegetation or artificial structures, this allows precipitation to directly reach the surface, whereas the hardened surface of the constructed land makes it difficult for precipitation to infiltrate. The average multi-year water yield of unutilized and constructed land reached 295.44 and 234.29 mm, respectively.

Grassland, arable land, and woodland exhibited medium water yield capacities. Although grassland was vegetated, water evaporation was relatively low, and the root system was shallow, with an average water yield of 214 mm. Arable land was usually artificially managed and improved, and the root systems of crops were more developed; therefore, it absorbed and utilized the water in the soil more efficiently, with an average water yield of 217.69 mm. Because woodland had exuberant branches and leaves with high evapotranspiration, the formation of a layer of withered material and humus on the surface played a positive role in water conservation, with an average water yield of 220.25 mm. Waters, despite being a storage space for water, were unable to absorb and utilize water, and the average water yield was 151.40 mm.

The water yield in Shanxi Province from 2000 to 2020 showed an N-shaped change, characterized by a small increase, then a decline and a subsequent increase (Figure 2). The average water yield in Shanxi Province in 2000 was 297.01 mm, and the high-value areas of water yield were mainly located in the Qin, Zhangwei, and Hutuo River Basins in the eastern part of Shanxi, owing to their abundant precipitation. However, 2015 was a very strong El Niño year; the province's precipitation was only 480.60 mm, the average temperature was 1 °C higher than normal, and, therefore, the water yield was low at only 265.27 mm. In contrast, 2020 was the subsequent year of El Niño, when the rainfall reached $877.2 \times 10^8$ $m^3$, with an average rain depth of 561.3 mm; thus, the year had a biased abundance of water [46]. Meanwhile, with the rapid development of the local society and economy and the popularization and implementation of the policy of "returning farmland to forests," the expansion of construction land has occupied a large amount of arable land and grassland. Furthermore, the increase in construction land from 2000 to 2020 was 106.22%, which has greatly increased the area of impermeable layers. Consequently, the average water yield of the province in 2020 reached 375.80 mm, which was higher than that of 2000 and 78.79 mm higher than the average water yield in 2000.

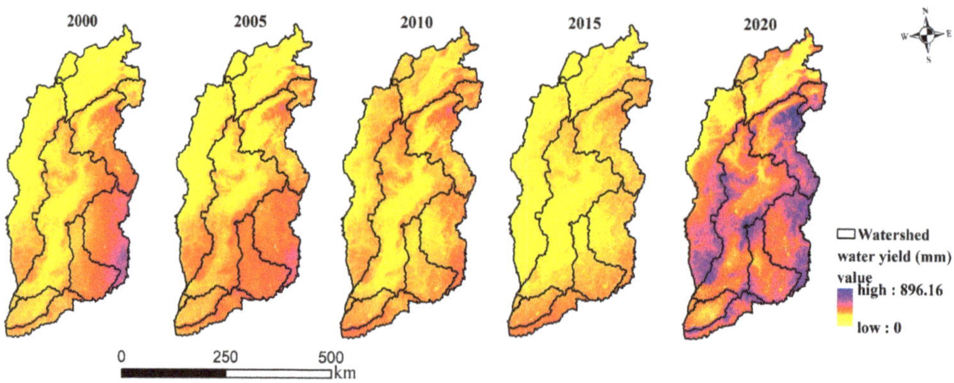


Figure 2. Water yield in Shanxi Province from 2000 to 2020.

### 3.1.2. Spatial Variations

Water yield is greatly influenced by climate and human activities. Affected by this, it shows the spatial differentiation characteristics are generally low in the middle and northwest and high in the southeast in terms of space (Table 4). The water yield in the central basin area is low, the topography of the basin is usually surrounded by surrounding mountains or mountains, and water replenishment is limited. Moreover, the basin's topography has a high rate of evaporation and transpiration because the surrounding mountains block the wind, making it possible for the humidity to increase, resulting in more evaporation and transpiration of water. In the northern region, Datong and Xinzhou had mines, and to ensure the dryness of the mines and associated workspaces, a large amount of groundwater was extracted, which led to the destruction of underground aquifers, a continuous decrease in water yield, and, thereby, a negative impact on the local water ecosystem. This encouraged the local area to actively promote the mine's ecological restoration and, thereby, the benign development of the regional ecology. The water yield in the western part of Shanxi Province is low, mainly because most of the area belongs to the arid climate zone, with a dry climate and lack of sufficient precipitation. The western region has high and complex terrain, more mountainous areas, and poor water storage conditions. The ecological environment in the western region is poor, and problems such as soil erosion are serious, affecting the maintenance and supply of water resources. In recent years, the Lvliang Mountain area has experienced abundant precipitation and vigorous vegetation growth. With the popularization of soil and water conservation projects, water-saving technology, and other measures, the high water yield area has shifted from the southeastern part of the Haihe River Basin to the Lvliang Mountain area in the Yellow River Basin.

**Table 4.** Water yield in different watersheds in Shanxi Province (all units in mm).

| Watershed | Year | 2000 | 2005 | 2010 | 2015 | 2020 |
|---|---|---|---|---|---|---|
| Yellow River Basin | Fen River Basin | 278.9 | 291.27 | 268.68 | 245.49 | 385.57 |
| | Yellow Tributary Basin | 281.75 | 300.44 | 315.34 | 298.05 | 362.45 |
| | Sushui River Basin | 281.44 | 319.17 | 321.27 | 327.17 | 335.06 |
| | Qin River Basin | 370.53 | 406.94 | 304.93 | 289.52 | 403.08 |
| Sea River Basin | Yongding River Basin | 213 | 233.73 | 283.2 | 250.43 | 267.09 |
| | Hutuo River Basin | 333.32 | 287.5 | 310.45 | 284.99 | 427.41 |
| | Zhangwei River Basin | 441.03 | 404 | 314.29 | 276.73 | 435.6 |
| | Daqing River Basin | 300.44 | 326.42 | 341.76 | 320.76 | 401.76 |

Located in the eastern part of Shanxi Province, the Haihe River Basin generally has a high level of urbanization, and land use is dominated by construction and industrial land, with an average annual water yield of 322.70 mm. The Zhangwei, Hutuo, and Daqing River Basins are situated on the windward slopes of the Taihang Mountain System, with high precipitation and low evaporation, with an average annual water yield of 328.73–374.33 mm.

The Yellow River Basin in the central and western parts of Shanxi Province is a semi-arid region, with land-use types dominated by cropland and grassland and an average water yield of only 319.35 mm. The Fen and Sushui River Basins are located on the leeward slopes of the Taihang and Taiyue Mountains, with topography dominated by basins, high temperatures, and strong evapotranspiration; their average water yields are 293.98 and 316.82 mm, respectively.

### 3.2. *Spatiotemporal Variation in Water Purification Services in Shanxi Province from 2000 to 2020*

The ability to remove nitrogen can be used to reflect the ability to purify water, and different land types have different abilities to absorb and remove nitrogen. Studies have shown that the annual average nitrogen loads of various land-use types are in the following order: cropland > construction land > unutilized land > grassland > woodland > waters. In agricultural activities, some pesticides and fertilizers are not absorbed by crops, which accumulate nitrogen on the surface, leading to a nitrogen load in the croplands of Shanxi Province of 1.64 kg/hm$^2$. During the urbanization process, a large amount of nitrogen oxides is generated from energy consumption, industrial production, and transportation, which is deposited into water bodies in the form of acid rain, and some enterprises directly discharge nitrogen-containing wastewater into the water bodies; this increased the nitrogen loading of construction land in Shanxi Province to 0.99 kg/hm$^2$. Because forests and grasslands have strong nitrogen-fixing capacities, their nitrogen loads were lower. In water, aquatic plants and plankton can absorb and utilize nitrogen in the water, whereas the water body can dilute the nitrogen via water flow and circulation. Consequently, the nitrogen load in the water bodies of Shanxi Province was only 0.01 kg/hm$^2$.

The nitrogen load in Shanxi Province is affected by a variety of factors, including agriculture, industry, urbanization, and the natural environment. Therefore, spatially, the nitrogen loading in Shanxi Province was high in the central part and low in the eastern and western parts (Figure 3). Areas with low water purification capacities were mainly distributed in the south-central part of the Yellow River Basin, the northern part of the Haihe River Basin, and the southeastern part of Shanxi Province. A large amount of discharged urban and industrial sewage and applied chemical fertilizers increased the nitrogen load and lowered the water purification capacity in the province. Areas with higher water purification capacities were mainly distributed in the southeastern part of the Yellow River Basin and the Lvliang Mountain area, which are mountainous and hilly, with large forest and grassland distributions.

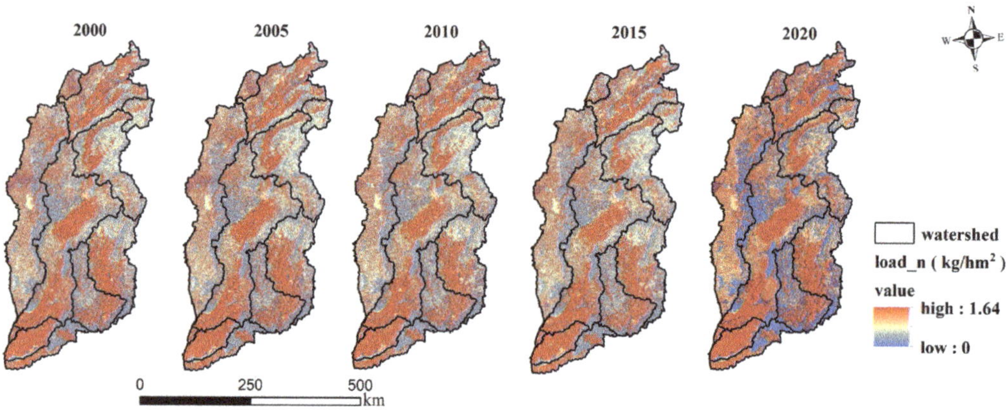


**Figure 3.** Nitrogen output load from 2000 to 2020.

During 2000–2020, the water purification capacity of the Haihe and Yellow River Basins increased overall. Among them, with the implementation of environmental protection and "returning farmland to forest" policies, the area with higher water purification capacity expanded significantly in the Lvliang Mountain area of the Yellow River Basin; however, in the central area of the Fen River Basin, owing to the increasing level of urbanization and subsequent increase in regional nitrogen loading, the areas with low water purification capacity are expanding into the north-central part of the Fenhe River Basin.

### *3.3. Spatial Matching Analysis of Water Yield and Purification Services*

Based on the Geoda spatial analysis tool, a spatial weight matrix was established, and the bivariate global Moran's I index for water yield and purification services in Shanxi Province was calculated (Table 5). The results show that a significant and expanding trade-off occurred between water yield and water purification services during 2000–2020.

**Table 5.** The global Moran's I index for water yield and water purification services.

| Norm | 2000 | 2005 | 2010 | 2015 | 2020 |
|---|---|---|---|---|---|
| Moran's I | −0.201 | −0.204 | −0.273 | −0.216 | −0.335 |
| *p*-value | 0.002 | 0.003 | 0.005 | 0.004 | 0.003 |

Analyzing the local spatial heterogeneity between water yield and purification services in Shanxi Province (Figure 4) revealed that the Hutuo River Basin, Zhangwei River Basin, southern part of the Yellow Tributary, and Fen River Basin were mainly "high-low" agglomeration areas. This region had abundant precipitation that was distributed across forested grasslands, and it currently represents the optimal area for water ecology in Shanxi Province.

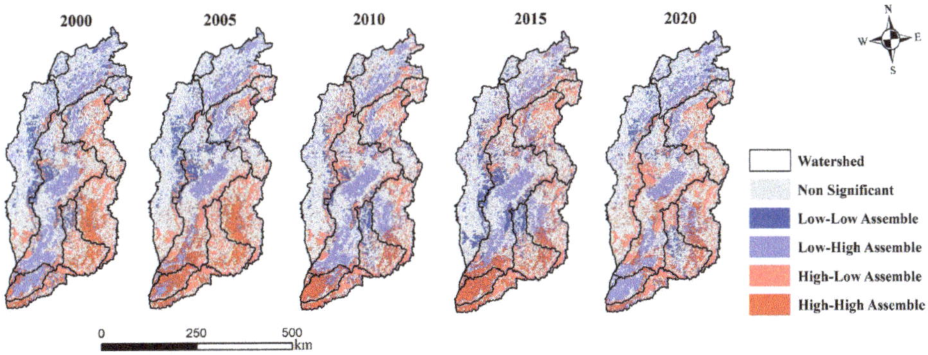


**Figure 4.** Bivariate local indicators of spatial association clustering of water yield and purification services in Shanxi Province.

The "low-high" agglomeration areas were mainly found to be distributed in the south-central part of the Yellow River Basin and the northern part of the Haihe River Basin, with high temperatures, high evaporation rates, and an abundance of arable land; this area is currently a low-value area for the quality of the water ecosystem in Shanxi Province. Under the influence of the increasing trend of local warming and drying, the significant increase in fertilizer application, and the uncontrolled discharge of nitrogen-containing wastewater, the "low-high" agglomeration area has a clear tendency to expand to the western part of the Zhangwei River Basin; therefore, special attention should be paid to the protection of the aquatic ecology in this area.

The "high-high" agglomeration areas were mainly distributed in the southern part of the Yellow River Basin and the southwestern part of the Haihe River Basin in Shanxi Province, where the arable land is widely spread and has rich precipitation, exhibiting high water yield and weak water purification capacity. However, over the past 20 years, with the promotion of the policy of returning farmland to forests, the local water purification capacity has been significantly improved, and "high-high" agglomeration areas have been decreasing.

The "low-low" agglomeration areas were mainly distributed in the eastern part of the basin of the tributaries of the Yellow River and the western part of the Qin River Basin, which are relatively small in area but have large forest and grassland distributions, and the vegetation has a strong ability to intercept precipitation. Notably, with the increase in precipitation in the Lvliang Mountain area, the local area showed a tendency to shift from the "low-low" catchment area to the "high-low" agglomeration area.

### 3.4. Characterization of Water Yield and Purification Services under Different Scenarios for 2030

By predicting the changes in water yield and nitrogen loading under the three scenarios (i.e., natural development, urban development, and ecological protection) for 2030, the following findings were obtained (Table 6).

**Table 6.** Average annual water yield and nitrogen loading in Shanxi Province from 2000 to 2020, and projections for 2030 under different scenarios.

| Ecosystems/Year | 2000 | 2005 | 2010 | 2015 | 2020 | Natural Development | Urban Development | Ecological Protection |
|---|---|---|---|---|---|---|---|---|
| Water yield (mm) | 297.01 | 299.39 | 294.06 | 265.27 | 375.81 | 372.19 | 380.53 | 368.22 |
| Nitrogen loading (kg/hm$^2$) | 0.97 | 0.96 | 0.95 | 0.94 | 0.95 | 0.91 | 0.94 | 0.85 |

Under the natural development scenario, Shanxi Province will continue with the development trend of the current land-use type. By 2030, the water yield capacity will change minimally, only 372.19 mm, which is slightly lower than that in 2020. Under the urban development scenario, construction land will be restricted from shifting to other land-use types but other land-use types can be transformed into construction land; therefore, by 2030, the area of construction land will have significantly increased compared with that of the other scenarios, whereas ecological land areas will decrease significantly, resulting in a water yield of 380.53 mm, a 1.3% increase compared with that in the year 2020. Under the ecological protection scenario, ecological environmental protection will be particularly emphasized by the government, which will lead to a significant increase in the area of ecological forests and grasslands, thereby increasing the intercepting capacity of vegetation for precipitation and increasing pollutants. The amount of water produced by 2030 is projected to decrease by 7.59 mm compared with that recorded in 2020.

With urban socioeconomic development and the improvement of wastewater treatment levels, the water purification capacity of Shanxi Province under the three scenarios will be higher in 2030 than it was in 2020, but the water purification capacities will be different under the different scenarios. The nitrogen loading under the natural development scenario will be 0.91 kg/hm$^2$, a decrease of 4.21% compared with that recorded in 2020. Under the urban development scenario, the nitrogen loading will be evidently higher than those in the other scenarios because of the significant increase in the area of construction land and the significant decrease in the areas of forest and grassland, which have strong absorption and removal capacities for nitrogen. Under the ecological protection scenario, the area of ecological forest and grassland will increase, and the ecological forest meadow will have an increased capacity to intercept pollutants and reach the water body. The capacity of the pollutants will be enhanced, and the nitrogen load reaching the water body is anticipated to decrease by 10.52%, or 0.85 kg/hm$^2$, compared with that recorded in 2020.

## 4. Conclusions and Discussion

### 4.1. Conclusions

Based on the InVEST model water yield and water purification module, the spatiotemporal changes in water yield and water purification in Shanxi Province from 2000 to 2020 and the correlation between the two were analyzed. Water yield and nitrogen loading under the three scenarios of natural development, urban development, and ecological protection by 2030 were predicted using the PLUS model. The results of this article are as follows:

1. The differences in the average water yield of the different land-use types in Shanxi Province were as follows: unutilized land > constructed land > grassland > cultivated land > woodland > waters. Temporally, the water yield of Shanxi Province from 2000 to 2020 can be characterized by an N shape. Affected by land-use change and climatic change, the lowest water yield (265.27 mm) occurred in 2015, and the highest (375.80 mm) was in 2020, showing a change of 110.53 mm; from 2000 to 2020, the water yield of Shanxi Province changed by 78.8 mm. Spatially, the water yield mainly exhibited characteristics of low spatial differentiation in the middle and northwest and high spatial differentiation in the southeast.
2. The water purification capacities of different land-use types were as follows: arable land > construction land > unutilized land > grassland > woodland > waters. Spatially, the areas with lower water purification capacities were mainly distributed in the south-central part of the Yellow River Basin and the northern and southeastern parts of the Haihe River Basin, whereas those with higher water purification capacities were mainly distributed in the southeastern part of the Yellow River Basin as well as in the Lvliangshan Mountain area. Due to the implementation of environmental protection policies and the improvement of the technical level, the nitrogen load was the highest (0.97 kg/hm$^2$) in 2000 because of its large arable land area and high usage

rate of pesticide fertilizers. In contrast, the nitrogen load was the lowest in 2015 (0.94 kg/hm$^2$).

3. From 2000 to 2020, a significant and expanding trade-off occurred between the water yield and water purification services in Shanxi Province. The Hutuo River Basin, Zhangwei River Basin, southern part of the Yellow Tributary Basin, and Fenhe River Basin were primarily "high-low" agglomeration areas. The "low-high" agglomeration areas were mainly distributed in the south-central part of the Yellow River Basin and the northern part of the Haihe River Basin.
4. By 2030, the urban development scenario yielded the most water (380.53 mm), whereas the ecological protection scenario yielded the least (368.22 mm). With urban socioeconomic development and sewage treatment level improvement, the water purification capacity of Shanxi Province in 2030 is expected to be higher than that in 2020 under all three scenarios; however, the ecological protection scenario had the strongest purification capacity, with a nitrogen load of only 0.85 kg/hm$^2$. In contrast, the urban development scenario had the weakest purification capacity (nitrogen load: 0.94 kg/hm$^2$).

### 4.2. Discussion

The results show that the land-use type in the study area under the natural development scenario develops along the historical benchmark, reflecting the changes in the water ecological environment under current conditions, thereby having a minimal impact on the water ecological environment. Under the urban development scenario, the research area focuses on economic development and urban construction. The water yield and nitrogen load change the most; hence, urban expansion should be reasonably controlled, and ecological land should be protected [47,48]. The water ecological environment of Shanxi Province is ideal under the ecological protection scenario, but economic growth becomes relatively slow. Therefore, optimizing land use, increasing investment in sewage treatment science and technology, using pesticides and fertilizers rationally, and implementing river basin zoning are necessary for economic development. Science and technology must be used to promote the construction of ecological civilization, ensure the safety of the water ecological environment in the river basin, and improve the water ecological environment.

In this study, the spatiotemporal changes and correlation between water yield and purification services in Shanxi Province from 2000 to 2020 were investigated. We chose 5 years as a time node; although this reflects the changes in regional water ecology in time and space, this may cause some spatiotemporal information on water yield and water purification to be omitted. Moreover, both precipitation and land-use changes can affect the water yield capacity of a region. As this study focused on analyzing the impact of land-use change on water yield capacity, future precipitation was assumed to remain unchanged. The three development scenarios were simulated to provide a certain reference for the rational adjustment of the land-use types and protection. In future studies, the dual impacts of precipitation and land use can be considered and combined with the representative concentration pathway and shared socioeconomic pathway scenarios proposed in CMIP6 to comprehensively predict changes in water yield and purification capacity in Shanxi and other similar critical regions.

Relevant studies have shown that warming and drying trends in the climate in Shanxi Province will become increasingly prominent in the future [49,50]. Shanxi Province should consider ecological, social, and economic factors when formulating relevant policies and action plans and seek the best balance to achieve comprehensive sustainability:

1. In the context of agricultural production, areas with low water yield in Shanxi Province, such as the Fenhe River Basin and the Sushui River Basin, should promote the popularization of water-saving technologies such as sprinkler and drip irrigation, improve the efficiency of water resource utilization, and optimize water resource management systems. Areas with wide agricultural distribution and low water purification capacity, such as the central and southern regions of the Yellow

River Basin, should also reduce the use of synthetic pesticides and chemical fertilizers and support organic agriculture.
2. In the context of ecosystem protection, areas with poor soil and water conservation, such as the cities of Jinzhong, Lvliang, and Yuncheng, should protect ecological forest belts (soil and water conservation forests), strengthen wind and sand control forests to reduce soil erosion and maintain soil moisture and nutrients, and protect water sources (wetlands, lakes, and rivers).
3. In the context of urban development, areas with a high level of urbanization, such as the cities of Taiyuan, Datong, and Changzhi, should rationally control the scale of urban construction to reduce the discharge of nitrogenous sewage. Sewage treatments should be strengthened, and the scale of use of reclaimed water should be expanded.

**Author Contributions:** Conceptualization, M.L. and S.L.; data curation, S.L.; formal analysis, S.L.; funding acquisition, M.L. and H.L.; project administration, M.L. and H.L.; resources, S.L.; software, S.L.; investigation, S.L.; visualization, J.Z.; writing—original draft, S.L.; writing—review and editing, M.L. and H.L. All authors have read and agreed to the published version of the manuscript.

**Funding:** This research was funded by the Natural Science Research Project of Shanxi Province (202103021224258), and the Key subject of Shanxi Federation of Social Sciences (SSKLZDKT2023036).

**Data Availability Statement:** Not applicable.

**Acknowledgments:** We thank the editors and reviewers for reviewing this article and look forward to your valuable comments on this article.

**Conflicts of Interest:** The authors declare no conflict of interest. The funders had no role in the design of the study; in the collection, analyses, or interpretation of data; in the writing of the manuscript; or in the decision to publish the results.

## References


1. Liu, J.; Li, J.; Qin, K. Changes in land-uses and ecosystem services under multi-scenarios simulation. *Sci. Total Environ.* **2017**, *586*, 522–526. [CrossRef]
2. Deng, X.; Zhao, C. Identification of Water Scarcity and Providing Solutions for Adapting to Climate Changes in the Heihe River Basin of China. *Adv. Meteorol.* **2014**, *2014*, 279173. [CrossRef]
3. Li, Y.; Mi, W.; Ji, L.; He, Q.; Yang, P.; Xie, S.; Bi, Y. Urbanization and agriculture intensification jointly enlarge the spatial inequality of river water quality. *Sci. Total Environ.* **2023**, *878*, 162559. [CrossRef] [PubMed]
4. Ban, Y.; Liu, X.; Yin, Z.; Li, X.; Yin, L.; Zheng, W. Effect of urbanization on aerosol optical depth over Beijing: Land use and surface temperature analysis. *Urban Clim.* **2023**, *51*, 101655. [CrossRef]
5. Chen, J.C.; Zhao, Z.; Wang, J.Y. Research on the essence and practical significance of the "Two Mountains Theory". *For. Econ.* **2020**, *42*, 3–13. [CrossRef]
6. Bojie, F.; Guoyi, Z.; Yongfei, B.; Changchun, S.; Jiyuan, L.; Huiyuan, Z.; Yihe, L.; Hua, Z.; Gaodi, X. Functions and ecological security of major terrestrial ecosystems in China. *Prog. Earth Sci.* **2009**, *24*, 571–576.
7. Dai, E.; Wang, X.; Zhu, J. Progress and trend prospect of ecosystem service trade-off/synergy research. *Adv. Earth Sci.* **2015**, *30*, 1250–1259.
8. Costanza, R.; Darge, R.; Groot, R. The value of the world's ecosystem services and natural capital. *Nature* **1997**, *387*, 253–260. [CrossRef]
9. Zhang, B.; Li, W.H.; Xie, G.D.; Xiao, Y. Water conservation function and measurement method of forest ecosystem. *J. Ecol.* **2009**, *28*, 6.
10. Cheng, X.; Chen, L.; Sun, R. An improved export coefficient model to estimate non-point source phosphorus pollution risks under complex precipitation and terrain conditions. *Environ. Sci. Pollut. Res.* **2018**, *25*, 20946–20955. [CrossRef]
11. Xie, Y.C.; Gong, J.; Qi, S.S. Spatial-temporal differentiation of water supply services in Bailongjiang River Basin based on InVEST model. *J. Nat. Resour.* **2017**, *32*, 1337–1347. [CrossRef]
12. Aryal, S.K.; Ashbolt, S.; Mcintosh, B.S. Assessing and Mitigating the Hydrological Impacts of Urbanisation in Semi-Urban Catchments Using the Storm Water Management Model. *Water Resour. Manag.* **2016**, *30*, 5437–5454. [CrossRef]
13. Nong, X.; Lai, X.; Chen, L.; Shao, D.; Zhang, C.; Liang, J. Prediction modelling framework comparative analysis of dissolved oxygen concentration variations using support vector regression coupled with multiple feature engineering and optimization methods: A case study in China. *Ecol. Indic.* **2023**, *146*, 109845. [CrossRef]
14. Wang, W.; Chen, L.; Shen, Z. Dynamic export coefficient model for evaluating the effects of environmental changes on non-point source pollution-Science Direct. *Sci. Total Environ.* **2020**, *747*, 141164. [CrossRef]

5. Zhang, F.P.; Li, X.J.; Feng, Q. Water conservation in the upper reaches of Heihe River Basin based on InVEST model. *Chin. Desert* **2018**, *38*, 9.
6. Ran, C.; Wang, S.; Bai, X. Trade-Offs and Synergies of Ecosystem Services in Southwestern China. *J. Environ. Eng. Sci.* **2020**, *37*, 669–678. [CrossRef]
7. Gao, J.; Li, F.; Gao, H. The impact of land-use change on water-related ecosystem services: A study of the Guishui River Basin, Beijing, China. *J. Clean. Prod.* **2015**, *163*, S148–S155. [CrossRef]
8. Redhead, J.W.; Stratford, C.; Sharps, K. Empirical validation of the InVEST water yield ecosystem service model at a national scale. *Sci. Total Environ.* **2016**, *569–570*, 1418–1426. [CrossRef] [PubMed]
9. Daneshi, A.; Brouwer, R.; Najafinejad, A. Modelling the impacts of climate and land use change on water security in a semi-arid forested watershed using InVEST. *J. Hydrol.* **2021**, *593*, 125621. [CrossRef]
0. Bejagam, V.; Keesara, V.R.; Sridhar, V. Impacts of climate change on water provisional services in Tungabhadra basin using InVEST Model. *River Res. Appl.* **2021**, *38*, 106–194. [CrossRef]
1. Li, Y.L.; He, Y.; Liu, W.Q.; Jia, L.P.; Zhang, Y.R. Evaluation and Prediction of Water Yield Services in Shaanxi Province, China. *Forests* **2023**, *14*, 229. [CrossRef]
2. Chen, Z.Y.; Yu, P.H.; Chen, Y.Y. Spatial-temporal evolution of water production and water purification services in the Han River Basin under shared socio-economic pathway. *Chin. J. Eco-Agric.* **2021**, *29*, 1800–1814. [CrossRef]
3. Li, Z.B.; Tao, Y.; Ou, W.X. Research on the relationship between supply and demand of aquatic ecological services in Taihu Lake Basin and multi-scenario assessment based on water quantity and water quality. *Acta Ecol. Sin.* **2023**, *5*, 2088–2100. [CrossRef]
4. Pan, H.H.; Wang, J.Q.; Du, Z.Q.; Wu, Z.T.; Zhang, H.; Ma, K.M. Spatiotemporal evolution of ecosystem services and its potential drivers in coalfields of Shanxi Province, China. *Ecol. Indic.* **2023**, *148*, 110109. [CrossRef]
5. Wang, J.; Li, Y.; Wang, S.; Li, Q.; Li, L.; Liu, X. Assessment of Multiple Ecosystem Services and Ecological Security Pattern in Shanxi Province, China. *Int. J. Environ. Res. Public Health* **2023**, *20*, 4819. [CrossRef] [PubMed]
6. Yang, Z.; Zhan, J.; Wang, C. Coupling coordination analysis and spatiotemporal heterogeneity between sustainable development and ecosystem services in Shanxi Province, China. *Sci. Total Environ.* **2022**, *836*, 155625. [CrossRef]
7. Liang, X.; Guan, Q.; Clarke, K.C. Understanding the drivers of sustainable land expansion using a patch-generating land use simulation (PLUS) model: A case study in Wuhan, China. *Comput. Environ. Urban. Syst.* **2021**, *85*, 101569. [CrossRef]
8. Yin, L.; Wang, L.; Li, T.; Lu, S.; Yin, Z.; Liu, X.; Zheng, W. U-Net-STN: A Novel End-to-End Lake Boundary Prediction Model. *Land* **2023**, *12*, 1602. [CrossRef]
9. Ferreira, M.R.; Almeida, A.M.; Quintela-Sabarís, C.; Roque, N.; Fernandez, P.; Ribeiro, M.M. The role of littoral cliffs in the niche delimitation on a microendemic plant facing climate change. *PLoS ONE* **2021**, *16*, e0258976. [CrossRef] [PubMed]
0. Gao, L.N.; Tao, F.; Liu, R.R. Multi-scenario simulation and ecological risk analysis of land use based on the PLUS model: A case study of Nanjing. *Sustain. Cities Soc.* **2022**, *85*, 104055. [CrossRef]
1. Li, J.; Yang, D.H.; Wu, F.Z. Dynamic simulation of land use change and carbon storage assessment in Kunming City based on PLUS and InVEST models. *Bull. Soil. Water Conserv.* **2023**, *43*, 378–387. [CrossRef]
2. Yang, Y.; Liu, L.; Zhang, P.; Wu, F.; Wang, Y.; Xu, C.; Kuzyakov, Y. Large-scale ecosystem carbon stocks and their driving factors across Loess Plateau. *Carbon Neutrality* **2023**, *2*, 5. [CrossRef]
3. Zhu, G.; Liu, Y.; Shi, P.; Jia, W.; Zhou, J.; Liu, Y.; Zhao, K. Stable water isotope monitoring network of different water bodies in Shiyang River basin, a typical arid river in China. *Earth Syst. Sci. Data* **2023**, *14*, 3773–3789. [CrossRef]
4. Wang, X.Z.; Wu, J.Z.; Wu, P.X. Spatial-temporal distribution and trade-off/synergy of water conservation, soil conservation and NPP services in Loess Plateau ecosystems, 2000–2015. *J. Soil Water Conserv.* **2021**, *35*, 114–121,+128. [CrossRef]
5. Su, C.H.; Wang, Y.L. Changes and driving factors of ecosystem services in the upper reaches of the Fenhe River Basin. *Acta Ecol. Sin.* **2018**, *38*, 7886–7898.
6. Wu, B.; Quan, Q.; Yang, S.; Dong, Y. A social-ecological coupling model for evaluating the human-water relationship in basins within the Budyko framework. *J. Hydrol.* **2023**, *619*, 129361. [CrossRef]
7. Budyko, M.I. *Climate and Life*; Academic Press: New York, NY, USA, 1974. [CrossRef]
8. Zhang, L.; Dawes, W.R.; Walker, G.R. Response of mean annual evapotranspiration to vegetation changes at catchment scale. *Water Resour. Res.* **2001**, *37*, 701–708. [CrossRef]
9. Li, W.; Zhao, Z.L.; Lv, S.S. Spatial and temporal differentiation of water purification function based on InVEST model. *J. Irrig. Drain.* **2022**, *41*, 105–113. [CrossRef]
0. Ou, Y.X.; Zhu, X.; He, Q.Y. Spatial interaction between urbanization and ecosystem services: A case study of Changzhutan urban agglomeration. *Acta Ecol. Sin.* **2019**, *39*, 12. [CrossRef]
1. Ren, Z.Z.; Chen, W.J.; Kang, H.T.; Li, X.; Zhang, X.; Wang, K. Research on threshold measurement method of influencing factors of safety vulnerability in traffic-intensive waters based on SD model. *Saf. Environ. Eng.* **2023**, *30*, 9–17. [CrossRef]
2. Zhang, F.; Zhan, J.; Zhang, Q. Impacts of land use/cover change on terrestrial carbon stocks in Uganda. *Phys. Chem. Earth Parts A/B/C* **2017**, *101*, 195–203. [CrossRef]
3. Zhang, M.F.; Liu, W.X.; Wang, J.N. Scenario simulation of ecosystem service value change in Dongguan section of Shima River Basin based on Clue-S model. *Bull. Soil. Water Conserv.* **2021**, *41*, 152–160. [CrossRef]
4. Yang, J.; Xie, B.P.; Zhang, D.G. Spatial-temporal variation of carbon storage in the Yellow River Basin based on InVEST and CA-Markov models. *Chin. J. Eco-Agric.* **2021**, *29*, 1018–1029. [CrossRef]

45. Ji, Q.Q.; Pan, Q.Q.; Wu, S.R. The spatial reconstruction of "three lives" and the impact of precipitation changes on wate production services in the Yellow River Basin of Shanxi Province. *Arid. Zone Res.* **2023**, *40*, 132–142. [CrossRef]
46. Peng, Y.Y.; Liu, Y.; Gao, Q.Q. Change characteristics of clouds in China in the summer of El Niño and their relationship wit precipitation. *Acta Meteorol. Sin.* **2022**, *80*, 701–720. [CrossRef]
47. Yin, Z.; Liu, Z.; Liu, X.; Zheng, W.; Yin, L. Urban heat islands and their effects on thermal comfort in the US: New York and Nev Jersey. *Ecol. Indic.* **2023**, *154*, 110765. [CrossRef]
48. Shang, M.; Luo, J. The Tapio Decoupling Principle and Key Strategies for Changing Factors of Chinese Urban Carbon Footprin Based on Cloud Computing. *Int. J. Environ. Res. Public. Health* **2021**, *18*, 2101. [CrossRef]
49. Zhang, C.L.; Zhao, J.B.; Niu, J.J. Study on Warming and Drying Climate of Shanxi Loess Plateau in Recent 50 Years. *J. Arid. Lan Resour. Environ.* **2008**, *22*, 70–74. [CrossRef]
50. Yao, Y.B.; Wang, Y.R.; Li, Y.H. Climate warming and drying of the Loess Plateau in China and its impact on the ecologica environment. *Resour. Sci.* **2005**, *27*, 146–152. [CrossRef]

water

Article

# Arid AREAS Water-Piled Photovoltaic Prevents Evaporation Effects Research

Jiamin Huang [1], Kebin Shi [1,*], Xingpeng Shi [2], Guocheng Hao [1] and Yimin Yang [1]

[1] College of Hydraulic and Civil Engineering Xinjiang, Xinjiang Agricultural University, 311 Nongda East Road, Urumqi 830052, China; H2842928692@outlook.com (J.H.)
[2] Xinjiang Water Conservancy and Hydropower Survey and Design Institute Co., Urumqi 830099, China; 13361649376@sina.cn
* Correspondence: xndsg@sina.com

**Abstract:** (1) Background: In arid and semi-arid reservoirs, water surface evaporation is the main method of water dissipation in order to inhibit the evaporation of water and enhance economic efficiency. The evaporation inhibition rate of water-piled PV at different times of the year is derived from the anti-evaporation test of water-piled PV, and a new idea is proposed for water conservation in plains reservoirs in arid areas. (2) Methods: The test was conducted by dividing the area into groups A and B, with and without PV panel shading. In situ observation and numerical calculation were used to measure the atmosphere's temperatures, test group, and PV module. The saturated water vapor pressure difference was then calculated according to Dalton's principle to analyze the economic benefits of water saving. (3) Results: Based on the test results, it was found that the shading of PV panels had a cooling effect on the water body, the PV module, and the atmosphere. Group A showed a 44.2% decrease in the saturation water vapor pressure difference compared to Group B. The maximum evaporation suppression rate of 40.2% was observed in July, while the minimum rate of 12.2% was observed in January. The average evaporation suppression rate for the entire year was 29.2%. By utilizing the annual water savings for agricultural irrigation, it is possible to cover 38 $hm^2$ of land and generate a revenue of 39,000 CNY. (4) Conclusions: The photovoltaic water cover can effectively reduce water evaporation and generate economic benefits.

**Keywords:** arid areas; evaporation; photovoltaic; water conservation

## 1. Introduction

Climate change is a pressing issue that affects all of humanity. The greenhouse effect, caused by harmful emissions, has far-reaching consequences [1,2]. However, reducing these emissions is a crucial step in mitigating the effects of climate change. In 2015, 192 countries came together in France to sign the Paris Agreement, which aimed to address the issue of carbon emissions [3]. One promising solution is the development of alternative energy sources like wind, solar, and geothermal, which can help to reduce carbon dioxide emissions significantly. Solar photovoltaic power generation is an up-and-coming energy source that is strongly supported by many countries worldwide. By installing photovoltaic, corresponding policy subsidies are given, and the cost of building photovoltaic power plants has been reduced by 85% compared to 2010. In 2020, new solar photovoltaic power generation accounted for over 50% of annual new energy generation, making it the primary contributor to new energy generation [4]. Although land PV covers a vast area, cleaning it is not easy, and wind and sand reduce power generation. Conversely, waterborne photovoltaics can be an excellent solution to this problem, providing the advantages of saving land area, improving power generation, reducing water evaporation, and the ability to combine it with hydroelectric power generation to improve economic efficiency [5,6]. Waterborne photovoltaics (WPV) are divided into floating photovoltaic (FPV) and pile-based photovoltaic (PPV), and water-based pile-based PV is mainly applied in areas where the water depth does not exceed 3 m. Prefabricated pipe pile-type concrete is used as the pile foundation, which is suitable for small lakes, rivers, reservoirs, artificial fishponds, and other environments.

Research on water resource utilization in arid and semi-arid areas has been a topic of great interest. The connection between human activity and the natural watershed in these areas is crucial. The socio-ecological coupling system of human–water relations can be described by using nD [7]. Some scholars have investigated the allocation of water resources and ecological water supply [8]. Their findings suggest that vegetation in the terminal lake area depends heavily on ecological water use [9]. Additionally, the benefits of using an arid area's ecology, landscape, and water resources can be evaluated and traded off. Predicting the evolution of lake boundaries can aid in protecting the arid area lake ecosystems [10]. Various methods have been proposed to address this concern, including utilizing palm fronds to cover the water surface. Full coverage of the water surface can reduce evaporation by up to 55%, while semi-coverage can decrease it by 26% [11]. Benzene sheets and other physical materials have also been tested for their efficacy in reducing evaporation rates. Results have shown that the more extensive the coverage, the more significant the reduction in evaporation [12]. Furthermore, studies have been conducted to examine the impact of land photovoltaic installation on evaporation. It has been observed that concentrated lighting reduces evaporation by 21%, while uniform lighting results in a 14% reduction. In contrast, no coverage leads to a 19% reduction in evaporation [13]. Currently, only a limited number of scholars have delved into the effects of FPV. Certain researchers have examined its influence on watersheds and formulated a mathematical model. Their findings indicate that as the coverage of FPV expands, so does the effectiveness of water conservation [14]. The installation of FPV in arid region reservoirs can increase power generation capacity by 58%, synergizing hydroelectric and photovoltaic power generation [15,16]. However, the studies mentioned above are solely based on theoretical models and have not been tested in the field. This raises concerns about the validity of the model as the installation environment and orientation of the reservoir can vary in different regions. Therefore, it is necessary to conduct field tests to verify the effectiveness of the model in real-world scenarios.

Floating photovoltaic (FPV) has been found to be an effective solution for reducing evaporation from water bodies. The rate of evaporation is influenced by temperature and inversely proportional to humidity levels, and through the adjustment of the angle of the photovoltaic (PV) tilt, it has been observed that a significant reduction in evaporation can be achieved. The implementation of single-axis tracking mechanisms or novel photosensitizers has the potential to induce varied levels of energy generation over the course of a year [17–19]. Despite the effectiveness of FPV, no relevant experiments have been conducted for waterborne photovoltaic (WPV), which holds practical and economic value in fishery–photovoltaic complementary fish ponds. In this study, we have analyzed and organized the results of experiments conducted on WPV. However, the mechanism of evaporation inhibition at different times has not been elaborated upon. This paper focuses on several research topics. Firstly, it aims to create a prototype observation model test. Then, it examines the primary factors that affect the temperature of water bodies due to PV panels, including the formation of saturated water vapor pressure difference and the day-to-day change process. Additionally, it briefly analyzes the evaporation inhibition rate of WPV on water bodies. Lastly, the paper explores the economic benefits of water saving with WPV technology.

## 2. Materials and Methods

### 2.1. Analysis of Solar Energy Resources at the Test Site

This testing site is located within the Tulufan Shengjintai Reservoir, situated in Xinjiang, China, at coordinates 42.50′36″ N for latitude and 89.15′50″ E for longitude. Meteonorm is a tool that provides information on rainfall, solar radiation, and more for a specific area. This region is representative of an arid area with a temperate continental climate characterized by an average annual precipitation of 150 mm. However, it receives an annual sunshine time of up to 3500 h, which is 1.25 times higher than that of Beijing at the same latitude. In Xinjiang, the total annual average solar radiation is 5800 MJ/m$^2$, which is ranked second in the country after Tibet and is 10–15% higher than in similar areas [20]. Comparing the total annual average horizontal solar radiation between the Turpan and Beijing regions, it is evident from Figure 1 that the former receives a significantly higher amount of solar radiation. Specifically, Turpan receives an annual solar radiation of 6061.8 kWh/m$^2$, while Beijing receives 5778.1 kWh/m$^2$, both at the same latitude. This discrepancy is particularly noticeable during July, the summer month when Turpan receives a maximum of 804.6 kWh/m$^2$ compared to Beijing's 542.5 kWh/m$^2$. During the winter season, the solar radiation levels in both regions are comparable. As Figure 2 exhibits, it is noteworthy that Turpan's highest single-day radiation is 30.73 kWh/m$^2$, compared to Beijing's 28 kWh/m$^2$. Consequently, the development of centralized PV and distributed PV in Turpan may prove beneficial [21].

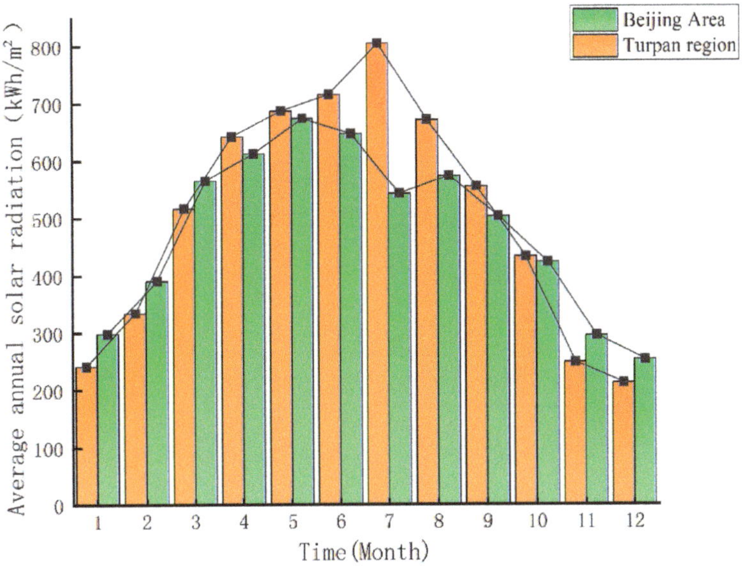


**Figure 1.** Comparison of the total annual average horizontal solar radiation between Turpan and Beijing.

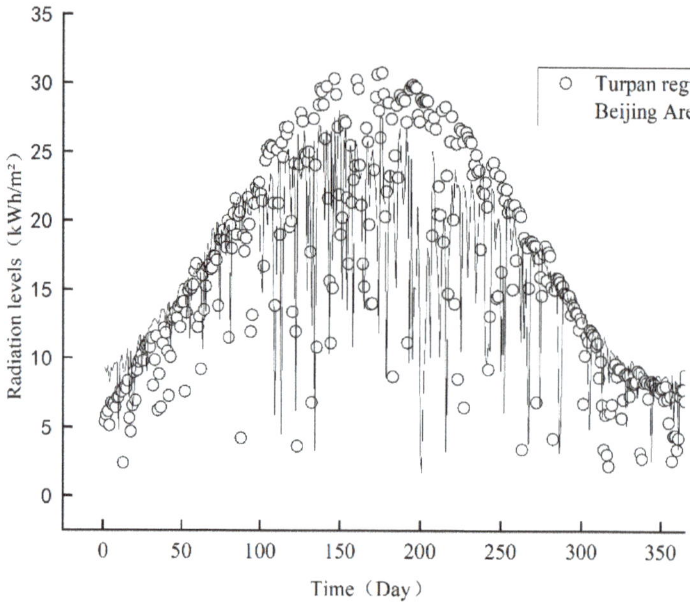


**Figure 2.** Daily variation curves of solar radiation in Turpan and Beijing.

### 2.2. *PV Array Inclination and Azimuth*

PV arrays require specific angles of azimuth and inclination to maximize solar radiation and power generation [22]. Solar PV arrays require a specific group spacing to prevent mutual shading, as shown in Figure 3. The time zone where the test site is located is East 8, and the shadows were modeled by Sketchup software, which allows for a group spacing of 2 m.

**Figure 3.** Pile photovoltaics on the water under the sun.

The test site is located in the northern hemisphere, and in order to maximize solar radiation, the PV array is fixed to face south. According to the photovoltaic system design aid software PVsyst, the optimal tilt angle for installing a photovoltaic system in this area is 36°, based on the principle of obtaining the minimum loss of solar radiation.

There are two main methods for calculating water surface evaporation from water bodies: the model calculation method and the in situ measurement method [23]. The current methods for calculating models include the Dalton model, Bowen ratio–energy balance method, Penman equation, water budget method, energy balance method, and Priestley–Taylor model. Measuring evapotranspiration in lakes and reservoirs is crucial for managing water balance in arid and semi-arid regions. The Dalton model, which is used to estimate evapotranspiration in open spaces, has a smaller error rate according to studies [24–26]; the calculation for this model is as follows:

$$E = f(u) \cdot (e_s - e_a) \tag{1}$$

where: $E$ is the water surface evaporation; $f(u)$ is the wind speed function; $e_s - e_a$ is the saturation water vapor pressure difference (VPD); $e_s$ is the water surface temperature corresponding to the saturation water vapor pressure; and $e_a$ is the saturation water vapor pressure in the air.

$$e_s(T_a) = 0.611 \cdot \exp\left(\frac{17.27 T_a}{T_a + 237.3}\right) \tag{2}$$

where: $T_a$ is the temperature of the air, $e_s$ is the water surface temperature corresponding to the saturation water vapor pressure, kPa, and $T_a$ is the temperature in the air.

$$e_a = \frac{RH}{100} e_s(T_a) \tag{3}$$

where: $e_a$ is the saturated water vapor pressure in the air, kPa and $RH$ is the relative humidity.

Therefore, we can obtain the following equation:

$$VPD = e_s(T_a)\left(1 - \frac{RH}{100}\right) = 0.611 \times \exp\left(\frac{17.27 T_a}{T_a + 237.3}\right) \times \left(1 - \frac{RH}{100}\right) \tag{4}$$

This paper quantifies the anti-evaporation effect of PPV on water using scale tests and actual measurement data, utilizing an in situ measurement method. The in situ measurement method was carried out using the E601 evaporator and was modified according to the test area [27–29]. The most commonly used method in China for measuring water surface evaporation is the evaporator pan, which is both economical and practical [27–30]. It is a cylindrical evaporator with a diameter of 20 cm and a height of 10 cm [31]. In this paper, the in situ measurement approach was employed to quantify the level of water evaporation from the designated site. The evaporation rate was monitored over the non-freezing period of one year in its natural state. As illustrated in Figure 4, the test configuration is showcased.

**Figure 4.** Water-based piled PV test diagram.

In accordance with the *"Surface Water Resources Investigation and Statistical Analysis Technical Rules"* test specification, the ideal method to measure water surface evaporation is by employing the E601 evaporator. Subsequently, the test data acquired from this measurement must be translated into the evaporation of the E601 evaporator. It is worth noting that the conversion coefficient may differ depending on the region. For this specific test area, the conversion coefficient utilized is 0.62 [32], and the calculation formula is as follows:

$$E_{601} = kE_{20} \tag{5}$$

where $k$ is the conversion factor; $E_{601}$ is the evaporation of large water bodies; and $E_{20}$ is the evaporation of evaporation pan evaporation.

For our research project, we meticulously selected two evaporator models that had measurements of $1.0 \times 1.0 \times 0.25$ m. We designated these models as A and B, respectively. In order to obtain precise and reliable results, we allocated Group A as the test group, while Group B served as the control group. We ensured that there was no possibility of leakage by lining both evaporators with polypropylene tarpaulin. In addition, we covered the exterior of each evaporator with a heat-resistant blanket to safeguard against solar radiation. For Group A, we strategically positioned solar photovoltaic panels that measured $1.04 \times 0.76$ m above the evaporator. These panels were tilted at a 36° angle and oriented southward to optimize their exposure to sunlight. Conversely, Group B did not have any form of shading.

In order to determine fluctuations in water levels on a daily basis, the water level stylus method is employed, which boasts a precision of $\pm 0.01$ mm. The handheld weather station is utilized to document changes in atmospheric temperature, wind speed, and humidity. The probe of the environmental monitoring system is utilized to oversee fluctuations in water temperature within both evaporators A and B. Furthermore, the CENTER-309 thermometer is employed to gauge temperature changes on the surface of the PV panel, which is installed above both evaporators. The same temperature and humidity recorder is situated above both evaporators at a height of 5 cm.

## 3. Results

### 3.1. Temperature Variation

Throughout the course of our experiment, we conducted a thorough analysis of atmospheric temperature for a full day. Our investigative approach incorporated detailed observations of the temperature fluctuations exhibited by both evaporators A and B. Additionally, we carefully monitored the temperature changes that occurred in the water body and PV module of evaporator A, as demonstrated by Figures 5 and 6. These findings have provided us with a comprehensive understanding of the various factors that contribute to atmospheric temperature and have allowed us to draw meaningful conclusions about the nature of these phenomena.

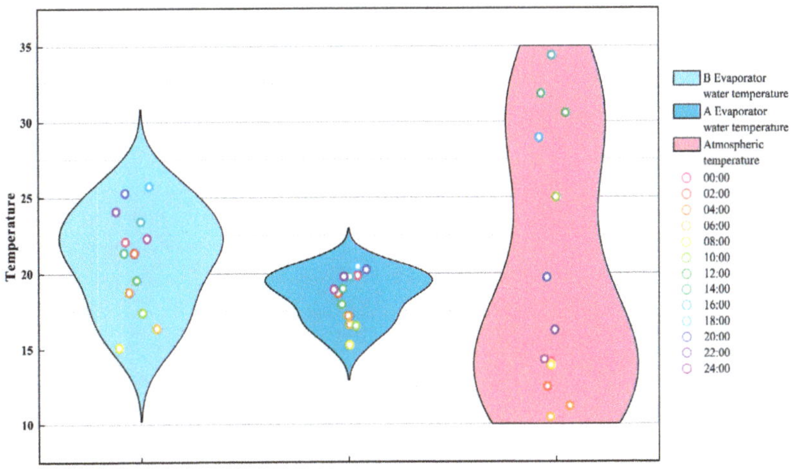


**Figure 5.** Daily variation of evaporators A, B, and atmospheric temperature.

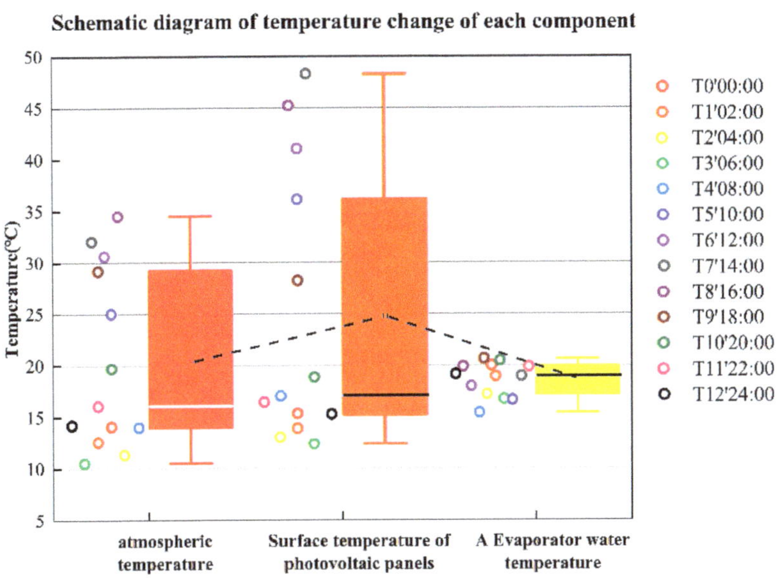


**Figure 6.** Schematic diagram of temperature change of each component.

Based on the information presented in Figure 5, it can be inferred that evaporator A has a more concentrated water temperature than evaporator B, and there is a significant temperature difference between the morning and evening atmospheric temperatures. The temperature in the atmosphere increases slowly after 8:00 a.m. The temperature of both evaporators is lower than that of the atmosphere, with Evaporator A exhibiting the lowest water temperature due to the shading provided by photovoltaic panels. This shading reduces the effect of solar radiation on the water surface temperature, thereby slowing the rate of water temperature increase and ultimately minimizing water evaporation. At 18:00, the temperature difference between the two evaporators peaked at 5.3 °C. From 18:00 to 8:00 the following day, the atmospheric temperature remained lower than the temperature of the evaporator due to the hindrance provided by the evaporator and photovoltaic panels to heat dissipation, leading to a slower rate of water temperature reduction. The testing site, situated in a desert area with mostly gravel, experiences rapid heating during the day and cooling at night, resulting in a significant temperature disparity between day and night. Consequently, the atmospheric temperature at night is lower than the temperature of the two evaporators. In comparison to Evaporator B, Evaporator A exhibits an average temperature that is 2.5 °C lower. Figure 6 portrays the temperature variation of the surface of the PV module, revealing that the solar panel's surface temperature is considerably higher than the water surface temperature inside Evaporator A. The temperature difference averages 6.1 °C, and the temperature difference can reach up to 29.3 °C at 14:00. The average surface temperature of the solar panel is 4.2 °C higher than the atmospheric temperature, with a maximum temperature difference of 16.2 °C. From 20:00 of the succeeding day to 8:00 of the following day, the surface temperature of the PV panel does not show a notable disparity from the atmospheric temperature. This is attributed to the primary components of PV panels, which are metal frames and monocrystalline silicon, resulting in a considerable temperature difference between the surface of the module during day and night due to quick rises and falls. Despite the high surface temperature of the PV panel, the water temperature inside the evaporator is not high. This is due to the absence of direct contact between the PV panel and the water surface. The heat mainly relies on air as a medium of transmission, and the heat transferred to the water surface only accounts for a part of this. The remaining heat is primarily dissipated into the atmosphere.

As previously explained, evaporation is influenced by various factors, with wind speed and saturated water vapor pressure difference being the primary ones. To determine the saturated water vapor pressure difference between the shaded conditions of test Group A's evaporator and the unshaded conditions of blank control Group B's evaporator, one can employ Equation (4). The calculation results are shown in Figure 7.

From 6:00 to 8:00, the saturated water vapor pressure difference between the shaded and unshaded groups was relatively similar. However, as the temperature rose between 8:00 and 18:00, the difference gradually increased due to greater evaporation from the water body. Furthermore, the presence of solar panels in the shaded group resulted in water vapor condensation between the photovoltaic panels and the water surface, further reducing the saturated water vapor pressure difference.

Between 18:00 and the following day at 6:00, the temperature decreased, and air humidity increased, leading to a gradual increase in the saturated water vapor pressure difference. By 18:00, the difference in saturated water vapor pressure between the sheltered and unsheltered groups had peaked, with the unsheltered group exhibiting a saturated water vapor pressure 1.12 kPa higher than that of the sheltered group. The daily averages of the saturated water vapor pressure difference between the evaporator in the sheltered Group A and the evaporator in the unsheltered Group B were 0.68 kPa and 1.22 kPa, respectively, and the saturated water vapor pressure difference in the sheltered group was reduced by 44.2% in comparison with that in the unsheltered group.

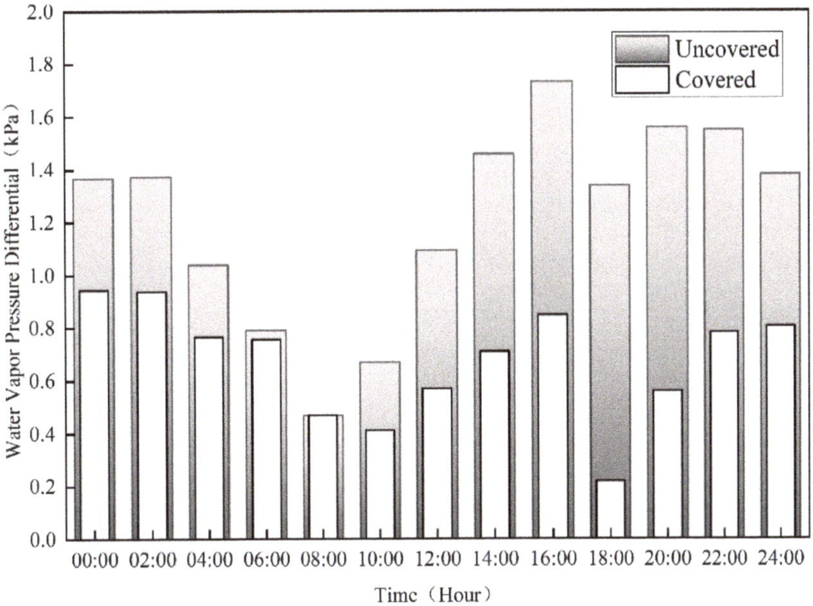


**Figure 7.** A, B evaporator daily saturation water vapor pressure difference.

### 3.2. Calculation of Evaporation Volume

The tested test data are shown in Figure 8 below:

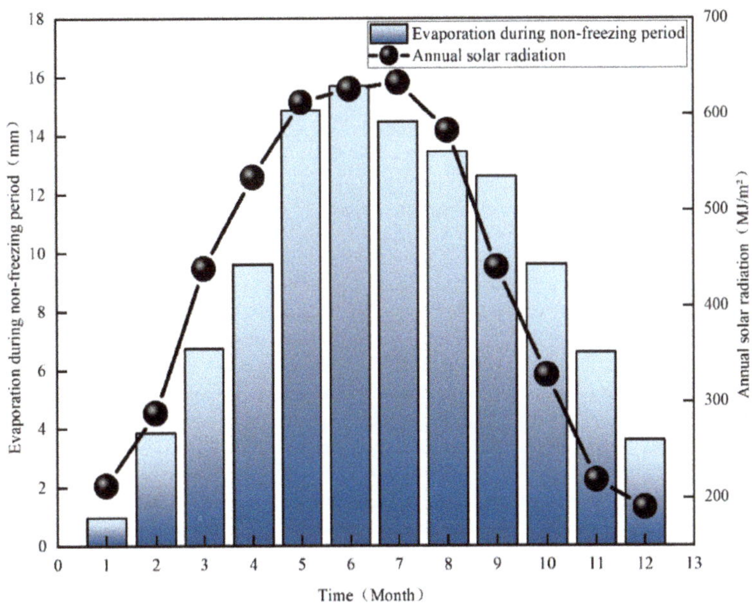


**Figure 8.** Schematic diagram of evaporation and average annual solar radiation in evaporation panes.

The evaporation suppression rate under solar PV panel shading is calculated as follows:

$$k = \frac{E_B - E_A}{E_B} \times 100\% \quad (6)$$

where $k$ is the evaporation inhibition rate, %; $E_B$ is the evaporation volume of evaporator B, mm; and $E_A$ is the evaporation volume of evaporator A, mm.

Figure 9 presents a detailed depiction of the evaporation process for evaporators A and B. The diagram showcases the monthly evaporation rates for both evaporators in the upper section. It highlights the difference between the monthly evaporation and the evaporation inhibition rates of A and B in the lower section.

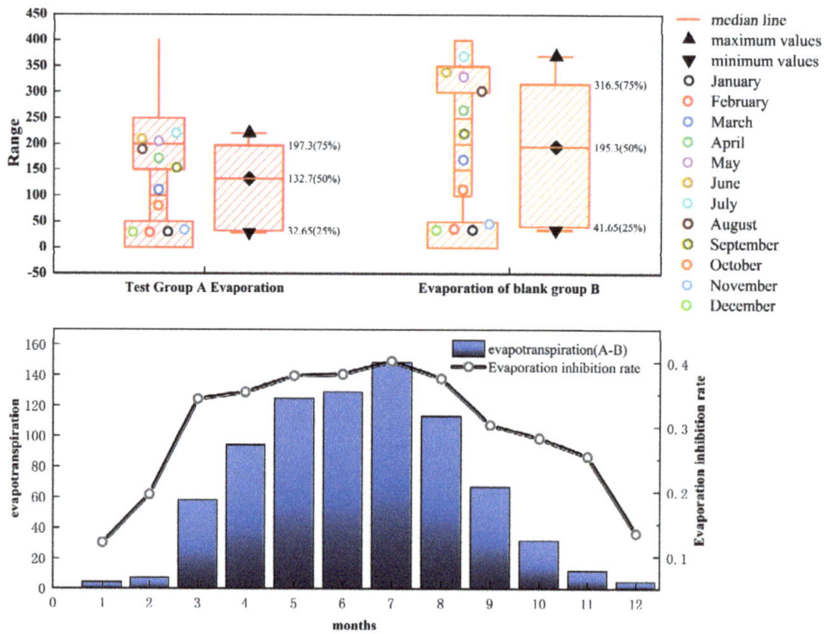


**Figure 9.** Evaporation and evaporation suppression rate under PV panel shading.

The diagram shows that evaporator A has a more concentrated and consistent monthly evaporation rate than evaporator B, which is more prone to fluctuations caused by air temperature and wind speed. During the winter season, both evaporators A and B have similar evaporation rates due to the water surface freezing, and the photovoltaic panels have minimal impact on the evaporation process. The chart indicates that both evaporators A and B experience an increasing trend in evaporation volume, followed by a decreasing trend, with the maximum volume reached in July. Evaporator A has a total evaporation volume of 221 mm, while evaporator B has an evaporation volume of 370 mm. The evaporation inhibition rate varies in different periods. In July and January, the evaporation inhibition rate can reach a maximum of 40.2% and a minimum of 12.2%, respectively. The average evaporation inhibition rate for the entire year is 29.2%. The shading of photovoltaic panels in summer reduces the temperature of the water surface, which decreases the water vapor pressure difference. As a result, moist air is not replaced by dry air between the water surface and the atmosphere, inhibiting the evaporation of water bodies.

### *3.3. Economic Benefit Analysis*

Based on the information provided about the project area, the reservoir has a total area of 31.6 hectares, making it a small irrigation-type reservoir. If the reservoir were completely covered with photovoltaic panels, it would be possible to calculate the economic benefits of water savings due to PV power generation and panel covering. Figure 10 shows a PPV system based on a reservoir. The specific parameters are shown in Table 1:

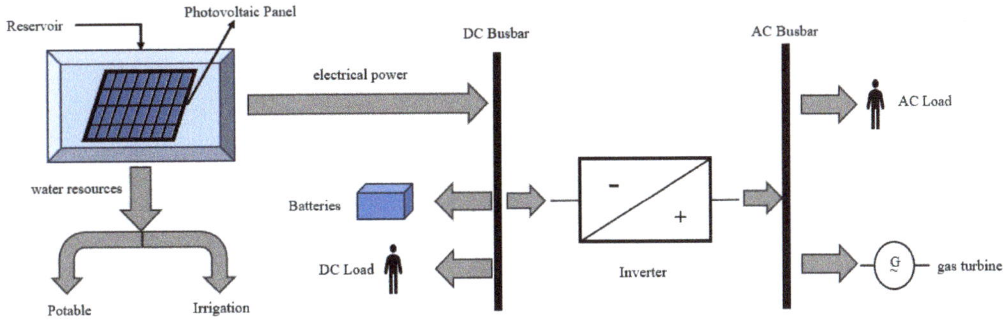


**Figure 10.** System Diagram for PPV.

**Table 1.** Reservoir parameters and installation PV parameters.

| Total Storage Volume | Total Reservoir Area | Photovoltaic Coverage Area | Photovoltaic Uncovered Areas | Average Annual Evaporation Suppression Rate | Evaporation Pan Conversion Factor |
|---|---|---|---|---|---|
| 1186.6 million $m^3$ | 31.6 $hm^2$ | 27.18 $hm^2$ | 4.46 $hm^2$ | 29% | 0.62 |

Based on the figure provided, the PV system produces DC power that can be utilized by DC loads, stored in batteries, or converted to AC power through an inverter for use by AC loads or a gas turbine. The PV panels used in the test have a power output of 160 w. With the coverage area shown in Table 1, a total of 388,285 PV panels can be installed. The photovoltaic panel used in the test was manufactured by model GHGN-150WDJBZ; each PV panel can generate 0.6 kWh of power per day (without considering losses from the inverter and other factors). This results in a daily power generation of 232,971 kWh, which translates to a revenue of 144,442 CNY per day.

According to the parameters shown in the table above, the evaporation volume of the evaporation pan can be converted into the evaporation volume of the standard evaporator, as shown in Figure 11.

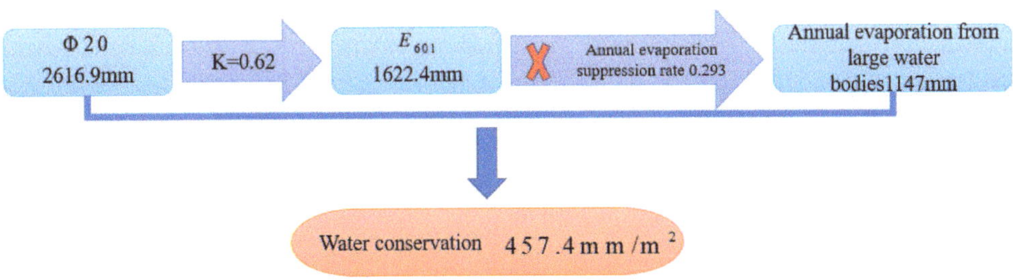


**Figure 11.** Water savings per square meter for large water bodies.

Based on the information provided above, the amount of evaporation varies for each month of the year due to the combined impact of wind speed, temperature, and relative humidity. This also affects the corresponding evaporation inhibition rate. By calculating the evaporation inhibition rate of evaporation panes for each month and converting it to the rate for large water bodies, we can determine the monthly evaporation amount for large water bodies. Please refer to Figure 12 for more details.

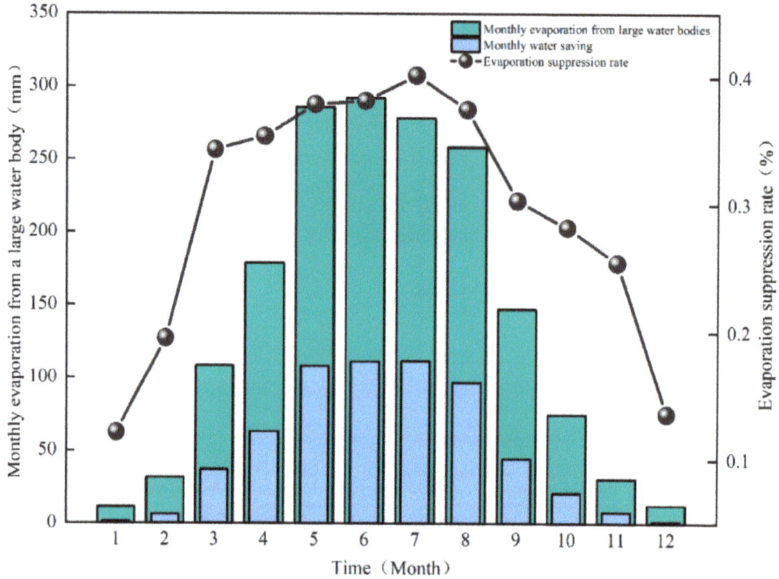


**Figure 12.** Schematic diagram of monthly evaporation from large water bodies.

Referring to Figure 12, the implementation of a photovoltaic water structure in a vast water body can preserve up to 610.81 mm of water per square meter every year. By utilizing the PV coverage area outlined in Table 1, the corresponding installation area can conserve 166,018 cubic meters of water annually. This amount is equivalent to 14% of the entire capacity of the reservoir. In the surrounding area, the primary crop is grapes, the reservoir water should be utilized primarily for irrigation purposes, the reservoir governs an impressive 267 hectares of irrigation land, and the water saved can be directed toward agricultural purposes, effectively serving an area of 38 hectares and bolstering economic gains. The reservoir sources its water from groundwater, boasting excellent quality with a soluble solids content (TDS) of 519 mg/L. If utilized for drinking water, the tap water in the region is priced at 0.23 CNY/m$^3$, providing an opportunity for revenue generation totaling 39,000 CNY.

## 4. Discussion

Land-based photovoltaics face the problems of extensive land area, wind and sand, and difficulty in cleaning [33,34], and the development of waterborne photovoltaics can effectively alleviate the increasingly tight land resources.

It has been found that using waterborne photovoltaic systems can effectively decrease water evaporation while increasing power generation capacity [15,35]. The principle of reducing water evaporation lies in the fact that PV panels block direct solar radiation and thus reduce the water surface temperature, which has been verified by different experimental and theoretical approaches [30,36]. However, the research approaches mentioned above did not systematically analyze the temperature of different layers of the water body, and future research directions could be aimed toward the continuous effect of water photovoltaic systems on the stratification effect of water temperature in large water bodies in different seasons. The above studies are based on floating photovoltaic systems but not on pile-based photovoltaic systems. Water-piled photovoltaic has a better spatial effect; the upper layer can implement photovoltaic power generation project, and the water surface can be combined with an anti-evaporation floating ball [37], which can achieve a better effect of preventing the evaporation of the water body, which has essential research value for arid areas.

In rural areas, an emerging trend is the combination of fishery and photovoltaic systems to generate additional income. By leveraging fish's distinct habits and ability to adjust to water temperatures, a layered aquaculture approach can be achieved in water bodies. Studies have shown that installing aquatic photovoltaic systems on fish ponds can bring significant economic gains [38,39], particularly in implementing this technology within desert regions; it has proven to be effective in addressing concerns surrounding aquaculture in arid areas, mitigating water body evaporation, and providing supplementary benefits [40], but insufficient research exists regarding the integration of waterborne photovoltaics and aquaculture. Further exploration is necessary to establish effective farming techniques and management practices incorporating waterborne photovoltaics. This is attributed to the impact that waterborne photovoltaics may have on water temperature, which can adversely influence the ability of diverse fish species to flourish in the water column. To ensure the durability of photovoltaic panels, the outer layer is sealed with toughened glass. To prevent moisture from seeping in, it is important to properly treat the location of the junction box and wiring. With proper treatment, the panels can be directly cleaned with water, making it easier and more cost-effective to maintain them. In the future, there will be research carried out on self-cleaning photovoltaic systems.

Research has indicated that the implementation of aquatic photovoltaic systems yields a substantial carbon reduction effect, resulting in the reduction of emissions by an impressive 3.3 million tons annually [18]. It can reduce the eutrophication of water bodies and reduce the concentration of chlorophyll and nitrate in water bodies [17,40]. However, the method mentioned above does not entirely cover vast water surfaces, and some research is relatively uniform with a limited observation duration. As the deployment of solar photovoltaic panels is regional and spatial, diverse areas may have varying inclinations and outcomes based on different arrangements and densities, which have not been adequately explored. To investigate the influence of water-based photovoltaic installations on water quality under distinct conditions, further research can examine the placement and spacing of photovoltaic arrays in sizeable water bodies.

This study examines the anti-evaporation effect of PPV. The findings indicate that these devices can effectively lower water temperature and reduce evaporation. However, the study only used model tests, and further research is needed to test the impact of pile photovoltaics installed at different heights and changes in water quality. Future research can explore these areas to broaden our understanding of this technology.

## 5. Conclusions

This paper focuses on investigating the anti-evaporation effect of PPV on the water body. Specifically, the study analyzes the impact on the water temperature and PV module temperature. The Dalton model is used to calculate the saturation water vapor pressure difference of the water body. Finally, the research derives the evaporation inhibition rate of the water body by waterborne photovoltaics. The test results are presented below:

Based on the test conducted, it was observed that the water temperature in evaporator A was significantly lower than that of evaporator B and the atmospheric temperature. During summer, the maximum evaporation inhibition rate reached 40.16%, while the lowest rate was 12.14% during winter. On average, the evaporation inhibition rate for the entire year was 29.3%. Furthermore, the economic benefits of preventing water evaporation using water pile-based PV were analyzed. If the water body's evaporation is used for agricultural irrigation, it can cover an area of 38 $hm^2$. On the other hand, if it is used for drinking water, it can generate a revenue of 30,000 CNY.

These test results indicate that the PV module blocks most of the solar radiation and that water pile-based PV can effectively reduce the water body's temperature, ultimately reducing the rate of evaporation. During summer, PPV had the highest evaporation inhibition rate, which was slightly lower during winter. Overall, the use of PPV in arid and semi-arid areas can help conserve water and produce good economic benefits.

**Author Contributions:** J.H.: Writing—original and final drafts, manipulating experiments, processing data, and analyses; K.S.: Supervising experiments, writing and revising papers; X.S.: Providing test equipment and software to guide data processing; G.H.: Data computation analysis, data collection; Y.Y.: Conducting a research and investigation process, specifically performing the experiments, data/evidence collection. All authors have read and agreed to the published version of the manuscript.

**Funding:** This research received no external funding.

**Data Availability Statement:** The data in the sub-paper were measured in personal experiments and are non-public.

**Conflicts of Interest:** The authors declare no conflict of interest.

## References


1. Semenov, S.M. Greenhouse Effect and Modern Climate. *Russ. Meteorol. Hydrol.* **2022**, *47*, 725–734. [CrossRef]
2. He, Y.; Gao, X.; Wang, Y. Sustainable Financial Development: Does It Matter for Greenhouse Gas Emissions? *Sustainability* **2022**, *14*, 5064. [CrossRef]
3. Zamarioli, L.H.; Pauw, P.; König, M.; Chenet, H. The climate consistency goal and the transformation of global finance. *Nat. Clim. Chang.* **2021**, *11*, 578–583. [CrossRef]
4. UN Environment Programme. Renewables 2021 Global Status Report [EB/OL]. Available online: https://go.exlibris.link/MtSvJ6Zz (accessed on 1 January 2021).
5. Liu, H.; Krishna, V.; Lun Leung, J.; Reindl, T.; Zhao, L. Field experience and performance analysis of floating PV technologies in the tropics. *Prog. Photovolt. Res. Appl.* **2018**, *26*, 957–967. [CrossRef]
6. Liu, L.; Sun, Q.; Li, H.; Yin, H.; Ren, X.; Wennersten, R. Evaluating the benefits of Integrating Floating Photovoltaic and Pumped Storage Power System. *Energy Convers. Manag.* **2019**, *194*, 173–185. [CrossRef]
7. Wu, B.; Quan, Q.; Yang, S.; Dong, Y. A social-ecological coupling model for evaluating the human-water relationship in basins within the Budyko framework. *J. Hydrol.* **2023**, *619*, 129361. [CrossRef]
8. Qiu, D.; Zhu, G.; Bhat, M.A.; Wang, L.; Liu, Y.; Sang, L.; Lin, X.; Zhang, W.; Sun, N. Water use strategy of *Nitraria tangutorum* shrubs in ecological water delivery area of the lower inland river: Based on stable isotope data. *J. Hydrol.* **2023**, *624*, 129918. [CrossRef]
9. Qiu, D.; Zhu, G.; Lin, X.; Jiao, Y.; Lu, S.; Liu, J.; Liu, J.; Zhang, W.; Ye, L.; Li, R.; et al. Dissipation and movement of soil water in artificial forest in arid oasis areas: Cognition based on stable isotopes. *Catena* **2023**, *228*, 107178. [CrossRef]
10. Yin, L.; Wang, L.; Li, T.; Lu, S.; Yin, Z.; Liu, X.; Li, X.; Zheng, W. U-Net-STN: A Novel End-to-End Lake Boundary Prediction Model. *Land* **2023**, *12*, 1602. [CrossRef]
11. Al Hassoun, S.A.; Al Shaikh, A.A.; Al Rehaili, A.M.; Misbahuddin, M. Effectiveness of using palm fronds in reducing water evaporation. *Can. J. Civ. Eng.* **2011**, *38*, 1170–1174. [CrossRef]
12. Jiang, H.; Tang, K.; He, X. Experimental Studies on Reduction of Evaporation from Plain Reservoirs in Drought Areas by Benzene Board Covering Technology. *J. Coast. Res.* **2015**, *73*, 177–182. [CrossRef]
13. Ali Abaker Omer, A.; Liu, W.; Li, M.; Zheng, J.; Zhang, F.; Zhang, X.; Osman Hamid Mohammed, S.; Fan, L.; Liu, Z.; Chen, F.; et al. Water evaporation reduction by the agrivoltaic systems development. *Sol. Energy* **2022**, *247*, 13–23. [CrossRef]
14. Bontempo Scavo, F.; Tina, G.M.; Gagliano, A.; Nižetić, S. An assessment study of evaporation rate models on a water basin with floating photovoltaic plants. *Int. J. Energy Res.* **2021**, *45*, 167–188. [CrossRef]
15. Gonzalez Sanchez, R.; Kougias, I.; Moner-Girona, M.; Fahl, F.; Jäger-Waldau, A. Assessment of floating solar photovoltaics potential in existing hydropower reservoirs in Africa. *Renew. Energy* **2021**, *169*, 687–699. [CrossRef]
16. Elshafei, M.; Ibrahim, A.; Helmy, A.; Abdallah, M.; Eldeib, A.; Badawy, M.; AbdelRazek, S. Study of Massive Floating Solar Panels over Lake Nasser. *J. Energy* **2021**, *2021*, 6674091. [CrossRef]
17. Abdelal, Q. Floating PV: An assessment of water quality and evaporation reduction in semi-arid regions. *Int. J. Low-Carbon Technol.* **2021**, *16*, 732–739. [CrossRef]
18. Ravichandran, N.; Ravichandran, N.; Panneerselvam, B. Floating photovoltaic system for Indian artificial reservoirs—An effective approach to reduce evaporation and carbon emission. *Int. J. Environ. Sci. Technol.* **2022**, *19*, 7951–7968. [CrossRef]
19. Zhang, J.; Zhong, A.; Huang, G.; Yang, M.; Li, D.; Teng, M.; Han, D. Enhanced efficiency with CDCA co-adsorption for dye-sensitized solar cells based on metallosalophen complexes. *Sol. Energy* **2020**, *209*, 316–324. [CrossRef]
20. Feng, G.; Li, W.; Han, Y.; Zhang, Y.; Zhang, D. Solar Energy Resources and Zoning in Xinjiang. *Renew. Energy* **2010**, *28*, 133–139.
21. Wei, G.; Jiapaer, G.; Hanmin, Y.; Liangliang, J.; Xiaofang, Z. Spatial and temporal distribution characteristic and division research of solar energy resources in southern Xinjiang. *Arid. Zone Geogr.* **2021**, *44*, 1665–1675.

- Wang, Z.; Fan, Y.; Wang, J.; Li, Q.; Cai, L.; Xue, J. Influence of tilt and azimuth on the characteristics of photovoltaic power generation in Zhangbei area. *Acta Energiae Solaris Sin.* **2022**, *43*, 73–79.
- Liu, X.; Yu, J.; Wang, P.; Zhang, Y.; Du, C. Lake Evaporation in a Hyper-Arid Environment, Northwest of China—Measurement and Estimation. *Water* **2016**, *8*, 527. [CrossRef]
- Abtew, W.; Obeysekera, J.; Iricanin, N. Pan evaporation and potential evapotranspiration trends in South Florida. *Hydrol. Process.* **2011**, *25*, 958–969. [CrossRef]
- Hojjati, E.; Ghorban, M.F.T.; Ozgur, K. Estimating evaporation from reservoirs using energy budget and empirical methods: Alavian dam reservoir, NW Iran. *Ital. J. Agrometeorol./Riv. Ital. Agrometeorol.* **2020**, *34*, 33.
- Wei, X.; Zhao, R.; Huang, D.; Zhao, Z. Water evaporation rate correlation based on Dalton model of enclosed space in frost-free refrigerator. *Case Stud. Therm. Eng.* **2021**, *28*, 101495. [CrossRef]
- Li, Z.; Chen, Y.; Shen, Y.; Liu, Y.; Zhang, S. Analysis of changing pan evaporation in the arid region of Northwest China. *Water Resour. Res.* **2013**, *49*, 2205–2212. [CrossRef]
- Shen, Y.; Liu, C.; Liu, M.; Zeng, Y.; Tian, C. Change in pan evaporation over the past 50 years in the arid region of China. *Hydrol. Process.* **2009**, *24*, 225–231. [CrossRef]
- Chen, D.; Gao, G.; Xu, C.; Guo, J.; Ren, G. Comparison of the Thornthwaite method and pan data with the standard Penman-Monteith estimates of reference evapotranspiration in China. *Clim. Res.* **2005**, *28*, 123–132. [CrossRef]
- Xing-Peng, S. Research on Anti-Evaporation and Water-Saving Efficiency of Floating Photovoltaic Power Generation Unit in Arid Area. *Water Conserv. Sci. Technol. Econ.* **2022**, *28*, 55–59.
- Han, K.; Shi, K.; Yan, X.; Lv, J.; Cheng, Y. Analysis of water saving efficiency of anti-evaporation in plain reservoirs in arid areas covered by PE floats. *J. Basic Sci. Eng.* **2020**, *28*, 376–385.
- Guo, J.; Chen, Z.; Xiao, F.; Wang, Y.; Yuan, L. Numerical analysis of wave loads on floating square array of photovoltaic power plant. *Acta Energiae Solaris Sin.* **2021**, *42*, 1–6.
- Yuan, W.; Lu, H.; Pan, T.; Ji, C.; Peng, X. Structural dynamics of floating photovoltaic power plants under the coupling effect of wind and waves. *Eng. J. Wuhan Univ.* **2021**, *54*, 7–13.
- Santos, F.R.D.; Wiecheteck, G.K.; Virgens Filho, J.S.D.; Carranza, G.A.; Chambers, T.L.; Fekih, A. Effects of a Floating Photovoltaic System on the Water Evaporation Rate in the Passaúna Reservoir, Brazil. *Energies* **2022**, *15*, 6274. [CrossRef]
- Majumder, A.; Innamorati, R.; Frattolillo, A.; Kumar, A.; Gatto, G. Performance Analysis of a Floating Photovoltaic System and Estimation of the Evaporation Losses Reduction. *Energies* **2021**, *14*, 8336. [CrossRef]
- Kewu, H.; Kebin, S.; Xinjun, Y.; Jianxun, L.; Yunpeng, Y. Study on evaporation inhibition rate of static water surface in plains reservoirs in arid areas under PE float coverage. *J. Water Resour. Water Eng.* **2017**, *28*, 235–239.
- Château, P.; Wunderlich, R.F.; Wang, T.; Lai, H.; Chen, C.; Chang, F. Mathematical modeling suggests high potential for the deployment of floating photovoltaic on fish ponds. *Sci. Total Environ.* **2019**, *687*, 654–666. [CrossRef]
- Zhang, J.; Liu, X.; Gu, Z.; Cheng, G.; Zhu, H. Ecological economic characteristics of fishery and light complementary and its development direction. *J. Fish. China* **2022**, *46*, 1525–1535.
- Vo, T.T.E.; Je, S.; Jung, S.; Choi, J.; Huh, J.; Ko, H. Review of Photovoltaic Power and Aquaculture in Desert. *Energies* **2022**, *15*, 3288. [CrossRef]
- Wang, Y.; Yu, H.; Yu, J. Factual analysis of observation of local ecological effects of PV on water surface. *Acta Energiae Solaris Sin.* **2022**, *43*, 38–44.

geosciences

MDPI

Article

# Using an Open-Source Tool to Develop a Three-Dimensional Hydrogeologic Framework of the Kobo Valley, Ethiopia

**Sisay S. Mekonen [1,2,*], Scott E. Boyce [1,3], Abdella K. Mohammed [4] and Markus Disse [1]**

1 School of Engineering and Design, Technical University of Munich, 80333 Munich, Germany; scott.boyce@tum.de or seboyce@usgs.gov (S.E.B.)
2 Faculty of Water Resources and Irrigation Engineering, Arba Minch University, Arba Minch P.O. Box 21, Ethiopia
3 U.S. Geological Survey, California Water Science Center, 4165 Spruance Rd., Suite 200, San Diego, CA 92101-0812, USA
4 Faculty of Hydraulic and Water Resources Engineering, Arba Minch University, Arba Minch P.O. Box 21, Ethiopia
* Correspondence: sisay.mekonen@tum.de

**Abstract:** Groundwater resource management requires understanding the groundwater basin's hydrogeology and would be improved with the development of a three-dimensional hydrogeologic framework model (HFM). A wide range of methods and software exist to quantify the extent, structure, and properties of geologic systems. However, most geologic software is proprietary and cost-prohibitive for use in developing countries. GemPy is a Python-based, open-source (no-cost) tool for generating three-dimensional geological models. This study uses available data and GemPy to develop the Kobo Valley Hydrogeologic Framework Model (KV-HFM), a three-dimensional HFM for Kobo Valley in northern Ethiopia, which is part of the East African Rift System. The KV-HFM is a conceptual model that comprises the hydrostratigraphy, structural features, and hydraulic properties of the Kobo Valley groundwater system. The limited data described the extent and altitude of the hydrostratigraphic units using the GemPy implicit potential–field interpolation. The KV-HFM showed the existence of an east-to-west, structural-based groundwater divide composed of volcanic rock and clay. This divide splits the catchment into two groundwater systems with limited interconnected flow. This study illustrates the use of open-source software for developing an HFM using sparse, existing geologic data.

**Keywords:** East Africa; Rift Valley; hydrogeologic framework; GemPy; groundwater management

## 1. Introduction

Groundwater is a crucial resource for urban and domestic water supply, irrigated agriculture, industry, and ecosystems. With the growing scarcity of surface water sources, developing nations are currently relying more on groundwater resources, particularly in rural parts of Africa. This paradigm shift results from groundwater becoming a strategic resource for economic growth, food security, poverty reduction, and groundwater suitability to adapt to climate change impacts on urban and rural livelihoods [1–3]. However, properly using and managing groundwater is difficult because of a lack of experience and knowledge [1,2]. Unlike surface water, groundwater systems are challenging to describe without a hydrogeologic framework model (HFM). HFM is essentially a conceptual model that helps develop a simplified path and framework for a numerical geologic model that embodies the concepts in a numerical representation of the geologic features and honors the geologic history of the region. The HFM helps groundwater resource management by describing the thickness and area extent of the aquifers, aquitards, and bedrock and embodies a 3D numerical regional estimation of the geologic framework. It also delineates faults, sedimentary facies, and related estimates of aquifer properties that result from different provenances.

The development of an HFM has evolved over time. Prior to the 1970s, the development of an HFM to represent the groundwater flow systems was accomplished using pencil and paper [4,5]. However, the approach did not fully represent complex terrains structurally or topographically. Detailed representations demanded maps that show details of lateral lithological contrasts for 3D modeling [6]. In order to close this gap, efforts were made to provide 3D information. For example, in the late 1970s, the Illinois State Geological Survey (ISGS) produced black-and-white maps and subsequently updated maps with color and patterns [7,8]. Yet even these maps did not sufficiently help users readily grasp the regional geologic features. As a result, efforts were again made to use the geographic information systems (GIS) approach to help speed up the production of colored maps to create cross-sections and line-and-dot patterns [9].

With the advance of computer technology, the development of HFM transitioned towards using advanced numerical-computer algorithms to create 3D geologic models that represent the subsurface structure and stratigraphy [4,10,11]. The layers that typically represent formations in the geologic model are delineated and estimated within the HFM (typically, formations above what is considered "bedrock") that provide an estimate of water-bearing layers that can be developed for groundwater supply. The availability of computer resources and 3D geological software has led to developing an HFM as an important step in developing and managing groundwater systems [10,12].

The recent development of HFMs started in the petroleum industry and included "sequence stratigraphy" to replicate the geologic history of a region and related layering, faults, folds, and associated properties. Most of the development of petroleum framework models occurred in offshore regions. The first onshore HFMs were sequence stratigraphic HFMs of terrestrial regions developed by the U.S. Geological Survey (USGS) for the Santa Clara Valley along with the Hydrostatrigraphic framework model of the Central Valley [13], which is a hybrid approach to the sequence approach. There are three types of HFMs: Sequence Stratigraphic Models (Example, Earthvision, etc.), Hydrostratigraphic Models (Example, Rockware, Surfer, LeapFrog, etc.), and Hybrid Elevation Layer Models (Example, ISATIS, FREEWARE, TPROGS, Arc-geostatistical analyst, etc.).

Modern studies have developed HFMs with computer and software resources. For example, the Pajaro Valley groundwater basin hydrogeologic framework was developed by combining a driller's log synthesis [14], where e-logs were used with resistance limits to delineate fractions of coarse and fine-grained material with hydrostratigraphic units [15]. A more recent use of driller's logs was combined with an access database script to delineate coarse-grained and fine-grained thicknesses to be kriged within hydrostratigraphic units [13,16]. Then, many studies emerged employing ground-based and aerial Transient Electromagnetic (EM). However, they use proprietary software and are not linked to lithology databases or well-based transient EM logs (dual induction logs) [17].

Three-dimensional geological models can play a vital role in understanding the stratigraphic framework of aquifer systems and are used to create hydrogeologic representations of a groundwater flow system [18,19]. Three-dimensional geological models can also help estimate the geometry and related volumes of stratigraphic units or subregions delineated as aquifers or aquitards [20–26]. For example, digital 3D HFMs were constructed for many parts of California [13,24,25,27–29]. These 3D HFMs defined the aquifer system geometry and subsurface lithologic characteristics for the subsurface hydrogeologic conceptualization of the aquifer system to be used in regional hydrologic and hydrogeologic modeling. With the increased computing power and advanced 3D numerical models, datasets such as drill logs and geo-electrical resistivity data are used as input to construct detailed 3D geological models of groundwater flow systems [20,23]. For example, drill log and vertical electrical sounding (VES) data were combined to construct 3D geological models of a groundwater basin and it was reported that the combined method improved the representation of the groundwater flow system and reduced model uncertainties [21,30–33].

There are various options for using software for building a geologic model. Typically, the choice for software is either to use open-source GIS software or commercial geological modeling software. Open-source GIS options such as QGIS and gdal [34], are examples of open-source GIS software that provide access to process earth observation raster and vector data for visualization but are 2D maps, complicating efforts to visualize fault networks, complex structures, or stratigraphic sequences. Examples of commercial software that are closed-source include EarthVision [35], Leapfrog [36], GOCAD [37], Petra [38], Rockware [39], and Hydro GeoAnalyst [40], provide advanced visualization and algorithms for developing geological models [41] but do not provide access to the source code implementation. Because commercial software is closed source, the implementation of the methods is unknown, other than from the provided documentation, and cannot be modified. This limits the utility to only what is provided by the software and prevents connecting to open-source libraries for machine learning and computational inference. As a result, the integration into other computational frameworks is limited. Often, commercial software is used to enhance groundwater management. However, the cost of such software is usually prohibitive for most areas in the world. This has led to the development of open-source and free software that constitutes a cornerstone for enhancing groundwater development and management [42,43].

To fill this gap, efforts have been made to develop a fully open-source software application to create complete 3D HFMs that are comparable to implementations in commercial packages. In this regard, the popularity of using open-source programming languages such as Python is emerging and playing a crucial role in facilitating scientific programming and script-based science. GemPy, for example, is a fully open-source and easy-to-use software presented recently to create 3D visualization of geological models [4,44]. To construct 3D geological models, GemPy relies on Python and is based on an explicit and implicit potential-field interpolation approach. This interpolation approach consists of fast and automated surface formation supported with manual framing and interpolating of a scalar function based on the cokriging of point data and structural orientations. However, there are also other common approaches to developing HFMs, such as the use of spline interpolators to estimate the geologic top and bottom layer surfaces [45] and the use of 2D ordinary kriging or cokriging to spatially estimate the vertically averaged properties [46].

Groundwater is a primary source of water for irrigation and domestic water supply in rural arid and semi-arid regions in sub-Saharan Africa. However, this resource is little studied and poorly understood, partly because of the scarcity of existing hydrogeological information in many regions of sub-Saharan Africa. As a result, existing studies on assessments of groundwater resources relied on remotely sensed data combined with modeling and missed detailed information at the catchment scale [47,48]. In this study, we developed a 3D HFM considering Kobo Valley in Northern Ethiopia as a case study to understand the groundwater basin's hydrogeology of the area, which is basic for sustainable groundwater resource evaluation and management based on field investigations and synthesizing Vertical Electrical Sounding (VES) measurements, driller's logs, pumping tests, groundwater levels, and land surface elevation data. This study aims to fill the existing gap of implicit modeling in an open-source tool to develop 3D HFM in geosciences for a region that has a shortage of surface water and utilizes the groundwater for similar developments. The groundwater resources have not been previously studied in this way and are potentially vulnerable to considerable future overdevelopment. Hence, the methodology developed can be used for similar purposes, contributing to sustainable groundwater development and management.

## 2. Materials and Methods

### 2.1. Description of Study Area

Kobo Valley is located in the Afar Depression, the western margin of the Main Ethiopian Rift, East African Rift. Geographically, the study area is located at 11°54′ to 12°24′ N and 39°20′ to 39°48′ E, with a total area of approximately 1544.24 km$^2$. The elevation ranges between 3975 m above sea level in the mountains and 1018 m above sea level in the Kobo Valley plain area (Figure 1) with reference to UTM zone 37N (WGS 84). The valley is bounded in the west by mountains of the western plateau and in the east by a chain of mountain terrain, the rift escarpment. The western catchment is mountainous that is covered by and composed of volcanic rocks and has about 1045 km$^2$ (68%), while the valley plain is about 499 km$^2$ (32%).

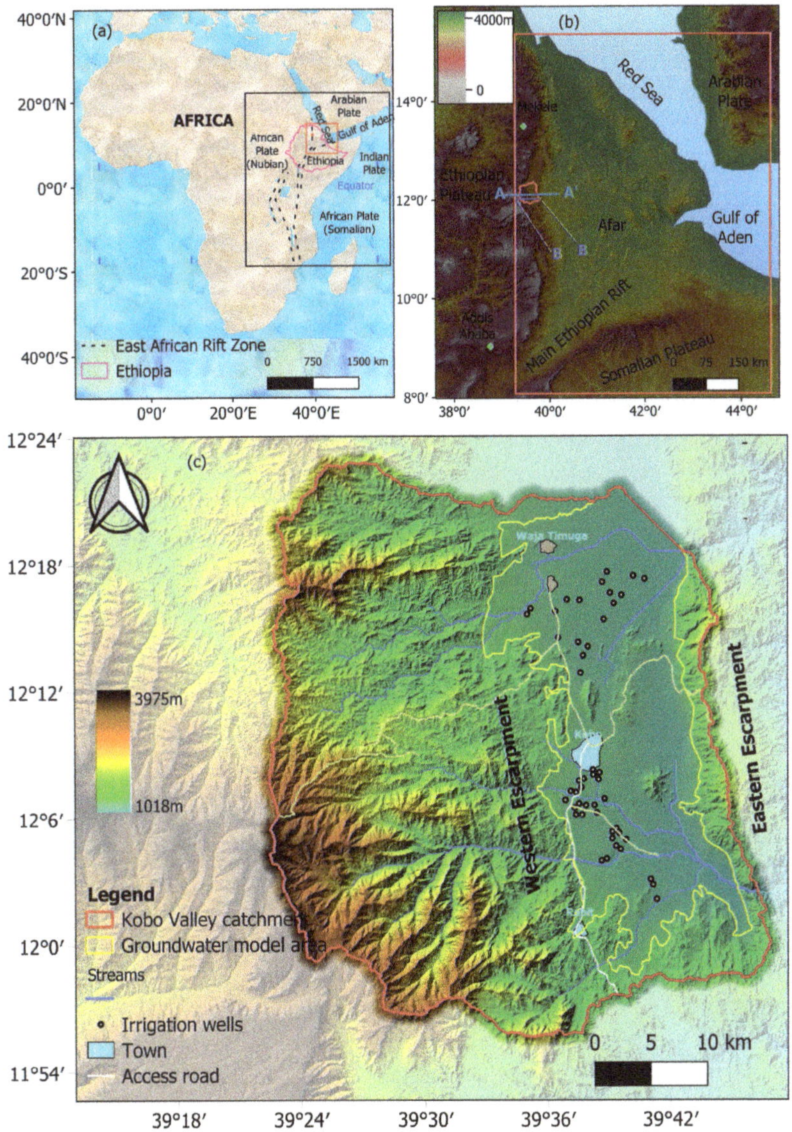


**Figure 1.** (**a**) East African Rift bordering the Red Sea and the Gulf of Aden. (**b**) Afar Depression location in the Main Ethiopian Rift, East Africa. (**c**) Kobo Valley catchment in the Afar Depression.

The Kobo Valley is a semi-arid catchment and has limited surface water resources [49]. The use of groundwater is expanding to meet the growing demand for irrigation [50].

The Kobo Grana Valley Irrigation (KGVI) project serves an area of 3100 hectares and was developed to support the local community as they strive to maintain food security [51]. Over the past two decades, KGVI has developed more than 100 public boreholes to provide water for irrigation and domestic water supply. A series of groundwater simulation models have been developed to understand the Kobo Valley water resources and plan for future development [52,53]. However, the lack of a detailed geologic layer formation led the studies to the assumption of homogeneous aquifer properties.

### 2.2. Geology of the Kobo Valley

The study area is found in the Western Margin of the Afar Depression, the region in East Africa that represents a key location for studying continental breakup [54–56]. Zwaan and Corti Zwaan, Corti [54] revealed that the south–south striking Afar Depression is still actively deforming and is characterized by NNW–SSE normal faulting and a series of marginal grabens (Figure 2). The Afar Depression forms a triangular depression near and partially below sea level between the Ethiopian and Somalian plateaus to the west and south and the Danakil and Ashia Blocks to the northeast and east (Figure 2).

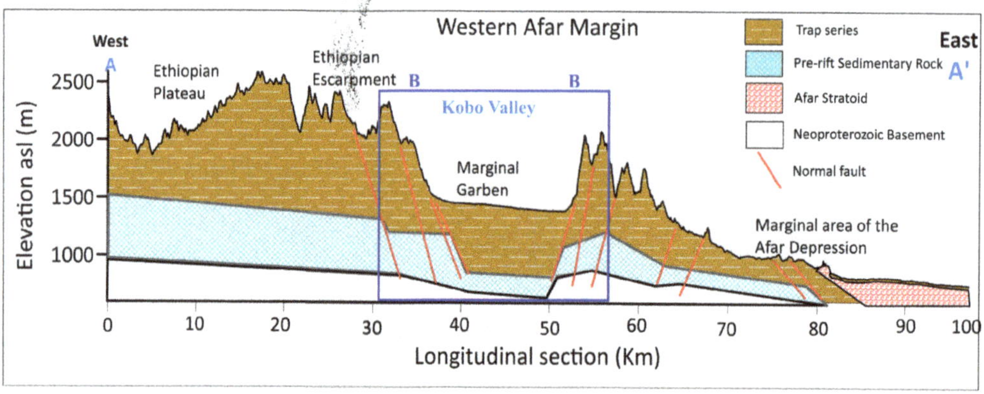


**Figure 2.** Longitudinal cross-section along Afar Depression (Section AA′ in Figure 1b), asl represents above sea level. It illustrates the structural style in the Western Afar Margin and is dominated by antithetic faulting (towards the Ethiopian Plateau) and associated marginal grabens. Image after [56,57]. The inset blue box—section BB shows the groundwater basin area (Section BB′ in Figure 1b).

The lithology of the Western Margin of the Afar Depression consists of Miocene sedimentary and volcanic infill (the Pliocene–Quaternary Stratoid units) [58]. This forms the most recent unit and covers the Afar Depression floor but has also accumulated in depressions along the Western Afar Margin and on the Ethiopian Plateau. Faulting along the margin is generally considered dominantly antithetic (i.e., dipping away from the Afar rift basin, e.g., [59], Figure 2). Furthermore, a series of faulted basins referred to as "marginal grabens" [54,60] align along the Western Afar Margin.

The formation of the geological structure is controlled by tectonic events that led to the development of the Ethiopian Rift System on the western side of the Afar Margin. Its origin is considered to be local tectonic development forming an intermountain trough. These events are characterized by tensional movements, which gave rise to fissural volcanism followed by block-faulting and tilting to form the escarpment zone, including marginal grabens. These marginal grabens are narrow, elongated depressions bounded on both sides by normal faults facing each other. The main axis of the trough runs in a north–south direction. The trough is formed in the west by the rift escarpment and east by the horst of the mountain ridge (Figure 1c). The frame is mainly composed of Tertiary volcanic rocks (Figure 3). The eastern and western ridges bounding the plain area are characterized by opposite dipping faults parallel to the plateau escarpments. The intermountain trough (Kobo Valley) is dominantly composed of poorly compacted sedimentary basin-fill deposits [57,61].

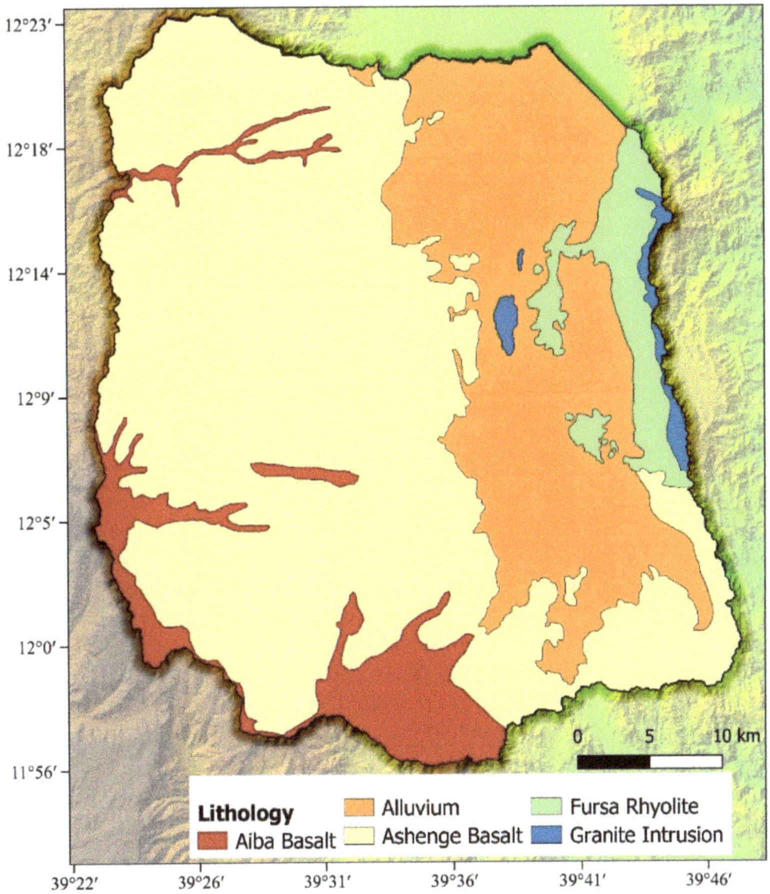


**Figure 3.** Geological map of Kobo Valley catchment [62] (Ethiopian Geological Study, 2012).

### 2.3. Data Collection and Processing

Hydraulic parameter data of aquifers in the Kobo Valley well field from 63 boreholes and driller's logs from 45 wells were collected from the office of Ethiopian Construction Design and Supervision Works Corporation [63] as well as 37-point VES data from the Ministry of Water, Irrigation and Energy (MWIE) office. The VES field data were collected by Metaferia Consulting Engineers [64]. The positions of the data points are validated using QGIS mapping and field visits made from January–February 2021. The data are also carefully checked for consistency in geological interpretation for each driller's log and the descriptions of the geology of the area from different published reports.

Geological data of the area were collected from the Ethiopian Geological Study [62] and the Ethiopian Construction Design & Supervision Works Corporation Office [63]. Earthquake data collected from the USGS Earthquake catalog (accessed on 9 August 2021; https://earthquake.usgs.gov/earthquakes/search/) also helped us visualize the ongoing tectonic activity in the area. Study area location preparation and mapping of the structures within the study area were carried out through analysis in QGIS 3.2 (Accessed on 27 January 2020; www.qgis.org), programming language Python, and IPI2win software also used for terrain and VES data analysis. Detailed (30 m resolution) Shuttle Radar Topography Mission (SRTM) and Advanced Spaceborne Thermal Emission and Reflection Radiometer (ASTER) digital topography data from NASA and METI (Accessed on 10 July 2020; https://earthexplorer.usgs.gov) provided an excellent basis for our mapping and processing using QGIS.

The final aim was to develop a three-dimensional hydrogeologic framework model (HFM), a conceptual model that comprised the hydrostratigraphy, structural features, and hydraulic properties that helped to understand the groundwater basin's hydrogeology for groundwater resource management in the area. We used the limited existing data in the study area to achieve these results and followed several steps. Figure 4 presents our conceptual workflow. The input data were VES measurements, driller's lithologic logs, pump tests, and land-surface altitude data. First, the input data were processed to extract the surface contact points, orientation measurements, and defined topological relationships (stratigraphic sequences and fault networks) and synthesized to create input data for the GemPy model. Then, the following three steps summarize the whole simulation hierarchically:

1. Create a digitized geospatial database from the input data that contains all the raw data, topological relationships, standardized projection, and spatial extent;
2. Define the spatial distribution of geological structures and discretize the 3D space regular grid geometry based on a potential-field interpolation method to define the spatial distribution of geological structures, such as layers, interfaces, and faults (computations of lithologic stratigraphic unit (LSU));
3. Discretize and visualize an interactive 3D geological model using Python fundamental plotting library;
4. Then, pre-process and analyze the driller's log data to check whether they are consistent with the defined geometry and to identify the information that the contacts bring about the possible positions of the surface deviations.

The GemPy-derived 3D HFM model performance was evaluated using the built-in functionality to compute forward gravity conserving the package's automatic differentiation and the concept of topology, a useful tool to describe adjacency relations in geological models, such as stratigraphic contacts or across-fault connectivity. As a final check, the generated aquifer profiles were mapped and evaluated with the drillhole profile data. Goodness of fit was qualitatively assessed by visual inspection and quantitatively assessed by statistical analysis of residuals (differences between the measured and calculated elevations).

### 2.4. VES Data

The VES point data were profiled into six lines from west to east (Figure 5). The raw data were collected using Terameter SAS 4000, which is the product of the ABEM Company according to Schlumberger electrode configuration. For each VES measurement point, the spreading was performed in the north–south direction in a way that the potential electrodes remained fixed for the defined current electrode, separating while the current electrodes were spread apart for each measurement. This arrangement is known as a Schlumberger Array. Thus, the VES points in each profile line were distanced at 1 km intervals in the west–east direction (Figure 5). The resistivity data were converted to a one-dimensional (1D) resistivity earth structure using IPI2win 3.1 software [65]. Then, using the IPI2win software, the apparent resistivity for each VES was plotted against half-electrode separation on double logarithmic paper [66,67]. The geophysical data were analyzed and interpreted for each VES point measurement. The specific resistivity values and the corresponding layers noted in each VES point were correlated along the west–east profile lines. A set of geoelectric sections was produced using the interpreted results of each VES point and correlating the values along the profile line. The inverted resistivity values were interpreted using the adopted resistivity values for earth materials [68–70].

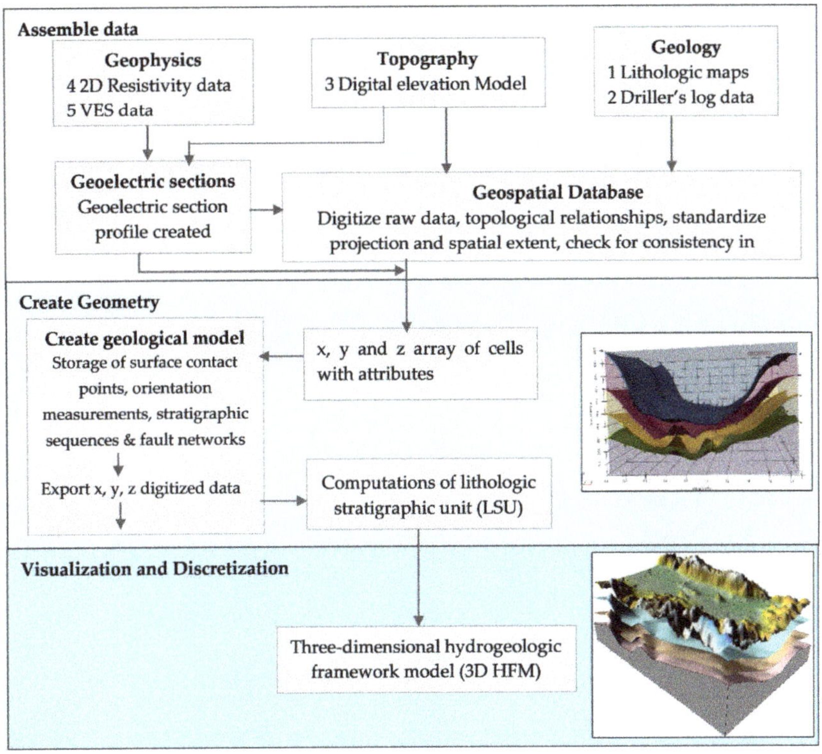


**Figure 4.** Conceptual workflow of the study.

*2.5. GemPy Modeling Approach*

GemPy is an open-source Python library for generating full 3D structural geological models based on an implicit potential-field interpolation approach [11,71]. The interpolation algorithm constructs 3D geological models, including fault networks, fault–surface interactions, and unconformities. This algorithm is applied in Python's programming language, using a Theano library for efficient code generation that directly executes on graphical processing units (GPUs).

The method was first introduced by [71] and is grounded on the mathematical principles of universal cokriging. Later, the method was updated by integrating it into Bayesian inference frameworks and advanced machine-learning [72] for stochastic geomodelling and Bayesian inversion, making efficient implementations of automatic differentiation in novel machine-learning frameworks [4]. To efficiently compute gradients and provide optimized compiled code, GemPy 1.0 was built on top of Theano libraries [73]. In addition, Pandas for data storage and manipulation [74], Visualization Toolkit (vtk) for interactive 3D visualization [75], Matplotlib [76], and NumPy for efficient numerical computations [77] were used.

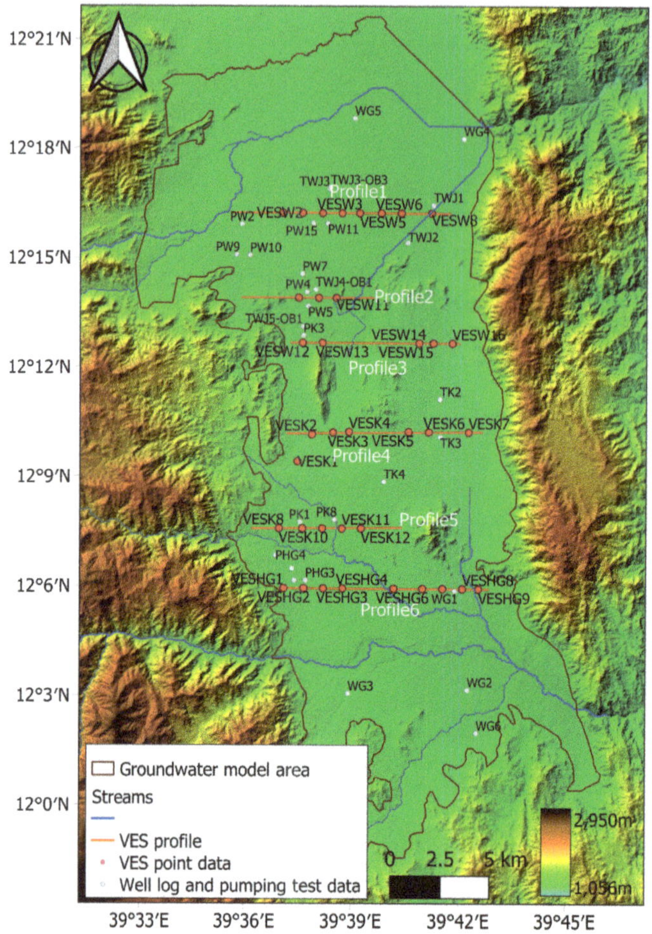


**Figure 5.** VES measurement data profiles arrangement from west to east at 1 km intervals. Topography is derived from ASTER data (30 m resolution).

In GemPy, the main method to generate the 3D geological models is the potential-field method developed by [71], which has been successfully deployed in the modeling software GeoModeller 3D [11]. The basic idea is to construct an interpolation function $Z(x_o)$ where $x$ is any point in the continuous 3D space $(x, y, z) \in R^3$ that describes the domain D as a scalar field. The gradient will follow the planar orientation of the stratigraphic structure throughout the volume. It means that every possible isosurface of the scalar field will represent every synchronal deposition of the layer (Figure 6). After creating the stratigraphic layers, the fault series is considered as layer formations with a "Fault_Series" representation as the key entry in the GemPy set_series dictionary.

In the final interpolation function $Z(x_o)$, $x_o$ refers to the estimated quantity for some integrable measure $p_o$. To characterize the scalar field interpolation, two types of parameters were used: (i) layer interface points $x_\alpha$ for the respective isosurfaces of interest; and (ii) the gradients of the scalar field, $x_\beta$, which are poles of the layer or normal vectors to the dip plane in geological terms. Accordingly, gradients are oriented perpendicular to the isosurfaces and located anywhere in space. The gradient of the scalar field is referred to as $\partial Z/\partial u$ with $u$ defined as any unit vector and its samples as $x_\beta$. A complete description of the core functionality of the GemPy model can be found in Varga and Schaaf Varga,

Schaaf [4] and additional references (See [71]). Appendix A contains the Python code that sets up the GemPy model presented in this paper, which utilizes the Kobo Valley datasets to generate a 3D visualization of the unknown hydrogeologic framework.

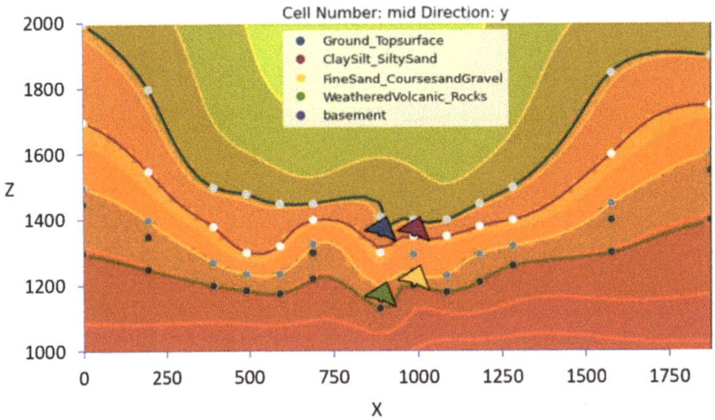


**Figure 6.** Scalar field for Kobo Valley. The input data are formed by the number of point data distributed in four layers ($x^1_{\alpha i}$, $x^2_{\alpha i}$, $x^3_{\alpha i}$ *and* $x^4_{\alpha i}$) and four arrows indicate the orientations of the layers ($x_{\beta j}$). An isosurface connects the interface points, and the scalar field is perpendicular to the gradient. Z is an elevation in the *y*-axis, and x is the longitudinal cross-section in the *x*-axis multiplied by 10.

### 2.6. Model Performance Evaluation

Obtaining field geological measurements is expensive and often cost-prohibitive for developing nations. Consequently, most field measurements are sparse in time and space. Geological models must address the uncertainty that results from sparse data to reach a reasonable level of confidence in the model [4,78,79]. The advantage of GemPy is that the software model is fully designed to be coupled with probabilistic frameworks. GemPy supports stochastic geological modeling for uncertainty analysis (e.g., Monte Carlo simulations, Bayesian inference), which helps consider uncertainties in the model input data and use additional secondary information in a Bayesian inference framework. For example, GemPy can be coupled with pymc3 [72] to build probabilistic graphical models [4,80]. GemPy uses the latest developments in uncertainty visualization for 3D structural geological modeling and geological inversion [81,82].

GemPy includes a built-in functionality to compute forward gravity conserving the package's automatic differentiation (AD). Topology helps to describe adjacency relations in geomodels, such as stratigraphic contacts or across-fault connectivity. GemPy can analyze the adjacency topology of its generated models using the topology compute method (See Appendix A (Figure A2)).

The generation of the stratigraphic geological layers using GemPy is formed by the number of points distributed in layers and orientations. The stratigraphic profiles were evaluated for goodness of fit using two approaches: visual inspection and statistical analysis of residuals. Visual inspection is a method of visually examining the map to see if the generated grid points with the model are a good representation of the original data. Residuals (difference between the measured value and an interpolated value) play an important role in interpolation characteristics; analysis is complete with a thorough examination of residuals [83]. Quantitative measures that can be used as goodness-of-fit statistics are performed, and a residual map is prepared to illustrate where the generated surface points are nearer or further away from the actual data of stratigraphic boundary elevations from drill logs. The mean absolute deviation and the standard deviation of the cross-validation residuals and the rank correlation between the measurements and the estimates are calculated for analysis. The coefficient of determination, $R^2$, is calculated by

$$R^2 = 1 - (SS_{res}/SS_{tot}) \tag{1}$$

where $SS_{res}$ = sum of the squares of the residuals, $SS_{tot}$ = sum of squares of the differences from the mean, $Sum(El_i - El_{mean})^2$, and El = the original elevation point data.

In addition to the built-in functionalities to the GemPy model, the model is checked by overlapping the generated 3D layer profiles with the well logger's data to check the discrepancy of the layer marked in the well logs with the model.

## 3. Results and Discussion

### 3.1. Driller's Log Lithology and VES Analysis

The geophysical data were interpreted for each VES measurement point. The specific resistivity values and the corresponding layers noted in each point were correlated with the west–east profile lines. The resistivity-survey data analysis showed that high resistivity values indicated greater sand and gravel content, whereas low resistivity values represented greater clay and silt content in the valley's deposits, which were crosschecked with the drillers' logs to confirm that low resistivity can also be caused by saline waters or poor water quality for coarse-grained layers. Six geoelectrical sections were produced, and four main layers were recognized along with the selected profiles (Figure 7) characterized as topsoil, clay and silty sand, sand and/or gravel, and weathered volcanic rock layers.

Profile lines with low resistivity segments due to clay deposition or salty water interpretation are compared to drillers' log data to dismiss any effects from saline waters that would give a false signal that could be identified as clay deposition towards the east. Generally, the geoelectric correlation sections showed vertical and lateral variations in the profile layers and thickness because of Kobo Valley's geomorphology.

The produced longitudinal sections (Figure 7) were constructed using the data collected at 1 km intervals. The created profile layers are not smooth because the layers were drawn by only connecting the points with a straight line. In transferring the raw data points to the GemPy model, this rough sketch was adjusted and smoothed with each layer orientation ($x_{\beta j}$) because an isosurface of the points connected the interface points and the scalar field perpendicular to the gradient. As explained in Section 2.2, the formation of the geological structure in the study area was controlled by tectonic events that affected the geophysical features of the valley plain and led to the development of uplands/remanent hills in the middle of Kobo Valley, which influenced the identified aquifer layer thickness to vary, as can be observed in Figure 7. For example, profile 3 of the generated geoelectric section showed that the layers are different in profiles and thickness to the west and east sides of the center, where the geomorphology shows upland areas/hills. This upland area is also coincident with faults, micro-seismic events, and a possible transform fault that separates the basin into two sub-basins.

To the east side of this section, either a layer of gravel is absent (profiles 2, 3, and 4 in Figure 7) or a thin layer of gravel is present (profiles 1, 5, and 6 in Figure 7). In the model, this thin layer is characterized as mainly clay. As a result, this area has a relatively thin layer of aquifer thickness. Generally, the produced longitudinal geoelectric sections show a clay layer beneath the thin topsoil layer. This clay layer has varying thicknesses throughout the whole valley. Furthermore, the sand and/or gravel and weathered volcanic-rock units follow from top to bottom as a third and fourth layer with varying thickness. Finally, the lowest layer below the weathered zone is the bedrock. The interpretation of each profile is given in Table A1 in Appendix B.

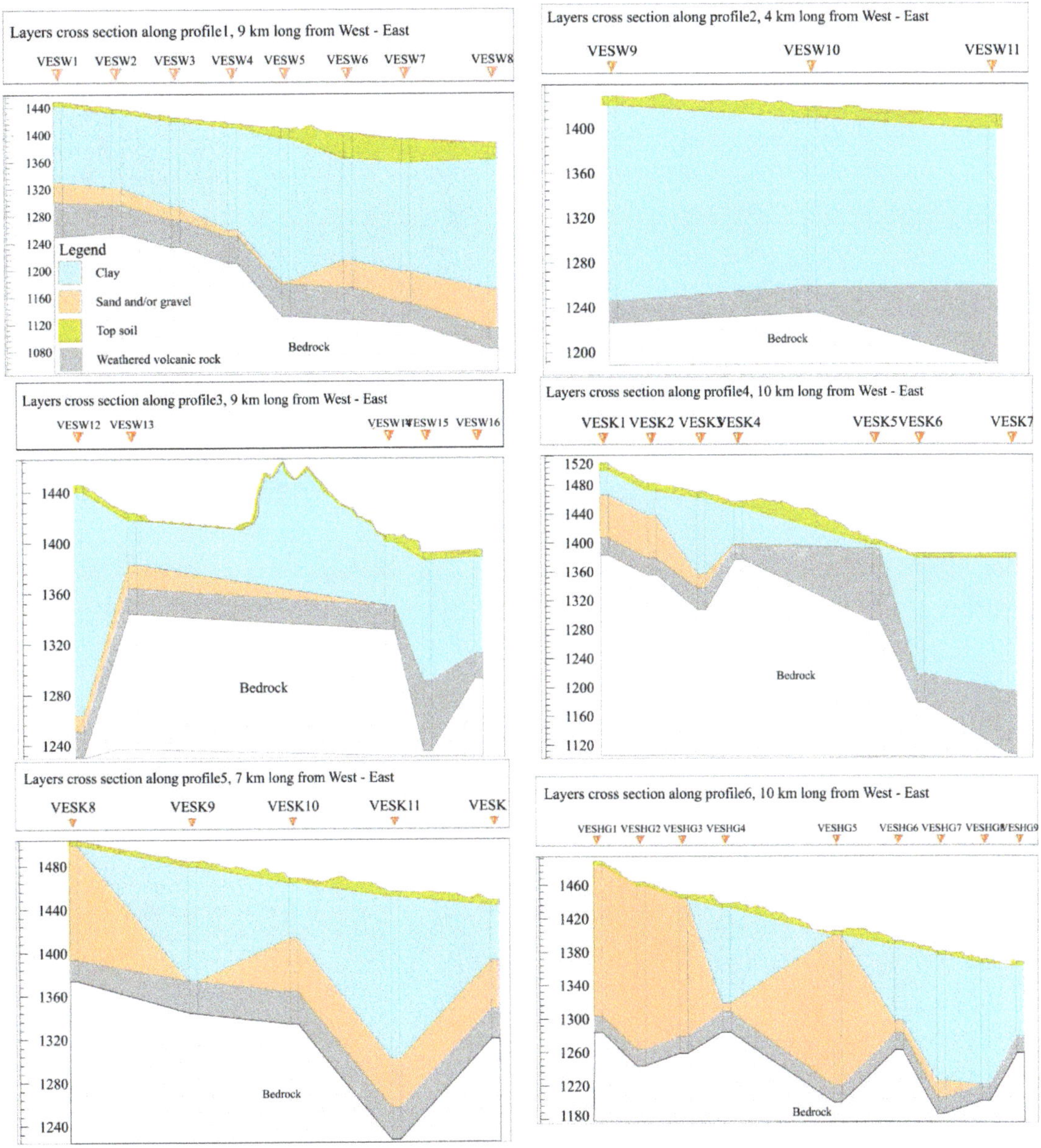


**Figure 7.** Produced longitudinal sections for the six profiles from west to east. $x$-axis represents longitudinal cross section from west to east in km and $y$-axis represents altitude of hydrogeologic units in m above sea level with reference to UTM zone 37N (WGS 84).

This study showed that the central part of the valley is thicker in sediment deposits and fine-grained material. At the same time, the sediment thickness around the eastern side from the center of the valley is mainly clay. The profile data are digitized to obtain the x, y, and z surface contact input data for the GemPy model and defined topological relationships (stratigraphic sequences and fault networks).

### 3.2. Three-Dimensional Hydrogeological Framework

The KV-HFM was developed by constructing a GemPy model using a GemPy python model object (object here is in reference to object-oriented programming). The model object uses a regular grid to interpolate the 3D geological model at any point in a 3D space. GemPy relies on Theano library for efficiency [73]. Theano_optimizer is used with the fast compile option before computing the model to generate Figure 8 (which uses PyVista, a 3D visualization library and mesh analysis in Python for 3D visualization).

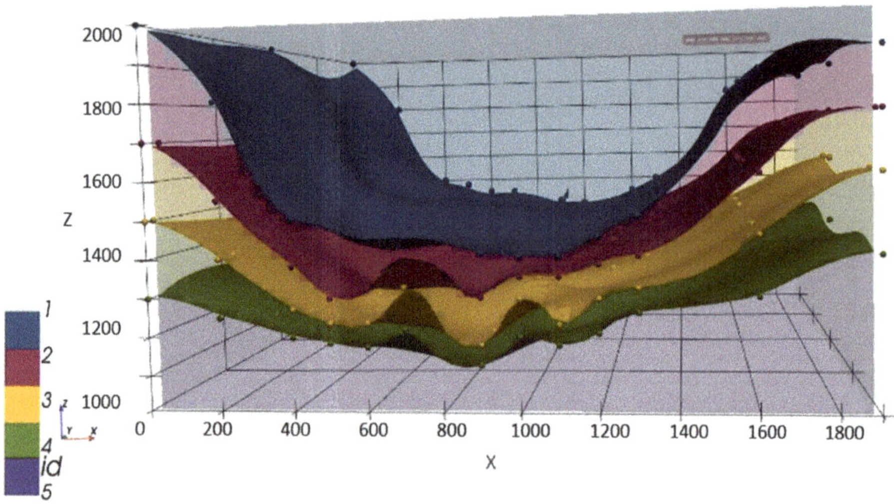


**Figure 8.** Three-dimensional visualization of Kobo Valley using GemPy geological model. *x*-axis has a scale of 1:10 and a vertical exaggeration of 10. Numbers 1 up to 4 represent the top surfaces of 1 = ground surface, 2 = clay silt and silty sand, 3 = fine sand, coarse sand, and gravel, 4 = weathered volcanic rocks (basalts), and 5 = hard rock.

The generated 3D GemPy model significantly enhances the visual interpretation and understanding of the valley's complex subsurface formations and geological profiles. This type of representation also helps non-professionals understand the subsurface profiles and complexity. The model enables each layer to be visualized individually using its spatial properties. Using this feature, the volume of each aquifer layer is calculated to estimate the water availability in each layer. This can be used to help quantify the groundwater resources in the region. Previous studies in Kobo Valley have been conducted with many generalized assumptions considering the homogenous thickness and the same structural settings throughout the catchment [50,53]. However, a GemPy model fills this knowledge gap by providing (1) the layer volume for the given geological settings; (2) visualization of the formation thickness distribution of each unit; and (3) the relations among the units (Figure 8, Figure 9, and Figure 10, respectively).

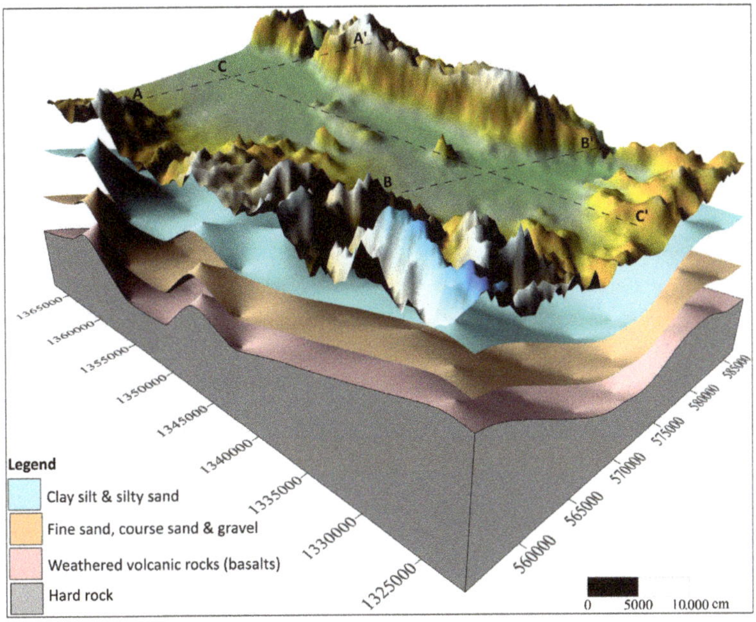


**Figure 9.** Three-dimensional geological map showing stratigraphic profiles of Kobo Valley. The vertical cross-section is exaggerated by a factor of 10.

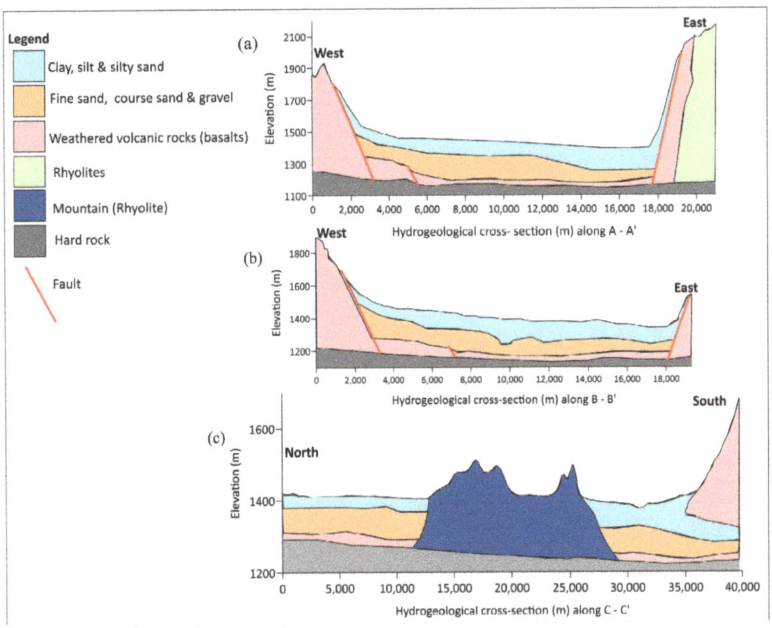


**Figure 10.** Hydrogeological longitudinal cross-section profile of Kobo Valley. (**a**) Cross-section A–A′ (northern part of Kobo Valley). (**b**) Cross-section B–B′ (southern part of Kobo Valley) from west to east. (**c**) Cross-section C–C′ from north to south.

The 3D model shows that the graben is bounded on both sides, from West and East normal faults facing each other. This fault bounding is also observable from topographical maps. A cross-sectional view of the model (see Figure 10a–c) for the profiles indicated in Figure 9 shows that the West and East frames are mountain ridges bounding the plain area (marginal grabens) that are mainly composed of volcanic rocks. The valley plain forms the main aquifer system.

The 3D HFM also provides a map of the distribution and thickness of the layers in the main aquifer system. The spatial distribution of the thickness of a layer determines its volume. Hence the model facilitates the estimation of the volume of a layer by considering the spatial distribution of its thickness (Table 1). This is the volume of the layers in the hydrogeologic stratigraphic unit (HSU), not the volume of water or the water that could be extracted from the HSU. The aquifer volume is estimated using the valley plain area of the main aquifer (alluvial part) of the valley using the difference between the isopach of the potentiometric surface and the top of the aquifer.

**Table 1.** Calculated volume of each individual layer of the study area.

| Hydrogeologic Stratigraphic Unit (HSU) | HSU Order | Volume of Unit in $m^3$ |
|---|---|---|
| Clay, silt, and silty sand | First (top) layer | $38.21 \times 10^9$ |
| Fine sand, coarse sand, and gravel | Second layer | $26.79 \times 10^9$ |
| Weathered volcanic rock (basalts) | Third layer [1] | $17.22 \times 10^9$ |

[1] The layer below the third is hard rock.

### 3.3. Uncertainty in the GemPy Model

The first evaluation made was a visual examination. The map was visually examined to see if the generated grid points with the GemPy model closely represented the original data. Next, the top surface of the 3D structural geological map generated using GemPy was compared with the digital elevation map (DEM) of the study area. Then, the minimum, maximum, mean, standard error, standard, and coefficient of determination ($R^2$) were calculated for the residuals from the differences between the known values of VES and the driller's logs with the estimates (Table 2).

**Table 2.** Summary of residual statistics.

| Minimum (m) | Maximum (m) | Mean (m) | SE (m) | SD (m) | $R^2$ | HSU | Description |
|---|---|---|---|---|---|---|---|
| −5.68 | 7.71 | 0.01 | 0.08 | 0.91 | 0.93 | HSU_1 | top of clay, silt, and silty sand layer |
| −7.52 | 7.84 | −0.58 | 0.60 | 6.76 | 0.95 | HSU_2 | bottom of clay, silt, and silty sand layer, and top of fine sand, coarse sand, and gravel layer |
| −9.68 | 7.71 | 0.01 | 0.10 | 1.01 | 0.90 | HSU_3 | bottom of fine sand, coarse sand, and gravel layer and top of basalts layer |
| −6.14 | 8.32 | 0.04 | 0.10 | 1.12 | 0.91 | Base | top of hard rock |

HSU = Hydrogeologic Stratigraphic Unit, SD = Standard Deviation, SE = Standard Error.

The residual's minimum and maximum values indicate the magnitudes of differences between the generated grid values from GemPy and the actual measured data points. For example, the absolute maximum value is 9.68 m for HSU_3, where the elevation of the HSU varies from 1094 m to 2621 m elevation above sea level with reference to UTM zone 37N (WGS 84). Hence, 9.68 m is less than a 1% variation. The mean differences between GemPy model-generated grid point elevation values and measured elevation values were less than 1% for all HSUs, demonstrating that the gridded values were reasonably close to the original data values for most of the grid. The standard deviation was 6.76 m. These residual values reflect the accuracy of the original data points compared to the gridded values. The large values of the coefficient of correlation ($R^2 > 0.90$) for all HSUs indicate that the model explains most of the variation observed in measured data points, but some patterns were detected in the residuals (Figure 11). Figure 11 demonstrates that the model closely matches the original data points except along the main axis of the trough frame, the rift escarpment, and the horst of the mountain ridges, where measured data were scarce. Overall, we concluded that the model performed well with the available data.

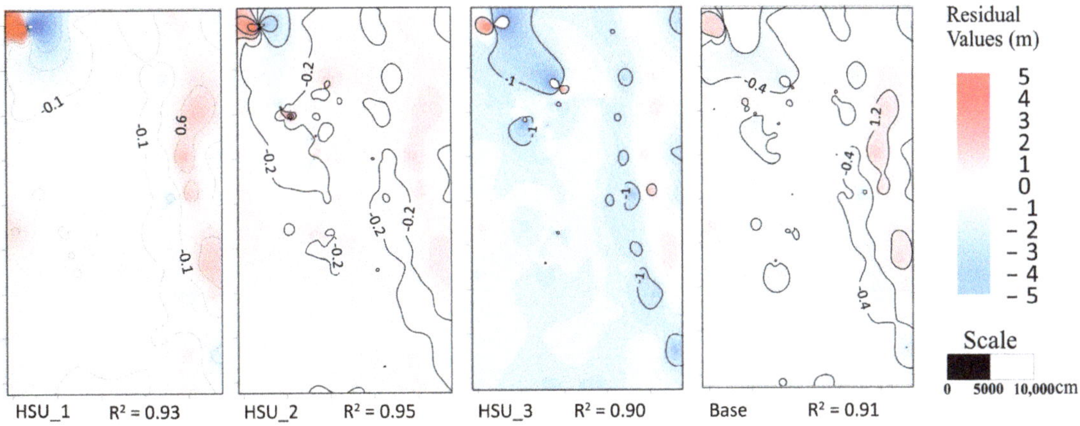


**Figure 11.** The residual map showing the difference between the original layer elevation value and the interpolated value for HSU_1, HSU_2, HSU_3, and Base.

To visualize the subsurface geologic unit extents and fault locations, fractures, and aquifer formations in GemPy, stratigraphic and unconformity connections were computed for each point in the grid (Figure 12a) to interpret the sparse field measurements. In the validation of the interpolation, it is assumed that the intrinsic error is zero and the validation is only for the error of the interpolation estimator. Well log data were used to evaluate the overlap variation with the layers generated by the GemPy model to minimize uncertainty in the input parameters and therefore in the model outcomes (Figure 12b).

### 3.4. Kobo Valley Aquifer and Sediment Layer Visualization

An aquifer is a saturated permeable geologic unit that can contain and transmit considerable quantities of water to wells and springs [84,85]. The thickness of Kobo Valley aquifer was determined from VES and drilling data. For this GemPy model, the Kobo Valley aquifer is assumed to be composed of sand, gravel, pebbles, and fractured volcanic rocks to obtain the spatial distribution of the aquifer thickness map. In addition, the driller's log and the geophysical survey data of the sub-surface material below the water table in the catchment were analyzed to assess the thickness of the aquifer. Generally, the aquifer thickness varied throughout Kobo Valley and increased from the center to the north and south (Figure 13a). The average aquifer thickness of the study area was 104 m. All layers were considered as penetrated by boreholes, including volcanic rocks, for the sediment thickness spatial distribution (Figure 13b). The sediment thickness to the eastern side of the center of Kobo Valley is mainly clay, and as a result, the aquifer is relatively thin in this area. The sediment thickness varied from a minimum of 120 m in the eastern central part to a maximum of 220 m in the western part, with an average sediment thickness of 150 m for the entire study area.

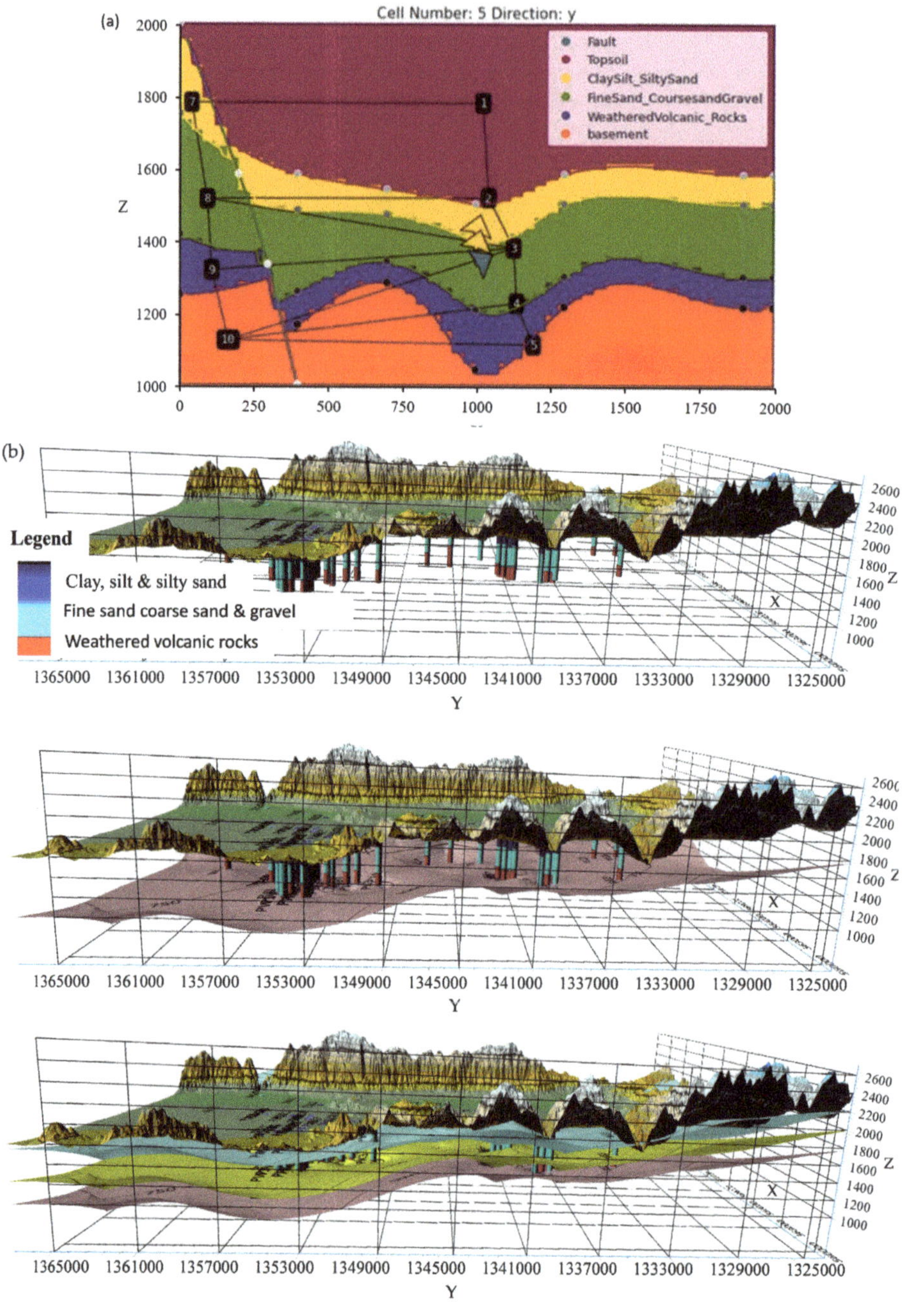


**Figure 12.** (**a**) Computed stratigraphic and unconformity connections at the western side of the valley at the fault section and models of special correlation. Z is an elevation in the *y*-axis, and x is the longitudinal cross-section in the *x*-axis multiplied by 10. (**b**) Driller's log lithology overlapped with interpolated aquifer profiles generated with GemPy. The X-axis, Y-axis, and Z-axis represent longitude, latitude, and elevation, respectively.

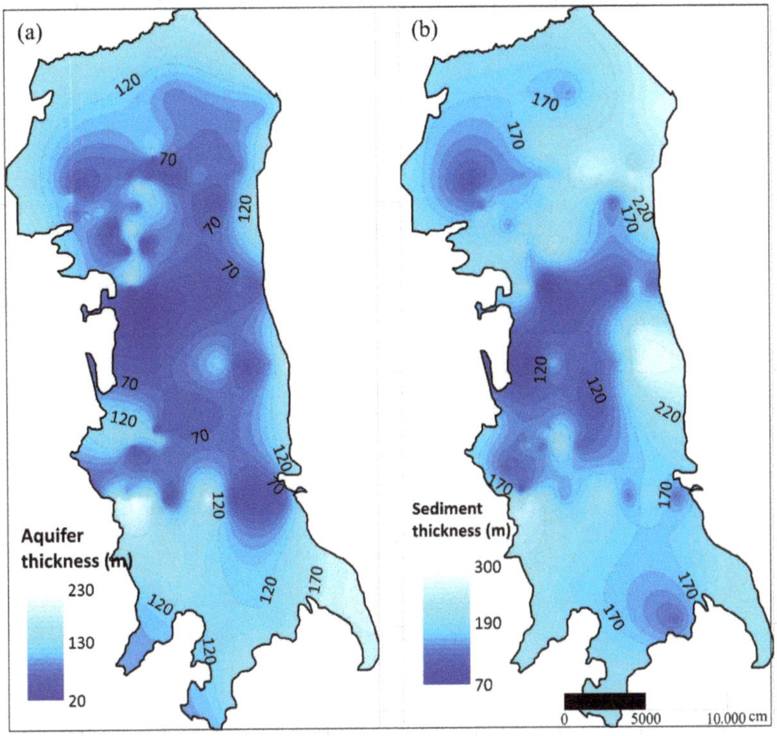


**Figure 13.** (**a**) Aquifer thickness and (**b**) sediment thickness distribution in Kobo Valley plain.

### 3.5. Hydraulic Properties of the Valley

The pertinent hydraulic properties of the aquifer were hydraulic conductivity, transmissivity, and specific yield. Hydraulic conductivity is the capacity of an aquifer to transmit water and is expressed as the volume of groundwater at the existing kinematic viscosity that will move in unit time under a unit hydraulic gradient through a unit area at a right angle to the direction of flow. Transmissivity is defined as the rate at which water of prevailing kinematic viscosity is transmitted through unit width of the aquifer under a unit hydraulic gradient and can be calculated by multiplying the hydraulic conductivity by the saturated thickness of the aquifer. Specific yield is defined as the ratio of the volume of water that drains because of gravity to the total volume of saturated aquifer [85–87].

Hydraulic conductivity, transmissivity, and specific yield are required inputs for most numerical groundwater simulation models and proper management of groundwater resources. A pumping test is the most common technique for estimating these parameter values. For this study, pumping test data from irrigation and water-supply wells were obtained from the MWIE office and were analyzed. The first hydraulic parameter analyzed was transmissivity as a product of hydraulic conductivity and saturated thickness. Irrigation and water-supply wells were used to determine the transmissivity distribution of the alluvial aquifer. Transmissivity data were determined from performed constant pumping tests, which were carried out uninterrupted for 3 days. The spatial distribution of the transmissivity of the Kobo Valley was plotted using QGIS and varied from 7.9 $m^2/day$ to 2500 $m^2/day$ with an average value of 467 $m^2/day$ (Figure 14b). The hydraulic conductivity of the Kobo Valley was also mapped, and values ranged from 0.1 m/day to 35 m/day (Figure 14a). Generally, as noted in Figure 14a, the western side of the central part of Kobo Valley has a low hydraulic conductivity zone, which indicates a fine deposit, and the western part of the alluvial aquifer is a high hydraulic conductivity zone ranging from 7.1 m/day to 33 m/day. The third parameter is specific yield, a material boundary that represents the effective (drainable) porosity of the unconfined sediments and the specific retention of the volumetric fraction of water that remains during a unit decline in the water table [88]. Specific yields obtained from the 63 wells were mapped, and observed values ranged from 0.06 to 0.3 with an average value of 0.22 (Figure 14c). As can be seen from the figure, the specific yield decreased from west to eastward. The minimum specific yield was observed on the eastern side of the central and northern part of Kobo Valley and was likely due to thick silt and clay aquifer materials. In contrast, the high specific yields observed at the western margin of the study area were likely due to the coarse aquifer material, as observed in Figure 14c.

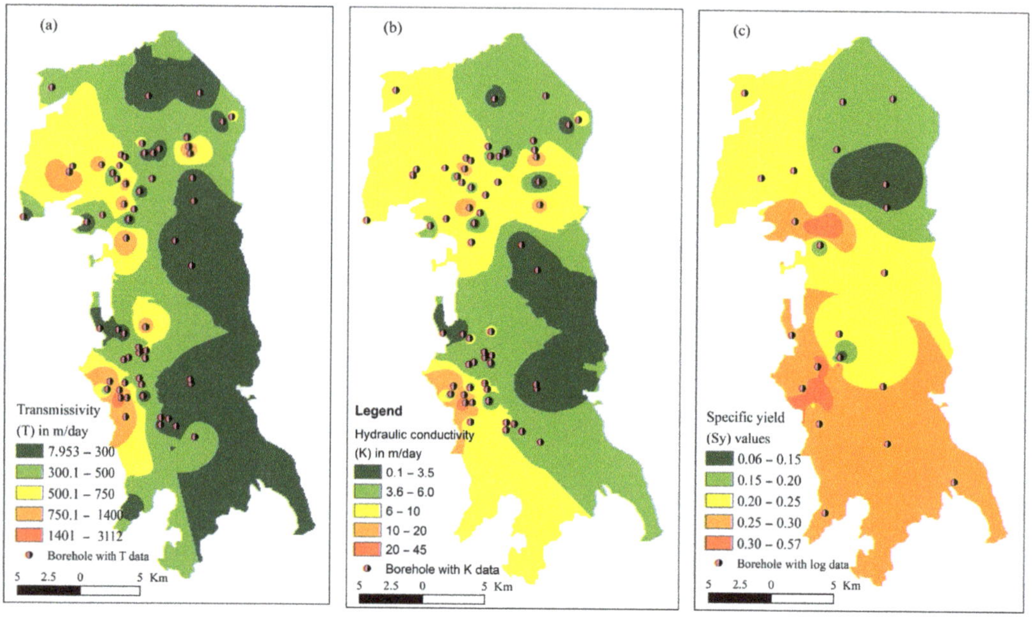


**Figure 14.** Spatial distribution of hydraulic parameters in Kobo Valley plain. (**a**) Hydraulic conductivity. (**b**) Transmissivity. (**c**) Specific yield.

### *3.6. Groundwater Flow System*

The groundwater-level contour map of Kobo Valley was generated from the GemPy model as shown in Figure 15a. Overlaying the groundwater-level contour map with the geomorphology of Kobo Valley (Figure 15b) and the hydrogeological longitudinal cross-section (Figure 10c), the data indicate a structural-based groundwater divide that divides the catchment into northern and southern groundwater systems that are prominently shown in the 3D HFM. The northern groundwater (north of the groundwater divide) flows in the northeast direction to the Selen Wuha outlet, whereas the southern groundwater (south of the groundwater divide) flows in the southeast direction to the Golina outlet (Figure 15b). Furthermore, the groundwater contour and the stream flow direction indicate that the groundwater systems have a hydraulic connection with the streams and rivers in Kobo Valley, and the flow direction is also influenced by the geomorphology and surface drainage of the area.

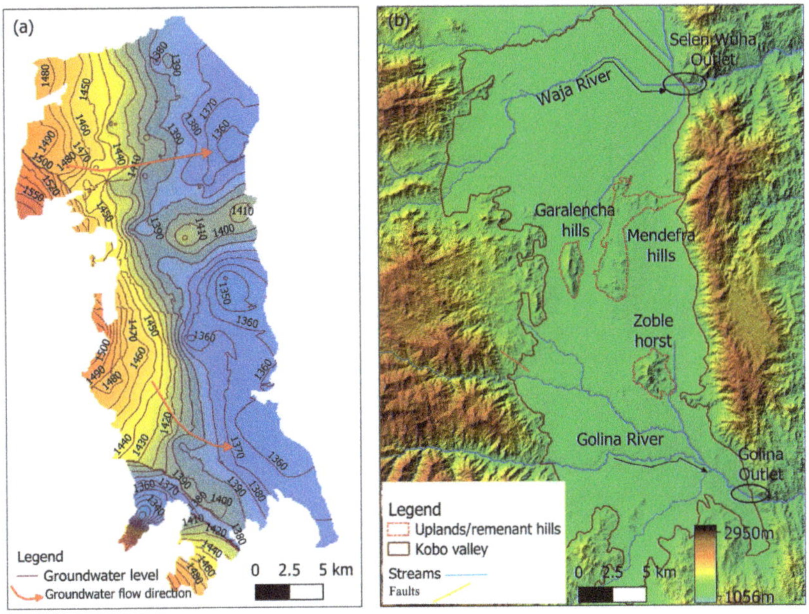


**Figure 15.** (**a**) Kobo Valley groundwater-level contour and groundwater flow system. (**b**) Kobo Valley shows the river flow outlet and remnant hills in the valley. Topography is derived from ASTER data (30 m resolution, accessed on 10 July 2020; https://earthexplorer.usgs.gov).

*3.7. Groundwater Storage in the Valley*

The developed 3D HFM and hydrogeological cross-section using GemPy (Figures 9 and 10) were analyzed and Kobo Valley's alluvial aquifer was identified as an unconfined aquifer with an impermeable hard rock as a base layer. Groundwater storage (GWS) is the groundwater in the pores of the alluvial aquifer and can be computed from the saturated aquifer thickness and specific yield of the aquifer [88–92].

$$GWS = (A * H)S_y \tag{2}$$

where A is the total area of the aquifer in meter square (m$^2$), H is saturated aquifer thickness (m), and $S_y$ is a specific yield of the aquifer. The aquifer thickness varies throughout Kobo Valley (detail shown in Section 3.2). Using the saturated aquifer thicknesses and the average specific yield of 0.22, groundwater storage in Kobo Valley was estimated to be 4132 MCM. This estimate is higher than the groundwater storage estimates of 3081 MCM reported by [93] and 2548 MCM reported by [94]. We believe that our estimation is more accurate because unlike the previous studies, we developed a three-dimensional hydrogeologic framework model (HFM) that accurately defined the hydrostratigraphy, structural features, and hydraulic properties of the Kobo Valley groundwater basin hydrogeology.

## 4. Conclusions

In this study, the GemPy model, a Python-based open-source tool, was used to develop a 3D HFM for Kobo Valley, part of the East African Rift System in northern Ethiopia. This work developed a simplified representation of a groundwater system to better understand the basin's groundwater hydrogeology and provide a tool that can be used for sustainable management of the groundwater system in the catchment. The developed model could help optimize the management of stratigraphic information; 3D visualization speeds up the process of stratigraphic setting evaluation, allowing for the identification of the existing geological aquifer layers in vertical and horizontal sections. The developed 3D HFM can help detect geological contacts, assess volumes and thicknesses, and evaluate geometric relationships between hydrostratigraphic units and their effects on the groundwater flow system of the basin. The 3D HFM, together with the geophysical and driller's lithologic logs and literature review, facilitated the development of a conceptual model describing the dynamics of the groundwater flow system in Kobo Valley.

For Kobo Valley, GemPy provided a new perspective for understanding the groundwater basin's hydrogeology and the development of a 3D HFM that could benefit sustainable groundwater management. Kobo Valley is an important source of fresh groundwater and has complex geological and structural settings. Estimation of the groundwater resource and modeling based on the generalized aquifer geometry may mislead the sustainable development of this aquifer. However, GemPy provides an easy, flexible, and interactive platform for incorporating the natural settings and complexity of the aquifer for the development and visualization of the 3D HFM of Kobo Valley. The groundwater storage in Kobo Valley was estimated as 4132 MCM; more accurate estimates can be used to improve groundwater resource management in Kobo Valley. The developed 3D HFM of Kobo Valley provides information to complement the development and management policy for sustainable groundwater extraction from Kobo Valley's alluvial aquifer.

The GemPy model also demonstrated the existence of remnant volcanic hills in the middle of Kobo Valley that act as a structural-based groundwater divide that splits the catchment into northern and southern groundwater systems. The acquired visualization and understanding of the subsurface structure were essential for the quantitative modeling of groundwater flow, and we recommend modeling and evaluating the two groundwater systems as two separate sub-basins.

We believe that GemPy fills the existing gap of implicit modeling in the open-source ecosystem in geosciences and offers a reliable and easy-to-use technology to generate complex models with only a few lines of code. The advancements described in this study maximize the computational capacity of present computing systems and have the potential to help improve groundwater management in Kobo Valley.

**Author Contributions:** Conceptualization, S.S.M. and S.E.B.; methodology, S.S.M.; software, S.S.M. and S.E.B.; validation, S.S.M., S.E.B. and M.D.; formal analysis, S.S.M.; data curation, S.S.M.; writing—original draft preparation, S.S.M.; writing—review and editing, S.S.M., S.E.B., A.K.M. and M.D.; supervision, S.E.B., A.K.M. and M.D. All authors have read and agreed to the published version of the manuscript.

**Funding:** This article was accomplished in the framework of the Home-Grown Scholarship Program funded by the governments of Ethiopia and the German Academic Exchange Service (DAAD). The project was implemented by Arba Minch University Water Research Center, Ethiopia (GOV/AMU/TH01/AWTI/WRRC/092014), in cooperation with the Chair of Hydrology and River Basin Management at the Technical University of Munich, Germany. In addition, this work was supported by the German Research Foundation (DFG) and the Technical University of Munich in the framework of the Open-Access Publishing Program.

**Data Availability Statement:** Data are contained within the article.

**Acknowledgments:** The authors are very grateful to the Chair of Hydrology and River Basin Management at the Technical University of Munich (TUM) and its Graduate School (TUM-GS) and Arba Minch University, Ethiopia, for the institutional services and facilities necessary to perform this study. The authors also thank the Office of Ethiopian Construction Design and Supervision Works Corporation (ECDSWC), the Ethiopian Ministry of Water, Irrigation and Energy (MWIE), and the Ethiopian Geological Study (EGS) for providing the necessary geologic data. Additionally, they acknowledge Randy Hanson, President and Hydrologist at One-Water Hydrologic for constructive comments about the content and methodology, Wayne Belcher from U.S. Geological Survey, and Nevada Water Science Center for the initial review of the journal article. They would also like to thank Nicole Parker for technical rhetoric comments or editorial comments.

**Conflicts of Interest:** The authors declare no conflict of interest.

**Disclaimer:** Any use of trade, firm, or product names is for descriptive purposes only and does not imply endorsement by the U.S. Government. This journal article has been peer reviewed and approved for publication consistent with USGS Fundamental Science Practices (https://pubs.usgs.gov/circ/1367/).

## Appendix A. GemPy Codes

```
import gempy as gp
# Importing auxiliary libraries
import numpy as np
import pandas as pd
import matplotlib.pyplot as plt
import os
# Main data management object containing
# Creating a model and importing the data from CSV-files and setting extent and resolution
geo_model = gp.create_model(Kobovalley_Geology')
gp.init_data(geo_model, [0, 2000., 0, 2000., 1000, 2000.], [100, 50, 100],
        path_o=data_path + "kobovalley_model_orientations.csv",
        path_i=data_path + "kobovalley_model_points.csv",
                default_values=True)
geo_model.surfaces
```

| | surface | series | order_surfaces | color | id |
|---|---|---|---|---|---|
| 0 | Ground_Topsurface | Default series | 1 | #015482 | 1 |
| 1 | ClaySilt_SiltySand | Default series | 2 | #9f0052 | 2 |
| 2 | FineSand_CoursesandGravel | Default series | 3 | #ffbe00 | 3 |
| 3 | WeatheredVolcanic_Rocks | Default series | 4 | #728f02 | 4 |
| 4 | basement | Basement | 1 | #443988 | 5 |

```
# Creating object with data prepared for interpolation and compiling
gp.set_interpolator(geo_model,
                        compile_theano=True,
                        theano_optimizer='fast_compile',
                        )
# Computing result
sol = gp.compute_model(geo_model)
# Plotting result: scalar field
gp.plot_2d(geo_model, show_data=True, show_scalar=True, figsize=(18,16))
# Plotting 3D result
geo3d = geo3d.plot_structured_grid()
grid3d[0].save('Kobo/kobogrid.vtk')
```

**Figure A1.** Python code to initiate GemPy model, import data, generate a single scalar field and plotting a section of regular grid and extracting surface points at the interfaces (Figure 6).

```
----
Add Code 1
# Importing auxiliary libraries
import numpy as np
import pandas as pd
import matplotlib.pyplot as plt
import os
from gempy.assets import topology as tp
import warnings
warnings.filterwarnings("ignore")
# Main data management object containing
# Creating a model and importing the data from CSV-files and order the formations (stratigraphic pile)
geo_model = gp.create_model("kobovalley_geology")
gp.init_data(
  geo_model, [0, 2000, 0, 20, 1000, 2200], [100, 10, 100],
  path_i=data_path+"kobovalley_model_interf.csv",
  path_o=data_path+"data_path + "kobovalley_model_data_fol.csv"
)
gp.map_stack_to_surfaces(
  geo_model,
    {
        "fault": "Fault",
        "Rest": ('Topsoil', 'ClaySilt_SiltySand', 'FineSand_CoursesandGravel', 'WeatheredVolcanic_Rocks')
    }
)
geo_model.set_is_fault(["fault"]);
gp.set_interpolator(geo_model)
sol = gp.compute_model(geo_model, compute_mesh=True)
#2-D Visualization of the Topology Graph
gp.plot_2d(geo_model, cell_number=[5])
gp.plot_2d(geo_model, cell_number=[5], show=False)
gp.plot.plot_topology(geo_model, edges, centroids, scale=True)
plt.show()
```

**Figure A2.** Unconformity connection analysis of a GemPy model (Figure 12a).

## Appendix B. Summary of VES Measurement Survey Interpretation

**Table A1.** Calculated volume of each individual layer of the study area.

| Profile | VES ID | Profile Layers and Lithology Identified |
|---|---|---|
| Profile1 | VESW1 to VESW8 | Top soil (1 to 33 m), clay layer 112 m at VESW1 to 210 m thick at VESW5, sandy/gravel layer (5 m at VESW5 to 59 m thick at VESW8), weathered volcanic 50 m at VESW1 and 48 m at VESW5, bed rock |
| Profile2 | VESW9 to VESW11 | Top soil (1 to 8 m), clayey layer (174 m at VESW9 and 140 at VESW11), thin sand/gravel layer, weathered volcanic 24 m at VESW10 and 68 m thick at VESW11, bed rock |
| Profile3 | VESW12 to VESW16 | Very thin top soil, clay layer ranges from 76 m at VESW16 to 176 m at VESW12, gravel layer of 13 m at VESW13 to 19 m at VESW14, weathered rock of 19 m at VESW15 and 56 m at VESW14, bed rock |
| Profile4 | VESK1 to VESK7 | Top soil (2 to 6 m), sand/gravel layer at the western half at VESK 2 and 3 thickness of 59 m and 20.9 and clay at the eastern half and weathered volcanic at the center. Clay layer on western half has thickness of 33.5 m at VESK2 and 104 m at VESK3. The eastern clay layer is 160 m at VESK6 and 183 m at VESK7, weathered zone 24 m at VESK2 and 88.8 m at VESk7, bed rock |
| Profile5 | VESK8 to VESK12 | The sandy/gravel layer thickness varies from 105 m at VESK8 to 45 m at VESK12, the clay layer filling the central and eastern part of the profile is 105 m to 149 m at VESK12, weathered zone has thickness of 20 to 30 m, bed rock |
| Profile6 | VESHG1 to VESHG9 | Thick clay layer of max 150 m at VESHG7 and 114 m at VESHG4, sand/gravel layer with maximum thickness of 195 m at VESHG2 and minimum thickness at VESHG4 (10 m), third layer above the fresh bed rock is the weathered zone of 30 to 40 m. |

## References


1. Altchenko, Y.; Villholth, K.G. Mapping irrigation potential from renewable groundwater in Africa—A quantitative hydrological approach. *Hydrol. Earth Syst. Sci.* **2015**, *19*, 1055–1067. [CrossRef]
2. Gaye, C.B.; Tindimugaya, C. Review: Challenges and opportunities for sustainable groundwater management in Africa. *Hydrogeol. J.* **2018**, *27*, 1099–1110. [CrossRef]
3. MacDonald, A.; Adelana, S. Groundwater research issues in Africa. *Appl. Groundw. Stud. Afr.* **2008**, *2008*, 6152.
4. Varga, M.d.L.; Schaaf, A.; Wellmann, F. GemPy 1.0: Open-source stochastic geological modeling and inversion. *Geosci. Model Dev.* **2019**, *12*, 1–32. [CrossRef]
5. Stumpf, A.J.; Keefer, D.A.; Turner, A.K. Overview and history of 3-D modeling approaches. In *Applied Multidimensional Geological Modeling: Informing Sustainable Human Interaction with the Shallow Subsurface*; Turner, A.K., Kessler, H., van der Muelen, M.J., Eds.; John Wiley & Sons: New York, NY, USA, 2022; pp. 95–112.
6. Van der Meulen, M.; Doornenbal, J.; Gunnink, J.; Stafleu, J.; Schokker, J.; Vernes, R.; van Geer, F.; van Gessel, S.; van Heteren, S.; van Leeuwen, R.; et al. 3D geology in a 2D country: Perspectives for geological surveying in the Netherlands. *Neth. J. Geosci.-Geol. En Mijnb.* **2013**, *92*, 217–241. [CrossRef]
7. Gross, D.L. *Geology for Planning in De Kalb County, Illinois. Champaign, IL: Illinois State Geological Survey*; Environmental Geology Notes 33; Pioneer Publishing Company: Oak Park, IL, USA, 1970; p. 26.
8. Hunt, C.S.A.; Kempton, J.P. *Geology for Planning in De Witt County, Illinois. Champaign, IL: Illinois State Geological Survey*; Environmental Geology Notes 83; Urbana publishing Company: Urbana, IL, USA, 1977; p. 42.
9. Berg, R.C.A.; Greenpool, M.R. *Stack-Unit Geologic Mapping: Color-Coded and Computer-Based Methodology*; State Geological Survey, Circular: Champaign, IL, USA, 1993; Volume 552, p. 11.
10. Jones, R.; McCaffrey, K.; Clegg, P.; Wilson, R.; Holliman, N.; Holdsworth, R.; Imber, J.; Waggott, S. Integration of regional to outcrop digital data: 3D visualisation of multi-scale geological models. *Comput. Geosci.* **2009**, *35*, 4–18. [CrossRef]
11. Calcagno, P.; Chilès, J.; Courrioux, G.; Guillen, A. Geological modelling from field data and geological knowledge. *Phys. Earth Planet. Inter.* **2008**, *171*, 147–157. [CrossRef]
12. Karlović, I.; Marković, T.; Vujnović, T.; Larva, O. Development of a Hydrogeological Conceptual Model of the Varaždin Alluvial Aquifer. *Hydrology* **2021**, *8*, 19. [CrossRef]
13. Faunt, C. Numerical Model of the Hydrologic Landscape and Groundwater Flow in California's Central Valley. In *Groundwater Availability of the Central Valley Aquifer*; USGS: Reston, VA, USA, 2009; pp. 121–212.
14. Hanson, R.T.; Schmid, W.; Faunt, C.C.; Lear, J.; Lockwood, B. *Integrated Hydrologic Model of Pajaro Valley, Santa Cruz and Monterey Counties, California*; Scientific Investigations Report; U.S. Geological Survey: Reston, VA, USA, 2014; p. 180.
15. Hanson, R.T.; Martin, P.; Koczot, K.M. *Simulation of Ground-Water/Surface-Water Flow in the Santa Clara-Calleguas Ground-Water Basin, Ventura County, California*; Water-Resources Investigations Report; U.S. Geological Survey: Sacramento, CA, USA, 2003.
16. Faunt, C.C.; Belitz, K.; Hanson, R.T. Development of a three-dimensional model of sedimentary texture in valley-fill deposits of Central Valley, California, USA. *Hydrogeol. J.* **2010**, *18*, 625–649. [CrossRef]
17. Knight, R.; Smith, R.; Asch, T.; Abraham, J.; Cannia, J.; Viezzoli, A.; Fogg, G. Mapping Aquifer Systems with Airborne Electromagnetics in the Central Valley of California. *Ground Water* **2018**, *56*, 893–908. [CrossRef]
18. Caruso, P.; Ochoa, C.G.; Jarvis, W.T.; Deboodt, T. A Hydrogeologic Framework for Understanding Local Groundwater Flow Dynamics in the Southeast Deschutes Basin, Oregon, USA. *Geosciences* **2019**, *9*, 57. [CrossRef]
19. Ben Saad, E.; Ben Alaya, M.; Taupin, J.-D.; Patris, N.; Chaabane, N.; Souissi, R. A Hydrogeological Conceptual Model Refines the Behavior of a Mediterranean Coastal Aquifer System: A Key to Sustainable Groundwater Management (Grombalia, NE Tunisia). *Hydrology* **2023**, *10*, 180. [CrossRef]
20. Lázaro, J.M.; Navarro, J.Á.S.; Gil, A.G.; Romero, V.E. 3D-geological structures with digital elevation models using GPU programming. *Comput. Geosci.* **2014**, *70*, 138–146. [CrossRef]
21. Cox, M.E.; James, A.; Hawke, A.; Raiber, M. Groundwater Visualisation System (GVS): A software framework for integrated display and interrogation of conceptual hydrogeological models, data and time-series animation. *J. Hydrol.* **2013**, *491*, 56–72. [CrossRef]
22. Brandenburg, J.P. *Geologic Frameworks for Groundwater Flow Models*; 2020: The Groundwater Project; Groundwater Project: Guelph, ON, Canada, 2020.
23. Raiber, M.; Webb, J.; Cendón, D.; White, P.; Jacobsen, G. Environmental isotopes meet 3D geological modelling: Conceptualising recharge and structurally-controlled aquifer connectivity in the basalt plains of south-western Victoria, Australia. *J. Hydrol.* **2015**, *527*, 262–280. [CrossRef]
24. Hanson, R.T. Hydrologic framework of the Santa Clara Valley, California. *Geosphere* **2015**, *11*, 606–637. [CrossRef]
25. Everett, R.R.; Gibbs, D.R.; Hanson, R.T.; Sweetkind, D.S.; Brandt, J.T.; Falk, S.E.; Harich, C.R. *Geology, Water-Quality, Hydrology, and Geomechanics of the Cuyama Valley Groundwater Basin, California, 2008–12*; Scientific Investigations Report; U.S. Geological Survey: Reston, VA, USA, 2013; p. 76.
26. Sweetkind, D.S.; Faunt, C.C.; Hanson, R.T. *Construction of 3-D Geologic Framework and Textural Models for Cuyama Valley Groundwater Basin, California*; U.S. Geological Survey Scientific Investigations Report 2013–5127; U.S. Geological Survey: Reston, VA, USA, 2013; p. 46.

27. Wentworth, C.M.; Jachens, R.C.; Williams, R.A.; Tinsley, J.C., III; Hanson, R.T. *Physical Subdivision and Description of the Water Bearing Sediments of the Santa Clara Valley, California*; Scientific Investigations Report; USGS: Reston, VA, USA, 2015; p. 84.
28. Sweetkind, D.S. *Three-Dimensional Hydrogeologic Framework Model of the Rio Grande Transboundary Region of New Mexico and Texas, USA, and Northern Chihuahua, Mexico*; Scientific Investigations Report; USGS: Reston, VA, USA, 2017; p. 61.
29. Belcher, W.R.; Sweetkind, D.S.; Faunt, C.C.; Pavelko, M.T.; Hill, M.C. *An Update of the Death Valley regional Groundwater Flow System Transient Model, Nevada and California*; Scientific Investigations Report; USGS: Reston, VA, USA, 2017.
30. Shishaye, H.A.; Tait, D.R.; Befus, K.M.; Maher, D.T.; Reading, M.J.; Jeffrey, L.; Tewolde, T.G.; Asfaw, A.T. Development of an improved hydrogeological and hydro-geochemical conceptualization of a complex aquifer system in Ethiopia. *Hydrogeol. J.* **2020**, *28*, 2727–2746. [CrossRef]
31. Bashir, I.Y.; Izham, M.Y.; Main, R. Vertical Electrical Sounding Investigation of Aquifer Composition and Its Potential to Yield Groundwater in Some Selected Towns in Bida Basin of North Central Nigeria. *J. Geogr. Geol.* **2014**, *6*, 60–69. [CrossRef]
32. Soomro, A.; Qureshi, A.L.; Jamali, M.A.; Ashraf, A. Groundwater investigation through vertical electrical sounding at hilly area from Nooriabad toward Karachi. *Acta Geophys.* **2019**, *67*, 247–261. [CrossRef]
33. Iserhien-Emekeme, R.; Ofomola, M.O.; Bawallah, M.; Anomohanran, O. Lithological Identification and Underground Water Conditions in Jeddo Using Geophysical and Geochemical Methods. *Hydrology* **2017**, *4*, 42. [CrossRef]
34. Jiang, Y.; Sun, M.; Yang, C. A Generic Framework for Using Multi-Dimensional Earth Observation Data in GIS. *Remote Sens.* **2016**, *8*, 382. [CrossRef]
35. Dynamic Graphics, Inc. EarthVision. 2020. Available online: http://www.dgi.com/earthvision/evmain.html (accessed on 10 June 2022).
36. ARANZ Geo Limited. Leapfrog3D. 2015. Available online: http://www.leapfrog3d.com/ (accessed on 18 May 2022).
37. GOCAD. Gocad Research Group Mira Geoscience. 2022. Available online: https://mirageoscience.com/mining-industry-software/gocad-mining-suite/ (accessed on 12 April 2022).
38. Petra. IHS Petra. 2022. Available online: https://www.spglobal.com/commodityinsights/en/ci/products/petra-geological-analysis.html (accessed on 24 July 2022).
39. Rockworks, Rockware, Inc. 2022. Available online: https://www.rockware.com (accessed on 10 July 2022).
40. HydroGeoAnalyst, Schlumberger Water Services. 2011. Available online: https://www.waterloohydrogeologic.com/products/hydro-geoanalyst/ (accessed on 2 August 2022).
41. Velasco, V.R.; Gogu, C.R.; Vázquez-Suñé, E.; Garriga, A.; Ramos, E.; Riera, J.; Alcaraz, M. The use of GIS-based 3D geological tools to improve hydrogeological models of sedimentary media in an urban environment. *Environ. Earth Sci.* **2013**, *68*, 2145–2162. [CrossRef]
42. Rossetto, R.; De Filippis, G.; Borsi, I.; Foglia, L.; Cannata, M.; Criollo, R.; Vázquez-Suñé, E. Integrating free and open source tools and distributed modelling codes in GIS environment for data-based groundwater management. *Environ. Model. Softw.* **2018**, *107*, 210–230. [CrossRef]
43. Bittner, D.; Rychlik, A.; Klöffel, T.; Leuteritz, A.; Disse, M.; Chiogna, G. A GIS-based model for simulating the hydrological effects of land use changes on karst systems—The integration of the LuKARS model into FREEWAT. *Environ. Model. Softw.* **2020**, *127*, 104682. [CrossRef]
44. Wellmann, F.; Caumon, G. *3-D Structural Geological Models: Concepts, Methods, and Uncertainties*; Elsevier: Amsterdam, The Netherlands, 2018; pp. 1–121.
45. Mitášová, H.; Mitáš, L. Interpolation by regularized spline with tension: I. Theory and implementation. *Math. Geol.* **1993**, *25*, 641–655. [CrossRef]
46. Matheron, G. Principles of geostatistics. *Econ. Geol.* **1963**, *58*, 1246–1266. [CrossRef]
47. MacDonald, A.M.; Bonsor, H.C.; Dochartaigh, B.É.Ó.; Taylor, R.G. Quantitative maps of groundwater resources in Africa. *Environ. Res. Lett.* **2012**, *7*, 024009. [CrossRef]
48. Cobbing, J.; Hiller, B. Waking a sleeping giant: Realizing the potential of groundwater in Sub-Saharan Africa. *World Dev.* **2019**, *122*, 597–613. [CrossRef]
49. Mekonen, S.S.; Boyce, S.E.; Mohammed, A.K.; Flint, L.; Flint, A.; Disse, M. Recharge Estimation Approach in a Data-Scarce Semi-Arid Region, Northern Ethiopian Rift Valley. *Sustainability* **2023**, *15*, 15887. [CrossRef]
50. Tadesse, N.; Nedaw, D.; Woldearegay, K.; Gebreyohannes, T. Groundwater Management for Irrigation in the Raya and Kobo Valleys, Northern Ethiopia. *Int. J. Earth Sci. Eng.* **2015**, *8*, 36–46.
51. Sisay Mengesha, G. Food Security Status of Peri-Urban Modern Small Scale Irrigation Project Beneficiary Female Headed Households in Kobo Town, Ethiopia. *J. Food Secur.* **2017**, *5*, 259–272. [CrossRef]
52. Ayenew, T.; GebreEgziabher, M.; Kebede, S.; Mamo, S. Integrated assessment of hydrogeology and water quality for groundwater-based irrigation development in the Raya Valley, northern Ethiopia. *Water Int.* **2013**, *38*, 480–492. [CrossRef]
53. Adane, G.W. *Groundwater Modelling and Optimization of Irrigation Water Use Efficiency to Sustain Irrigation in Kobo Valley, Ethiopia*; UNESCO-IHE Institute for Water Education: Delft, The Netherlands, 2014.
54. Zwaan, F.; Corti, G.; Keir, D.; Sani, F. A review of tectonic models for the rifted margin of Afar: Implications for continental break-up and passive margin formation. *J. Afr. Earth Sci.* **2019**, *164*, 103649. [CrossRef]
55. Beyene, A.; Abdelsalam, M.G. Tectonics of the Afar Depression: A review and synthesis. *J. Afr. Earth Sci.* **2005**, *41*, 41–59. [CrossRef]

6. Corti, G.; Bastow, I.D.; Keir, D.; Pagli, C.; Baker, E. Rift-Related Morphology of the Afar Depression. In *Landscapes and Landforms of Ethiopia*; Springer Science and Business Media: Dordrecht, The Netherlands, 2015; pp. 251–274.
7. Barberi, F.; Santacroce, R. The Afar Stratoid Series and the magmatic evolution of East African rift system. *Bull. de la Société Géologique de Fr.* **1980**, *S7-XXII*, 891–899. [CrossRef]
8. Stab, M.; Bellahsen, N.; Pik, R.; Quidelleur, X.; Ayalew, D.; Leroy, S. Modes of rifting in magma-rich settings: Tectono-magmatic evolution of Central Afar. *Tectonics* **2016**, *35*, 2–38. [CrossRef]
9. Corti, G. Continental rift evolution: From rift initiation to incipient break-up in the Main Ethiopian Rift, East Africa. *Earth-Sci. Rev.* **2009**, *96*, 1–53. [CrossRef]
0. Zwaan, F.; Corti, G.; Sani, F.; Keir, D.; Muluneh, A.A.; Illsley-Kemp, F.; Papini, M. Structural Analysis of the Western Afar Margin, East Africa: Evidence for Multiphase Rotational Rifting. *Tectonics* **2020**, *39*, e2019TC006043. [CrossRef]
1. Hammond, J.O.S.; Kendall, J.-M.; Stuart, G.W.; Keir, D.; Ebinger, C.; Ayele, A.; Belachew, M. The nature of the crust beneath the Afar triple junction: Evidence from receiver functions. *Geochem. Geophys. Geosyst.* **2011**, *12*. [CrossRef]
2. EGS, Ethiopian Geological Study, Government Document. 2012. Available online: https://docplayer.net/133114738-Geological-survey-of-ethiopia.html (accessed on 22 July 2022).
3. ECDSWC, Ethiopian Construction Design and Supervision Works Corporation, Government Document. 2021. Available online: https://waterpip.un-ihe.org/ethiopian-construction-design-and-supervision-works-corporation (accessed on 22 July 2022).
4. MCE, M.C.E. Hydrogeological and Geophysical Investigation Report of Kobo—Girana irrigation project by Metaferia Consulting Egneers. Government Document. 2009. Available online: https://www.metaferia.com/portfolio-4-columns-no-space/irrigation-agro-industry/ (accessed on 22 July 2022).
5. Program, I.W. *Program for Vertical Electrical Sounding Curves 1-D Interpreting along a Single Profile*; Department of geophysics, Geological Faculty, Moscow University: Moscow, Russia, 2000.
6. Ibuot, J.; Akpabio, G.; George, N. *A Survey of the Repository of Groundwater Potential and Distribution Using Geoelectrical Resistivity Method in Itu Local Government Area (L.G.A)*; Open Geosciences: Akwa Ibom State, Southern Nigeria, 2013; Volume 5.
7. Okoyeh, E.I.; Akpan, A.E.; Egboka, B.C.E.; Okeke, H.I. An Assessment of the Influences of Surface and Subsurface Water Level Dynamics in the Development of Gullies in Anambra State, Southeastern Nigeria. *Earth Interact.* **2014**, *18*, 1–24. [CrossRef]
8. González-Álvarez, I.; Ley-Cooper, A.Y.; Salama, W. A geological assessment of airborne electromagnetics for mineral exploration through deeply weathered profiles in the southeast Yilgarn Cratonic margin, Western Australia. *Ore Geol. Rev.* **2016**, *73*, 522–539. [CrossRef]
9. Loke, M.H. 2-D and 3-D Electrical Imaging Surveys. 2021. Available online: https://www.researchgate.net/publication/264739285_Tutorial_2-D_and_3-D_Electrical_Imaging_Surveys (accessed on 22 July 2022).
0. Cyril, A.G. Interpretation of Geolectric Pseudo Section and Seismic Refraction Tomography with Borehole Logs Carried out across a Functional Borehole at Garaje-Kagoro Area of Kaduna Northwestern Nigeria. *NIPES J. Sci. Technol. Res.* **2020**, *2*, 124. [CrossRef]
1. Lajaunie, C.; Courrioux, G.; Manuel, L. Foliation Fields and 3D Cartography in Geology: Principles of a Method Based on Potential Interpolation. *Math. Geol.* **1997**, *29*, 4. [CrossRef]
2. Salvatier, J.; Wiecki, T.V.; Fonnesbeck, C. Probabilistic programming in Python using PyMC3. *PeerJ Comput. Sci.* **2016**, *2*, e55. [CrossRef]
3. Theano Development Team. *A Python Framework for Fast Computation of Mathematical Expressions*; Montreal Institute for Learning Algorithms (MILA), Université de Montréal: Montréal, QC, Canada, 2016.
4. Mckinney, W. pandas: A Foundational Python Library for Data Analysis and Statistics, Python for High Performance and Scientific Computing. *ResearchGate* **2011**, *14*, 1–9.
5. Schroeder, W.; Martin, K.; Lorensen, B. The Visualization Toolkit an Object-Oriented Approach to 3D Graphics. *Kitware* **2004**, *2004*.
6. Hunter, J.D. Matplotlib: A 2D graphics environment. *Comput. Sci. Eng.* **2007**, *9*, 90–95. [CrossRef]
7. Walt, S.V.D.; Colbert, S.C.; Varoquaux, G. The NumPy Array: A Structure for Efficient Numerical Computation. *Comput. Sci. Eng.* **2011**, *13*, 22–30. [CrossRef]
8. Lark, R.; Mathers, S.; Thorpe, S.; Arkley, S.; Morgan, D.; Lawrence, D. A statistical assessment of the uncertainty in a 3D geological framework model. *Br. Geol. Surv.* **2013**, *124*, 946–958.
9. Wellmann, J.F.; Horowitz, F.G.; Schill, E.; Regenauer-Lieb, K. Towards incorporating uncertainty of structural data in 3D geological inversion. *Tectonophysics* **2010**, *490*, 141–151. [CrossRef]
0. Koller, D.; Friedman, N. *Probabilistic Graphical Models: Principles and Techniques Adaptive Computation and Machine Learning*; The MIT Press: Cambridge, MA, USA, 2009.
1. Wellmann, J.F.; Regenauer-Lieb, K. Uncertainties have a meaning: Information entropy as a quality measure for 3-D geological models. *Tectonophysics* **2012**, *526–529*, 207–216. [CrossRef]
2. Schweizer, D.; Blum, P.; Butscher, C. Uncertainty assessment in 3-D geological models of increasing complexity. *Solid Earth* **2017**, *8*, 515–530. [CrossRef]
3. Martin, J.; Adana, D.D.R.D.; Asuero, A.G. Fitting Models to Data: Residual Analysis, a Primer. In *Uncertainty Quantification and Model Calibration*; InTechOpen Publisher: London, UK, 2017.

84. Bear, J. *Hydraulics of Groundwater, McGraw-Hill Series in Water Resources and Environmental Engineering*; McGraw-Hill: New York NY, USA, 1979. Available online: https://www.perlego.com/book/110730/hydraulics-of-groundwater-pdf (accessed on 22 July 2022).
85. Freeze, R.A.; Cherry, J.A. *Groundwater*; Prentice-Hall Inc.: Englewood Cliffs, NY, USA, 1979; Volume 7632, p. 604.
86. Todd, D.K. Groundwater Hydrology; Agrosy Publishing: 1959. Available online: https://old.amu.ac.in/emp/studym/9999412€ .pdf (accessed on 22 July 2022).
87. Heath, R.C. Basic ground-water hydrology. In *Water Supply Paper*; USGS: Reston, VA, USA, 1983; p. 91.
88. Scanlon, B.R.; Longuevergne, L.; Long, D. Ground referencing GRACE satellite estimates of groundwater storage changes in the California Central Valley, USA. *Water Resour. Res.* **2012**, *48*. [CrossRef]
89. Wahyuni, S.; Oishi, S.; Sunada, K. The Estimation of the Groundwater Storage and Its Distribution in Uzbekistan. *Proc. Hydraul. Eng.* **2008**, *52*, 31–36. [CrossRef]
90. Evans, S.W.; Jones, N.L.; Williams, G.P.; Ames, D.P.; Nelson, E.J. Groundwater Level Mapping Tool: An open source web application for assessing groundwater sustainability. *Environ. Model. Softw.* **2020**, *131*, 104782. [CrossRef]
91. Bhanja, S.N.; Rodell, M.; Li, B.; Saha, D.; Mukherjee, A. Spatio-temporal variability of groundwater storage in India. *J. Hydrol.* **2017**, *544*, 428–437. [CrossRef]
92. Todd, D.K.; Mays, L.W. *Groundwater Hydrology*, 3rd ed.; John Wiley & Sons, Inc.: New York, NY, USA, 2005.
93. ECDSWC. *Kobo Chefa Groundwater Resource Evaluation, Assessment and Test Wells Drilling Supervision Project, Volume-I: Updating the Groundwater Potential Evaluation of Kobo Area*; Government Document; Addis Ababa, Ethiopia, 2018. Available online: https://waterpip.un-ihe.org/ethiopian-construction-design-and-supervision-works-corporation (accessed on 22 July 2022).
94. MCE. *Metaferia Consulting Engineers*; Hydrogeological Investigation Report; Ministry of Water Resources of Ethiopia: Addis Ababa, Ethiopia, 2009.

water

MDPI

*Article*

# Evaluation of Various Forms of Geothermal Energy Release in the Beijing Region, China

**Zebin Luo [1], Mingbo Yang [2,*], Xiaocheng Zhou [3,*], Guiping Liu [2], Jinlong Liang [4], Zhe Liu [5], Peixue Hua [2], Jingchen Ma [5], Leyin Hu [2], Xiaoru Sun [2], Bowen Cui [2], Zhiguo Wang [2] and Yuxuan Chen [2]**

1 School of Emergency Management, Xihua University, Chengdu 610039, China; zebin_l@mail.xhu.edu.cn
2 Beijing Earthquake Agency, Beijing 100080, China; liuguiping@bjseis.gov.cn (G.L.); huapeixue@bjseis.gov.cn (P.H.); huleyin@bjseis.gov.cn (L.H.); sunxr@bjseis.cn (X.S.); cuixie2001@bjseis.gov.cn (B.C.); wzg@bjseis.gov.cn (Z.W.); chenyx@bjseis.cn (Y.C.)
3 United Laboratory of High-Pressure Physics and Earthquake Science, Institute of Earthquake Forecasting, China Earthquake Administration CEA, Beijing 100036, China
4 College of Earth Science, Chengdu University of Technology, Chengdu 610059, China; richardljl04@aliyun.com
5 Beijing Institute of Geo-Engineering, Beijing 100048, China; liuzhedante@126.com (Z.L.); 13910656366@126.com (J.M.)
* Correspondence: yangmb2008@163.com (M.Y.); zhouxiaocheng188@163.com (X.Z.)

**Abstract:** The energy inside the Earth can not only be released outward through earthquakes and volcanoes but also can be used by humans in the form of geothermal energy. Is there a correlation between different forms of energy release? In this contribution, we perform detailed seismic and geothermal research in the Beijing area. The results show that the geothermal resources in Beijing belong to typical medium-low temperature geothermal resources of the sedimentary basin, and some areas are controlled by deep fault activities (e.g., Xiji geothermal well (No. 17)). The heat sources are upper mantle heat, radioactive heat in granite, and residual heat from magma cooling. The high overlap of earthquakes and geothermal field locations and the positive correlation between the injection water and earthquakes indicate that the exploitation and injection water will promote the release of the earth's energy. The energy releases are partitioned into multiple microearthquakes, avoiding damaging earthquakes ($M_L \geq 5$) due to excessive energy accumulation. Therefore, the exploitation of geothermal resources may be one way to reduce destructive earthquakes. Furthermore, the use of geothermal resources can also reduce the burning of fossil energy, which is of great significance in dealing with global warming.

**Keywords:** geothermal; earthquake forecasting; global warming; hot spring; Beijing; Zhangjiakou-Bohai fault

## 1. Introduction

The interior of the earth is filled with energy, which originates from the magma and the decay of radioactive materials. The energy can be released into the shallow surface or atmosphere in various ways. They can be fierce and destructive, like earthquakes and volcanoes, or relatively gentle, like hot springs. The difference is that earthquakes and volcanoes represent disasters, while hot springs are clean energy that can be used by humans. It is worth noting that they all originate from the release of energy inside the Earth. Is there a correlation between the different forms of energy release?

Unlike earthquakes and volcanoes, geothermal resources can be used by humans in a gentle way. Geothermal resources are considered one of the ways to combat global warming. The exploitation of geothermal resources has always been the focus of attention [1–11]. According to statistics, the geothermal energy reserves in the upper crust (3–10 km) are $1.3 \times 10^{27}$ J. In the World Energy Association's "Energy and Sustainability Challenges" report published in 2000, geothermal energy ranked first among all renewable energy sources [12]. However, induced earthquakes have been observed at various production stages of geothermal energy extraction, including initial injection of geothermal working fluid during stimulation, withdrawal of working fluid from geothermal reservoirs, reinjection of working fluid after heat extraction, and post-well closure [13]. Along with the disturbance of the crustal stress state, some exploitation projects have induced sizable earthquakes, even causing significant disasters and social problems [14–19]. For example, Soultz-Sous-Forêts in France [20], Basel in Switzerland [21], and Pohang in South Korea [22]. Therefore, exploring the mechanism of earthquakes induced by geothermal energy mining has been a hot topic in the world.

Efforts have long been made to mitigate or even eliminate induced earthquakes, whether from geothermal or oil and gas extraction. Induced seismicity is often perceived as an unsolicited and uncontrollable side effect of geothermal development [16,23–31]. But in fact, in most cases of induced seismicity, many events usually have magnitudes smaller than $M_L = 3$ and hence without economic consequences [15,32]. Seismicity triggered by fluid injection-induced earthquakes are still natural earthquakes, and their energy still comes from the Earth itself. Therefore, we put forward a conjecture: the total amount of Earth's energy is fixed, and earthquakes and geothermal are different forms of energy release. Is it possible to reduce the energy released by earthquakes by increasing the energy released by geothermal development?

To test the assumption, we choose the Beijing area for seismic and geothermal research. There are abundant geothermal resources in the Beijing area. Statistically, from 1971 to 2013, the total amount of geothermal resources exploitation quantity in Beijing is $2.87 \times 10^8$ m$^3$, and the injection water is $3.02 \times 10^7$ m$^3$ (Data from Beijing Geological Archive). In addition, there is a complete seismic network in the Beijing area, with detailed records of earthquakes ($M_L \geq 1$) since 1970. Therefore, Beijing is a natural laboratory for studying the relationship between seismic activity and geothermal energy.

## 2. The Study Area

The North China Craton (NCC) is one of the ancient cratons in the world [33]. It is bounded by the Central Asian orogenic belt in the north and the Qinling—Dabbe orogenic belt in the south. The basement rocks of the NCC consist of biotite-hornblende gneisses and Trondhjemite, Tonalite, Granodiorite (TTG) [34]. Overlying sedimentary layers with a thickness of several thousand meters, mainly carbonate rocks and clastic rocks. During the Yanshan tectonic period, the NCC experienced destruction and thinning, accompanied by a series of volcanic tectonic processes [35–39].

Beijing is located in the northern margin of NCC, high in the northwest and low in the southeast. Tectonic movement is active in the area [40]. The main faults include the Yanqing Fault, Dahuicang Fault, Liangxiang Fault, and Zhangjiakou-Bohai Fault (Figure 1). It is a seismic activity zone in eastern China. In history, there has been the $M_L$7.8 Tangshan earthquake (28 July 1978), the $M_L$7.4 Bohai earthquake (18 July 1969), and the $M_L$8.0 Shanhe-Pinggu earthquake (2 September 1679) [41].

Beijing area is enriched in geothermal resources [42]. At present, there are 10 geothermal fields: (1) Yanqing, (2) Xiaotangshan, (3) Houshayu, (4) Northwest district, (5) Tianzhu, (6) Lisui, (7) Southeast district, (8) Shuangqiao, (9) Liangxiang and (10) Fengheying geothermal field (Figure 1) [43]. The total geothermal resources are about $9.94 \times 10^{16}$ KJ, equivalent to $3.39 \times 10^9$ t of standard coal (Data from Beijing Geological Archive).

The terrestrial heat flow in Beijing ranges from 16.45 to 383.97 mW/m$^2$, with an average value of 65.95 mW/m$^2$. The geothermal gradient in the central area of the geothermal field is more than 3.0 °C/100 m. In some areas (Figure 1b, 5: Tianzhu and 6: Lishui), the geothermal gradient is more than 5 °C/100 m. The geothermal resources in Beijing belong to the medium and low-temperature hot water, and the temperature range is 25.0–118.5 °C. The geothermal water is Na-$HCO_3 \cdot SO_4$ type water with a salinity between 500 and 700 mg/L, with a high content of F and $SiO_2$, containing a small amount of trace elements, which is used for medical treatment and health care.

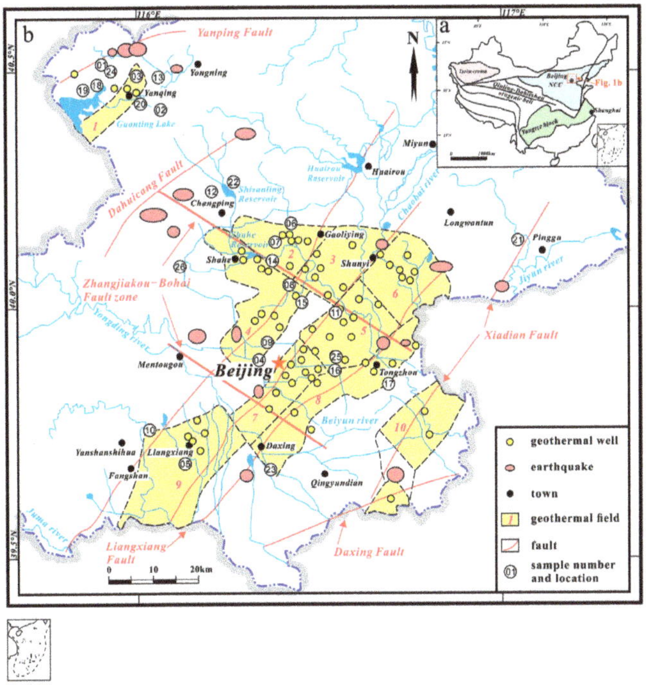


**Figure 1.** (**a**) A simple map of China. (**b**) Schematic map showing the distribution of geothermal fields and location of sampling points in the Beijing area, modified after Liu et al. [43]. 1: Yanqing, 2: Xiaotangshan, 3: Houshayu, 4: Northwest district, 5: Tianzhu, 6: Lishui, 7: Southeast district, 8: Shuangqiao, 9: Liangxiang and 10: Fengheying geothermal field. The size of the symbol of the earthquake label indicates the magnitude of the earthquake.

Beijing's climate is a warm, temperate, semi-humid, semi-arid monsoon climate, with an average annual temperature of 9 to 19 °C and annual precipitation of 600 mm. The seasonal distribution of precipitation is very uneven, with 80% of the annual precipitation concentrated in summer. The natural rivers of Beijing run through five major river systems from west to east: the Juma River, the Yongding River, the Beiyun River, the Chaobai River, and the Jiyun River. Most of them originated from the northwest mountain, meandered through the plain to the southeast, and finally merged into the Bohai Sea at the Haihe River.

## 3. Sampling and Analytical Methods

### 3.1. Geothermal Water Samples Collection and Analysis

Twenty-six samples of water were collected in Beijing, including hot springs and geothermal wells. All samples were analyzed for anions, cations, trace elements, hydrogen, and oxygen isotopes at the Beijing Institute of Geology of the Nuclear Industry. Detailed sample collection and testing methods can be found at Luo et al. [44]. In short, the waters were collected in a 50 mL clear polyethylene bottle, and the pH and temperature were recorded. Two water samples need to be collected at each geothermal water sampling site, one with ultrapure Nitric acid for cation analysis and the other for hydrogen and oxygen isotopes and anion analysis. Each sample is filtered with a 0.45 μm filter membrane before being tested. The cation and anion were analyzed by Dionex ICS-900 ion chromatograph (Thermo Fisher Scientific Inc., Bremen, Germany), and SiO2 was analyzed by inductively coupled plasma emission spectrometer Optima-5300 DV (PerkinElmer Inc., Waltham, MA, USA). $HCO_3^-$ and $CO_3^{2-}$ was determined by acid-base titration with a ZDJ-100 potentiometric titrator. Trace elements were analyzed by Element XR ICP-MS. Multielement standard solutions (IV-ICPMS 71A, IV-ICP-MS 71B and IV-ICP-MS 71D, iNORGANIC VENTURES) were used for quality control (the analytical error margin of major cations and trace elements were less than 10%). MAT 253 was used to analyze hydrogen and oxygen isotopes (reported as δD and $\delta^{18}O$ relative to Vienna Standard Mean Ocean Water (V-SMOW)).

### *3.2. Geothermal Gas Samples Collection and Analysis*

Between April 2022 and April 2023, we collected geothermal gas samples five times at the No. 17 geothermal well. 500 mL glass bottles were used to collect gas by drainage gas collection method [44]. During transportation and storage, glass bottles are kept sealed to prevent contamination by air. The chemical composition of geothermal gas samples was measured using the Agilent Macro 490 portable gas chromatograph with a measurement accuracy of better than 5%. He concentration in hot spring gas samples was analyzed using the Noblesse noble gas isotope mass spectrometer by the Northwest Institute of Eco-Environmental Resources, Chinese Academy of Sciences.

## **4. Results and Discussion**

### *4.1. Hydrochemistry of Geothermal Waters*

The physical properties and chemical and isotopic compositions of geothermal waters are shown in Table S1. The temperature of water varies from 13 to 92 °C. In this study, we divided the samples into three groups according to the sampling sites (Figure 2). The first group of geothermal waters was distributed in the Beijing urban area, and the second group of geothermal waters was collected in the Yanqing basin. In particular, we also classified the geothermal water in group 3, which is similar to that in group 1 in terms of collection location but obviously different from that in group 1 in terms of hydrochemical characteristics. This will be discussed in detail below.

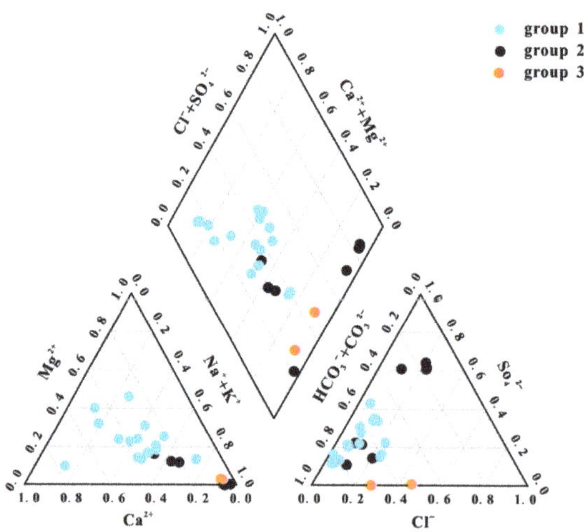


**Figure 2.** Piper diagram of geothermal waters in Beijing. These waters are Na·Ca·Mg-$HCO_3$, Na-$SO_4$, and Na-Cl types.

The $\delta^{18}O$ and δD of waters of Beijing are −16.2‰ to −9.6‰ and −92.4‰ to −69.2‰ respectively, which is close to the local meteoric water line (LMWL) of the Beijing δD = 7.0181 $\delta^{18}O$ + 3.5231 ($R^2$ = 0.86, n = 36) (Figure 3) [45], suggesting they originated in meteoric. Group 2 is more enriched in light isotope composition than groups 1 and 3. The $\delta^{18}O$ value of a few waters went off the LMWL, indicating that the isotopic exchange of $^{18}O$ occurs during the water-rock reaction.

From Figure 2 and Table S1, the geothermal waters are Na·Ca·Mg-$HCO_3$ (group 1), Na-$SO_4$·$HCO_3$ (groups 2) and Na-Cl·$HCO_3$ (groups 3) types. Groups 2 have significantly higher concentrations of $Na^+$ (84.50–151 mg/L) but lower $Ca^{2+}$ (3.16–46.90 mg/L) and $Mg^{2+}$ (0.04–14.8 mg/L) than group 1($Na^+$ (5.41–135 mg/L), $Ca^{2+}$ (19.20–57.50 mg/L) and $Mg^{2+}$ (2.30–37.10 mg/L)), which may reflect the reaction between groundwater and silicate rocks (Figure 4). It is consistent with the fact that group 2 waters are located in the granite thermal reservoir of the Yanqing basin. Analogously, there were also differences between group 2 and group 3. The anions of group 2 are $HCO_3^-$ (17.70–303 mg/L) and $SO_4^{2-}$ (29.70–177 mg/L), while Group 3 contains more $Cl^-$ (225–325 mg/L) and $HCO_3^-$ (648–1022 mg/L). In the Paleogene period, the gypsum salt layer was widely distributed in the North China Plain [46]. The elevated concentration of $SO_4^{2-}$ could be caused by the dissolution of sulfate minerals, such as anhydrite ($CaSO_4$) and mirabilite ($Na_2SO_4$). In addition, the high concentrations of $Cl^-$ in geothermal water probably originated from brine or the mixing with a deep fluid [44,47,48]. However, in the Beijing area, the brine has almost no effect on $Cl^-$. Because group 3 is well water located near the fault zone. The depth of 3588 m has exceeded the thickness of the sedimentary and reached the top of the magmatic batholith. Moreover, the $Cl^-$ versus $Na^+$ also indicates that the fluid is non-brined (Figure 5). Therefore, deep $Cl^-$-rich magmatic fluids rise along faults and then mix with groundwater to form group 3 of geothermal waters with high $Cl^-$ concentration.

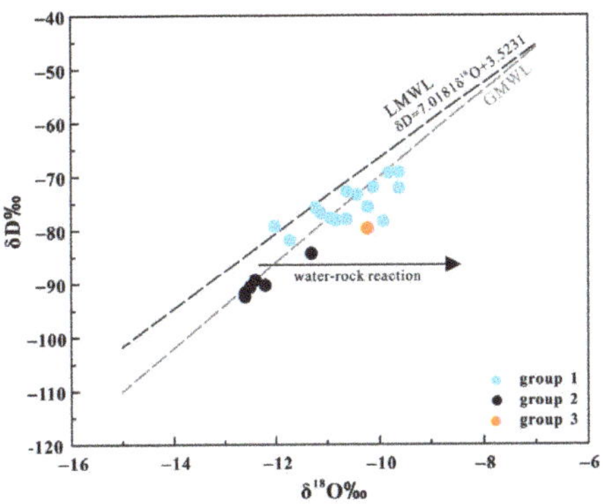


**Figure 3.** δD and $\delta^{18}O$ (relative to V-SMOW) values for waters collected from the Beijing area. The GMWL is a global meteoric water line [49]. The LMWL is a local meteoric water line [45]. Arrows indicate enhanced water-rock reactions.

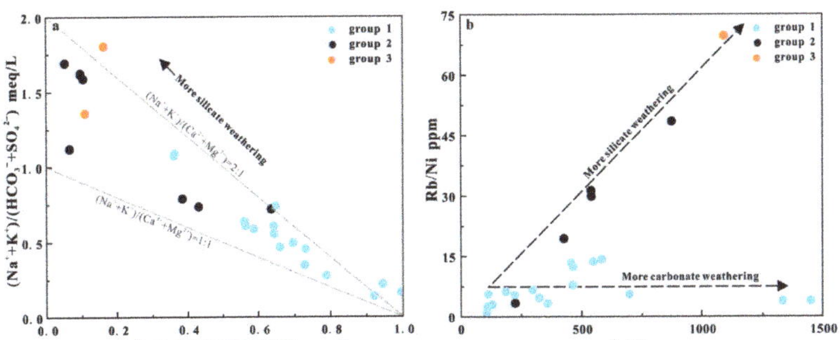


**Figure 4.** $(Na^+ + K^+)/(HCO_3^- + SO_4^{2-})$meq/L versus $(Ca^{2+} + Mg^{2+})/(HCO_3^- + SO_4^{2-})$ meq/L (**a**) and Rb/Ni ppm versus Sr/Ni ppm (**b**) for geothermal waters of Beijing area. Group 1 is characterized by the reaction of carbonate rock with water, while groups 2 and 3 are characterized by the reaction of silicate rock with water.

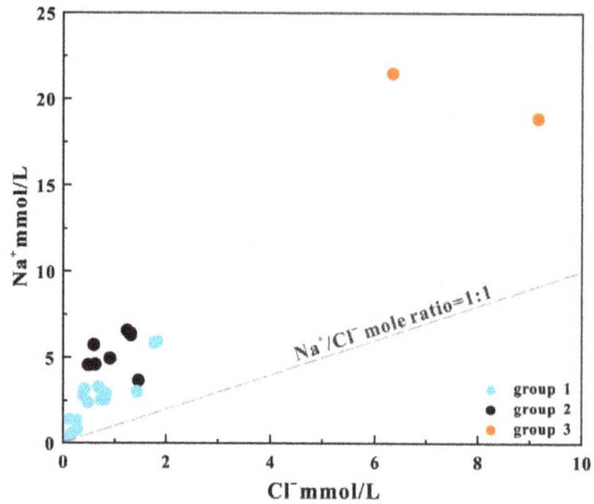


**Figure 5.** $Na^+$ versus $Cl^-$ for Beijing area geothermal waters.

The sedimentary layer in the North China Plain is several thousand meters high, including carbonate and clastic rocks, which well explains the high $Ca^{2+}$ and $Mg^{2+}$ concentrations in group 1 (Table S1) [42,43,46]. Relatively, the sedimentary layer in the granite thermal reservoir of the Yanqing basin is thinner, and the $Ca^{2+}$ and $Mg^{2+}$ concentrations in group 2 are lower. It can be seen from Figure 4 that the weathering characteristics of the silicate rocks of group 2 geothermal waters are significantly greater than those of group 1. Rubidium (Rb) occurs preferentially in K-containing minerals, while Strontium (Sr) occurs preferentially in Ca-containing minerals, and Nickel (Ni) is an extremely compatible element. Using Ni as the regional background value to normalize Rb and Sr, the source of ions in geothermal water can be distinguished. As can be seen from Figure 4, groups 2 and 3 are characterized by the reaction of silicate rocks with water, while group 1 is characterized by the reaction of carbonate rocks with water.

Carbonate rocks, including limestone and dolomite, can be further distinguished by the variation of $Ca^{2+}$ and $Mg^{2+}$ content. The $Mg^{2+}/Ca^{2+}$ molar ratio of the geothermal water in the dolomite area is near one, while it is much lower than one in the limestone area [50]. Meanwhile, the $Na^+/Ca^{2+}$ molar ratio can distinguish the carbonate rock and silicate rock area. As shown in Figure 6, the geothermal water in Beijing mainly comes from dolomite and silicate rock, or a mixture of them, with almost no contribution from limestone.

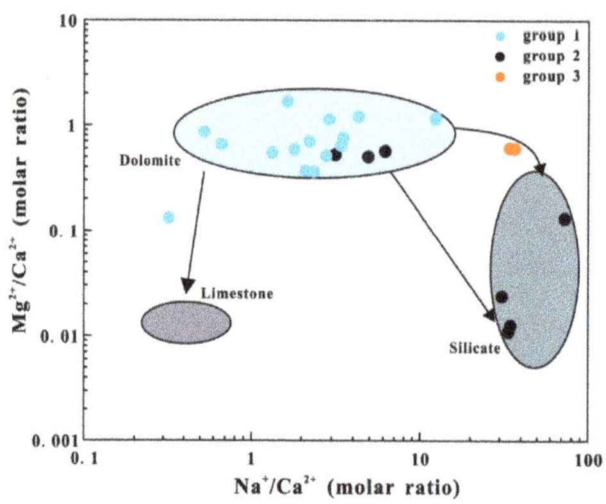


**Figure 6.** $Mg^{2+}/Ca^{2+}$ versus $Na^+/Ca^{2+}$ for geothermal waters of the Beijing area. The Dolomite, silicate, and limestone areas are from [50].

### 4.2. Characteristics of Heat Reservoir

The geothermal resources in Beijing belong to typical medium-low temperature geothermal resources of the sedimentary basin. The temperature varied from 25.0 to 118.5 °C, and the heat reservoirs are carbonate rocks [42]. The good thermal conductivity of carbonate rock results in a geothermal gradient in the study area (3–3.5 °C/100 m). All geothermal waters are plotted in the immature water field or partially equilibrated or mixed water (Figure S1). Therefore, pay attention to the applicability of the geothermometer when selecting the temperature scale. Previous studies have shown that the accuracy of the Na-Li geothermometer is higher than that of other thermometers in the carbonate rock region [44,51]. Therefore, the heat storage temperature of geothermal water in the study area was estimated by Na-Li geothermometer, and the results are shown in Table S2. Furthermore, quartz thermometers are also used as a reference [52]. The reservoir temperature and circulation depths of geothermal waters in the Beijing area calculated based on Na-Li and $SiO_2$ geothermometers are 65–240 °C and 1592–6597 m, respectively [44].

### 4.3. Origin of High He, $H_2$, and $CH_4$ Concentrations in Geothermal Gases

Yang et al. [40] observed that the No. 17 geothermal well has high concentrations of $H_2$ (330 ppm), He (5993 ppm), and $CH_4$ (volume ratio = 27.6%), and indicated that it may contain important information. Therefore, we have made a more in-depth study of No. 17 geothermal wells. Five samples were collected from the No. 17 geothermal well from April 2022 to April 2023. The chemical compositions of the geothermal gas samples are shown in Table S3 and Figure 7. $N_2$ and $CH_4$ account for more than 93% of No. 17 geothermal wells. He concentration (4243–6049 ppm) is significantly higher than other geothermal gases in the Beijing area (150–1851 ppm). What is the genesis of these high abnormal concentrations of these gas components? We will discuss this in detail below.

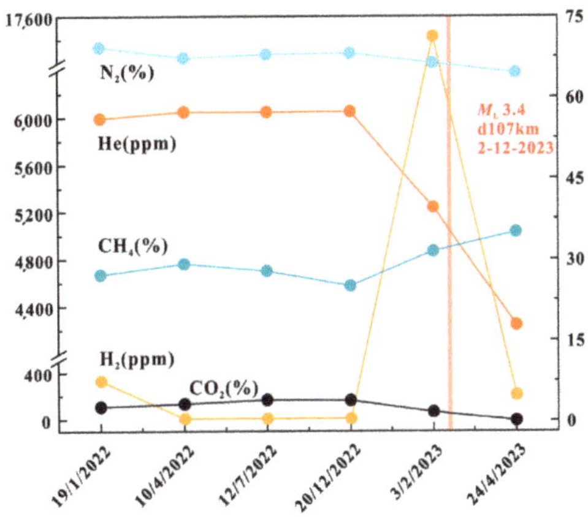


**Figure 7.** Characteristics of gas composition with time in Xiji (No. 17) geothermal well. Data of 19 January 2022 from Yang et al. [40].

#### 4.3.1. He

As discussed earlier, the No. 17 geothermal water belongs to Group 3 and has a high $Cl^-$ concentration, which probably reflects the intensity of deep fluid activity. Helium is also a geochemical indicator of tectonic activity and earthquakes. Both tectonic activity and earthquakes release large amounts of He [44,53]. However, He suddenly descends into No. 17 geothermal water in response to a significant earthquake, which is different from the traditional understanding [54]. In fact, even if the He concentration was reduced from 6049 ppm to 4243 ppm, the He concentration at the No. 17 geothermal water was still much higher than that of other geothermal water in the Beijing area. The reason for the decrease in He concentration before the earthquake may be that the faulting activity leads to the mixing of more air, diluting He in the geothermal gas. Therefore, the high He concentration of No. 17 geothermal gas should originate from deep fluid activity.

#### 4.3.2. $CH_4$

Most of the world's methane is biotic $CH_4$, which is produced either by microbial processes or by thermogenic degradation of organic matter in sedimentary rocks [55,56]. However, there is another origin of $CH_4$, called abiotic $CH_4$. It can formed by chemical reactions that do not directly include organic matter [57–64]. Abiotic $CH_4$ is extremely important in a wide range of scientific fields, including the origin of life, hydrocarbon synthesis, astrobiology, and planetary exploration [56].

During the Paleogene, oil shales were widely distributed in the Beijing area [46]. Does the high concentration of $CH_4$ in No. 17 geothermal gas originate from the thermogenic degradation of oil shales? Yes, but not entirely! Because well No. 17 had already cut through the sediment (3588 m, Table S1), and both $Cl^-$ and He indicate that No. 17 geothermal is polluted by deep fluid. In addition, deep tectono-magmatic activity, magma cooling, and gas-water-rock reactions can produce abiotic $CH_4$ [56]. Hence, the gas of the No. 17 geothermal well should be coming from deeper and contain abiotic $CH_4$. Although the $\delta^{13}CH_4$ of No. 17 is −36.4 [40], which shows the characteristics of biotic $CH_4$ [56], it may be a mixed value. The mixture of abiotic $CH_4$ from deep and biotic $CH_4$ released by thermogenic degradation of oil shales formed geothermal gas No. 17.

What is the genesis of abiotic $CH_4$ in geothermal gas No. 17? We propose that there are three ways:

(1) Carbonate reacts with water in the presence of Fe (500–1500 °C) [56]:

$$8FeO + CaCO_3 + 2H_2O = 4Fe_2O_3 + CH_4 + CaO \quad (1)$$

(2) $CO_2$ evolution to $CH_4$ during magma cooling (<500 °C) [56]:

$$CO_2 + 2H_2O = CH_4 + O_2 \quad (2)$$

(3) The Sabatier reaction (25–500 °C) [56]:

$$CO_2 + 4H_2 = CH_4 + 2H_2O \quad (3)$$

The sedimentary layer in the NCC is several thousand meters high, including carbonate and clastic rocks, which can provide sufficient $CaCO_3$ for (1). The (2) benefits from magmatic rocks produced by extensive Yanshanian magmatic activity [35,36,39]. The high concentration of $H_2$ and $CO_2$ in the geothermal gas of No. 17 provided the conditions for (3).

#### 4.3.3. $H_2$

A large number of experiments and natural observations have shown that hydrogen can be produced by faulting movements [54,65–69]. The origin of $H_2$ is usually attributed to a chemical reaction between crushed silicate minerals and water (e.g., serpentinization produces molecular hydrogen (4)) [67,70], which enables $H_2$ to reflect the activity of the fault to a certain extent [40,54,71].

$$(Mg, Fe)_2SiO_2 + H_2O \rightarrow Mg_2Si_2O_5(OH)_4 + Fe_3O_4 + H_2 \quad (4)$$

Olivine + fluid→serpentine + magnetite + hydrogen

Hydrogen content varies greatly in No. 17 geothermal gas. In particular, in the 3 February 2023 sample, a concentration of 17,426 ppm of $H_2$ was recorded, which probably reflects a precursory pulse of seismic activity (Figure 7). Sure enough, on 12 February 2023, the ninth day after the signal was detected, 11 earthquakes were detected in the same place; the maximum magnitude was $M_L 3.4$, at a maximum depth of 14 km (Table S4).

Beijing area is located on the northern margin of NCC. During the Yanshan tectonic period, the NCC experienced destruction and thinning under the influence of Pacific subduction [33,35,36,39]. The resulting magmatic rocks and fault zones provide the material sources and ascending channels for hydrogen generation. The hydrogen production is controlled by the activity of the fault zone. Therefore, the $H_2$ concentration in No. 17 geothermal gas can be used for monitoring fault activity and earthquake warnings.

*4.4. Geothermal Water Cycle Model and Genesis of Geothermal Field*

As discussed above, the geothermal water in the Beijing area can be divided into three groups. Group 1 is located in the sedimentary area, dominated by $Na{\cdot}Ca{\cdot}Mg\text{-}HCO_3$, and group 2 is located in the silicate rock area, dominated by $Na\text{-}SO_4{\cdot}HCO_3$. In particular, although group 3 is located in the sedimentary area, the depth of 3588 m has exceeded the thickness of the sedimentary and reached the top of the magmatic rock batholith, so that group 3 has the characteristics of deep fluid with high $Cl^-$, He, $H_2$ and $CH_4$. Combined with geochemical and isotopic composition, we propose that the geothermal water in the Beijing area originated from atmospheric precipitation. The precipitation flows into the ground along the fault and reacts with the surrounding rock while being heated. Eventually, they go up well along the fault to form hot springs (Figure 8). The geothermal resources in Beijing belong to typical medium-low temperature geothermal resources of the sedimentary basin, and some areas are controlled by deep fault activity (e.g., Xiji geothermal well (No. 17)). The heat sources are upper mantle heat, radioactive heat in granite and magmatic cooling residual heat. The heat reservoir is carbonate rock.

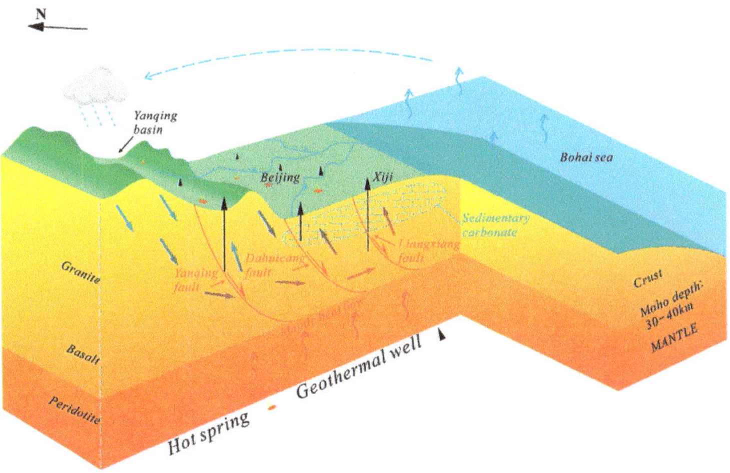


**Figure 8.** The water cycle model of the geothermal waters and gases in the Beijing area. The geothermal water in the Beijing area originated from atmospheric precipitation. The precipitation flows into the ground along the fault and reacts with the surrounding rock while being heated. Eventually, upwell along the fault to form hot springs.

*4.5. The Promotion of Geothermal Resources to Promote the Earth's Energy Release*

The way the earth releases energy can be geothermal energy or earthquakes. A large number of studies on oil and gas extraction, wastewater treatment, and geothermal exploitation have shown that fluids can promote seismic activity [14,15,17,23–32]. The geothermal development in the Beijing area includes the extraction and injection of water. So, what is the relationship between geothermal fluid activity and earthquakes?

We have been collecting earthquake records in the Beijing area since 1970. Considering that earthquakes with smaller magnitudes may not have been recorded due to insufficient coverage area of the seismic network in the early stage, we only conducted statistics on earthquakes with magnitudes above $M_L$ 2, and the results are shown in Figure 9. The earthquakes are distributed near the Zhangjiakou-Bohai fault zone, which is similar to the geothermal field (Figure 1). The magnitude is mainly 2–4, while earthquakes above 4 are rare (Figure 9). Since 1970, the number of earthquakes in Beijing has shown a slow upward tendency, but the total amount of energy released by earthquakes has not increased significantly (Figure 10). This reflects the fact that in the absence of a significant change in fault activity, the rise in the number of earthquakes results in less energy being released each time, i.e., a smaller magnitude. In fact, since 2013, the fault activity in the Beijing area has been weakening [41]. Exploitation of geothermal resources, on the other hand, has steadily increased. Statistically, from 1971 to 2013, the total amount of geothermal resources developed in Beijing was $2.87 \times 10^8$ m$^3$, and the injection water was $3.02 \times 10^7$ m$^3$. Subsequently, the annual production has been maintained at $600–800 \times 10^4$ m$^3$/y (Figure 10).

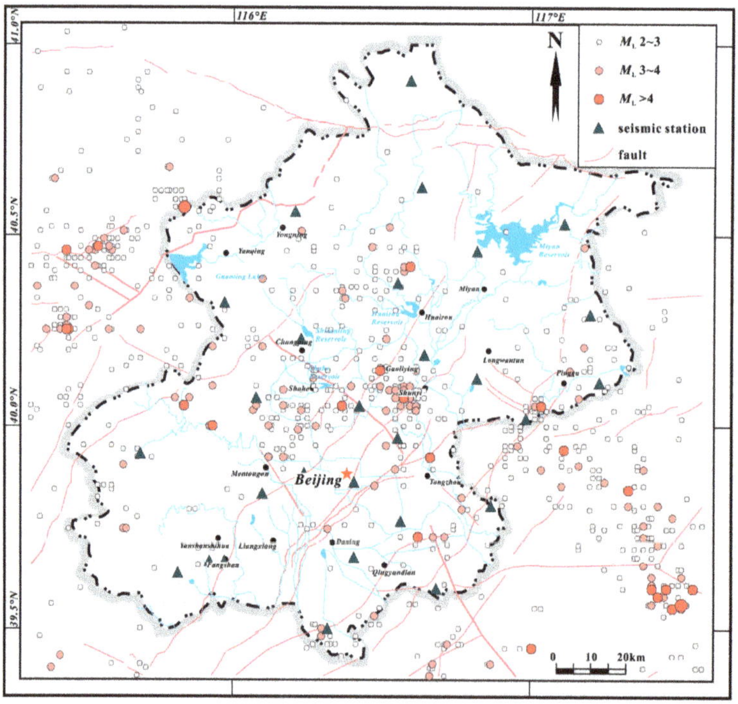


**Figure 9.** Earthquake records from 1970 in the Beijing area. The triangle shows the distribution of seismic stations, and their locations are from the China Earthquake Administration.

Significantly, the high overlap of the earthquake and geothermal field location and the positive correlation between injection water and earthquakes indicates that geothermal resource development will promote the occurrence of earthquakes (Figures 9, 10 and S2). The extraction and injection of water change the fluid pressure of the geothermal water, which leads to a change in rock stress, releasing its elastic potential energy and triggering earthquakes [24,26]. Due to the continuous exploitation of geothermal water, the elastic potential energy of the rock cannot be accumulated excessively, which effectively reduces the occurrence of destructive earthquakes. Therefore, we can reduce the occurrence of destructive earthquakes by rational use of geothermal resources.

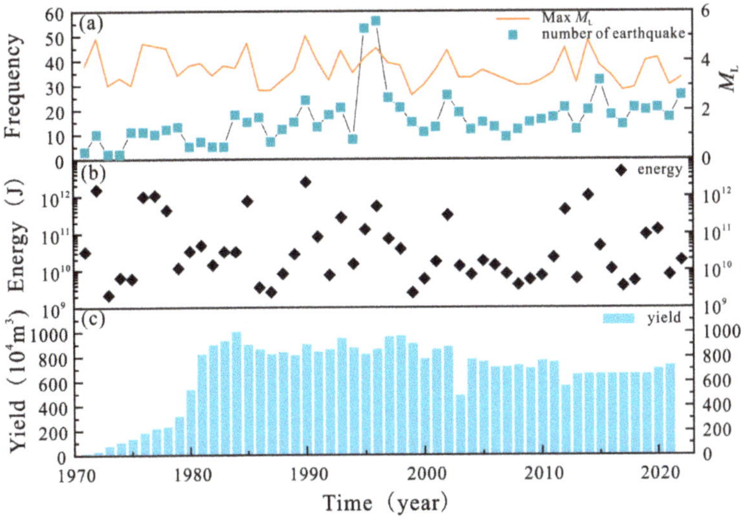


**Figure 10.** Temporal variations of earthquake frequency (time) (**a**), energy release (J) (**b**), and geothermal production ($10^4$ $m^3$) (**c**). The conversion formula of magnitude and energy: lgE = 4.8 + 1.5 M, E is energy (J). M is magnitude ($M_L > 2$), earthquake data from the China Earthquake Administration. Geothermal production data from the Beijing Hydrogeological Engineering Team (2014–2019 are estimates).

## 5. Conclusions and Outlook

In this contribution, we perform a detailed elemental and isotopic analysis of geothermal waters and gases collected from the Beijing area. By integrating geochemical results of geothermal waters and gases, we propose that the geothermal resources in Beijing belong to typical medium-low temperature geothermal resources of the sedimentary basin, and some areas are controlled by deep fault activity (e.g., Xiji geothermal well (No. 17)).

The $H_2$ and $CH_4$ in the geothermal water/gas of the No. 17 geothermal well are sensitive to deep structural activities. By monitoring the elements and isotopes of geothermal well No. 17, the deep fluid activities can be reflected and thus forewarn earthquakes.

The extraction and injection of water will promote the release of Earth's energy. The energy is differentiated into multiple releases and avoids the excess accumulation of one-time energy, resulting in damaging earthquakes ($M_L \geq 5$). On the one hand, the exploitation of geothermal resources may be one way to reduce destructive earthquakes; on the other hand, the utilization of geothermal resources can reduce the consumption of fossil energy, which is of great significance for tackling global warming.

We propose that the exploitation of geothermal resources may be one of the means to reduce destructive earthquakes. However, given the complex thermal structure of the Earth's crust, the conversion mechanism between geothermal and seismic energy release is not known. Geothermal water links to earthquakes and to earthquake stress release are not established (no physical robust, statistical, or quantitative analyses). In addition, the study area is limited. Therefore, the contribution of this paper is that we provide a new research idea for earthquake and geothermal research, and more in-depth and systematic research is needed in the future.

**Supplementary Materials:** The following supporting information can be downloaded at: https://www.mdpi.com/article/10.3390/w16040622/s1, Table S1: Physical properties, hydrogeochemistry and isotopic compositions of geothermal waters from Beijing area; Figure S1: Na-K-Mg ternary diagram of geothermal waters in Beijing area; Table S2: Temperature results obtained with empirical chemical geothermometers (values in °C) and depths (m) of origin for Beijing area geothermal waters; Table S3: Geothermal gases chemistry compositions of Xiji geothermal well (No. 17) from Beijing area; Table S4: Earthquake records at 12 February 2023 in Beijing area; Figure S2: Relationship between volume of injection water ($10^4$ $m^3$) and number of earthquakes ($>M_L = 1$) in Beijing area.

**Author Contributions:** Conceptualization, Z.L. (Zebin Luo) and X.Z.; methodology, Z.L. (Zebin Luo); software, Z.L. (Zebin Luo); validation, X.Z. and J.L.; formal Analysis, X.Z.; investigation, M.Y., Z.L. (Zhe Liu), P.H., J.M., L.H., X.S., B.C., Z.W. and Y.C.; resources, M.Y.; data Curation, M.Y.; writing—Original Draft, Z.L. (Zebin Luo); writing—Review and Editing, Z.L. (Zebin Luo); supervision, X.Z.; project Administration, M.Y. and G.L.; funding acquisition, M.Y. All authors have read and agreed to the published version of the manuscript.

**Funding:** This work is supported by the surface technology project of Beijing earthquake agency (NO.BJMS-2024005) and Key Laboratory of Ionic Rare Earth Resources and Environment, Ministry of Natural Resources of the People's Republic of China.

**Data Availability Statement:** Water and gas of hot springs data were measured by experimental instruments. Seismic records and geothermal resource data will be made available on request.

**Acknowledgments:** We would like to thank the Associate Editor and anonymous reviewers for their constructive comments, suggestions, and corrections. We also thank Lincheng Jiang and Tian Liang for the discussion.

**Conflicts of Interest:** The funders had no role in the design of the study, in the collection, analyses, or interpretation of data, in the writing of the manuscript, or in the decision to publish the results.

## References


1. Alam, B.Y.C.S.S.S.; Itoi, R.; Taguchi, S.; Saibi, H.; Yamashiro, R. Hydrogeochemical and isotope characterization of geothermal waters from the Cidanau geothermal field, West Java, Indonesia. *Geothermics* **2019**, *78*, 62–69. [CrossRef]
2. Bahri, F.; Saibi, H.; Cherchali, M.-E.-H. Characterization, classification, and determination of drinkability of some Algerian thermal waters. *Arab. J. Geosci.* **2011**, *4*, 207–219. [CrossRef]
3. Elbarbary, S.; Zaher, M.A.; Saibi, H.; Fowler, A.-R.; Saibi, K. Geothermal renewable energy prospects of the African continent using GIS. *Geotherm. Energy* **2022**, *10*, 8. [CrossRef]
4. Saibi, H.; Batir, J.F.; Pocasangre, C. Hydrochemistry and geothermometry of thermal waters from UAE and their energetic potential assessment. *Geothermics* **2021**, *92*, 102061. [CrossRef]
5. Saibi, H.; Ehara, S. Temperature and chemical changes in the fluids of the Obama geothermal field (SW Japan) in response to field utilization. *Geothermics* **2010**, *39*, 228–241. [CrossRef]
6. Aboulela, H.; Amin, A.; Lashin, A.; El Rayes, A. Contribution of geothermal resources to the future of renewable energy in Egypt: A case study, Gulf of Suez-Egypt. *Renew. Energy* **2021**, *167*, 248–265. [CrossRef]
7. El-Rayes, A.E.; Arnous, M.O.; Aboulela, H.A. Hydrogeochemical and seismological exploration for geothermal resources in South Sinai, Egypt utilizing GIS and remote sensing. *Arab. J. Geosci.* **2015**, *8*, 5631–5647. [CrossRef]
8. Fan, Y.; Pang, Z.; Liao, D.; Tian, J.; Hao, Y.; Huang, T.; Li, Y. Hydrogeochemical Characteristics and Genesis of Geothermal Water from the Ganzi Geothermal Field, Eastern Tibetan Plateau. *Water* **2019**, *11*, 1631. [CrossRef]
9. Gu, X.M.; Zhang, Q.L.; Cui, Y.L.; Shao, J.L.; Xiao, Y.; Zhang, P.; Liu, J.X. Hydrogeochemistry and Genesis Analysis of Thermal and Mineral Springs in Arxan, Northeastern China. *Water* **2017**, *9*, 61. [CrossRef]
10. Deng, J.; Lin, W.; Xing, L.; Chen, L. The Estimation of Geothermal Reservoir Temperature Based on Integrated Multicomponent Geothermometry: A Case Study in the Jizhong Depression, North China Plain. *Water* **2022**, *14*, 2489. [CrossRef]
11. Yu, M.; Tian, X.; Zhang, H.; Li, J.; Wang, L.; Zhang, Z.; Lin, H.; Yang, X. Hydrogeochemical Characteristics of Geothermal Water in Ancient Deeply Buried Hills in the Northern Jizhong Depression, Bohai Bay Basin, China. *Water* **2023**, *15*, 3881. [CrossRef]
12. Yin, X.; Jiang, C.; Zhai, H.; Zhang, Y.; Jiang, C.; Lai, G.; Zhu, A.; Yin, F. Review of induced seismicity and disaster risk control in dry hot rock resource development worldwide. *Chin. J. Geophys. Chin. Ed.* **2021**, *64*, 3817–3836.
13. Okamoto, K.; Yi, L.; Asanuma, H.; Okabe, T.; Abe, Y.; Tsuzuki, M. Triggering processes of microseismic events associated with water injection in Okuaizu Geothermal Field, Japan. *Earth Planets Space* **2018**, *70*, 15. [CrossRef]
14. Brodsky, E.E.; Lajoie, L.J. Anthropogenic Seismicity Rates and Operational Parameters at the Salton Sea Geothermal Field. *Science* **2013**, *341*, 543–546. [CrossRef] [PubMed]
15. Gruenthal, G. Induced seismicity related to geothermal projects versus natural tectonic earthquakes and other types of induced seismic events in Central Europe. *Geothermics* **2014**, *52*, 22–35. [CrossRef]

6. Guglielmi, Y.; Cappa, F.; Avouac, J.P.; Henry, P.; Elsworth, D. Seismicity triggered by fluid injection–induced aseismic slip. *Science* **2015**, *348*, 1224–1226. [CrossRef]
7. Trutnevyte, E.; Wiemer, S. Tailor-made risk governance for induced seismicity of geothermal energy projects: An application to Switzerland. *Geothermics* **2017**, *65*, 295–312. [CrossRef]
8. Li, C.; Zhou, X.; Yan, Y.; Ouyang, S.; Liu, F. Hydrogeochemical Characteristics of Hot Springs and Their Short-Term Seismic Precursor Anomalies along the Xiaojiang Fault Zone, Southeast Tibet Plateau. *Water* **2021**, *13*, 2638. [CrossRef]
9. Wang, B.; Zhou, X.; Zhou, Y.; Yan, Y.; Li, Y.; Ouyang, S.; Liu, F.; Zhong, J. Hydrogeochemistry and Precursory Anomalies in Thermal Springs of Fujian (Southeastern China) Associated with Earthquakes in the Taiwan Strait. *Water* **2021**, *13*, 3523. [CrossRef]
0. Majer, E.L.; Baria, R.; Stark, M.; Oates, S.; Bommer, J.; Smith, B.; Asanuma, H. Induced seismicity associated with enhanced geothermal systems. *Geothermics* **2007**, *36*, 185–222. [CrossRef]
1. Haering, M.O.; Schanz, U.; Ladner, F.; Dyer, B.C. Characterisation of the Basel 1 enhanced geothermal system. *Geothermics* **2008**, *37*, 469–495. [CrossRef]
2. Kim, K.-H.; Ree, J.-H.; Kim, Y.; Kim, S.; Kang, S.Y.; Seo, W. Assessing whether the 2017 $M_W$ 5.4 Pohang earthquake in South Korea was an induced event. *Science* **2018**, *360*, 1007–1009. [CrossRef]
3. Block, L.V.; Wood, C.K.; Yeck, W.L.; King, V.M. The 24 January 2013 M-L 4.4 Earthquake near Paradox, Colorado, and Its Relation to Deep Well Injection. *Seismol. Res. Lett.* **2014**, *85*, 609–624. [CrossRef]
4. Che, Y.; Yu, J. Influence and controlling of fluid in the crust on earthquake activity. *Recent Dev. World Seismol.* **2014**, *8*, 1–9, (In Chinese with English Abstract).
5. Ellsworth, W.L. Injection-Induced Earthquakes. *Science* **2013**, *341*, 142. [CrossRef] [PubMed]
6. Evans, D.M. Denver area earthquakes and the Rocky Mountain Arsenal disposal well. *Mt. Geol.* **1966**, *3*, 23–36.
7. Horton, S. Disposal of Hydrofracking Waste Fluid by Injection into Subsurface Aquifers Tiggers Earthquake Swarm in Central Arkansas with Potential for Damaging Earthquake. *Seismol. Res. Lett.* **2012**, *83*, 250–260. [CrossRef]
8. Keranen, K.M.; Savage, H.M.; Abers, G.A.; Cochran, E.S. Potentially induced earthquakes in Oklahoma, USA: Links between wastewater injection and the 2011 Mw 5.7 earthquake sequence. *Geology* **2013**, *41*, 699–702. [CrossRef]
9. Keranen, K.M.; Weingarten, M.; Abers, G.A.; Bekins, B.A.; Ge, S. Sharp increase in central Oklahoma seismicity since 2008 induced by massive wastewater injection. *Science* **2014**, *345*, 448–451. [CrossRef] [PubMed]
0. Kim, W.Y. Induced seismicity associated with fluid injection into a deep well in Youngstown, Ohio. *J. Geophys. Res. Solid Earth* **2013**, *118*, 3506–3518. [CrossRef]
1. McGarr, A. Maximum magnitude earthquakes induced by fluid injection. *J. Geophys. Res. Solid Earth* **2014**, *119*, 1008–1019. [CrossRef]
2. Gaucher, E.; Schoenball, M.; Heidbach, O.; Zang, A.; Fokker, P.A.; Wees, J.D.V.; Kohl, T. Induced seismicity in geothermal reservoirs: A review of forecasting approaches. *Renew. Sustain. Energy Rev.* **2015**, *52*, 1473–1490. [CrossRef]
3. Gao, S.; Rudnick, R.L.; Yuan, H.L.; Liu, X.M.; Liu, Y.S.; Xu, W.L.; Ling, W.L.; Ayers, J.; Wang, X.C.; Wang, Q.H. Recycling lower continental crust in the North China craton. *Nature* **2004**, *432*, 892–897. [CrossRef] [PubMed]
4. Zhao, G.C.; Sun, M.; Wilde, S.A.; Li, S.Z. Late Archean to Paleoproterozoic evolution of the North China Craton: Key issues revisited. *Precambrian Res.* **2005**, *136*, 177–202. [CrossRef]
5. Ling, M.X.; Li, Y.; Ding, X.; Teng, F.Z.; Yang, X.Y.; Fan, W.M.; Xu, Y.G.; Sun, W.D. Destruction of the North China Craton Induced by Ridge Subductions. *J. Geol.* **2013**, *121*, 197–213. [CrossRef]
6. Wu, F.; Xu, Y.; Gao, S.; Zheng, J. Lithospheric thinning and destruction of the North China Craton. *Acta Petrol. Sin.* **2008**, *24*, 1145–1174, (In Chinese with English Abstract).
7. Zhu, R.; Chen, L.; Wu, F.; Liu, J. Timing, scale and mechanism of the destruction of the North China Craton. *Sci. China-Earth Sci.* **2011**, *54*, 789–797. [CrossRef]
8. Zhu, R.; Xu, Y.; Zhu, G.; Zhang, H.; Xia, Q.; Zheng, T. Destruction of the North China Craton. *Sci. China-Earth Sci.* **2012**, *55*, 1565–1587, (In Chinese with English Abstract). [CrossRef]
9. Zhu, R.X.; Xu, Y.G. The subduction of the west Pacific plate and the destruction of the North China Craton. *Sci. China-Earth Sci.* **2019**, *62*, 1340–1350. [CrossRef]
0. Yang, M.B.; Liu, G.P.; Liu, Z.; Ma, J.C.; Li, L.W.; Wang, Z.G.; Hua, P.X.; Xing, L.T.; Sun, X.R.; Han, K.Y.; et al. Geochemical characteristics of geothermal and hot spring gases in Beijing and Zhangjiakou Bohai fault zone. *Front. Earth Sci.* **2022**, *10*, 933066. [CrossRef]
1. Chen, F.; Guo, L.; Zhang, Z. Research on activity of Zhangjiakou-Bohai fault zone based on GPS observations. *Seismol. Geol.* **2020**, *42*, 95–108, (In Chinese with English Abstract).
2. Wang, G.; Zhang, W.; Lin, W.; Liu, F.; Zhu, X.; Liu, Y.; Li, J. Research on formation mode and development potential of geothermal resources in Beijing-Tianjin-Hebei region. *Geol. China* **2017**, *44*, 1074–1085, (In Chinese with English Abstract).
3. Liu, K.; Wang, S.; Sun, Y.; Cui, W.; Zhu, D. Characteristics and regionalization of geothermal resources in Beijing. *Geol. China* **2017**, *44*, 1128–1139, (In Chinese with English Abstract).
4. Luo, Z.; Zhou, X.; He, M.; Liang, J.; Li, J.; Dong, J.; Tian, J.; Yan, Y.; Li, Y.; Liu, F.; et al. Earthquakes evoked by lower crustal flow: Evidence from hot spring geochemistry in Lijiang-Xiaojinhe fault. *J. Hydrol.* **2023**, *619*, 129334. [CrossRef]
5. Zhai, Y.Z.; Wang, J.S.; Zhang, Y.; Teng, Y.G.; Zuo, R.; Huan, H. Hydrochemical and isotopic investigation of atmospheric precipitation in Beijing, China. *Sci. Total Environ.* **2013**, *456*, 202–211. [CrossRef] [PubMed]

46. Wang, X.X.; Zhou, X.R.; Li, S.H.; Zhang, N.D.; Ji, L.; Lu, H. Mechanism Study of Hydrocarbon Differential Distribution Controlled by the Activity of Growing Faults in Faulted Basins: Case Study of Paleogene in the Wang Guantun Area, Bohai Bay Basin, China. *Lithosphere* **2022**, *2021*, 7115985. [CrossRef]
47. Guo, Q.H. Hydrogeochemistry of high-temperature geothermal systems in China: A review. *Appl. Geochem.* **2012**, *27*, 1887–1898. [CrossRef]
48. Pan, S.; Kong, Y.; Wang, K.; Ren, Y.; Pang, Z.; Zhang, C.; Wen, D.; Zhang, L.; Feng, Q.; Zhu, G.; et al. Magmatic origin of geothermal fluids constrained by geochemical evidence: Implications for the heat source in the northeastern Tibetan Plateau. *J. Hydrol.* **2021**, *603*, 126985. [CrossRef]
49. Craig, H. Isotopic Variations in Meteoric Waters. *Science* **1961**, *133*, 1702–1703. [CrossRef]
50. Han, G.L.; Liu, C.Q. Water geochemistry controlled by carbonate dissolution: A study of the river waters draining karst-dominated terrain, Guizhou Province, China. *Chem. Geol.* **2004**, *204*, 1–21. [CrossRef]
51. Fouillac, C.; Michard, G. Sodium/lithium ratio in water applied to geothermometry of geothermal reservoirs. *Geothermics* **1981**, *10*, 55–70. [CrossRef]
52. Fournier, R.O. Chemical geothermometers and mixing models for geothermal systems. *Geothermics* **1977**, *5*, 41–50. [CrossRef]
53. Zhou, X.; Wang, W.; Chen, Z.; Yi, L.; Liu, L.; Xie, C.; Cui, Y.; Du, J.; Cheng, J.; Yang, L. Hot Spring Gas Geochemistry in Western Sichuan Province, China after the Wenchuan Ms 8.0 Earthquake. *Terr. Atmos. Ocean. Sci.* **2015**, *26*, 361. [CrossRef]
54. Zhou, X.C.; Yan, Y.C.; Fang, W.Y.; Wang, W.L.; Shi, H.Y.; Li, P.F. Short-Term Seismic Precursor Anomalies of Hydrogen Concentration in Luojishan Hot Spring Bubbling Gas, Eastern Tibetan Plateau. *Front. Earth Sci.* **2021**, *8*, 586279. [CrossRef]
55. Beaudry, P.; Stefansson, A.; Fiebig, J.; Rhim, J.H.; Ono, S. High temperature generation and equilibration of methane in terrestrial geothermal systems: Evidence from clumped isotopologues. *Geochim. Cosmochim. Acta* **2021**, *309*, 209–234. [CrossRef]
56. Etiope, G.; Lollar, B.S. Abiotic methane on earth. *Rev. Geophys.* **2013**, *51*, 276–299. [CrossRef]
57. Etiope, G.; Schoell, M. Abiotic Gas: Atypical, But Not Rare. *Elements* **2014**, *10*, 291–296. [CrossRef]
58. Etiope, G.; Schoell, M.; Hosgormez, H. Abiotic methane flux from the Chimaera seep and Tekirova ophiolites (Turkey): Understanding gas exhalation from low temperature serpentinization and implications for Mars. *Earth Planet. Sci. Lett.* **2011**, *310*, 96–104. [CrossRef]
59. Etiope, G.; Tsikouras, B.; Kordella, S.; Ifandi, E.; Christodoulou, D.; Papatheodorou, G. Methane flux and origin in the Othrys ophiolite hyperalkaline springs, Greece. *Chem. Geol.* **2013**, *347*, 161–174. [CrossRef]
60. Klein, F.; Grozeva, N.G.; Seewald, J.S. Abiotic methane synthesis and serpentinization in olivine-hosted fluid inclusions. *Proc. Natl. Acad. Sci. USA* **2019**, *116*, 17666–17672. [CrossRef]
61. McDermott, J.M.; Seewald, J.S.; German, C.R.; Sylva, S.P. Pathways for abiotic organic synthesis at submarine hydrothermal fields. *Proc. Natl. Acad. Sci. USA* **2015**, *112*, 7668–7672. [CrossRef] [PubMed]
62. Proskurowski, G.; Lilley, M.D.; Seewald, J.S.; Fruh-Green, G.L.; Olson, E.J.; Lupton, J.E.; Sylva, S.P.; Kelley, D.S. Abiogenic hydrocarbon production at Lost City hydrothermal field. *Science* **2008**, *319*, 604–607. [CrossRef] [PubMed]
63. Reeves, E.P.; Fiebig, J. Abiotic Synthesis of Methane and Organic-Compounds in Earth's Lithosphere. *Elements* **2020**, *16*, 25–31. [CrossRef]
64. Sciarra, A.; Saroni, A.; Etiope, G.; Coltorti, M.; Mazzarini, F.; Lott, C.; Grassa, F.; Italiano, F. Shallow submarine seep of abiotic methane from serpentinized peridotite off the Island of Elba, Italy. *Appl. Geochem.* **2019**, *100*, 1–7. [CrossRef]
65. Di Martino, R.M.R.; Camarda, M.; Gurrieri, S.; Valenza, M. Continuous monitoring of hydrogen and carbon dioxide at Mt Etna. *Chem. Geol.* **2013**, *357*, 41–51. [CrossRef]
66. Hirose, T.; Kawagucci, S.; Suzuki, K. Mechanoradical $H_2$ generation during simulated faulting: Implications for an earthquake-driven subsurface biosphere. *Geophys. Res. Lett.* **2011**, *38*, L17303. [CrossRef]
67. Kameda, J.; Saruwatari, K.; Tanaka, H. $H_2$ generation in wet grinding of granite and single-crystal powders and implications for $H_2$ concentration on active faults. *Geophys. Res. Lett.* **2003**, *30*, 2063. [CrossRef]
68. Wakita, H.; Nakamura, Y.; Kita, I.; Fujii, N.; Notsu, K. Hydrogen release: New indicator of fault activity. *Science* **1980**, *210*, 188–190. [CrossRef]
69. Zhou, X.; Du, J.; Chen, Z.; Cheng, J.; Tang, Y.; Yang, L.; Xie, C.; Cui, Y.; Liu, L.; Li, Y.; et al. Geochemistry of soil gas in the seismic fault zone produced by the Wenchuan Ms 8.0 earthquake, southwestern China. *Geochem. Trans.* **2010**, *11*, 5. [CrossRef]
70. Huang, R.; Shang, X.; Zhao, Y.; Sun, W.; Liu, X. Effect of Fluid Salinity on Reaction Rate and Molecular Hydrogen ($H_2$) Formation during Peridotite Serpentinization at 300 °C. *J. Geophys. Res. Solid Earth* **2023**, *128*, e2022JB025218. [CrossRef]
71. Skelton, A.; Andren, M.; Kristmannsdottir, H.; Stockmann, G.; Morth, C.-M.; Sveinbjoernsdottir, A.; Jonsson, S.; Sturkell, E.; Gudorunardottir, H.R.; Hjartarson, H.; et al. Changes in groundwater chemistry before two consecutive earthquakes in Iceland. *Nat. Geosci.* **2014**, *7*, 752–756. [CrossRef]

atmosphere

MDPI

*Article*

# An Assessment of the Coupled Weather Research and Forecasting Hydrological Model on Streamflow Simulations over the Source Region of the Yellow River

Yaling Chen [1,2,3], Jun Wen [1,*], Xianhong Meng [2], Qiang Zhang [1,2,3], Xiaoyue Li [1], Ge Zhang [1,4] and Run Chen [1]

1 Key Laboratory of Plateau Atmosphere and Environment, College of Atmospheric Sciences, Chengdu University of Information Technology, Chengdu 610225, China
2 Key Laboratory of Land Surface Process and Climate Change in Cold and Arid Regions, Northwest Institute of Eco-Environment and Resources, Chinese Academy of Sciences, Lanzhou 730000, China
3 University of Chinese Academy of Sciences, Beijing 100049, China
4 College of Natural Resources and Environment, Northwest A&F University, Yangling 712100, China
* Correspondence: jwen@cuit.edu.cn

**Abstract:** The Source Region of the Yellow River (SRYR), renowned as the "Water Tower of the Yellow River", serves as an important water conservation domain in the upper reaches of the Yellow River, significantly influencing water resources within the basin. Based on the Weather Research and Forecasting (WRF) Model Hydrological modeling system (WRF-Hydro), the key variables of the atmosphere–land–hydrology coupling processes over the SRYR during the 2013 rainy season are analyzed. The investigation involves a comparative analysis between the coupled WRF-Hydro and the standalone WRF simulations, focusing on the hydrological response to the atmosphere. The results reveal the WRF-Hydro model's proficiency in depicting streamflow variations over the SRYR, yielding Nash Efficiency Coefficient (NSE) values of 0.44 and 0.61 during the calibration and validation periods, respectively. Compared to the standalone WRF simulations, the coupled WRF-Hydro model demonstrates enhanced performance in soil heat flux simulations, reducing the Root Mean Square Error (RMSE) of surface soil temperature by 0.96 K and of soil moisture by 0.01 $m^3/m^3$. Furthermore, the coupled model adeptly captures the streamflow variation characteristics with an NSE of 0.33. This underscores the significant potential of the coupled WRF-Hydro model for describing atmosphere–land–hydrology coupling processes in regions characterized by cold climates and intricate topography.

**Keywords:** the Source Region of the Yellow River; WRF-Hydro; the atmosphere–land–hydrology coupling processes; streamflow

## 1. Introduction

Water, energy, and heat fluxes, along with the interactions between the atmosphere, land surface, and hydrology, constitute a complex nexus [1,2]. Within the water cycle of the whole Earth Climate System, the land surface hydrological processes serve as a crucial link, connecting atmospheric water components (e.g., precipitation, evapotranspiration, water vapor transport), terrestrial surface water (e.g., rivers, lakes, glacial meltwater, snow meltwater, surface runoff), groundwater (e.g., baseflow, subsurface runoff, soil water), and ecological water (vegetation water). This interconnected system provides feedback to weather and climate by regulating land–atmosphere energy and water cycle processes. Therefore, a nuanced comprehension of hydrological cycle processes at the interface of the atmosphere and the land surface in mesoscale river basins holds vital importance for ecological preservation and the overarching regulation of water resources [3].

The Source Region of the Yellow River (SRYR), located in the Tibetan Plateau (TP) hinterland, falls within the continental semi-arid climate zone with complex climatic conditions exhibiting a temperature rise rate of 0.48 °C/(10a)$^{-1}$ and precipitation increase of 7.6 mm/(10a)$^{-1}$ [4]. This region is highly susceptible to climate change and ecological shifts, characterized by numerous alpine lakes and wetlands, regarding it as a vital area for East Asia and global climate change [5]. Known as the "Yellow River Water Tower", it encompasses approximately 16.2% of the Yellow River Basin's total area. The streamflow, dominated by precipitation and evaporation, plays a crucial role as the primary flow-producing area and water conservation area for the middle and upper reaches of the Yellow River [6]. The historical continuity of the Yellow River civilization can be attributed, in part, to the SRYR's stable ecological environment and water supply [7]. However, contemporary challenges emerge due to global climate change, human activities, and the uneven distribution of regional water resources, leading to transformative shifts in water conservation elements such as glaciers, permafrost, and grasslands. This has resulted in a heightened frequency of extreme meteorological and hydrological events, including rainstorms, blizzards, droughts, and floods escalating the spatiotemporal distribution uncertainties of precipitation and hydrology in watersheds. Consequently, the sustainable development of the ecological environment and social economy over the SRYR encounters formidable challenges [8,9].

With the rapid development of high-resolution Earth System Models, the significance of land surface variability in simulations garners increasing attention [10,11]. Currently, studies on the coupled atmosphere–hydrology processes predominantly use regional climate models (RCMs) or combine land surface models (LSMs) with hydrological models to investigate the intricate interaction between climate change and hydrological cycle processes [12]. Key methodological tools encompass satellite remote sensing, data assimilation, error correction, and various downscaling approaches (statistical downscaling and dynamic downscaling). Prominent models such as the Weather Research and Forecasting model (WRF), Community Land Model (CLM), Community Noah Land Surface Model with Multi-Parameterization Options (Noah-MP), and Soil and Water Assessment Tool (SWAT) are deployed, emphasizing the impact of climate change and human activities on hydrological cycle processes [13,14]. Nevertheless, there are still large uncertainties regarding the RCMs in climate change simulations and projections, especially for precipitation simulations in regions characterized by complex terrain. It is found that these uncertainties are mainly associated with the physical parameterization schemes and the lateral forcing of the RCMs [15,16]. In addition, previous studies concentrate on the influence of climate change on the hydrological process, adopting a single-directional linkage of "atmospheric circulation change–regional precipitation change–land hydrological change". This approach falls short in accurately depicting the feedback between land surface and hydrological processes on regional climate, thereby affecting the simulation accuracy of hydrological processes in watersheds [17].

The Weather Research and Forecasting Model Hydrological modeling system (WRF-Hydro) stands as a high-resolution distributed land–atmosphere coupled model, developed by the National Center for Atmospheric Research (NCAR), with the primary aim of improving surface, subsurface, and river water redistribution and facilitating the coupling of atmospheric and hydrological models [18]. This model can operate as a standalone land surface hydrological model or can be coupled with an atmospheric model (such as WRF) to achieve a two-way feedback process between the atmosphere and land surface. Distinguishing from traditional land surface hydrological models, the WRF-Hydro model is explicitly designed to provide continuous spatially gridded information on soil temperature and moisture, evapotranspiration, water and heat exchange fluxes, and runoff [19,20]. Notably, the WRF-Hydro model has demonstrated success in numerous coupled atmosphere–hydrology studies [2,21]. Senatore et al. [22] found superior simulation performances for precipitation, surface runoff, and surface fluxes with the coupled WRF-Hydro model compared to the WRF model in the Crati River Basin. Gu et al. [20] thought that radar data assimilation can effectively improve flood simulations at small- and medium-sized basins based on the coupled WRF-Hydro model. Li et al. [23] concluded that the coupled WRF-Hydro model improved soil moisture and precipitation simulations and has potential in simulating and projecting streamflow over the Source Region of the Three Rivers. However, the great overestimation of precipitation by the coupled model leads to the fact that reproducing daily streamflow with the coupled model remains a challenge in complex terrain areas.

The main aim of this study is to investigate the feedback of land surface hydrological processes on precipitation in a large-sized basin with complex subsurface and climatic conditions based on a high-resolution coupling model. Meanwhile, based on the experimental data from Ngoring Lake and Maqu stations, the surface and hydrological parameters and parameterization schemes were modified. The scientific question addressed in this study is whether the atmosphere–land–hydrology coupling simulation after optimizing the parameterization schemes can effectively characterize the land–atmosphere interaction processes over the SRYR. To address this question, this research is structured as follows. The study area and data are arranged in Section 2. Section 3 introduces the methodology and experimental designs. A comparison between the standalone WRF and the coupled WRF-Hydro simulations follows. And then, the characteristics of coupled simulated streamflow are analyzed. Finally, the Discussion and Conclusion are presented to provide a summary and perspective of this study in Section 5 and Section 6, respectively.

## 2. Study Area and Data

### 2.1. Study Area

The SRYR, situated between 32.12° and 35.48° N and 95.50° and 103.28° E, lies in the northeastern part of the TP. The region encompasses a total catchment area of $1.22 \times 10^5$ km$^2$ and boasts a mean elevation of 4000.0 m, as displayed in Figure 1, which also serves as the configuration for the nested domains in the WRF and coupled WRF-Hydro. Positioned in the periphery influenced by the East Asian monsoon, the SRYR falls within the plateau cold climate zone, characterized by an annual mean temperature approaching 0.0 °C and annual mean precipitation ranging between 300.0 and 500.0 mm [24]. Recognized as the primary flow-producing area and water-conserving area in the middle and upper reaches of the Yellow River, the SRYR is referred to as the "Yellow River Water Tower", taking the Tangnaihai hydrological site as the basin outlet [25]. The region predominantly comprises high mountains, plains, and hills, featuring expansive lakes, notably the largest plateau freshwater lakes in China, Zhaling Lake and Ngoring Lake [17]. The terrain is undulating, characterized by alpine meadow grassland and alpine wetland as the primary land use types. The hydrological environment influences soil composition, primarily loam and sandy loam with a coarse texture, and a widespread distribution of seasonal frozen soil.

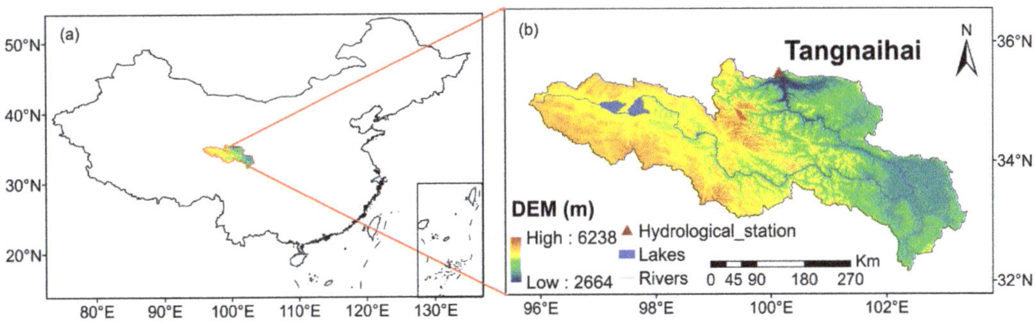


**Figure 1.** The geographic location of the Source Region of the Yellow River (**a**) and the distribution of the hydrological site (**b**).

### 2.2. Data

This study utilizes the daily streamflow data from the Tangnaihai hydrological site, generously provided by the Yellow River Water Conservancy Bureau, spanning the period of 2012–2013. Additionally, turbulent heat fluxes, top-layer soil temperature, and soil moisture data sourced from the Ngoring Lake site [26,27] and Maqu site [28,29] from the National Cryosphere Desert Data Center are incorporated. Quality inspection is carried out for the observational data, and the data with obvious errors are processed to ensure that there are not significantly incorrect data in the deployed data.

The WRF-Hydro model necessitates a substantial volume of input data, including meteorological driving data, underlying surface data, and river network data. The meteorological driving data are mainly composed of seven variables: downward longwave and shortwave radiation, surface pressure, specific humidity, air temperature, near-surface wind speed, and precipitation rate. Given the scarcity and non-uniform distribution of meteorological observation sites over the SRYR, significant challenges arise in model driving. To address this, the Global Land Data Assimilation System (GLDAS) data, jointly developed by the National Aeronautics and Space Administration (NASA) and the National Center of Environmental Prediction (NCEP), prove invaluable. With a temporal resolution of three hours and a spatial resolution of $0.25° \times 0.25°$, GLDAS integrates the ground and satellite observation data, demonstrating great applicability over the SRYR [30].

In addition, the quality of precipitation data is critical to streamflow simulation, emerging as the most sensitive factor affecting streamflow variations. Therefore, a high-quality precipitation product is of considerable significance to serve as the forcing data for the WRF-Hydro model. The China Meteorological Forcing Dataset (CMFD) is a high-spatial-temporal-resolution ($0.10° \times 0.10°$) gridded meteorological driving dataset which was developed for studying land surface processes in China [31]. The dataset integrates a variety of reanalysis, satellite remote sensing, and site observation data and is widely used in climate change and numerical simulations. The CMFD precipitation data and GLDAS non-precipitation data form the ultimate driving dataset for the model through the bilinear interpolation method [32].

The initial and boundary conditions for the WRF and coupled WRF-Hydro models are from the Final Operational Global Analysis (FNL), whose spatial and temporal resolutions are 6 h and $1.00° \times 1.00°$, respectively (https://doi.org/10.5065/D6M043C6, accessed on 23 November 2022). The model requisites, such as vegetation type, land use type, soil type, and other land surface information are sourced from the WRF Pre-processing System (WPS). The default soil type is substituted with the soil type dataset from Beijing Normal University (BNU), renowned for its heightened accuracy within China [33]. High-resolution river network data are from the United States Geological Survey (USGS) Hydrological data and maps based on SHuttle Elevation Derivatives at multiple Scales (HydroSHEDS), with a resolution of 90.0 m chosen to extract accurate river network information. The details of the aforementioned data are shown in Table 1.

**Table 1.** An overview of the research data.

| Category | Data Type | Temporal and Spatial Resolutions | Variables |
|---|---|---|---|
| Climate | CMFD | 3 h; $0.10° \times 0.10°$ | Precipitation |
| | GLDAS | 3 h; $0.25° \times 0.25°$ | Temperature, wind speed, solar radiation, downward longwave radiation, pressure, specific humidity |
| | FNL | 6 h; $1.00° \times 1.00°$ | Initial and boundary conditions |
| Hydrology | Site | 1 d | Streamflow |
| Eddy covariance | Site | 30 min | Water/heat flux, soil temperature and moisture |
| Topography | HydroSHEDS | $90.0\ m \times 90.0\ m$ | Digital Elevation Model |

## 3. Methodology

### 3.1. The Numerical Model

#### 3.1.1. WRF Model

The WRF is a sophisticated non-hydrostatic mesoscale numerical weather prediction model, renowned for its capability to accommodate various mesoscale and small-scale atmospheric and hydrological processes in numerical simulation research [34]. There are numerous physical parameterization schemes for the WRF model, and the applicability of schemes is different due to the variations in the weather, climate, and underlying surface conditions. The microphysical process scheme and cumulus scheme have a particularly significant impact on precipitation simulations. In addition, the quality of different reanalysis data also directly affects the effectiveness of precipitation simulation [21,22]. By testing different microphysical process and cumulus parameterization schemes using four different types of reanalysis data, it was found that the FNL data and the Thompson and Grell–Devenyi (G-D) schemes have the best applicability over the Source Region of the Yellow River (not shown in the manuscript). The Advanced Research WRF (ARW) model (version 4.1.2) for both the standalone WRF and the coupled WRF-Hydro model was employed in this research.

#### 3.1.2. Noah-MP Model

The Noah-MP model, derived from the Noah Land Surface Model (LSM) with substantial enhancements, is a notable advancement [35,36]. It provides multiple parameterization options for the key biogeophysical processes, featuring a distinct vegetation canopy, a two-stream radiation transfer approach, and a short-term dynamic vegetation scheme. Additionally, updates to the frozen soil scheme within the groundwater model and the snow model significantly influence streamflow simulation [37]. In this study, the Noah-MP model was selected as the land surface process module of the WRF and the WRF-Hydro model.

#### 3.1.3. WRF-Hydro Model

The WRF-Hydro model, developed as a hydrological extension package for WRF, is a new generation of a distributed hydrometeorological forecasting system with a physical basis and multi-scale and multi-parameter schemes. Serving as a linkage between the large-scale regional climate model and the refined hydrological model, it employs the LSM (Noah/Noah-MP) as a bridge. The model enhances the land surface hydrological process, focusing on the spatial redistribution of land surface water, groundwater, and river water. It demonstrates proficiency in quantitatively studying the water–heat exchange process between the atmosphere and land surface [18]. Comprising five modules, namely, surface overland flow, saturated subsurface flow, channel, reservoir routing, and the conceptual baseflow module, the WRF-Hydro model computes quasi-3D subsurface flow, accounting for both vertical and horizontal water exchange. This research utilized WRF-Hydro system version 5.1.1, and a comprehensive model description is available in [18].

### 3.2. Experimental Designs

#### 3.2.1. The Parameterization Schemes in the WRF and Coupled WRF-Hydro Model

The Lambert Projection, featuring a central longitude and latitude of 99.50° E and 33.75° N, is applied in the model, incorporating two-way nested domains with horizontal resolutions of 25 km and 5 km, respectively. The vertical structure encompasses 40 levels, reaching a 50 hPa pressure top, utilizing a time step of 100 s in the outer domain. Continuous runs are initialized with lateral atmospheric boundary conditions provided by the Final Operational Global Analysis (FNL) data from the National Centers for Environmental Prediction (NCEP), as outlined in Table 1. The physics parameterization schemes for the selected WRF domains are listed in Table 2, with the cumulus parameterization exclusively employed in the outer domain [22]. Notably, routing processes at a resolution of 500.0 m are executed solely on the innermost domain within the coupled WRF-Hydro model. The simulation spans from 1 March 2013 to 1 September 2013 UTC, with the initial two months allocated as spin-up time and the remainder for analysis.

**Table 2.** Physical options of WRF and coupled WRF-Hydro model.

| Physics Process | Parameterization | Reference |
|---|---|---|
| Microphysics | Thompson | [38] |
| Cumulus parameterization | G-D | [39] |
| Planetary boundary layer | MYNN2 | [40] |
| Land surface | Noah-MP | [34] |
| Longwave radiation | RRTMG | [41] |
| Shortwave radiation | RRTMG | [41] |

#### 3.2.2. The Calibration of Sensitivity Parameters in the Offline WRF-Hydro Model

Before analyzing the effects of the land–hydrological processes on the atmosphere simulation, the WRF-Hydro model is run in an offline/uncoupled way to calibrate relevant sensitivity parameters and evaluate its efficacy in simulating streamflow. Hydrological model parameters serve as the reflections of the underlying surface characteristics, and variations in default parameters' applicability across different basins are noteworthy. In terms of the WRF-Hydro model, prior studies have categorized the sensitivity parameters governing streamflow processes into those controlling streamflow distribution and water volume and those regulating flood peaks and flood hydrographs [20]. A stepwise manual approach is adopted in calibrating the sensitivity parameters, following previous WRF-Hydro studies [42]. Given the steep slope of the SRYR, distinct from that of the Daihe River Basin, the surface retention depth (RETDEPRTFAC) is set as 0.0, and only four parameters, including the saturated soil moisture (SMCMAX), runoff infiltration parameter (REFKDT), channel Manning roughness (MannN), and overland flow roughness parameter (OVROUGHRTFAC), were calibrated within reasonable ranges to select the optimal parameter value. The calibration process considered the daily streamflow variations at the Tangnaihai hydrological station during the rainy season from 1 June to 30 September in 2012 [43].

Additionally, the water–heat exchange process is of vital importance to the understanding of the atmosphere–land–hydrology interaction process, influencing the land–surface water cycle process by modulating the evapotranspiration process. Relevant studies indicate that the default parameterization schemes of the Noah-MP model exhibit an underestimation of latent heat (LE) and an overestimation of sensible heat (H) in the alpine grassland area [44]. To rectify the issue of H overestimation, the Chen97 scheme for a sensible heat transfer coefficient is employed, while the Jarvis canopy stomatal resistance scheme enhances vegetation transpiration, improving simulated LE and achieving a more balanced distribution of heat flux between LE and H. Other parameterization scheme options applied in this study remain consistent with the default settings in the uncoupled WRF-Hydro model.

#### 3.2.3. The Interpolation Method

In this study, a bilinear interpolation method was used to harmonize the data resolution to $0.1^{\circ} \times 0.1^{\circ}$ to analyze the model's performances for depicting the distributions of the meteorological and hydrological elements over the Source Region of the Yellow River. The bilinear interpolation is a method used to estimate the value of a function at a point, given its values at the surrounding points. The detailed calculation formula can be found in [33].

#### 3.2.4. Evaluation Index

To evaluate the model's simulation performance, various metrics including the Nash Efficiency Coefficient (NSE), Root Mean Square Error (RMSE), Correlation Coefficient

(R), and Relative Deviation (BIAS) are employed in this research. The NSE is used to evaluate the performance of streamflow, which can penalize large errors in the simulations and give a better measure of how well the model simulations match the observational data. The NSE ranges from −1 to 1, where a value close to 1 indicates that the experiment has a better performance. The RMSE is a commonly used metric to evaluate the accuracy of the model by quantifying the differences between the observed and simulated values. A lower RMSE indicates a better performance, and it is a widely accepted metric for regression calculation. As for R, it is a measure of the linear correlation between two data sets. It ranges from −1 to 1, where 1 indicates a perfect positive linear relationship, −1 indicates a perfect negative linear relationship, and 0 indicates no linear relationship. The BIAS can be useful for understanding the relative performance of different experiments, where the magnitude of the simulation errors can vary greatly depending on the scale of the observational data. The calculation formula and optimal value of each evaluation index are shown in our previous study [45].

#### 3.2.5. The Applicability of the WRF-Hydro Model

After a 2-month spin-up period, the uncoupled WRF-Hydro was calibrated from 1 June to 30 September in 2012, based on daily streamflow from the Tangnaihai hydrological site. As depicted in Figure 2, the simulated streamflow closely aligns with the observation, exhibiting consistency between flood and precipitation hydrographs. During the calibration period, the simulated and measured streamflow achieves an R of 0.84, NSE of 0.44, RMSE of 465.61 $m^3 \cdot s^{-1}$, and BIAS of −11.44%. In the validation period, R is 0.81, the NSE is 0.61, RMSE is 351.36 $m^3 \cdot s^{-1}$, and BIAS is −10.21%. However, the model tends to underestimate peak flows during flood season and presents some unrealistic peak flows, indicating potential issues with the base flow model and precipitation [22]. Nevertheless, the WRF-Hydro model demonstrates the ability to produce a relatively realistic hydrological regime over the SRYR; hence, the calibrated parameters are used for comparing WRF and coupled WRF-Hydro simulations.

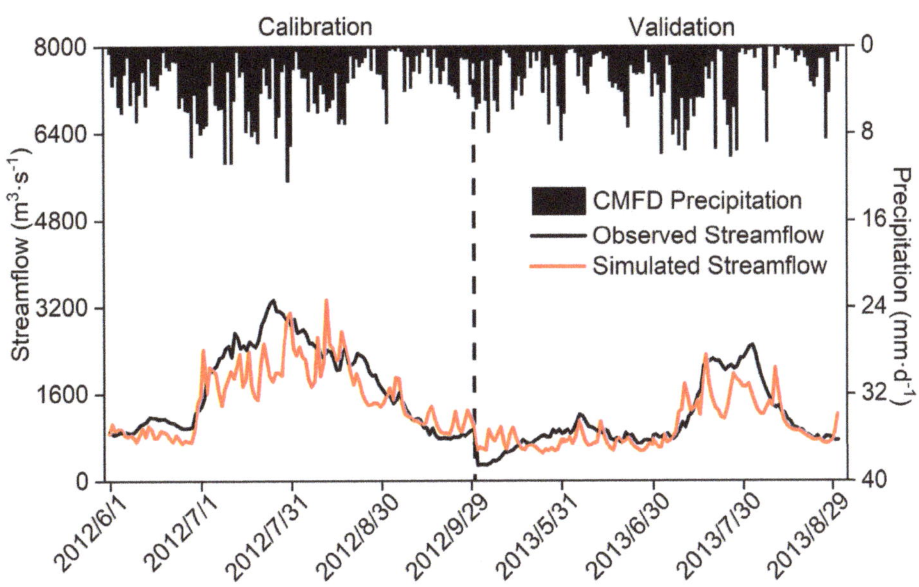


**Figure 2.** The variation in simulated and observed daily streamflow (units: $m^3 \cdot s^{-1}$) over the Source Region of the Yellow River during the calibration and validation periods. The black dotted line in correspondence with 30 September 2012 splits the calibration and validation periods.

## 4. Results

The land surface hydrological cycle constitutes a crucial Earth System process. Climate change exerts influence on the global water cycle, eliciting diverse responses at the regional scale. Simultaneously, variations in the land surface process contribute to the regional and catchment-scale modification of water resource distribution and runoff [4]. Based on the coupled WRF-Hydro model, the impact of climate change on land surface hydrological processes and the feedback of the surface water cycle to precipitation are comprehensively considered, and the characteristics of the coupled atmosphere–land–hydrology process and streamflow variations over the SRYR during the rainy season of 2013 are also explored.

### *4.1. The Temporal Variation in Hydrometeorological Elements*

Given the sparse and uneven distribution of monitoring sites over the SRYR, reanalysis products are employed in this study to examine the variation in the hydrometeorological elements. Previous studies [44,45] have demonstrated the suitability of CMFD precipitation data over the SRYR, while GLDAS data, encompassing temperature and runoff, are recognized for effectively characterizing climate change and water cycle processes over the SRYR. Therefore, CMFD and GLDAS are utilized as reference datasets (denoted as Reference) for analyzing the regional mean simulation results in the following study.

Figure 3 shows the time series of regional mean meteorological and hydrological elements. It exhibits that both the WRF and coupled WRF-Hydro models can effectively capture the evolution characteristics of these elements over the SRYR. Compared with the standalone WRF model, the coupling process results in a slight increase in the wet deviation of precipitation, with an average RMSE of 2.51 mm. However, it enhances the simulation of temperature and downward longwave and shortwave radiation to a certain extent. The simulations of surface pressure, specific humidity of 2 m, and wind speed near the ground exhibit minimal differences between the two experiments. Notably, the coupling process elevates soil moisture levels by accounting for the terrestrial vertical and lateral flow of soil water in three-dimensional space, consequently leading to larger simulated values for evapotranspiration and precipitation. Table 3 depicts the evaluation indices of diverse meteorological and hydrological variables simulated by WRF and WRF-Hydro models over the SRYR. Both models, especially WRF-Hydro, demonstrate good overall performance in simulating various hydrometeorological elements, with high correlation coefficients and generally low RMSE values.

**Table 3.** The evaluation indices for hydrometeorological elements' simulations.

| Variables | Model | R | RMSE |
|---|---|---|---|
| Precipitation | WRF | 0.80 | 2.50 |
| | WRF-Hydro | 0.81 | 2.51 |
| Temperature | WRF | 0.96 | 1.45 |
| | WRF-Hydro | 0.96 | 1.2 |
| Downward longwave radiation | WRF | 0.90 | 20.43 |
| | WRF-Hydro | 0.90 | 20.01 |
| Downward shortwave radiation | WRF | 0.76 | 63.14 |
| | WRF-Hydro | 0.77 | 61.27 |
| Surface pressure | WRF | 0.98 | 1.04 |
| | WRF-Hydro | 0.98 | 1.08 |
| Specific humidity | WRF | 0.93 | 0.002 |
| | WRF-Hydro | 0.91 | 0.002 |
| Wind speed | WRF | 0.72 | 0.01 |
| | WRF-Hydro | 0.75 | 0.03 |

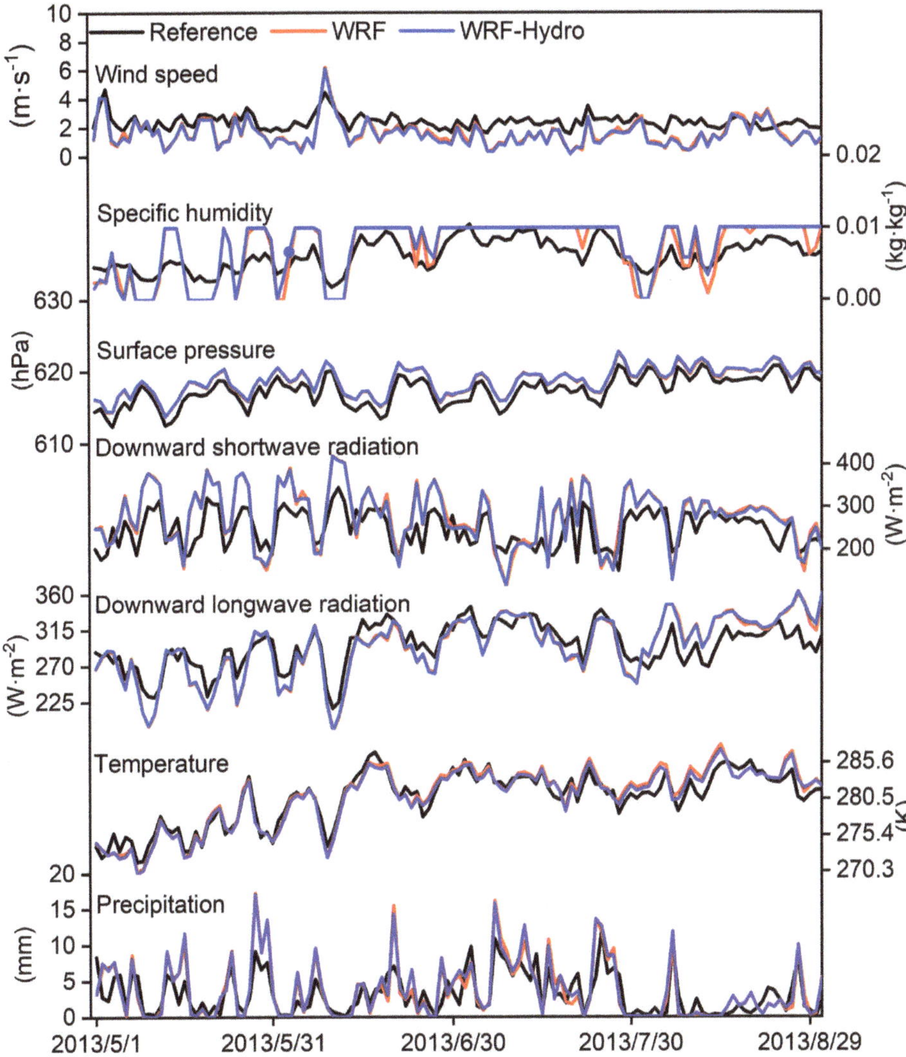


**Figure 3.** The time series of the mean meteorological and hydrological elements simulated by the WRF and coupled WRF-Hydro models from 1 May 2013 to 31 August 2013.

The land–atmosphere water and heat exchange processes exert influence on surface soil moisture through alterations in evapotranspiration, consequently impacting the surface energy and hydrological cycle [46]. To ensure data comparability, the typical sunny/cloudy days at Ngoring Lake (lakeside underlying surface) and Maqu (grass underlying surface) sites are selected based on characteristic downward total radiation patterns throughout the entire simulation period. The screening methods for typical sunny/cloudy days are consistent with the work of Zhang et al. [47].

Figure 4 illustrates the diurnal variation characteristics of the LE and H simulated by both the WRF and coupled WRF-Hydro models on typical sunny days, where the land–atmosphere water and heat exchange processes are notably active. The results at the Ngoring Lake site show that the coupled WRF-Hydro model produces LE and H values more consistent with measurements, resulting in reduced RMSE values of 29.91 W·m$^{-2}$ and

31.06 W·m$^{-2}$ compared to WRF simulations. Conversely, at the Maqu site, the coupled process amplifies the deviation in LE simulations, attributed to the overestimation of evapotranspiration, while effectively mitigating the issue of overestimated H.

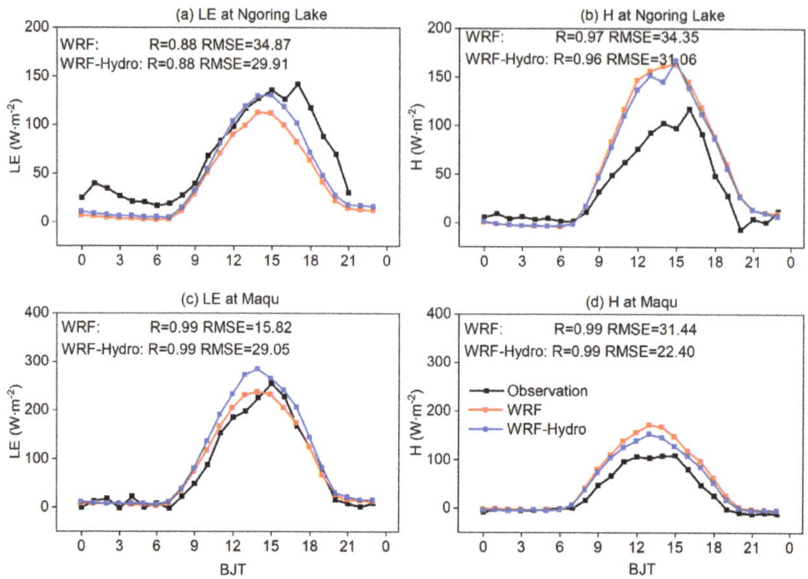


**Figure 4.** Comparison of turbulent fluxes (units: W·m$^{-2}$) between WRF-Hydro/WRF and observations on typical sunny days at Ngoring Lake (**a**,**b**) and Maqu (**c**,**d**) sites.

Figure 5 demonstrates the diurnal variation characteristics of the LE and H on typical cloudy days, characterized by more complicated weather conditions and physics processes. During these days, the agreement between turbulent flux simulation results and observations is less favorable compared to typical sunny days. The simulations at both sites indicate that the coupled process can reduce the RMSE in LE and H, with a mean RMSE of 29.48 W·m$^{-2}$ and 26.55 W·m$^{-2}$, respectively. Overall, the coupled WRF-Hydro simulations exhibit enhancements in simulating surface heat flux variables, attributed to the incorporation of lateral terrestrial water flow in the hydrological process.

Furthermore, soil temperature and moisture play an important role in influencing land surface evaporation and groundwater processes, directly or indirectly affecting the land–hydrology process. The Taylor diagram [48] provides a visual assessment of model performance by comparing the bias and correlation of model simulations to observations. It highlights which models perform well in simulating the climate variable and which ones may require improvement or further calibration. Therefore, the analysis of top-layer soil temperature and moisture at the Maqu site is conducted by Taylor diagrams in Figure 6, leveraging available observed data. The results display a commendable agreement between the simulated and observed soil temperature, albeit with a notable discrepancy in August, attributed to deviations in downward shortwave radiation and temperature. The coupling process reduces the RMSE of soil temperature simulation (from 5.18 to 4.22 K), indicating improved accuracy through the comprehensive consideration of soil water content variation. In addition, the WRF-Hydro model exhibits a prolonged soil moisture memory compared to standalone WRF, resulting in significantly higher simulated soil moisture values in WRF-Hydro, attributable to the inclusion of subsurface lateral flow in the WRF-Hydro model. Nevertheless, the simulated top-layer soil moisture in both experiments does not align with observations and struggles to capture the response of soil moisture to precipitation.

Furthermore, the uncertainty introduced by the site-specific results necessitates verification through spatial distribution analysis.

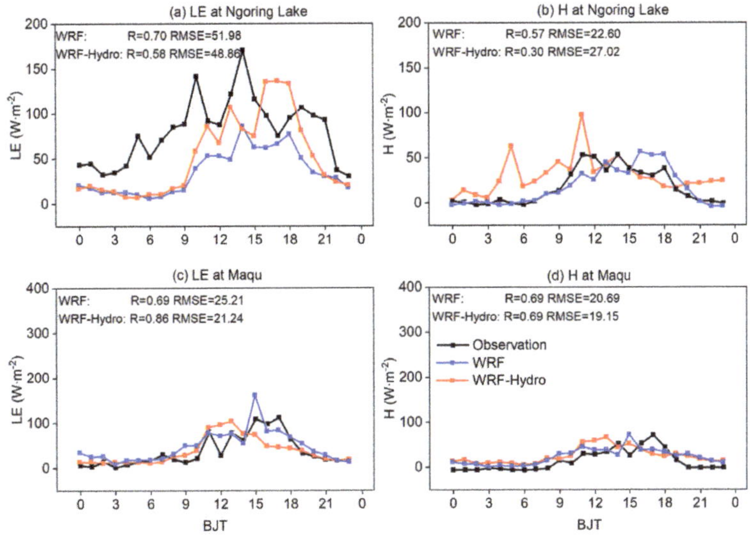


**Figure 5.** As in Figure 4, but for the typical cloudy days.

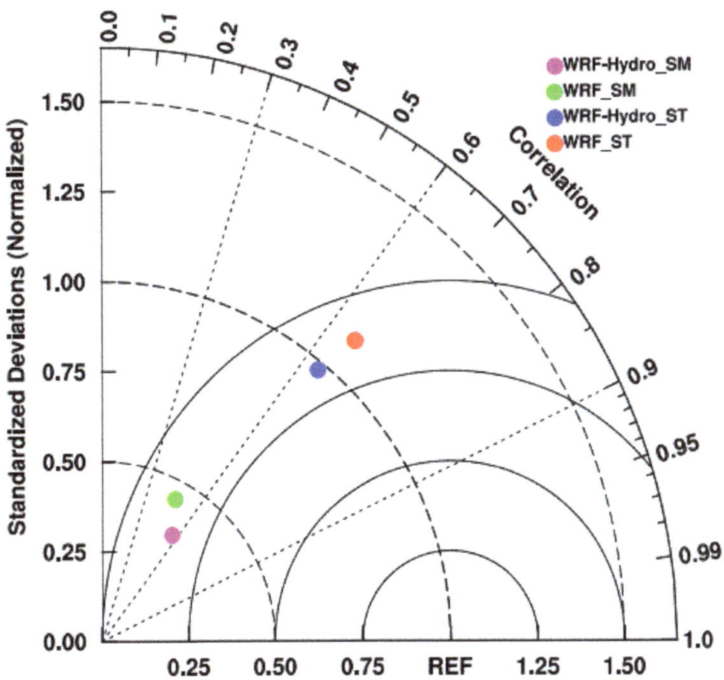


**Figure 6.** Taylor diagrams of correlation coefficients and standard deviations for daily soil temperature and soil moisture at Maqu site among simulations with observations from 1 May 2013 to 1 September 2013.

*4.2. The Spatial Distribution of Hydrometeorological Elements*

Moreover, the spatial distribution of the accumulated precipitation for the CMFD reference data (denoted as Reference), WRF, and coupled WRF-Hydro, along with their discrepancies, is displayed in Figure 7. The precipitation patterns underscore a pronounced reliance on topography, showing a decreasing trend from southeast to northwest over the SRYR. Both WRF and coupled WRF-Hydro demonstrate enhanced capabilities in capturing precipitation distribution characteristics. However, both simulations exhibit a considerable wet bias, particularly in the Jiuzhi and Maqu areas, coupled with a dry bias in the southeastern SRYR. The coupled WRF-Hydro, in contrast to the standalone WRF, incorporates subsurface lateral flow considerations, resulting in increased soil moisture and a more rational spatial distribution of soil water. This, in turn, engenders a feedback effect on precipitation, contributing to a mean wet bias of 16.63 mm compared to WRF simulations.

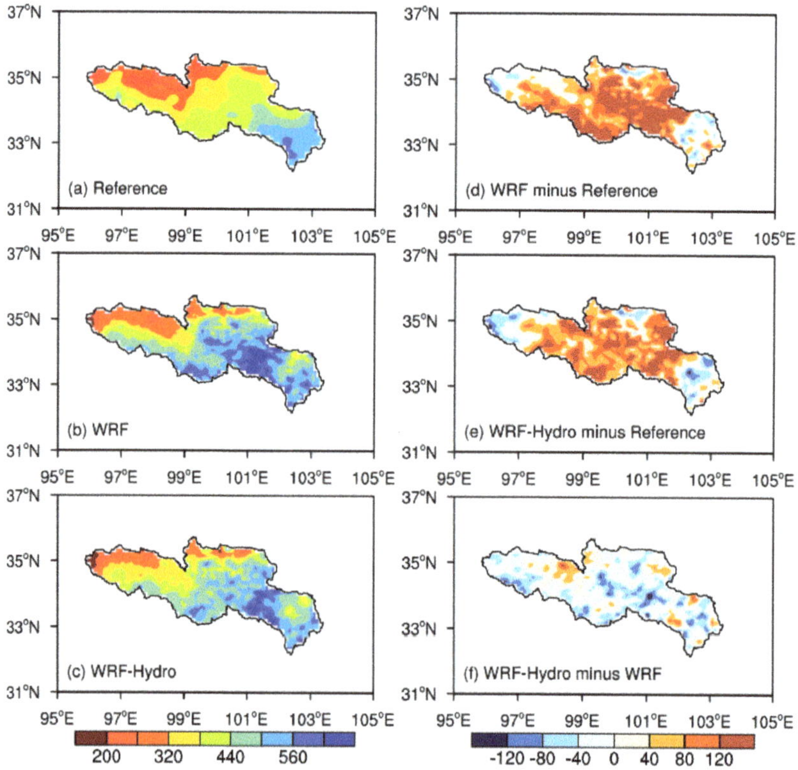


**Figure 7.** The spatial distribution of the accumulated precipitation (units: mm) in the time interval 1 May 2013 to 31 August 2013 with (**a**) observation, (**b**) the standalone WRF simulation, (**c**) the coupled WRF-Hydro simulation, and difference maps for (**d**) WRF minus observation, (**e**) WRF-Hydro minus observation, and (**f**) WRF-Hydro minus WRF.

Concerning temperature, the spatial patterns of the mean 2 m air temperature during the simulated period are shown in Figure 8. The temperature spatial distribution presents gradient characteristics, with elevated temperatures observed in flat regions and lower temperatures in alpine areas. Both experiments adeptly capture the temperature distribution characteristics. On the whole, the simulated temperature tends to be relatively higher, particularly in the northeast area of the SRYR. The regional mean bias is 0.44 K for the standalone WRF and 0.15 K for the coupled WRF-Hydro, indicating a slight reduction in the mean temperature deviation in the coupled simulation.

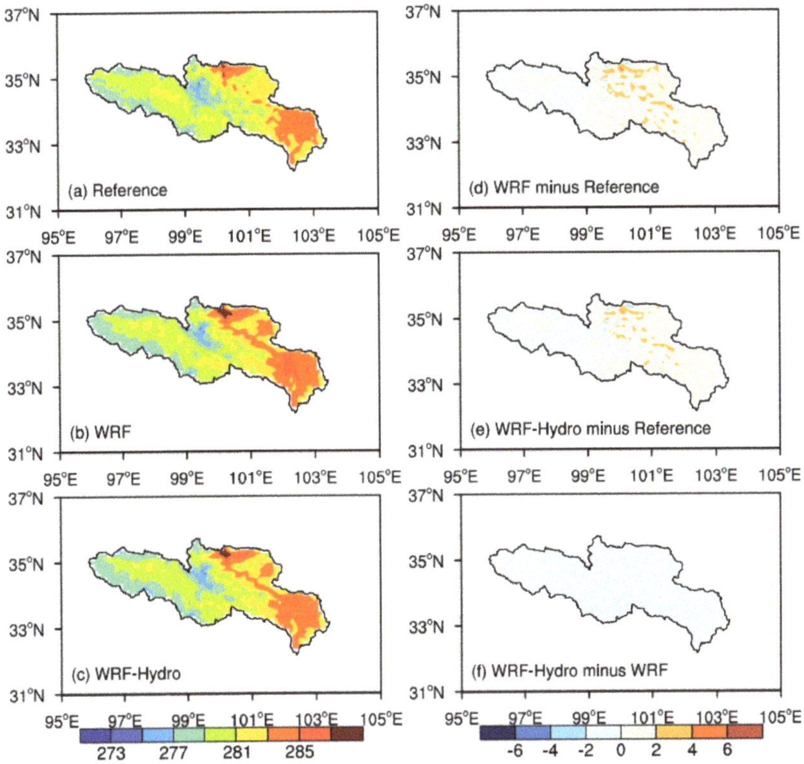


**Figure 8.** As in Figure 7, but for the mean temperature (units: K).

Soil temperature serves as an important parameter in the land surface process, providing a direct reflection of the thermal state of the land. The fluctuations in soil temperature have a consequential impact on the movement and phase transitions of surface soil water, consequently influencing the surface hydrological cycle [49]. Simultaneously, soil moisture plays a vital role in land–atmosphere interactions. Functioning as a reservoir for heat and moisture, soil moisture exhibits remarkable memory, retaining information from weeks to months, and subsequently influencing atmospheric conditions through the remembrance of preceding atmospheric perturbations.

The spatial distribution characteristics of top-layer soil temperature (Figure 9a–c) and moisture (Figure 9d–f) during the 2013 rainy season over the SRYR are analyzed in Figure 9. Given the high altitude and substantial diurnal temperature fluctuations over the TP, the mean temperature is lower than that of the inland areas, with lower soil temperature closer to the plateau's interior. Both simulations adeptly illustrate the characteristic of lower temperatures in the lake area compared to the surroundings. The coupled WRF-Hydro reduces the simulated surface soil temperature, exhibiting a cold deviation of 1.07 K, with potential implications for atmospheric water vapor convergence via land–atmosphere interactions. Regarding soil moisture, both experiments show wet centers in the Zhaling Lake and Ngoring Lake areas. The coupled WRF-Hydro model, influenced by terrestrial lateral water and soil moisture redistribution processes, demonstrated a more reasonable spatial distribution of soil water content over the study region. The simulations of soil moisture from WRF-Hydro significantly exceed those of WRF, presenting a wet deviation of 0.02 $m^3/m^3$. On the whole, the areas surrounding the two lakes in the SRYR function as cold and wet centers during the simulation period, and the coupled simulations aptly capture the variation characteristics of soil temperature and moisture.

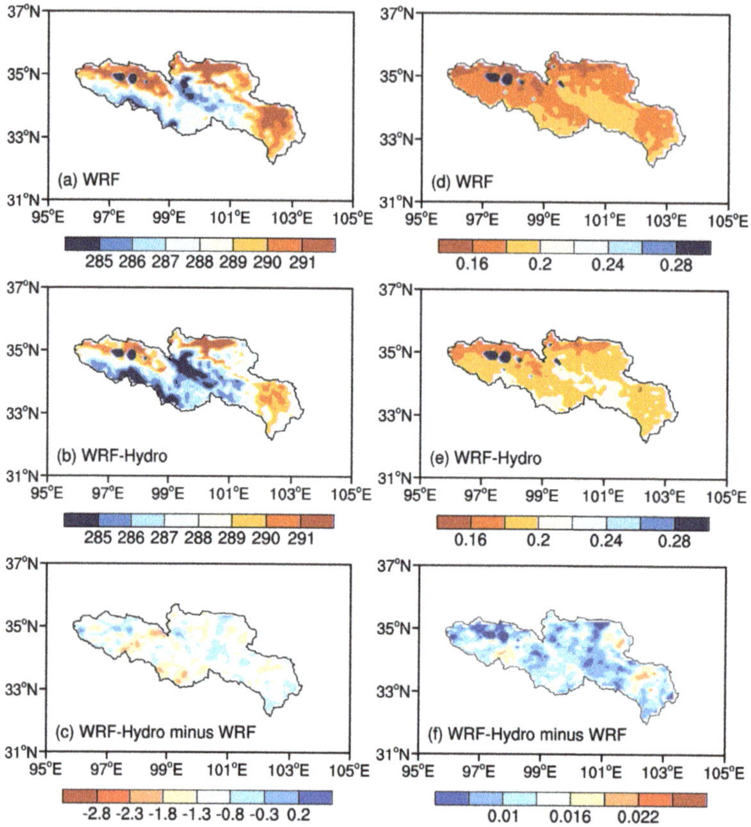


**Figure 9.** The spatial distribution of the mean top-layer (0–10 cm) soil temperature (units: K) in the time interval 1 May 2013 to 31 August 2013 with (**a**) the standalone WRF simulation, (**b**) the coupled WRF-Hydro simulation, and the difference map for (**c**) WRF-Hydro minus WRF; (**d**–**f**) are the same as (**a**–**c**), but for top-layer soil moisture (units: $m^3/m^3$).

### *4.3. The Time Series of the Streamflow Simulated by the Coupled Model*

The time series depicting the coupled simulated streamflow for the 2013 rainy season, where direct meteorological site observations are not required, is presented in Figure 10. The coupled model adeptly captures the temporal variation in observed hydrographs, exhibiting an R of 0.77. However, reproducing daily streamflow with the coupled model poses a challenge, yielding an NES of 0.33 and RMSE of 458.85 $m^3 \cdot s^{-1}$. The substantial overestimation of coupled simulated streamflow, particularly in peak flow reproduction, is a notable limitation. This performance degradation primarily stems from the WRF-Hydro model's extreme sensitivity to precipitation data quality, where the RMSE of precipitation is merely 2.51 mm, contrasting with the considerably higher RMSE of streamflow. Furthermore, this discrepancy may arise from distinct frequencies of Noah-MP invocation in uncoupled calibration and coupled runs. In uncoupled simulations, Noah-MP is typically invoked at the physical time step of the hydrological model, while in coupled simulations, it is invoked at the physical time step of the WRF model. This leads to more water traversing down-slope or entering the channel before infiltration occurs again, contributing to elevated streamflow [22].

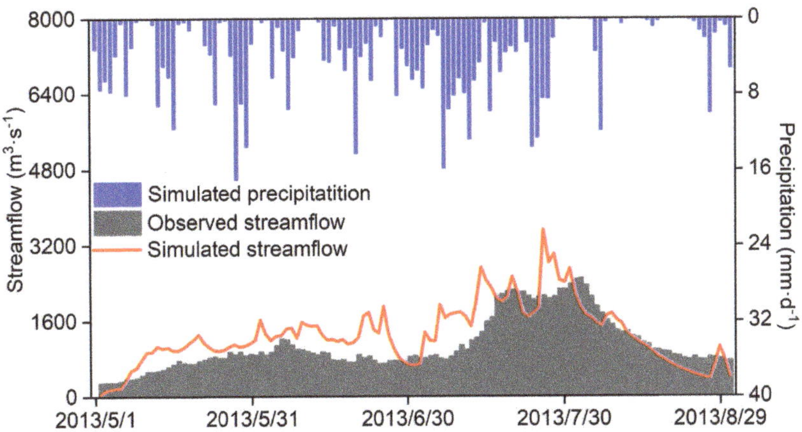


**Figure 10.** The observed and coupled WRF-Hydro simulated streamflow (units: $m^3 \cdot s^{-1}$) for the period 1 May 2013 to 31 August 2013.

## 5. Discussion

Based on the standalone WRF and coupled WRF-Hydro models, a high-resolution atmosphere–land–hydrology coupling model was constructed with various data over the SRYR. In this study, the key variables associated with atmosphere–land–hydrology coupling processes were compared and analyzed to evaluate the streamflow simulation capability of the coupled model. The results reveal that the WRF-Hydro model successfully reproduces the daily streamflow, yet refinement is warranted for a more nuanced depiction of the hydrograph, particularly regarding underestimation and steep changes during flood peaks. This limitation underscores the need for an in-depth exploration of groundwater and soil water content dynamics to explain their influences on streamflow. Future research endeavors should encompass a comprehensive analysis of land surface water cycle processes to enhance our understanding.

In addition, a significant overestimation of streamflow is observed in the coupled model. To explore the factors influencing this performance degradation, a sensitivity analysis is conducted by using different combinations of atmospheric forcing data and the uncoupled WRF-Hydro model (not shown in the manuscript). The results show that the coupled simulated precipitation data introduces slight deviations cumulatively leading to a substantial error in streamflow simulation (RMSE of 2.51 mm for precipitation and 458.85 $m^3 \cdot s^{-1}$ for streamflow). While using the coupled simulated non-precipitation data and CMFD precipitation, the simulated streamflow is close to the observations, which suggests that accurate precipitation is recommended as the forcing data for streamflow simulations over the SRYR. Presently, reproducing daily streamflow over the Tibetan Plateau with the coupled model remains a challenge [25,50]. Studies have indicated that advanced data assimilation methods have the potential to enhance precipitation forecasting accuracy, thus fostering advancements in atmosphere–land–hydrology simulations [20]. Incorporating satellite and radar data into the WRF model in future studies offers a potential avenue for improving precipitation and streamflow simulation and projection.

The inaccurate estimation of initial soil moisture is also a limitation for streamflow simulation. Studies found that streamflow is sensitive to early rainfall, which may influence the reconstruction of later flood peaks [20]. The observational sites over the SRYR were usually established on flat terrain with moist soils and rivers passing through, where the lateral flow of soil water flows in, not out. As a result, improvements in initial soil water content accuracy help capture streamflow over this region.

In future studies, the consideration of improving the accuracy of initial soil moisture and precipitation data will not only help the WRF-Hydro coupled model to simulate streamflow but also serve as an effective tool in interdisciplinary flood simulation studies.

## 6. Conclusions

This study conducts a comparative analysis of key variables associated with the coupled atmosphere–land–hydrology processes over the SRYR during the 2013 rainy season. The primary focus is on investigating the impacts of climate change on land surface and water cycle processes, as well as the feedback of the land surface hydrological cycle to precipitation. The following main conclusions have been drawn:

(1) The uncoupled WRF-Hydro model effectively characterizes the variability of streamflow over the SRYR basin, demonstrating an R of 0.84 and an NSE of 0.44 during the calibration period and an R of 0.81 and an NSE of 0.61 during the validation period.
(2) Both the standalone WRF and coupled WRF-Hydro models indicate reasonable performance in reproducing variables associated with atmosphere–land–hydrology processes over the SRYR. The consideration of soil water lateral flow in the coupling process significantly reduces biases in water–heat exchange fluxes, soil temperature, and soil moisture simulations, with mean RMSE values of 32.27 $W \cdot m^{-2}$, 24.91 $W \cdot m^{-2}$, 4.22 K, and 0.06 $m^3/m^3$, respectively.
(3) The coupled model success captures the streamflow variation. Nevertheless, reproducing daily streamflow with the coupled model remains a challenge, yielding an NSE of 0.33 and an RMSE of 458.85 $m^3 \cdot s^{-1}$.

The main findings in this work indicate that the WRF-Hydro model has the potential to reproduce the interaction processes between the land surface and atmosphere in the complex terrain areas. For a future perspective, accurate initial soil moisture and precipitation estimation by incorporating advanced data assimilation methods is supposed to improve streamflow simulations and interdisciplinary flood simulation studies.

**Author Contributions:** Conceptualization, Y.C. and J.W.; methodology, Y.C., J.W., and Q.Z.; software, J.W.; validation, Y.C. and X.M.; formal analysis, Y.C.; data curation, X.M.; writing—original draft preparation, Y.C.; visualization, Y.C., Q.Z., X.L., G.Z., and R.C.; supervision, J.W.; funding acquisition, J.W. All authors have read and agreed to the published version of the manuscript.

**Funding:** This research has been jointly supported by the National Natural Science Foundation of China (Grant 42375032) and the Scientific Research Project of Chengdu University of Information Technology (Grant KYTZ201821).

**Institutional Review Board Statement:** Not applicable.

**Informed Consent Statement:** Not applicable.

**Data Availability Statement:** The daily streamflow data of Tangnaihai hydrological station can be downloaded at http://www.yrcc.gov.cn , and the turbulent heat fluxes and soil temperature and moisture data of the Ngoring Lake station from the National Cryosphere Desert Data Center (http://www.ncdc.ac.cn). The CMFD data are provided by the National Tibetan Plateau Data Center (https://doi.org/10.11888/AtmosphericPhysics.tpe.249369.fle). The GLDAS data are from https://ldas.gsfc.nasa.gov/gldas/.

**Acknowledgments:** The authors would like to thank the National Cryosphere Desert Data Center and National Meteorological Information Center for providing the precious observational data. Moreover, we acknowledge the developers of the open-source models used for this research.

**Conflicts of Interest:** The authors declare no conflicts of interest.

## References

1. Arnault, J.; Wagner, S.; Rummler, T.; Fersch, B.; Bliefernicht, J.; Andresen, S.; Kunstmann, H. Role of Runoff-Infiltration Partitioning and Resolved Overland Flow on Land-Atmosphere Feedbacks: A Case Study with the WRF-Hydro Coupled Modeling System for West Africa. *J. Hydrometeorol.* **2016**, *17*, 1489–1516. [CrossRef]

Fersch, B.; Senatore, A.; Adler, B.; Arnault, J.; Mauder, M.; Schneider, K.; Völksch, I.; Kunstmann, H. High-resolution fully-coupled atmospheric-hydrological modeling: A cross-compartment regional water and energy cycle evaluation . *Hydrol. Earth Syst. Sci.* **2020**, *24*, 2457–2481. [CrossRef]

Milly, P.C.D.; Dunne, K.A.; Vecchia, A.V. Global pattern of trends in streamflow and water availability in a changing climate. *Nature* **2005**, *438*, 347–350. [CrossRef] [PubMed]

Meng, X.H.; Chen, H.; Li, Z.G.; Zhao, L.; Zhou, B.R.; Lvy, S.H.; Deng, M.S.; Liu, Y.M.; Li, G.W. Review of Climate Change and Its Environmental Influence on the Three-River Regions. *Plateau Meteor.* **2020**, *39*, 1113–1143.

Wu, G.X.; Mao, J.Y.; Duan, A.M.; Zhang, Q. Recent progressing in the study on the impacts of Tibetan Plateau on Asian summer climate. *Acta Meteorol. Sin.* **2004**, *62*, 528–549.

Tian, D.X.; Tian, S.M.; Jiang, S.Q.; Dong, X.N.; Li, Z.W.; Zhang, L. Research process of the evolition of runoff in the Source Area of the Yellow River. *Yellow River* **2020**, *42*, 90–95.

Zhang, A.; Li, T.J.; Fu, W.; Wang, Y.T. Model simulation of flood season runoff in the headwaters of the Yellow River Basin using satellite-ground merged precipitation data. *J. Abbr.* **2017**, *25*, 1–16.

Milly, P.C.D.; Wetherald, R.T.; Dunne, K.A.; Delworth, T.L. Increasing risk of great floods in a changing climate. *Nature* **2002**, *415*, 514–517. [CrossRef] [PubMed]

Ji, P.; Yuan, X. High-resolution land surface modeling of hydrological changes over the Sanjiangyuan Region in the eastern Tibetan Plateau: 2. Impact of climate and land cover change. *J. Adv. Model. Earth Syst.* **2018**, *10*, 2829–2843. [CrossRef]

0. Clark, M.P.; Fan, Y.; Lawrence, D.M.; Adam, J.C.; Bolster, D.; Gochis, D.J.; Hooper, R.P.; Kumar, M.; Leung, L.R.; Mackay, D.S.; et al. Improving the representation of hydrologic processes in Earth System Models. *Water Resour. Res.* **2015**, *51*, 5929–5956. [CrossRef]

1. Tang, Q.H.; Liu, X.C.; Li, Z.; Yun, X.B.; Zhang, X.J.; Yu, Q.; Li, J.; Zhang, Y.Y.; Cui, H.J.; Sun, S.A.; et al. Integrated water systems model for terrestrial water cycle simulation. *Adv. Earth Sci.* **2019**, *34*, 115–123.

2. Kruk, N.S.; Vendrame, I.F.; Chou, S.C. Coupling a Mesoscale Atmospheric Model with a Distributed Hydrological Model Applied to a Watershed in Southeast Brazil. *J. Hydrol. Eng.* **2012**, *18*, 58–65. [CrossRef]

3. Cuo, L.; Zhang, Y.X.; Gao, Y.H.; Hao, Z.C.; Cairang, L.S. The impacts of climate change and land cover/use transition on the hydrology in the upper Yellow River Basin, China. *J. Hydrol.* **2013**, *502*, 37–52. [CrossRef]

4. Sheng, M.Y.; Lei, H.M.; Jiao, Y.; Yang, D.W. Evaluation of the runoff and river routing schemes in the Community Land Model of the Yellow River Basin. *J. Adv. Model. Earth Syst.* **2017**, *9*, 2993–3018. [CrossRef]

5. Bergant, K.; Belda, M.; Halenka, T. Systematic errors in the simulation of european climate (1961–2000) with RegCM3 driven by NCEP/NCAR reanalysis. *Int. J. Climatol.* **2007**, *27*, 455–472. [CrossRef]

6. Flocas, H.A.; Hatzaki, M.; Yolika, K.; Anagnostopoulou, C.; Kostopoulou, E.; Giannakopoulos, C.; Kolokytha, C.; Tegoulias, K. Ability of RCM/GCM couples to represent the relationship of large scale circulation to climate extremes over the Mediterranean region. *J. Clim. Res.* **2011**, *46*, 197–209. [CrossRef]

7. Wen, J.; Lan, Y.C.; Su, Z.B.; Tian, H.; Shi, X.K.; Zhang, Y.; Wang, X.; Liu, R.; Zhang, T.T.; Kang, Y.; et al. Advances in observation and modeling of land surface process over the Source Region of the Yellow River. *Adv. Earth Sci.* **2011**, *26*, 575–585.

8. Gochis, D.J.; Chen, F. *Hydrological Enhancements to the Community Noah Land Surface Model (No. NCAR/TN-454+STR)*; University Corporation for Atmospheric Research: Boulder, CO, USA, 2020.

9. Gharamti, M.E.; McCreight, J.L.; Noh, S.J.; Hoar, T.J.; RafieeiNasab, A.; Johnson, B.K. Ensemble streamflow data assimilation using WRF-Hydro and DART: Novel localization and inflation techniques applied to Hurricane Florence flooding. *Hydrol. Earth Syst. Sci.* **2021**, *25*, 5315–5336. [CrossRef]

0. Gu, T.W.; Chen, Y.D.; Gao, Y.F.; Qin, L.Y.; Wu, Y.Q.; Wu, Y.Z. Improved streamflow forecast in a small-medium sized river basin with coupled WRF and WRF-Hydro: Effects of radar data assimilation. *Remote Sens.* **2021**, *13*, 3251. [CrossRef]

1. Zhang, Z.; Arnault, J.; Wagner, S.; Laux, P.; Kunstmann, H. Impact of lateral terrestrial water flow on land-atmosphere interactions in the Heihe River Basin in China: Fully coupled modeling and precipitation recycling analysis. *J. Geophys. Res. Atmos.* **2019**, *124*, 8401–8423. [CrossRef]

2. Senatore, A.; Mendicino, G.; Dochis, D.J.; Yu, W.; Yates, D.N.; Kunstmann, H. Fully coupled atmosphere-hydrology simulations for the central Mediterranean: Impact of enhanced hydrological parameterization for short and long time scales. *J. Adv. Model. Earth Syst.* **2015**, *7*, 1693–1715. [CrossRef]

3. Li, G.W.; Meng, X.H.; Blyth, E.; Chen, H.; Shu, L.L.; Li, Z.G.; Zhao, L.; Ma, Y.S. Impact of Fully Coupled Hydrology-Atmosphere Processes on Atmosphere Conditions: Investigating the Performance of the WRF-Hydro Model in the Three River Source Region on the Tibetan Plateau, China. *Water* **2021**, *13*, 3409. [CrossRef]

4. Ji, P.; Yuan, X.; Ma, F.; Pan, M. Accelerated hydrological cycle over the Sanjiangyuan region induces more streamflow extremes at different global warming levels. *Hydrol. Earth Syst. Sci.* **2020**, *24*, 5439–5451. [CrossRef]

5. Zheng, H.X.; Zhang, L.; Liu, C.M.; Shao, Q.X.; Fukushima, Y. Changes in stream flow regime in headwater catchments of the Yellow River basin since the 1950s. *Hydrol. Process.* **2006**, *21*, 886–893. [CrossRef]

6. Meng, X.H.; Lyu, S.H. Observation Data of Turbulent Flow at Lakeside Observation Points in Erling Lake Basin. 2013. Available online: http://www.ncdc.ac.cn (accessed on 17 December 2022 ).

7. Meng, X.H.; Lyu, S.H.; Li, Z.G.; Ao, Y.H.; Wen, LJ.; Shang, L.Y.; Wang, S.Y.; Deng, M.S; Zhang, S.B.; Zhao, L.; et al. Dataset of comparative observations for land surface processes over the semi-arid alpine grassland against alpine lakes in the Source Region of the Yellow River. *Adv. Atmos. Sci.* **2023**, *40*, 1142–1157. [CrossRef]

28. An Y.Y.; Meng, X.H.; Zhao, L.; Li, Z.G.; Lyu, S.H.; Ma, Y.T. Evaluation the applicability of albedo products of GLASS, MODIS and GlobAlbedo under the alpine meadow over the Qinghai-Tibetan Plateau. *Plateau Meteor.* **2019**, *38*, 88–100.
29. An, Y.Y.; Meng, X.H.; Zhao, L.; Li, Z.G.; Wang, S.Y.; Shang, L.Y.; Chen, H.; Lyu, S.H.; Li, G.W.; Ma, Y.S. Performance of GLASS and MODIS Satellite Albedo products in diagnosing Albedo variations during different time scales and special weather conditions in the Tibetan Plateau. *Remote Sens.* **2008**, *12*, 2456. [CrossRef]
30. Li, X.; Gao, Y.H.; Wang, W.Z.; Lan, Y.C.; Xu, J.W.; Li, K. Climate changes and applicability of GLDAS in the headwater of the Yellow River Basin. *Adv. Earth Sci.* **2014**, *29*, 531–540.
31. He, J.; Yang, K.; Tang, W.J.; Lyu, H.; Qin, J.; Chen, Y.Y.; Li, X. The first high-resolution meteorological forcing dataset for land process studies over China. *Sci. Data* **2020**, *7*, 25. [CrossRef]
32. Dmitriev, V.; Ovchinnikov, V. Interpolation in real method spaces. *Geoderma* **1979**, *246*, 794–797.
33. Shangguan, W.; Dai, Y.J.; Liu, B.Y.; Ye, A.Z.; Yuan, H. A soil particle-size distribution dataset for regional land and climate modelling in China. *Geoderma* **2012**, *171–172*, 85–91. [CrossRef]
34. Skamarock, W.C.; Klemp, J.B. A time-split nonhydrostatic atmospheric model for weather research and forecasting applications. *J. Comput. Phys.* **2008**, *227*, 3465–3485. [CrossRef]
35. Niu, G.-Y.; Yang, Z.-L.; Mitchell, K.E.; Chen, F.; Ek, M.B.; Barlage, M.; Kumar, A.; Manning, K.; Niyogi, D.; Rosero, E.; et al. The community Noah land surface model with multiparameterization options (Noah-MP): 1. Model description and evaluation with local-scale measurements. *J. Geophys. Res. Atmos.* **2011**, *116*, D12109. [CrossRef]
36. Yang, Y.Z.; Yuan, H.L.; Yu, W. Uncertainties of 3D soil hydraulic parameters in streamflow simulations using a distributed hydrological model system. *J. Hydrol.* **2011**, *567*, 12–24. [CrossRef]
37. Niu, G.Y.; Yang, Z.L. Effects of frozen soil on snowmelt runoff and soil water storage at a continental scale. *J. Hydrometeorol.* **2006**, *7*, 937–952. [CrossRef]
38. Thompson, G.; Field, P.R.; Rasmussen, R.M.; Hall, W.D. Explicit forecasts of winter precipitation using an improved bulk microphysics scheme. Part II: Implementation of a new snow parameterization. *Mon. Weather Rev.* **2008**, *136*, 5095–5115. [CrossRef]
39. Grell, G.A.; Devenyi, D. A generalized approach to parameterizing convection combining ensemble and data assimilation techniques. *Geophys. Res. Lett.* **2002**, *29*, 38-1–38-4. [CrossRef]
40. Nakanishi, M.; Niino, H. An improved mellor-yamada level-3 model: Its numerical stability and application to a regional prediction of advection fog. *Bound. Layer Meteorol.* **2006**, *119*, 397–407. [CrossRef]
41. Iacono, M.J.; Delamere, J.S.; Mlawer, E.J.; Shephard, M.W.; Clough, S.A.; Collins, W.D. Radiative forcing by long-lived greenhouse gases: Calculations with the AER radiative transfer models. *J. Geophys. Res. Atmos.* **2008**, *113*, D13. [CrossRef]
42. Yucel, I.; Onen, A.; Yilmaz, K.K.; Gochis, D.J. Calibration and evaluation of a flood forecasting system: Utility of numerical weather prediction model, data assimilation and satellite-based rainfall. *J. Hydrol.* **2015**, *523*, 49–66. [CrossRef]
43. Wang, J.; Liu, D.W.; Tian, S.N.; Hu, Y.H.; Ma, J.L.; Wang, L.X. Coupling analysis of short-term weather and runoff in an arid lake basin of China. *Reg. Sustain.* **2021**, *2*, 264–279. [CrossRef]
44. Ye, D.; Zhang, S.W.; Wang, F.Y.; Mao, F.P.; Yang, X.X. The applicability of different parameterization schemes in semi-arid region based on Noah-MP land surface model. *Chin. J. Atmos. Sci.* **2017**, *41*, 189–201.
45. Chen, Y.L.; Wen, J.; Yang, C.G.; Long, Y.P.; Li, G.W.; Jia, H.J.; Liu, Z. Analysis on the applicability of different precipitation products and WRF-Hydro model over the Source Region of the Yellow River. *Chin. J. Atmos. Sci.* Available online: http://www.dqkxqk.ac.cn/dqkx/dqkx/article/abstract/2022057B (accessed on 17 December 2022 ). [CrossRef]
46. Jia, D.Y.; Wen, J.; Zhang, T.T.; Xi, J.J. Response of soil water content and soil thermal conductivity on precipitation in Loess Plateau. *Plateau Meteor.* **2014**, *33*, 712–720.
47. Zhang, Q.; Wen, J.; Zhang, T.T.; Yang, Y.T.; Wu, Y.Y.; Zhang, G.; Liu, W.H.; Chen, Y.L.; Yan, Z.T. The impacts of roughness length on the simulation of land-atmosphere water and heat exchanges over the Yarlung Zangbo Grand canyon region. *Front. Earth Sci.* **2023**, *10*, 2296–6463. [CrossRef]
48. Taylor, K.E. Summarizing multiple aspects of model performance in a single diagram. *J. Geophys. Res. Atmos.* **2001**, *106*, 7183–7192. [CrossRef]
49. Zhang, H.X.; Yuan, N.M.; Ma, Z.G.; Huang, Y. Understanding the soil temperature variability at different depths: Effects of surface air temperature, snow cover, and the soil memory. *Adv. Atmos. Sci.* **2021**, *38*, 493–503. [CrossRef]
50. Jia, S.F.; Liang, Y.; Zhang, S.F. Discussion on evaluation of natural runoff in the Yellow River Basin. *Water Resour. Prot.* **2022**, *38*, 33–38+55.

atmosphere

MDPI

·ticle

# ·oundary Layer Height and Trends over the Tarim Basin

·kida Salam [1], Qing He [2,*], Alim Abbas [3,*], Tongwen Wu [4], Jie Zhang [4], Weihua Jie [4] and Junjie Liu [5]

1 Kezhou Meteorological Administration, Kezhou 845350, China; akidasalam@163.com
2 Institute of Desert Meteorology, China Meteorological Administration, Ürümqi 830002, China
3 College of Resources and Environment, Xinjiang Agricultural University, Ürümqi 830052, China
4 Earth System Modeling and Prediction Centre, China Meteorological Administration, Beijing 100081, China; twwu@cma.gov.cn (T.W.); jiezhang@cma.gov.cn (J.Z.); jiewh@cma.gov.cn (W.J.)
5 Anhui Climate Center, Hefei 230031, China; jieagle@126.com
* Correspondence: qinghe@idm.cn (Q.H.); alimabbas@xjau.edu.cn (A.A.)

**Abstract:** This study aimed to examine the spatio-temporal variations in the atmospheric boundary layer height (ABLH) over the Tarim Basin (TB). Monthly ABLH data from the ERA-Interim dataset from January 1979 to December 2018 were used. Periodicity analysis and the Mann–Kendall Abrupt Changes test were employed to identify the change cycle and abrupt change year of the boundary layer height. The Empirical Orthogonal Function (EOF) method was utilized to determine the spatial distribution of the boundary layer height, and the RF method was used to establish the relationship between the ABLH and influencing factors. The results demonstrated that the highest values of ABLH (over 1900 m) were observed in the middle parts of the study area in June, and the ABLH exhibited a significant increase over the TB throughout the study period. Abrupt changes in the ABLH were also identified in 2004, as well as in 2-, 5-, 9-, and 15-year changing cycles. The first EOF ABLH mode indicated that the middle and northeast regions are relatively high ABLH areas within the study area. Additionally, the monthly variations in ABLH show a moderately positive correlation with air temperature, while exhibiting a negative correlation with air pressure and relative humidity.

**Keywords:** atmospheric boundary layer height; surface air temperature; Taklamakan Desert; abrupt change

## 1. Introduction

Solar radiation and its daily fluctuations play a crucial role in the exchange of heat fluxes between the Earth's surface and the atmosphere. However, these heat fluxes are primarily limited to a shallow layer near the land surface known as the atmospheric boundary layer (ABL) [1]. The ABL directly influences various factors such as water vapor, heat, and pollutants between the land surface and the free atmosphere [2–6], thereby impacting atmospheric and weather-scale adjustments [7,8]. Additionally, the ABL also plays a significant role in extreme climate events [9]. As the lowest part of the atmosphere, the ABL is greatly influenced by the characteristics of the land surface [10].

The thickness of the ABL, referred to as the ABLH, varies from a few meters to several kilometers [11]. It depends on factors such as atmospheric system types, surface fluxes, and land cover [12,13]. The ABLH has significant implications for air quality, as well as for various environmental issues such as heat transmission, land surface modeling, air pollution, and drought [2–4,9,14–23].

Previous research on ABLH has predominantly focused on calculation methods [2,4,11, 24,25], influencing factors [13], and changing characteristics [26,27]. However, most studies have been limited to specific stations or short time scales. With regard to global warming, the tropopause height has shown an upward trend [28]. Therefore, it is worth investigating whether the ABLH exhibits a similar trend. In 2013, Zhang et al. [29] evaluated ABLH trends in Europe and found that daytime boundary layer heights at most stations significantly increased during all four seasons. Similarly, Zhang et al. [23] reported a significant upward trend in the average ABLH in the arid and semi-arid regions of East Asia from 1900 to 2015. Additionally, Darand et al. [30] indicated an upward trend in the ABLH over Iran.

Arid and semi-arid areas cover approximately 30% of the Earth's surface and are highly vulnerable to the impacts of climate change. However, there is a lack of studies on the ABLH in these regions due to insufficient observational data and meteorological measurements. Therefore, the main objective of this study is to investigate the temporal and spatial variations in ABLH over the Tarim Basin, which will serve as a foundation for future studies on the impact of the ABL on climate.

## 2. Data

Situated in the southern region of Xinjiang in northwestern China, the Taklamakan Desert (TB) covers an area of $53 \times 10^4$ km$^2$ [31]. As the second largest shifting desert globally, the Taklamakan Desert experiences minimal precipitation and high evapotranspiration rates. This region falls under a continental arid climate and serves as a significant source of sand–dust storms in China (Figure 1).

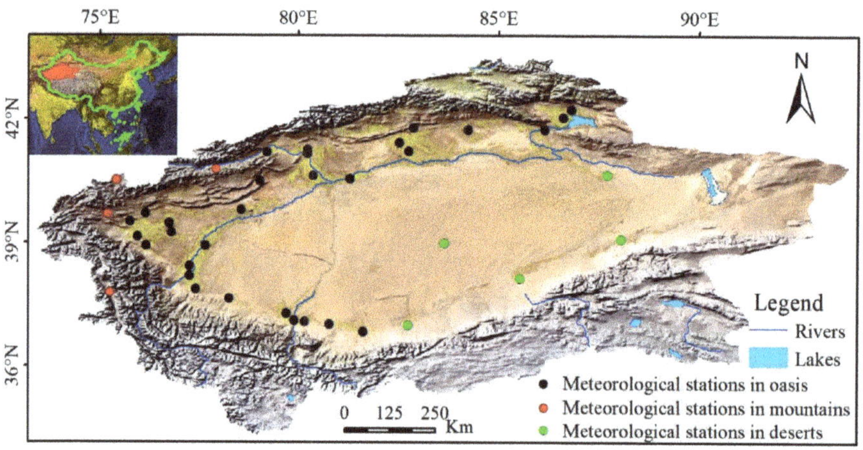


**Figure 1.** The territory of the study area. Located in northwestern China, lies between Tianshan Mountain, Kunlun, and Altun Mountain. The black, green, and red circles represent stations in oases, desert, and mountainous terrain, respectively.

For this study, we utilized the ERA Interim ABLH dataset, which offers a spatial resolution of 0.125° spanning from January 1979 to December 2018. The satellite data (GLAS boundary layer height) are typically 200–400 m higher than the ERA interim over oceans, but smaller-scale and global patterns of ABL height exhibit similar characteristics [32,33]. Additionally, the ERA Interim dataset has been validated worldwide [9] when compared to observational radiosoundes. ERA Interim data have been widely applied in many academic studies and have become some of the most important data in the field of atmospheric science in the past few years. The research and validation of these data have been widely recognized, and their accuracy and reliability have been confirmed in many studies. Moreover, these data have identified that the deviation between the boundary layer height reanalysis data (ERA-interim) and the measured data is relatively small [34–36]. The dataset is freely available online https://www.ecmwf.int/en/forecasts/datasets/archive-datasets/reanalysis-datasets/era-interim (accessed on 5 April 2020).

To supplement our analysis, we incorporated data from 39 weather stations during the period of 1979–2018. Specifically, we utilized ground-based monthly mean air temperature (1.5 m) (Mean air TEM), maximum air temperature (Max air TEM), minimum air temperature (1.5 m $\pm$ 5 cm) (Min air EM), air pressure (1.2 m), and relative humidity

(1.5 m). Figure 1 displays the locations of these weather stations. It is important to note that the meteorological data are part of the synoptic observation program and consist of two series: 8892 data points in the result part, and 59 data points in the study area part. Xinjiang Meteorological Administration provided the data, which underwent stringent quality control procedures before release.

## 3. Methodology

To uncover the evolutionary characteristics and influential factors of ABLH in the Tarim Basin, this study employs a range of methodologies including trend analysis, abrupt change analysis, wavelet analysis, Emperial Orthogonal Function (EOF) analysis, random forest model, and other techniques (Table 1).

**Table 1.** The main research methods used in this paper.

| Methods | Characteristics | Purposes |
|---|---|---|
| Linear regression method | Identifies the continuity of long-term trends | Revealing the changing trend characteristics of continuous meteorological data |
| Morlet wavelet analysis | Simultaneously analyzes time and frequency characteristics | Determine the period of data change |
| Abrupt changes test | Abnormal recognition ability | Monitoring in abrupt change point of detection data |
| EOF method | Decomposes spatial and temporal principal components of data, | Determine the main characteristics of data distribution |
| Random forest model | Based on Decision Tree Ensemble Model Feature Importance, Prediction Accuracy Prediction Analysis | Determine the importance of meteorological factors on the height of the boundary layer |

### 3.1. The Linear Regression Method

Linear regression is one of the main methods used to test the changing trend and can express the changing trend of variables in a time series. The equation is as follows:

$$Y = a_0 + a_1 t \tag{1}$$

where Y is the precipitation; t is the time; $a_0$ is the regression constant; $a_1$ is the regression coefficient; and $a_1 \times 10$ is the changing trend rate of per decade.

### 3.2. Morlet Wavelet Analysis

In this study, Morlet wavelet analysis is applied to display the periodic change of ABLH; it results in a number of wavelet coefficients. The corresponding wavelet family involves sub-wavelets that are generated from the basic wavelet function $\psi(t)$ shown as follows [37]:

$$\psi_{a,b}(t) = |a|^{-\frac{1}{2}}\psi\left(\frac{t-b}{a}\right) a, b \in R, a \neq 0 \tag{2}$$

where $\psi_{a,b}$ *(t)* is the sub-wavelet, and parameters a and b denote the scale factor and the horizontal shift, respectively.

For any function *f (t)* $\in L^2$ (R), its WT is expressed as:

$$W_f(a,b) = |a|^{-\frac{1}{2}} \int_{-\infty}^{+\infty} f(t)\Upsilon^*(\frac{t-b}{a})dt \tag{3}$$

where $W_f$ *(a,b)* is wavelet coefficient. According to the wavelet coefficient, the wavelet variance is computed according to the following Equation (4).

$$Var(a) = \int_{-\infty}^{+\infty} \left|W_f(a,b)\right|^2 db \tag{4}$$

Wavelet analysis decomposed the signal series on a time scale, and the time–frequency field change can be clearly observed and distinguished.

### 3.3. Mann–Kendall Test of Abrupt Changes

In this study, the time series data were assumed to be steady and independent. Variables (i.e., X = {x1, x2, . . . , xn}) shows no change because of the null hypothesis predicting no trends in the data. For data point xi, ni is calculated by the number of data points that exceed xi. The Mann–Kendall statistic ni is calculated as [38]:

$$E(d_k) = \frac{k(k-1)}{4}, 2 \leq k \leq n \tag{5}$$

$$\text{Var}(d_k) = \frac{k(k-1)(2k+5)}{72}, 2 \leq k \leq n \tag{6}$$

The standard value of $d_k$ is computed by:

$$u(d_k) = \frac{d_k - E(d_k)}{\sqrt{\text{Var}(d_k)}}, 2 \leq k \leq n \tag{7}$$

Given that u(d1) = 0, all $u(d_k)$ will result in a curve, UF. A retrograde $u(d_k)$ is expressed in Equation (8).

$$u'(d_k) = -u(d_{k'}) \quad k' = n+1-k, \ \ 2 \leq k \leq n \tag{8}$$

Given that u(d1) = 0, all $u(d_k)$ will establish a curve, UB. The intersection points of UF and UB are located between the confidence lines when abrupt climate change occurs.

### 3.4. EOF Method

The EOF method is used to calculate orthogonal functions, representing spatio-temporal components. Each component may represent a changing characteristic of the variable [39]. The EOF method provides the spatio-temporal change patterns of the variable [40].

### 3.5. Random Forest Model

The random forest (RF) model is capable of processing diverse data and can be effectively applied in data collection. RF selection can be utilized to identify the most significant variables for regression by selecting a reduced set of partition variables, and its output represents the average value derived from all decision-making trees. The RF model employs multivariate sorting to determine the variables and provide insights into their relative importance [41]. In this study, ABLH served as the dependent variable.

## 4. Results

### 4.1. Long-Term Mean of ABLH

In winter, the lowest ABLH was observed in January and December with a value of 100 m in the southwestern mountain area of the TB. In February, with the increase in air temperature, the ABLH increased by 200 m in comparison to January. In March, an ABLH of 1800 m was observed in the southeast and central area due to high levels of air dryness. As the air temperature gradually rose in spring, the ABLH increased, with the lowest value (400 m) observed in the north, northwest, and west part of the TB, mainly in mountain areas, indicating a major role of high latitude. The spatial distribution of ABLH in May is similar to that in April, but the value has increased. The highest value (1700 m) was observed in May in the central part of the basin, which is mainly influenced by low altitude, desert climates, low moisture, and high air temperature.

In June, the ABLH reached over 1900 m due to changes in air temperature, but it significantly decreased in September. In October, the ABLH was above 1000 m. The ABLH in the center of the desert decreased by 1400 m in comparison to its peak height of 1900 m, resulting in a decline of 500 m. This downward trend continued in November and

December. In November, the decrease in ABLH was particularly rapid, with ABLH values less than 200 m in most regions except for a small area. The lowest ABLH was observed in December over the entire area, with a value of less than 100 m.

The minimum ABLH (600 m) is observed in the southwest and northwestern parts of the TB, which were influenced by high latitude and high humidity. Meanwhile, the highest ABLH value (1000 m) was observed in the center of the desert, attributed to high air temperature and the absence of vegetation. The ABLH gradually decreased from the center to the surrounding areas of the basin, influenced by terrain and temperature (Figure 2).

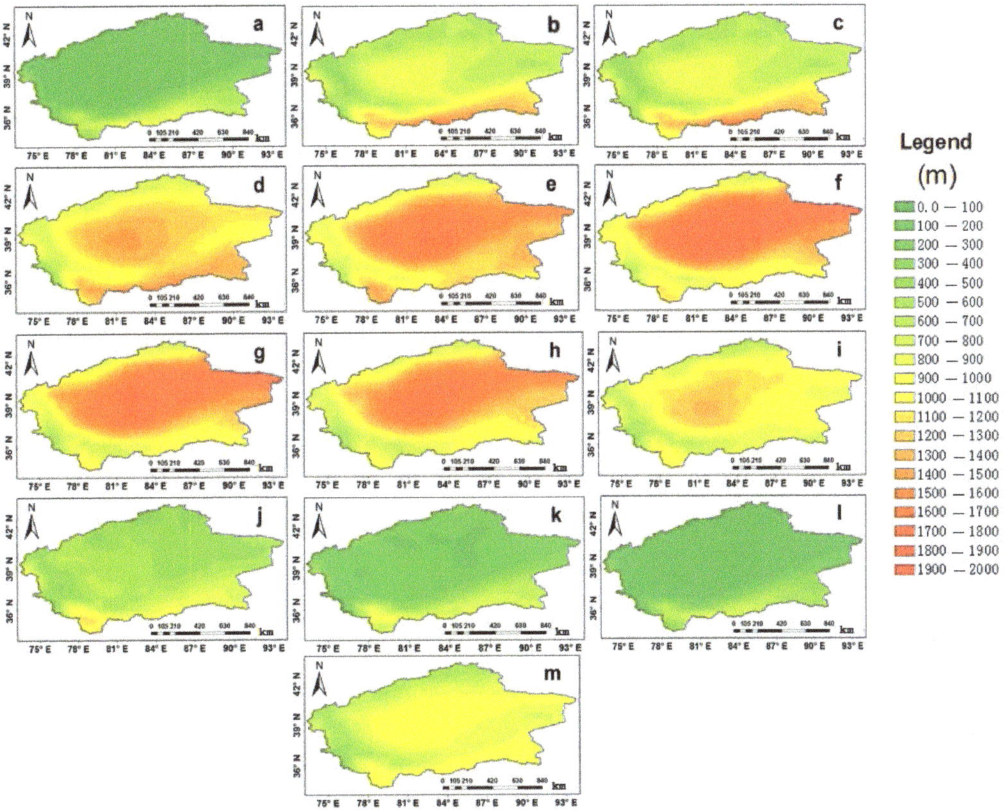


**Figure 2.** The long-term annual mean of ABLH for (**a**) January, (**b**) February, (**c**) March, (**d**) April, (**e**) May, (**f**) June, (**g**) July, (**h**) August, (**i**) September, (**j**) October, (**k**) November, (**l**) December, and (**m**) annually.

## *4.2. Trends in ABLH*

The trend analysis of the Atmospheric Boundary Layer Height (ABLH) data from ECMWF for the period 1979–2018 revealed both upward and downward trends, represented by the colors red, yellow, and green. Figure 3 illustrates the monthly trends of ABLH in the Tarim Basin. In January, the annual ABLH tendency rate ranged from −30 to 60 m/10 a. The lowest value of −30 m/10 a was observed in the eastern and southeastern parts of the basin, while the highest value of 60 m/10 a was observed in the eastern and southwestern parts. Most areas in the basin showed an upward trend, with a tendency rate of approximately 10 m/10 a. In February, the annual ABLH tendency rate ranged from −50 to 50 m/10 a. The lowest value of −50 m/10 a was observed in the southern part of the basin, while the highest value of 50 m/10 a was observed in the southwestern part. Similar to January, most areas showed an upward trend. In March, the annual ABLH tendency rate ranged from −110 to 70 m/10 a. The lowest value of −110 m/10 a was observed in the southern part, mainly in mountainous areas, while the highest value of 70 m/10 a was observed in the western part. Significant high-value centers were found in the south, southwest, and west, indicating an upward trend in most areas. For April, the ABLH tendency rate ranged from −10 to 130 m/10 a. The lowest value of −10 m/10 a was observed in the northern and northeastern parts of the basin, while the highest value of 130 m/10 a was observed in the southern part. Significant high-value centers were present in the west, with a low-value center in the east of the desert. Most parts of the study area showed an upward trend in ABLH. In May, the ABLH tendency rate ranged from −30 to 70 m/10 a. The lowest value of −30 m/10 a was observed in the northern, eastern, and western parts of the basin, while the highest value of 70 m/10 a was observed in the southern part. Significant low-value centers were found in the east, west, and north, with a high-value center in the south of the desert. The height of the boundary layer showed an increasing trend in most parts of the study area. In June, the ABLH tendency rate ranged from −30 to 60 m/10 a. The lowest value of −30 m/10 a was observed in the eastern and southwestern parts of the basin, while the highest value of 60 m/10 a was observed in the northeastern and southeastern parts. There was a dominant upward trend in the height of the boundary layer, while the surrounding mountains showed a downward trend. In July, the ABLH tendency rate ranged from −30 to 100 m/10 a. The lowest value of −30 m/10 a was observed in the surrounding mountains of the basin, while the highest value of 100 m/10 a was observed in the northeastern and central parts of the desert. There is an evident zonal distribution from the center of the desert to the surrounding regions.

The plain areas were dominated by an upward trend in ABLH, whereas the surrounding mountains showed a downward trend. In August, the ABLH tendency rate ranged from −60 to 110 m/10 a; the lowest value of −60 m/10 a was observed in the eastern part of the basin, while the highest value of 110 m/10 a was observed in the western part. There is an obvious zonal distribution from the low and high-value centers to the surrounding regions. The eastern and southeastern parts of the basin demonstrated a significant downward trend, while the western and southwestern parts showed a noticeable upward trend. The zonal distribution of either an upward or downward trend is highly pronounced. In September, the ABLH tendency rate ranged from −70 to 50 m/10 a. The lowest value of −50 m/10 a was observed in the eastern part of the basin, while the highest value of 50 m/10 a was observed in the northern and western parts. There is an evident zonal distribution and two high-value centers. The north and northwest parts of the basin exhibited a significant upward trend, while the east and southeast showed a distinct downward trend. The zonal distribution of either an increasing or decreasing trend is highly apparent. In October, the ABLH tendency rate ranged from −30 to 70 m/10 a. The lowest value of −30 m/10 a was observed in the eastern, southeastern, and southern parts of the basin, while the highest value of 70 m/10 a was observed in the southwestern part. The zonal distribution is not very pronounced. Most regions in the basin were dominated by ABLH tendency rates of 10 m/10 a and 20 m/10 a, with no clear zonal distribution. In November, the ABLH tendency rate ranged from −40 to 80 m/10 a. The lowest value of −40 m/10 a was observed in the eastern part of the basin, while the highest value of 80 m/10 a was observed in the southwestern part. There was a zonal distribution, but it is not very evident. Most regions in the basin were dominated by ABLH tendency rates of 10 m/10 a and 20 m/10 a. In December, the ABLH tendency rate ranged from −20 to 70 m/10 a. The lowest value of −20 m/10 a was observed in the southern part of the basin, while the highest value of 70 m/10 a was observed in the southwestern part. Most regions in the basin were dominated by an ABLH tendency rate of 20 m/10 a.

The ABLH tendency rate (Figure 3m) ranges from −20 to 40 m/10 a. The lowest value of −20 m/10 a was observed in the eastern part of the basin, while the highest value of 40 m/10 a was observed in the western part. Most regions of the basin were dominated by ABLH tendency rates of 10 m/10 a and 20 m /10 a. The increasing trend was dominant in most regions, displaying evident zonal distribution. Figure 4 depicts the annual average time series of ABLH during the period of 1979–2018, showing an upward trend with an increase of approximately 30 m per decade.

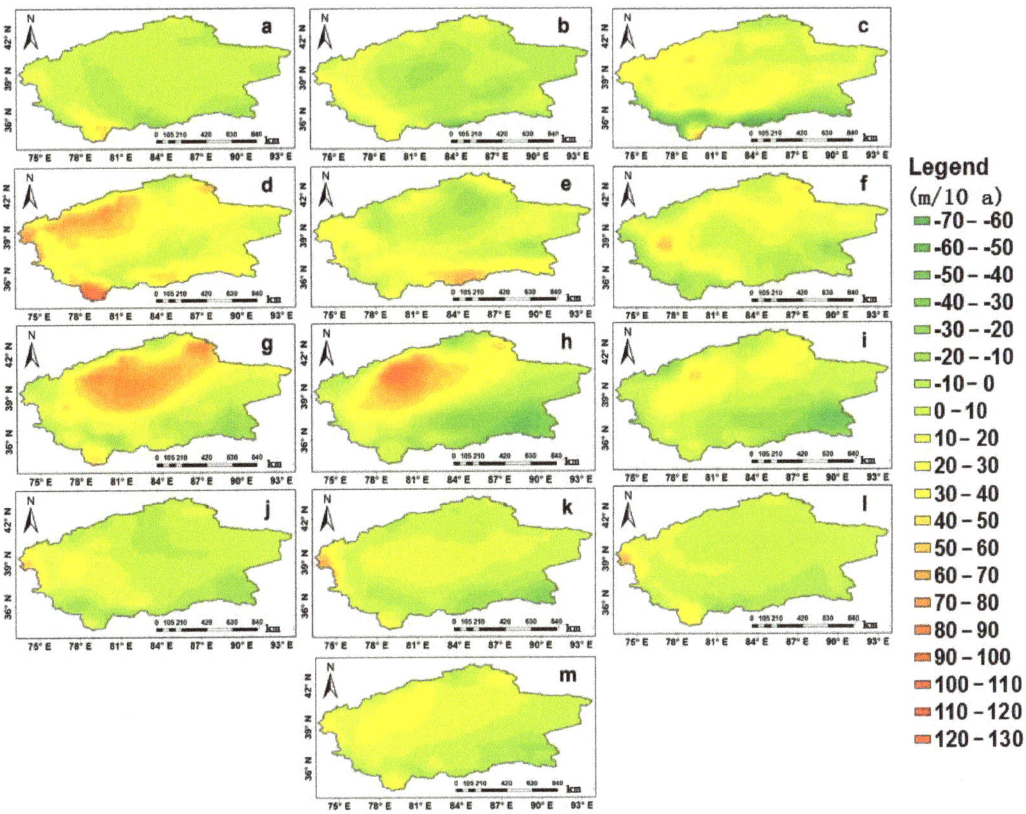


**Figure 3.** The trends of ABLH for (**a**) January, (**b**) February, (**c**) March, (**d**) April, (**e**) May, (**f**) Jun, (**g**) July, (**h**) August, (**i**) September, (**j**) October, (**k**) November, (**l**) December, and (**m**) annually.

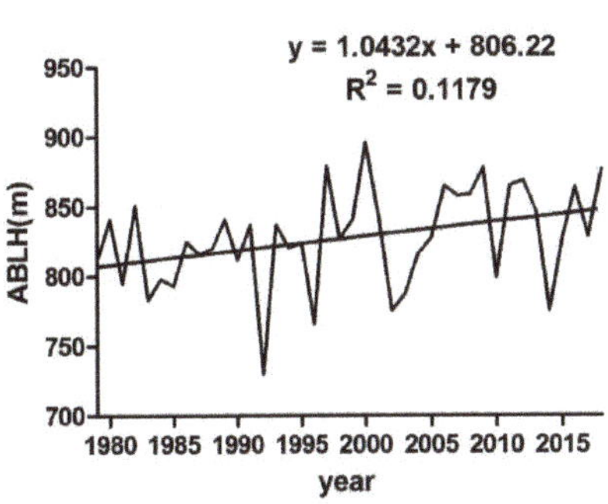


**Figure 4.** Annual area-averaged time series of ABLH.

*4.3. Periodicity Analysis and the Mann–Kendall Abrupt Changes Test*

Wavelet analysis was employed to detect periodic changes in ABLH. Figure 5 illustrates the wavelet variances of ABLH from 1979 to 2018. In this plot, a positive real part corresponds to the annual average ABLH during a high (increasing) period, while a negative real part indicates that the annual average ABLH belongs to a low (decreasing) period. Figure 5 shows a real-line contour plot of the Morlet wavelet coefficient of the annual average ABLH in the Tarim basin. It reveals four main cycles of ABLH: 2-, 5-, 9-, and 15-year cycles. Among these cycles, the time scale of approximately 15 years corresponds to the most significant variance-extreme value, followed by 9 and 5 years. The annual changes of about 2 years are too rapid and relatively insignificant.

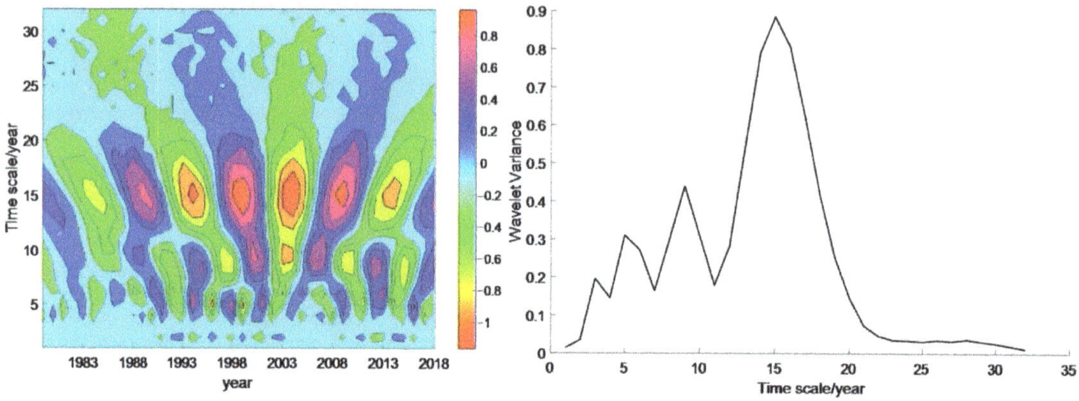


**Figure 5.** Distribution of wavelet frequency and wavelet variance of ABLH.

The periodic oscillations with a characteristic time scale of 2 years have undergone alternating changes over 13 periods. These oscillations were weak during 1979–1990 and 2004–2008, while they were more pronounced from 1991 to 2003 and from 2009 to 2016. They consist of seven periods with a negative real part center (decreasing) and 6 periods with a positive real part center (increasing). On the other hand, the periodic oscillations with a characteristic time scale of 5 years have undergone alternating changes over 21 periods. These oscillations were weak during 1979–1990, but became more evident from 1990 to 2018. The study reveals the presence of 11 consecutive periods characterized by negative real part center (decreasing) and 11 consecutive periods with positive real part center (increasing). The periods showing a decreasing trend are 1981–1982, 1984–1985, 1988–1990, 1992 −1993, 2002–2003, 2006–2008, 2010–2011, and 2014–2015. Conversely, the periods exhibiting an increasing trend are 1982–1983, 1986–1987, 1990–1991, 1994–1995, 1997–1998, 2000–2001, 2004–2005, and 2012–2013. The presence of any clear cycles in the remaining periods is not apparent. The periodic oscillations, with a characteristic time scale of 9 years, have experienced alternating changes over 14 periods. The periodic oscillation was weak during 1979–1989 and it was more obvious from 1989 to 2018. The oscillation pattern observed in this study exhibits a characteristic time scale of 15 years, with alternating changes occurring over nine periods. Among these cycles, the periodic oscillation is most prominent and displays a relatively stable and intense pattern. Specifically, there are seven periods characterized by a negative real part center (decreasing), and seven periods characterized by a positive real part center (increasing). The decreasing periods span from 1979 to 1981, 1987 to 1991, 2001 to 2007, and 2011 to 2016. On the other hand, the increasing periods occur from 1981 to 1983, 1987 to 1990, 1997 to 2001, 2007 to 2011, and 2017 to 2018.

The M-K abrupt change method was utilized to demonstrate the abrupt change year in ABLH within the study area. Figure 6 illustrates the year of abrupt change as tested by the M-K method. The findings indicate that an abrupt change was observed in 2004. The factors contributing to the abrupt changes in ABLH in this specific study area remain unknown, necessitating further investigations into the interplay between land use and the spatial distribution of air temperature in order to address this unresolved issue.

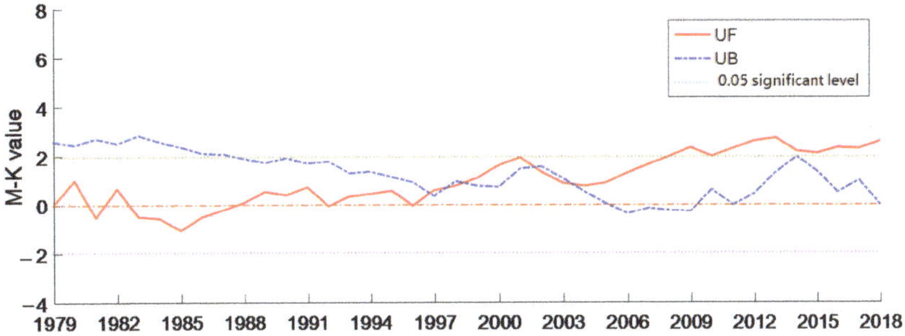


**Figure 6.** Annual average BLH Mann–Kendall abrupt change test from 1979 to 2018.

### *4.4. EOF Analysis of ABLH*

The analysis of Empirical Orthogonal Function (EOF) modes revealed that the first four modes accounted for 42.3%, 15.6%, 10.9%, and 5.7% of the total variation in the ABLH data, respectively, explaining a combined variability of 74.5% (Figure 7). The spatial pattern of the first EOF mode of ABLH was primarily observed in the northeast part of the region, indicating similar changes in the southwest and northeast areas of the basin. This distribution suggests that the ABLH is relatively low in the southwest and relatively high in the northeast, which may be influenced by air temperature and terrain.

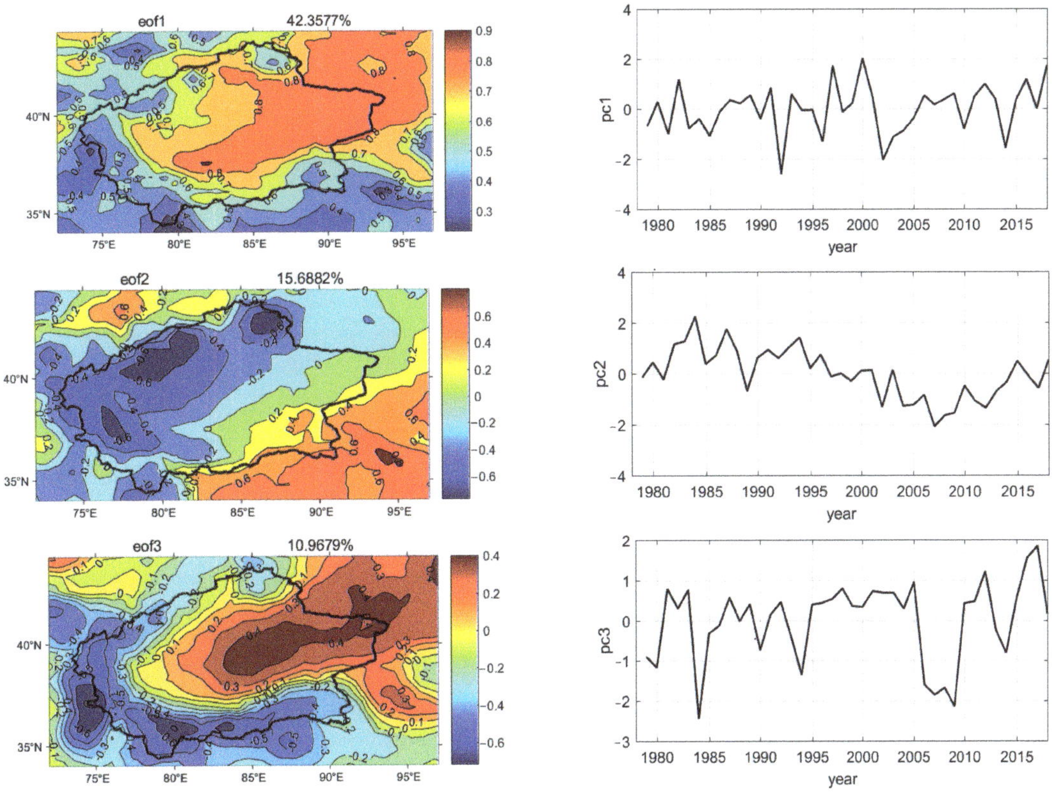


**Figure 7.** *Cont.*

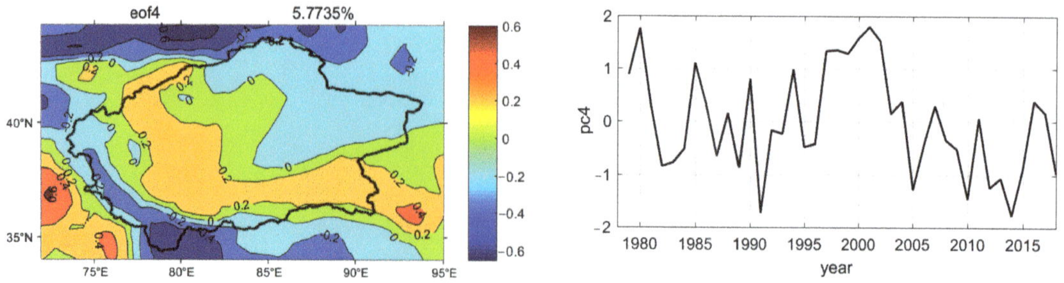


**Figure 7.** Spatial patterns and temporal amplitudes of EOF modes for air temperature time series (°C).

Furthermore, the time series analysis revealed an upward trend since 2003, with intermittent positive amplitudes in several years. The most significant changes occurred during the period 1995–2000, while the weakest changes were observed in 1992. This indicates a rapid downward trend in ABLH from 2002 to 2009, with the minimum ABLH recorded in 1992.

The second EOF mode explains 15.6% of the total variance, representing another important spatial distribution of boundary layer changes. Positive loadings were primarily observed in the eastern portion of the study area, while the weakest loadings were found in the northeast and west. Prior to 2000, the loadings were mostly positive, but turned negative after 2000. Notably, the highest value was observed in 1984, whereas the lowest value was recorded in 2006.

The third mode accounted for 10.9% of the total change, with the strongest positive loadings observed in the center of the desert and negative loadings in the surrounding regions of the Taklamakan Desert. The strongest negative value was recorded in 1984, while the strongest positive value was observed in 2012.

Similarly, the fourth mode explained 5.77% of the total change, with positive loadings in the southwest and negative loadings in the northeast. The strongest negative value was seen in 1992, while the most positive value was recorded in 2002.

In summary, the modal distribution of ABLH values indicates higher values in the northeast and lower values in the southwest, with high-value centers primarily situated in the northeast of the basin and the middle of the desert. This spatial pattern corresponds to the distribution of air temperature and land surface temperature, with the northeast regions of the basin characterized by lower average air temperatures and corresponding lower boundary layer heights in mountainous areas [42].

### *4.5. Relation between ABLH and Meteorological Factors*

The movement of air within the atmospheric boundary layer (ABL) is heavily influenced by ground friction and primarily depends on the thermal and dynamic effects of the ground surface. The variation in thickness of this layer is related to the speed of the outer airflow, its own meteorological conditions, and underlying surface conditions such as terrain, topography, buildings, and vegetation. Changes in land–atmosphere conditions can lead to changes in the ABL height (ABLH), with higher air temperatures resulting in a higher ABLH [30]. Latitude, solar radiation, topography, and underlying surface type are the major factors affecting the ABLH. We applied five indicators (i.e., mean air temperature, maximum air temperature, minimum air temperature, air pressure, relative humidity) to study the principal factors influencing changes in ABLH. The RF model was employed to quantitatively analyze the selected indicators, and the results are shown in Figure 8a.

Air temperature is the most fundamental and direct factor in meteorological conditions that affect the ABLH. High air temperatures increase the heat capacity of the atmosphere, leading to an increase in the ABL and subsequently increasing the ABLH. During the day, the ground is heated by solar radiation, causing an increase in ground temperature. This results in hot air rising and a strong convective motion. In this case, the boundary layer often exhibits characteristics such as decreasing temperature, increasing wind speed, and decreasing humidity. Conversely, at night, the ground releases heat, causing the temperature to gradually decrease, and the air no longer produces convective motion. As a result, the boundary layer begins to stabilize, exhibiting characteristics such as increasing temperature, decreasing wind speed, and increasing humidity. This phenomenon is commonly known as the nighttime stable layer. The correlation coefficients of mean, maximum, and minimum air temperatures were high, at 0.95, 0.94, and 0.95, respectively, with significant positive correlations.

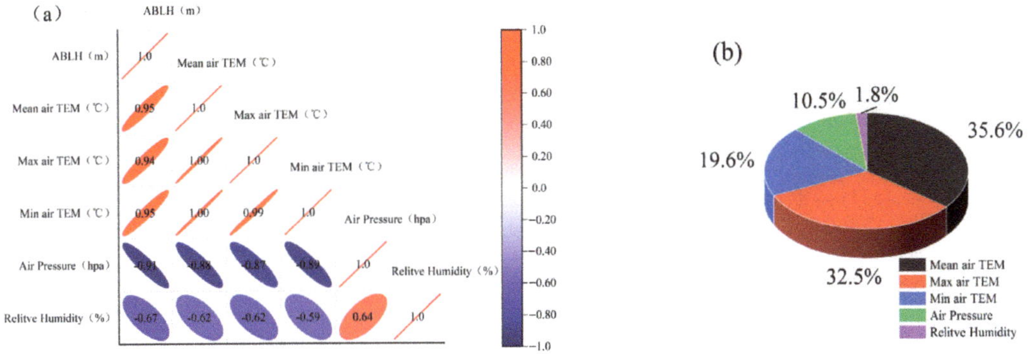


**Figure 8.** (**a**) Random forest (RF) model simulation of ABLH and (**b**) correlation between FD and influencing factors in the Northern Xinjiang (NX).

The air pressure and atmospheric composition also have a certain impact on the structure of the boundary layer. Changes in air pressure affect the pressure gradient force, gravity, and inertia force of the air, thereby altering the flow field and temperature field. Humidity is another important factor affecting the structure of the boundary layer. Changes in humidity can affect its stability, as well as heat and water vapor exchange within the boundary layer, and chemical reaction processes in the atmosphere. Under humid conditions, water vapor enhances the condensation and precipitation processes in the atmosphere, thus affecting the vertical distribution and dynamic characteristics of the boundary layer. The correlation coefficients of air pressure and relative humidity with ABLH were 0.91 and 0.67, respectively, exhibiting a noticeable negative correlation.

Furthermore, wind speed, precipitation, and altitude also greatly affect the evolution of ABLH. Due to ground friction, wind speed near the surface gradually decreases, forming a wind speed gradient layer. This phenomenon is often observed as varied wind speeds at different heights, such as kites experiencing stronger wind speeds at higher altitudes. Within the boundary layer, the vertical gradient of wind speed is also important, with the magnitude of this gradient determining the dynamic characteristics and degree of mixing. Precipitation significantly impacts the height of the boundary layer, with increased rainfall intensity causing a decrease in boundary layer height. The development history and height of the stable boundary layer are also related to altitude. In high-altitude areas, the stable boundary layer has a longer development history and higher height compared to low-altitude areas.

There is a certain correlation between the ABLH and atmospheric pollution. Local accumulation of atmospheric pollutants under conditions of weak wind speed, low boundary layer height, and low ventilation can easily cause moderate to severe pollution. Air pollution usually has a reducing effect on the height of the boundary layer. Continuous pollution emissions can lead to an increase in the concentration of pollutants in the atmosphere, causing a decrease in the heat and water vapor released within the boundary layer, thereby making the boundary layer more stable and reducing its height. In addition, certain pollutants may affect the thermal characteristics and radiation balance of the atmosphere, leading to changes in the temperature distribution and height of the boundary layer, which usually results in a decrease in the height of the boundary layer. Therefore, atmospheric pollution usually reduces the height of the boundary layer [43].

## 5. Conclusions and Discussion

This study examined the spatio-temporal distribution and influencing factors of the atmospheric boundary layer height (ABLH) in the Tarim Basin from January 1979 to December 2018. The results showed that the ABLH was higher (over 1900 m) in the middle parts of the study area in June, which was associated with higher air temperatures. The ABLH demonstrated a significant increasing trend across different seasons, which was consistent with the findings of Zhang et al. [18] in Europe and those of Mohammad et al. [30] in Iran.

The analysis also revealed that the highest upward trend rates of ABLH (120–130 m/decade) occurred in August over the western part of the basin, while the surrounding mountain regions experienced a downward trend. Overall, the annual ABLH exhibited a downward trend in most parts of the region, with the highest upward trend rate being approximately 30–40 m per decade.

Furthermore, the Morlet wavelet analysis identified four main cycles of the annual ABLH: 2, 5, 9, and 15 years. Among these cycles, the time scale of approximately 15 years corresponded to the most significant variance extreme value, followed by 9 and 5 years. The annual changes of about 2 years were comparatively fast and relatively insignificant. The M-K method also detected an abrupt change in the ABLH in 2004.

The first four Empirical Orthogonal Function (EOF) modes of ABLH explained approximately 74.5% of the total variance. Specifically, the first EOF mode indicated that the middle and northeast regions of the study area were characterized by a relatively high ABLH.

This correlation analysis showed that between the ABLH and mean, maximum, and minimum air temperature in the Tarim Basin demonstrated positive correlations, with correlation coefficients of 0.95, 0.94, and 0.95, respectively. Conversely, air pressure and relative humidity exhibited negative correlations, with correlation coefficients of 0.91 and 0.67, respectively. These findings are consistent with previous studies conducted in Europe [19], China [21,24], East Asia, and North Africa, which reported the sensitivity of ABLH to air temperature [23].

**Author Contributions:** A.S.: Data curation, Formal analysis, Investigation, Methodology, Software, Visualization, Writing—original draft. Q.H.: Conceptualization, Project administration, Supervision. A.A.: Methodology, Software, Writing—review and editing. T.W.: Data curation, Validation, Writing—review and editing. J.Z.: Project administration, Resources, Writing—review and editing. W.J.: Project administration, Resources, Writing—review and editing. J.L.: Project administration, Resources, Writing—review and editing. All authors have read and agreed to the published version of the manuscript.

**Funding:** This work was supported by the National Natural Science Foundation of China (Grant no. 42230608) and Tianshan Talent Project: 2023TSYCJC0068.

**Institutional Review Board Statement:** Not applicable.

**Informed Consent Statement:** Not applicable.

**Data Availability Statement:** Data are contained within the article.

**Acknowledgments:** We would like to thank the Xinjiang Meteorological Administration (XMA) for providing the meteorological data.

**Conflicts of Interest:** The authors declare that they have no known competing financial interests or personal relationships that could have appeared to influence the work reported in this paper.

## eferences


Stull, R.B. *Meteorology for Scientists and Engineers*, 2nd ed.; Brooks/Cole/Thomson: Pacific Grove, CA, USA, 2000; 502p.

Couvreux, F.; Guichard, F.; Austin, P.H.; Chen, F. Nature of the Mesoscale Boundary Layer Height and Water Vapor Variability Observed 14 June 2002 during the IHOP_2002 Campaign. *Mon. Weather Rev.* **2009**, *137*, 414–432. [CrossRef]

Schmid, P.; Niyogi, D. A method for estimating planetary boundary layer heights and its application over the ARM Southern Great Plains Site. *J. Atmos. Ocean. Technol.* **2012**, *29*, 316–322. [CrossRef]

Compton, J.C.; Delgado, R.; Berkoff, T.A.; Hoff, R.M. Determination of Planetary Boundary Layer Height on Short Spatial and Temporal Scales: A Demonstration of the Covariance Wavelet Transform in Ground-Based Wind Profiler and Lidar Measurements. *J. Atmos. Ocean. Technol.* **2013**, *30*, 1566–1575. [CrossRef]

Feng, X.; Wu, B.; Yan, N. Amethod for deriving the boundary layer mixing height from MODIS atmospheric profile data. *Atmosphere* **2015**, *6*, 1346–1361. [CrossRef]

Yang, Y.H.; Liu, C.H.; Jimy, D. Evaluation of two typical PBL parameterization schemes based on large-eddy simulation result. *Plateau Meteorol.* **2016**, *35*, 17–180. [CrossRef]

McGrath-Spangler, E.L. The impact of a boundary layer height formulation on the GEOS-5 model climate. *J. Geophys. Res. Atmos.* **2016**, *121*, 3263–3275. [CrossRef]

Ma, L.M.; Bao, X.W. Parametrization of planetary boundary-layer height with helicity and verification with tropical cyclone prediction. *Bound.-Layer Meteorol.* **2016**, *160*, 569–593. [CrossRef]

Von, E.A.; Teixeira, J. A planetary boundary layer height climatology derived from ECMWF reanalysis data. *J. Clim.* **2013**, *26*, 6575–6590.

Garratt, J. Review: The atmospheric boundary layer. *Earth Sci. Rev.* **1664**, *37*, 89–134. [CrossRef]

Molod, A.; Salmun, H.; Dempsey, M. Estimating Planetary Boundary Layer Heights from NOAA Profiler Network Wind Profiler Data. *J. Atmos. Ocean. Technol.* **2015**, *32*, 1545–1561. [CrossRef]

Liu, S.; Liang, X.-Z. Observed Diurnal Cycle Climatology of Planetary Boundary Layer Height. *J. Clim.* **2010**, *23*, 5790–5807. [CrossRef]

Leventidou, E.; Zanis, P.; Balis, D.; Giannakaki, E.; Pytharoulis, I.; Amiridis, V. Factors affecting the comparisons of planetary boundary layer height retrievals from CALIP-SO, ECMWF and radiosondes over Thessaloniki, Greece. *Atmos. Environ.* **2013**, *74*, 360–366. [CrossRef]

Marsik, F.J.; Fischer, K.W.; McDonald, T.D.; Samson, P.J. Comparison of Methods for Estimating Mixing Height Used during the 1992 Atlanta Field Intensive. *J. Appl. Meteorol.* **1995**, *34*, 1802–1814. [CrossRef]

Basha, G.; Ratnam, M.V. Identification of atmospheric boundary layer height over a tropical station using high-resolution radiosonde refractivity profiles: Comparison with GPS radio occultation measurements. *J. Geophys. Res.* **2009**, *114*, D16101. [CrossRef]

Wang, X.Y.; Wang, K.C. Estimation of atmospheric mixing layer height from radiosonde data. *Atmos. Meas. Tech.* **2014**, *7*, 1701–1709. [CrossRef]

Dai, C.; Wang, Q.; Kalogiros, J.A.; Lenschow, D.H.; Gao, Z.; Zhou, M. Determining Boundary-Layer Height from Aircraft Measurements. *Bound.-Layer Meteorol.* **2014**, *152*, 277–302. [CrossRef]

Zhang, Y.; Gao, Z.; Li, D.; Li, Y.; Zhang, N.; Zhao, X.; Chen, J. On the computation of planetary boundary-layer height using the bulk Richardson number method. *Geosci. Model Dev.* **2014**, *7*, 2599–2611. [CrossRef]

Coen, M.C.; Praz, C.; Haefele, A.; Ruffieux, D.; Kaufmann, P.; Calpini, B. Determination and climatology of the planetary boundary layer height above the Swiss plateau by in situ and remote sensing measurements as well as by the COSMO-2 model. *Atmos. Meas. Tech.* **2014**, *14*, 13205–13221.

Guo, J.; Miao, Y.; Zhang, Y.; Liu, H.; Li, Z.; Zhang, W.; He, J.; Lou, M.; Yan, Y.; Bian, L.; et al. The climatology of planetary boundary layer height in China derived from radiosonde and reanalysis data. *Atmos. Chem. Phys.* **2016**, *16*, 13309–13319. [CrossRef]

Seibert, P.; Beyrich, F.; Gryning, S.-E.; Joffre, S.; Rasmussen, A.; Tercier, P. Review and intercomparison of operational methods for the determination of the mixing height. *Atmos. Environ.* **2000**, *34*, 1001–1027. [CrossRef]

Seinfeld, J.H.; Pandis, S.N. *Atmospheric Chemistry and Physics: From Air Pollution to Climate Change*, 2nd ed.; Wiley Blackwell: Hoboken, NJ, USA, 2006; 1152p.

Zang, Z.; Wang, W.; Cheng, X.; Yang, B.; Pan, X.; You, W. Effects of boundary layer height on the model of ground-level PM2.5 concentrations from AOD: Comparison of stable and convective boundary layer heights from different methods. *Atmosphere* **2017**, *8*, 104. [CrossRef]

Seidel, D.J.; Ao, C.O.; Li, K. Estimating climatological planetary boundary layer heights from radiosonde observations: Comparison of methods and uncertainty analysis. *J. Geophys. Res. Atmos.* **2010**, *115*, D16113. [CrossRef]

Bachour, D.; Perez-Astudillo, D. Boundary layer height measurements over Doha using Lidar. *Energy Procedia* **2014**, *57*, 1086–1091. [CrossRef]

Sawyer, V.; Li, Z. Detection, variations andintercomparison of the planetary boundary layer depth from radiosonde, lidar and infrared spectrometer. *Atmos. Environ.* **2013**, *79*, 518–528. [CrossRef]

Patil, M.N.; Patil, S.D.; Waghmare, R.T. Planetary boundary layer height over the Indian subcontinent during extreme monsoon years. *J. Atmos. Solar-Terr. Phys.* **2013**, *92*, 94–99. [CrossRef]

28. Randel, W.J.; Wu, F.; Gaffen, D.J. Interannual variability of the tropical tropopause derived from radiosonde data and NCEP reanalyses. *J. Geophys. Res.* **2000**, *105*, 15509–15523. [CrossRef]
29. Zhang, Y.; Seidel, D.J.; Zhang, S. Trends in planetary boundary layer height over Europe. *J. Clim.* **2013**, *26*, 10071–10076. [CrossRef]
30. Darand, M.; Zandkarimi, F. Identification of atmospheric boundary layer height and trends over Iran using high-resolution ECMWF reanalysis dataset. *Theor. Appl. Climatol.* **2019**, *137*, 1457–1465. [CrossRef]
31. Hao, X.; Li, W. Oasis cold island effect and its influence on air temperature: A case study of Tarim Basin, Northwest China. *J. Arid Land* **2016**, *8*, 172–183. [CrossRef]
32. Liu, J.; Huang, J.; Chen, B.; Zhou, T.; Yan, H.; Jin, H.; Huang, Z.; Zhang, B. Comparisons of PBL heights derived from CALIPSO and ECMWF reanalysis data over China. *J. Quant. Spectrosc. Radiat. Transf.* **2015**, *153*, 102–112. [CrossRef]
33. Palm, S.P.; Benedetti, A.; Spinhirne, J. Validation of ECMWF global forecast model parameters using GLAS atmospheric channel measurements. *Geophys. Res. Lett.* **2005**, *32*, L22S09. [CrossRef]
34. Dee, D.P.; Uppala, S.M.; Simmons, A.J.; Berrisford, P.; Poli, P.; Kobayashi, S.; Andrae, U.; Balmaseda, M.A.; Balsamo, G.; Bauer, P.; et al. The ERA-Interim reanalysis: Configuration and performance of the data assimilation system. *Q. J. R. Meteorol. Soc.* **2011**, *137*, 553–597. [CrossRef]
35. Hersbach, H.; Bell, B.; Berrisford, P. *ERA5 Monthly Averaged Data on Pressure Levels from 1979 to Present*; ECMWF: Reading, UK, 2018.
36. Poli, P.; Hersbach, H.; Tan, D.; Dee, D.; Thépaut, J.-N.; Simmons, A.; Peubey, C.; Laloyaux, P.; Komori, T.; Berrisford, P.; et al. The Data Assimilation System and Initial Performance Evaluation of the ECMWF Pilot Reanalysis of the 20th Century Assimilating Observations up to 2010. *J. Clim.* **2016**, *29*, 6851–6872.
37. Hu, C.; Xu, Y.; Han, L.; Yang, L.; Xu, G. Long-term trends in daily precipitation over the yangtze river delta region during 1960–2012, eastern china. *Theor. Appl. Climatol.* **2016**, *125*, 131–147. [CrossRef]
38. Jones, J.R.; Schwartz, J.S.; Ellis, K.N.; Hathaway, J.M.; Jawdy, C.M. Temporal variability of precipitation in the Upper Tennessee valley. *J. Hydrol.* **2015**, *3*, 125–138. [CrossRef]
39. Larson, M.; Capobianco, M.; Janses, H.; Rozynski, G.; Southgate, H.N.; Stive, M.; Wijnberg, K.M.; Hulscher, S. Analysis and modeling of field data on coastal morphological evolution over yearly and decadal time scales, Part 1: Background and linear techniques. *J. Coast. Res.* **2003**, *19*, 760–775.
40. Jollife, I.T. *Principal Component Analysis*, 2nd ed.; Springer: New York, NY, USA, 2002. [CrossRef]
41. Hong, Y.; Chen, S.; Liu, Y.; Zhang, Y.; Yu, L.; Chen, Y.; Liu, Y.; Cheng, H.; Liu, Y. Combination of fractional order derivative and memory-based learning algorithm to improve the estimation accuracy of soil organic matter by visible and near-infrared spectroscopy. *Catena* **2019**, *174*, 104–116. [CrossRef]
42. Abbas, A.; He, Q.; Jin, L.; Li, J.; Salam, A.; Lu, B.; Yasheng, Y. Spatio-Temporal Changes of Land Surface Temperature and the Influencing Factors in the Tarim Basin, Northwest China. *Remote Sens.* **2021**, *13*, 3792. [CrossRef]
43. Jia, J.; Yang, P.Q.; Jiang, H.M. Analysis of the variation characteristics of daily maximum boundary layer height in Urumqi and its relationship with air quality. *J. Meteorol. Environ.* **2019**, *35*, 6.

water

Article

# Landscape Drivers of Floods Genesis (Case Study: Mayo Mizao Peri-Urban Watershed in Far North Cameroon)

Lucas Bouba [1], Pierre-Alain Ayral [2,*] and Sophie Sauvagnargues [3]

1 Faculty of Sciences, The University of Maroua, Maroua P.O. Box 814, Cameroon; iss.bouba.lucas@gmail.com
2 ESPACE, UMR 7300, Aix Marseille University, Côte d'Azur University, Avignon University, CNRS, Cévennes Location, 410 Chemin de Boissières, 30380 Saint-Christol-Lez-Alès, France
3 Risk Sciences Laboratory, IMT Mines Ales, 30319 Ales, France; sophie.sauvagnargues@mines-ales.fr
* Correspondence: pierre-alain.ayral@cnrs.fr

**Abstract:** Landscape has significant effects on hydrological processes in a watershed. In the Sudano-Sahelian area, watersheds are subjected to a quick change in landscape patterns due to the human footprint, and the exact role of the actual landscape features in the modification of the hydrological process remains elusive. This study tends to assess the effects of landscape on the genesis of the runoff in the Mayo Mizao watershed. To achieve this goal, 62 infiltration tests were performed at different points and depths (5 cm and 20 cm) using the double-ring method and the Porchet method. The results show that the combination of many factors (soil type, land use, and farming practices) can guide the hydraulic conductivity behavior of soils. For example, at 5 cm depths, clayey-evolved soils, such as vertisols and halomorphic soils, inhibit infiltration, as opposed to non-evolved mineral soils, such as lithosols and clayey-sandy soils. However, at 20 cm depths, gray soils with halomorphic tendencies followed by vertisols have a low sensitivity to infiltration, as opposed to soils derived from loose materials and halomorphic soils. For a given soil type, rainfed crops are the primary land use that runs against infiltration. However, the effect of tillage varies according to the soil type. Finally, given the extent of vertisols and halomorphic soils in the Far North region of Cameroon in general, and in the Mayo Mizao watershed in particular, and regarding the increase in cultivated areas, a probable reduction in the infiltration capacity of soils in this region is to be expected in the medium term. The results of this study can be used as a basis for land-use planning and sustainable watershed management in semi-arid tropical zones.

**Keywords:** saturated hydraulic conductivity; infiltration tests; landscape feature; runoff; flood hazard

## 1. Introduction

The temporal evolution of the number of floods that occurred between 1900 and 2023 reveals an upward trend, marked by a resurgence in the 1950s [1]. The increase in the number of flood occurrences could be associated with the various changes to which hydrological catchments have always been subject. Watersheds exhibit great variability in landscape structure (topography, geology, soils, occupation, and land management). This variability, in turn, controls the distribution of precipitation into different components: soil moisture, runoff, evapotranspiration, infiltration, and groundwater flow [2]. Runoff and floods occur when one or a combination of the above factors runs against infiltration [3]. Given the ecological changes affecting our environment, a close relationship has been established between landscape structure and flooding [4]. In the Sahelian area, major environmental changes that have occurred in recent decades [5] are likely to have led to profound changes in the landscape, raising the risk of flood disasters. Indeed, the Sahelian population, whose way of life has long remained most dependent on natural resources, has led to a high ecological footprint on the environment [6], with a significant change in the landscape features. This phenomenon seems to be the cause of the decrease in the soil's infiltration capacity [7–10]. Sahelian population growth has been accompanied by urban sprawl [11] and a high demand for livelihoods [12,13]. Thus, with the aim of attempting to satisfy the needs of the population, the cropland's size and urban area increased while new farming practices were developed [14]. The combination of many factors will then have an impact on hydrosystem functioning, increasing runoff and river flows [15]. People will then have to face more environmental hazards. In Far North Cameroon, several disastrous flooding events have been recorded [16]. The geohistorical analysis performed by Bouba et al. [17] enabled the identification of approximately 21 large-scale flood disasters between 1977 and 2011. The Maroua urban area (the greatest city in Far North Cameroon) appeared to be the most affected by the flood disaster. Therefore, with the aim of improving the resilience of populations, it is crucial to facilitate the comprehension of the origin of this hazard. On the one hand, the results obtained by Bouba et al. [17] highlighted that, despite the increase in rainfall since 1990, the occurrence of flood hazards in Far North Cameroon is far from being a direct consequence of the rainfall pattern. On the other hand, some studies have linked the vegetation cover evolution with human activities [18–22]. Their results showed a loss of vegetation cover and a development of cropland. However, change in landscape and landform is a rapid phenomenon, while accommodation as a response to the hydro-environmental consequences of these changes is slow or non-existent. Thus, identifying the implications of landscape on hydro-environmental processes in a watershed is a key that can help predict the watershed's hydrological response and anticipate runoff and flash flood hazards. What incidence has the actual landscape on the saturated hydraulic conductivity of soils in the Mayo Mizao catchment? How sensitive is the Mayo Mizao watershed to runoff and flooding under different landscape conditions? The aim of this study was to assess the impacts of soil type, land use, and farming practices on the field-saturated hydraulic conductivity of soils in the Mayo Mizao watershed. To achieve this goal, a field experiment based on the determination of soil permeability was carried out.

## 2. Materials and Methods

Two main components of the methodology are presented: the field campaign method and the data analyzing processing methods.

### 2.1. Presentation of the Mayo Mizao Watershed

The Mayo Mizao watershed is located at 10.58° N; 14.20° E (Figure 1). It is actually a sub-basin of the Mayo Kaliao River, the main tributary of the Mayo Tsanaga River, which is also a tributary of the Logone River. The Logone River is a tributary of the Chari, which is the main source for providing water to Lake Chad [23].

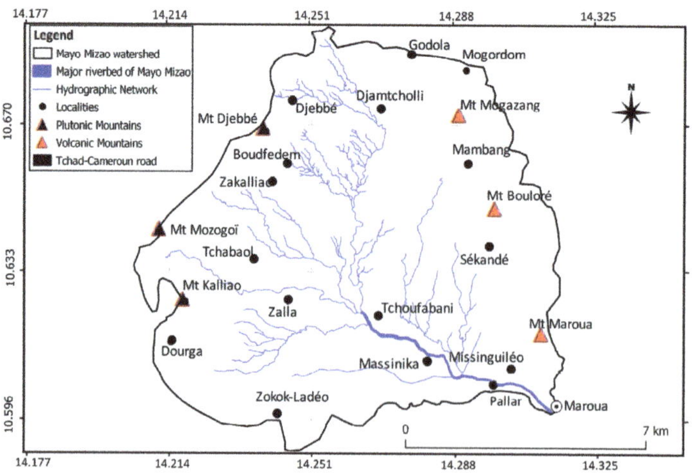


Figure 1. Location of the Mayo Mizao watershed.

The Mayo Mizao watershed size is 100 km$^2$. It is surrounded to the west by granitic mountains that form the Precambrian basement [24]. These mountains have altitudes ranging from 520 m (Mt Kaliao) to 640 m (Mt Djebbé) (Figure 1). Toward the east side, there is Mt Bouloré (700 m) and Mt Mogazang (700 m), made of volcanic and volcano-sedimentary rocks. The hypsometric characteristics of the watershed show, from upstream to downstream, three land surfaces: a surface of altitude greater than 440 m, a surface of altitude between 440 m and 420 m, and a surface of altitude less than 420 m. The slope varies in the same direction, from 3‰ to 16‰. The Gravelus compactness index (Kc = 1.21) is close to 1, giving the watershed a high compactness. The surface area is essentially covered by vertisols and gray soils with a halomorphic tendency. Vegetation that is mostly made of shrubs and grasses is characterized by a strong ecological footprint [20]. In terms of rainfall, the Mayo Mizao watershed area receives an annual rainfall of 800 mm. Between 2015 and 2019, daily rainfall varied from 1 mm to 170 mm, often generating severe flooding due to runoff. This was the case on 23 June 2018, when rainfall of 104 mm gave rise to significant floods observed in various localities of the watershed area.

### 2.2. Field Data Collection Process

Infiltration is the process of water entry into the soil. The saturated hydraulic conductivity $K_s$ is an essential parameter for infiltration because if the soil is saturated and homogeneous, it represents the limit value for the infiltration rate. Hydraulic conductivity can be determined in the field or in the laboratory. In this study, the determination of $K_s$ was performed on the field by applying two methods: the double-opened ring method used at a very shallow depth (5 cm) and the Porchet method (PM) at a depth of 20 cm. Both of these methods have a transient regime. They were requested because of their simplicity, accuracy, complementarity, and low implementation costs [25]. For each given test point, a double-ring method and a Porchet method were assigned. The 20 cm depth was chosen because it is considered to be a zone that is influenced by farming practices (plowing). According to Collinet and Lafforge [26], there is no clear relationship between soil infiltration capacity and the organization of their internal structures.

#### 2.2.1. Double-Opened Ring Method: Device and Operation

In order to overcome equipment shortages and achieve our goal in this study, the double-ring infiltrometer device was designed with care using rigid steel. It was modeled with two rings of 22 cm in diameter for the inner ring and 38 cm in diameter for the outer ring. The equipment was set up in the field by sinking the double ring 5 cm deep into the ground to prevent lateral water flow during the experiment. At the same time, the device was fitted with a levelogger for the automatic measurement of water level into the inner ring and two Frequency Domain Reflectrometry (FDR) probes that controlled variation in soil water content during the experiment. The two sensors were calibrated on the basis of measurements taken at a time resolution of 5 min. Once the device was installed, the levelogger and the FDR probes were switched on. The experiment was carried out by uniformly spreading around 400 mL of water inside the double ring every 30 s for 15 min to achieve a complete saturation of the ground. Depending on the soil conditions, the duration of any trial could take from 1 to 4 h.

#### 2.2.2. Porchet Method

The Porchet method was chosen because of its advantage in measuring permeability in multilayer soils [26]. The principle of the method consists of digging a hole, filling it with water, and measuring the temporal variation in water level. In this study, the hole was 20 cm deep and 10 cm in diameter. The use of pvc (PolyVinyl Chloride) pipe prevented the risk of the hole wall collapsing, which could distort the results of the experiment. Monitoring the decrease in the water level contained in the hole made it possible to calculate the hydraulic conductivity value by using empirical formulas.

### 2.2.3. Field Trial Deployment Method

Two main factors directed the deployment of the test points across the watershed area:

- The combination of soil type, land use, and farming practices;
- The steeper slope line of the watershed, which takes into account the different hydrogeomorphological aspects associated with the structural organization of the long profile shaped by erosion and alluvial deposits.

Indeed, in the first case, it is important to note that all soil types do not have the same occupation, and all cultivated plots do not have the same agricultural techniques. The location of any test point will, therefore, depend on all these factors. In the second case, the topography, which is defined by the longitudinal profile of the Mayo Mizao watershed, reflects a precise pedological organization shaped by water erosion (on upstream slopes) and sedimentation (downstream). The material removed from the slopes was deposited progressively, according to size. In this case, it may be noted that in the upstream of the watershed, soils are not very evolved and consist of coarse materials (products of physical alteration, nodules) that have been more or less reconstituted. In the middle stream, soils are clayey and well-reconstituted (evolved soils). In the downstream, there are colluvial deposits made up of fine, more or less powdery materials (silty-clay texture). Therefore, it seems worthwhile to take into account this structural organization of the landscape in the strategy aimed at defining the location of a field test point.

The number of test points was theoretically planned at first. Thus, for each land use class affected by any soil type, we assigned a test point (A in Table 1), and a total of 31 test points were required for each method. However, farming practices that are dependent on the will of the farmers can change on the same plot during the same rainy season. During the period of our field campaign, five farming practices were identified: rainfed crops plowed up with a machine; rainfed crops plowed up with a hoe; slash-and-burn crops; irrigated crops that are practiced in wetlands; fallowing (Figure 2). By taking into account this component of landscape (except irrigated land and fallowing), some of the soil types received more test points than others (B in Table 1) because of the difference observed in their agricultural development. The above criteria have resulted in a spatial distribution of test points in the field, as defined in Figure 3.

**Table 1.** Distribution of the number of test points performed in the Mayo Mizao watershed area.

| N° | Soils Types | Number of Test Points | | Identity of Test Points |
|---|---|---|---|---|
| | | A | B | |
| 0 | Bare rock (Granite) | 0 | 0 | Untested materials |
| I | Lithosols on andesite (sandy-clay texture) | 4 | 3 | $P_{24}$; $P_{25}$; $P_{27}$ |
| II | Stony soils derived from loose materials (Mokoya series) | 3 | 4 | $P_{13}$; $P_{14}$; $P_{15}$; $P_{22}$ |
| III | Soils derived from loose materials with a sandy-clay texture (Doyang series) | 2 | 2 | $P_{20}$; $P_{23}$ |
| IV | Gravelly soils derived from sandy materials | 4 | | Untested materials |
| V | Soils derived from loose material made of gray sand (Kodeck series) | 3 | 2 | $P_{21}$; $P_{31}$ |
| VI | Sandy-clay soils on the pediment, tending toward dark soils | 4 | 2 | $P_{26}$; $P_{28}$. |
| VII | Gray soils with halomorphic tendencies | 4 | 3 | $P_{19}$; $P_{29}$; $P_{30}$. |
| VIII | Vertisols («Hardé») | 4 | 6 | $P_1$; $P_{10}$; $P_{11}$; $P_{16}$;$P_{17}$; $P_{18}$ |
| IX | Undifferentiated halomorphic soils | 4 | 9 | $P_2$; $P_3$; $P_4$; $P_5$; $P_6$; $P_7$; $P_8$; $P_9$; $P_{12}$ |
| | Total | 31 | 31 | 31 |

Note: (A = theoretical planning; B = actual planning).

**Figure 2.** Main farming practices identified in the Mayo Mizao watershed.

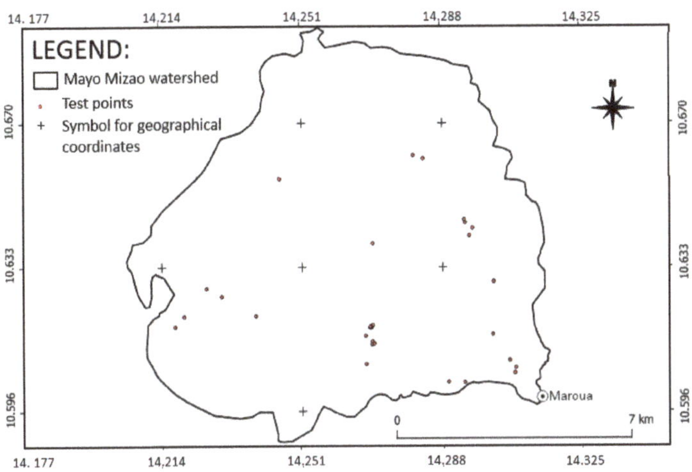


**Figure 3.** Number and spatial distribution of the test points through the watershed surface area.

### 2.3. Data Processing and Analysis

The data processing methods included two essential steps, namely the calculation of hydraulic conductivity values based on empirical formulas and analysis of these.

#### 2.3.1. The Calculation of the Saturated Hydraulic Conductivity Values

The saturated hydraulic conductivity values ($K_s$) are obtained by applying the empirical formulas to the raw data collected in the field when the vertical flow of water becomes constant, and the soil has reached its saturation state [25].

In the case of the double-ring test, we know that the flow inside the inner ring is considered to be constant. The quantity of infiltrated water is given by the equation below:

$$Q = V/t = (A1 \cdot \Delta h)/t \quad (1)$$

In Equation (1), $\Delta h/t$ is the expression of the speed of infiltration. Therefore, Darcy's law can be applied to the vertical flow descending inside the inner ring, considering that at the depth $Z_w$, the hydraulic load is null due to the influence of atmospheric pressure. In this case, the hydraulic gradient is given as follows:

$$i = (h'' * '' + Zw)/Zw = 1 + h * /Zw, \tag{2}$$

where h∗ is the hydraulic load applied inside the inner ring. The relationship of Darcy can be written as follows:

$$v = ki \tag{3}$$

Equating the two equations of the flow inside the inner ring leads to obtaining the formula of hydraulic conductivity, also called the permeability coefficient:

$$K_s = \Delta h/it = \Delta hZw/(t(h * +Zw)) \tag{4}$$

Considering Equation (4) above, we can note the following:
$Z_w$ (cm) = the depth at which the double ring was sunk into the ground;
H (cm) = hydraulic load;
A ($cm^2$) = the surface of the inner ring;
$\Delta h$ (cm) = the variation of water level during the time "t" taken by the experiment;
V ($cm^3$) = $A1 \cdot \Delta h$ = the volume of the infiltrated water.

In the case of the Porchet test, we can note that at the initial time $t_1$ (s), the water level in the hole is $h_1$ (cm). Then, at the final time, $t_2$ (s), the water level in the hole becomes $h_2$ (cm). The flow rate of water into the soil is given as follows:

$$Q = K_s SI, \tag{5}$$

with
Q ($cm^3/s$) = flow rate of water flow into the soil;
$K_s$ (cm/s) = saturated hydraulic conductivity;
I = driving slope;
S ($cm^2$) = wet section (bottom surface of the hole + side surface).

According to the Darcy theory,

$$I = h/l, \tag{6}$$

where l (cm) is the wet length. In this experiment, the driving slope I was not known, and we considered that I = 1, which is close to reality because it increases when the preliminary humectation is long [25].

In this case, Formula (5) becomes the following:

$$Q = K_s \cdot S \tag{7}$$

S, which represents the wet area of the hole, is expressed as follows:

$$S = \pi R^2 + 2\pi Rh, \tag{8}$$

where S ($cm^2$) = moistened surface of the hole, R (cm) = the section of the hole, h (cm) = wet height, and $\pi = 3.14$.

By integrating Equation (8) with Equation (7), we obtained the following:

$$Q = 2\pi K_s R\ (h + R/2) \tag{9}$$

On the other hand, the expression of the flow rate of water through the ground is the following:

$$Q = -\pi R2dh/dt \tag{10}$$

By equaling Equations (9) and (10), we obtained Equation (11) as follows:

$$-\pi R^2 dh/dt = 2\pi K_s R\ (h + R/2) \tag{11}$$

By transforming Equation (11), we obtained the Equation (12) as follows:

$$-R/Ks\ ((dh/R)/2 + h) = dt \tag{12}$$

By integrating (12), this became the following:

$$-R/2Ks \int_{h1}^{h2} \frac{\frac{dh}{R}}{2} + R = \int_{t1}^{t2} dt \tag{13}$$

Equation (12) can also be explained as follows:

$$-R/2Ks \left| \ln\left(\frac{R}{2} + h\right) \right|_{h1}^{h2} = t2 - t1 \tag{14}$$

Finally, Equation (15) represents the expression of the saturated hydraulic conductivity that governs the Porchet method:

$$K_s = R/2(t2 - t1)\ln(h1 + R/2)/(h2 + R/2) \tag{15}$$

For both the double-ring and Porchet methods, the $K_s$ values were corrected for temperature since the tests were performed in a warm environment where the temperature conditions were different from the reference temperature. For the Sudano-Sahelian environment, we considered that the reference temperature for this study was 20 °C, which corresponds to a dynamic water viscosity of 1 Centipoise and a correction factor of 0.66 (Table 2). The temperature correction is given by Equation (16):

$$K_{sTref} = K_T = n_T / n_{Tref} \tag{16}$$

with

$n_T / n_{Tref}$ : temperature correction factor that has no unit;

$K_{sT}(cm/s)$: saturated hydraulic conductivity measured at the field conditions;

$K_{sT_{ref:}}(cm/s)$: saturated hydraulic conductivity corresponding to the corrected temperature.

**Table 2.** Variation in field temperature correction factors according to reference temperature.

| $T_{ref}$ (°C) | 2 | 4 | 5 | 6 | 8 | 10 | 12 | 14 | 16 | 18 | 20 |
|---|---|---|---|---|---|---|---|---|---|---|---|
| Dynamic water viscosity (Centipoise) | 1.67 | 1.57 | 1.52 | 1.47 | 1.39 | 1.33 | 1.24 | 1.2 | 1.11 | 1.06 | **1** |
| $\frac{n_T}{n_{T_{ref}}}$ | 1.1 | 1.03 | 1 | 0.97 | 0.91 | 0.86 | 0.81 | 0.8 | 0.73 | 0.7 | **0.66** |

2.3.2. Data Analysis

The data analysis was based on graphical methods of data representation and then on statistical methods of data description, visualization, and comparison. Graphical data comparison methods allowed us to identify landscape conditions that were favorable or not favorable to infiltration. In addition, the use of descriptive statistics allowed us to characterize the distribution of saturated hydraulic conductivity values ($K_{sd}$ and $K_{sp}$). Thus, parameters such as mean, standard deviation, and extreme values were performed. Following this, the application of the Boxplot method made it possible to assess the dispersion of $K_{sd}$ and $K_{sp}$ values around their means and to then visualize more easily the so-called 'extreme values', which are more likely to be outliers. In addition, the normality of the distributions of $K_s$ values was verified by applying the Shapiro–Wilk test to the data series, and the Wilcoxon test was used to compare the two matched samples of field data ($K_{sd}$ and $K_{sp}$).

## 3. Results

### 3.1. Lithology, Land Use Type, and Hydraulic Conductivity ($K_s$) of Soils in the Mayo Mizao Watershed

The Mayo Mizao watershed contains nine classes of soil (Figure 4). Among them, vertisols and gray soils with a halomorphic tendency occupy nearly 70% of the watershed surface area. The eastern and western borders of the watershed are bordered by outcrops of crystalline rocks. To the east, there are volcanic formations, mainly olivine basalt and dolerite. To the west, we find plutonic formations consisting mainly of anatexis granite, syntectonic granite, or migmatite [24].

The land use features are given from the African land cover database in 2015. Four main land use classes can be, therefore, identified: rainfed crops, mixed crops/vegetation, mixed vegetation/crops and closed to open landscape (Figure 5).

With regard to soil hydraulic conductivity values, the table below (Table 3) presents the statistical distribution of $k_s$ ($K_{sd}$, $K_{sd}$). At a depth of 5 cm, $K_{sd}$ varies from 4 mm/h to 150 mm/h, with a mean of 36 mm/h and a standard deviation of 34 mm/h. However, at a depth of 20 cm, $K_{sp}$ data range from 0.05 mm/h to 416 mm/h, with a mean of 38 mm/h and a standard deviation of 92 mm/h. Thus, we can see that there is a strong dispersion of $K_s$ values around the means, especially with regard to data from the Porchet method ($K_{sd}$). Between the surface (5 cm) and the depth (20 cm), there is heterogeneity in hydraulic conductivity data, suggesting more extreme values of $K_{sp}$.

**Table 3.** Statistical distribution of hydraulic conductivity values for the Mayo Mizao watershed.

| Ground Depth | Variable | Number of Test Points | Range | Minimum | Maximum. | Mean | Std. Deviation |
|---|---|---|---|---|---|---|---|
| 5 cm | $K_{sd}$ (mm/h) | 31 | 146 | 4 | 150 | 36 | 34 |
| 20 cm | $K_{sp}$ (mm/h) | 31 | 416 | 0 | 416 | 38 | 92 |

### 3.2. Impact of Soil Type on the Variability of $K_s$ in the Mayo Mizao Watershed

The balance between infiltration and runoff depends on the soil type due to its structure [27] and its surface conditions [28–30]. In the Mayo Mizao watershed, the results of this study show that regardless of the measurement technique, vertisols are the lithological formation that has the highest resistance to infiltration. Resistance to infiltration is also remarkably observed in soils derived from loose materials with a sandy-clay texture, in gray soils with a halomorphic tendency, and in undifferentiated halomorphic soils (Figure 6). The impermeability of the soils mentioned above could be justified firstly by their low organic matter content and secondly by their grain size, most of which is dominated by fine materials, sometimes adding a high proportion of sandy materials [28,31]. It should be noted that the structural stability of soil depends on its intrinsic characteristics, which could make it more vulnerable to rainfall aggressiveness and, hence, to crusting. Therefore, since these soil types cover more than 75% of the Mayo Mizao watershed surface area, their implication in the genesis of runoff and flooding should rightly be considered in this study.

In addition, we were given the opportunity to compare the degree of permeability observed between the surface (5 cm) and depth (20 cm) using the Wilcoxon test. Before doing so, it is important to check the normality of the distribution of $K_s$ values by applying the Shapiro–Wilk test. Thus, to the null hypothesis $H_0$ (samples follow a normal distribution), we opposed the alternative hypothesis $H_a$ (samples do not follow a normal distribution). Regarding the results in (Table 4), we accept that the alternative hypothesis was retained for each of the two samples, as the calculated *p*-value (0.0001) was below the significance threshold $\alpha = 0.05$. This means that the distributions of the two samples ($K_{sd}$ and $K_{sp}$) did not follow a normal distribution and that the risk $\lambda$ of rejecting the null hypothesis $H_0$ when it is true was less than 1%.

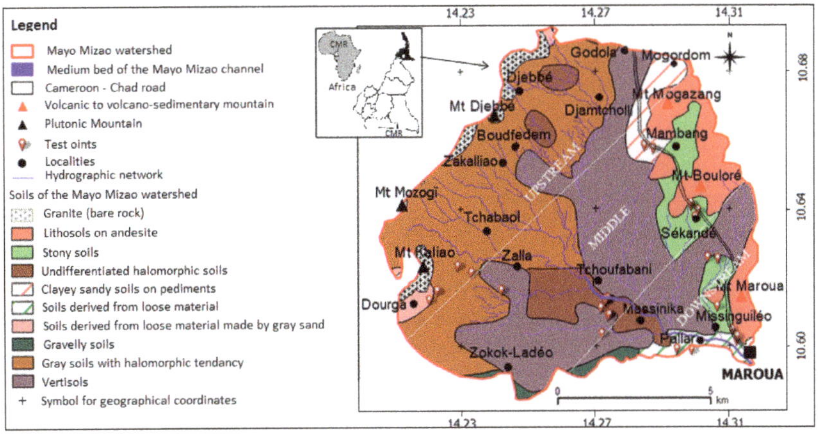


**Figure 4.** Soil class distribution in the Mayo Mizao watershed [31].

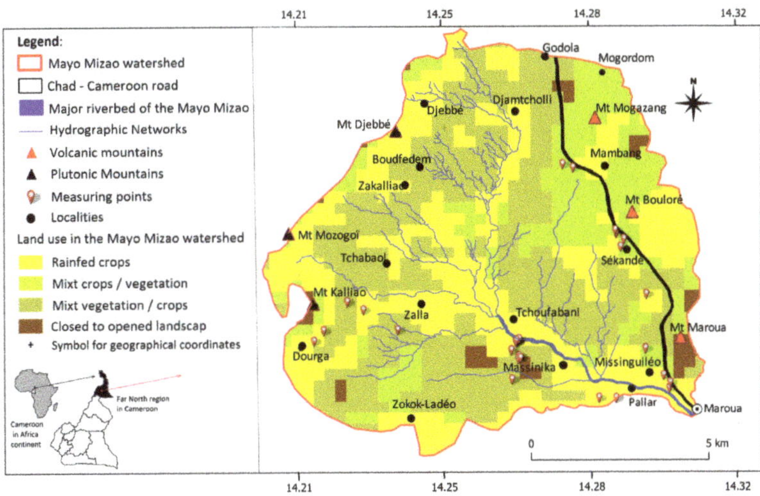


**Figure 5.** Distribution of the main land use classes in the Mayo Mizao watershed (http://2016 africalandcover20m.esrin.esa.int/ accessed on 11 August 2016).

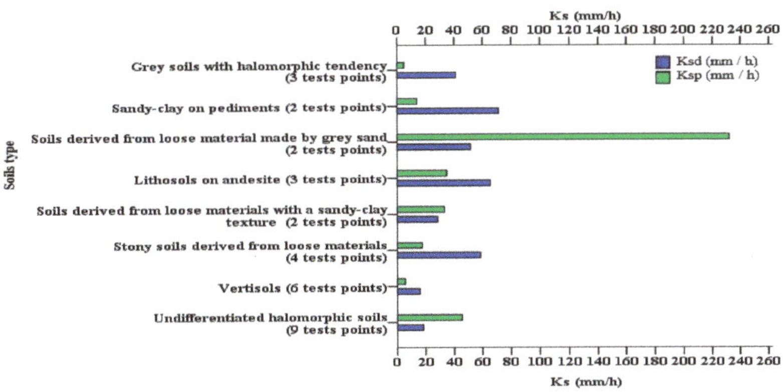


**Figure 6.** Variability of $K_S$ values according to the soil type in the Mayo Mizao watershed.

**Table 4.** Results of the Shapiro–Wilk test.

| Statistic Parameters | $K_{sd}$ (mm/h) | $K_{sp}$ (mm/h) |
|---|---|---|
| W | 0.780 | 0.432 |
| *p*-value | <0.0001 | <0.0001 |
| Significant level (α) | 0.05 | 0.05 |
| Risk of rejecting the true hypothesis (λ) | 0.01% | 0.01% |

According to the Wilcoxon test itself, the principle consists of testing the null hypothesis $H_0$ (the distribution of the two samples is not significantly different) against the $H_a$ hypothesis (the distribution of the two samples is significantly different). The results show that the alternative hypothesis $H_a$ (the two distributions are significantly different) was supported, and the risk λ of these results being considered true when they are false was less than 1.07% for the first sample ($K_{sd}$) and less than 4.89% for the second sample ($K_{sp}$). The results of this test (Table 5) mean that for the same type of soil, hydraulic conductivity values at the topsoil are different, with some obtained at depth. The variability of permeability at the soil surface area and depth is then under the control of different factors. It may be considered that the difference in the hydrodynamic behavior of soils is due to the organization and the structure of the soil layers, generally shaped by the nature of the soil, the land use type, and the cultivation practices.

**Table 5.** Results of the Wilcoxon test applied to $K_{sd}$ and $K_{sp}$ values.

| Statistic Parameters | $K_{sd}$ (mm/h) | $K_{sp}$ (mm/h) |
|---|---|---|
| V | 23 | 349 |
| Expectation | 15,500 | 248 |
| Variation (V) | 7750 | 2,603,875 |
| *p*-value | 0.011 | 0.049 |
| Significant level (α) | 0.05 | 0.05 |
| Risk of rejecting the true hypothesis (λ) | 1.07% | 4.89% |

Analysis of the results in (Figure 7) shows that, for the majority of individual test points, the soils have high surface permeability and low depth permeability ($K_{sd} > K_{sp}$), with the exception of the following test points: 150813P4, 150901P7, 150901P8, 160309P13, 160317P17, 160930P20, 161001021, and 170320P31, where $K_{sd} < K_{sp}$. This result may not always be true when considering the weighted average of $K_s$ values obtained from test points carried out on the same soil type (Figure 6). The exceptions mentioned above are linked, on the one hand, to an abundance of granular material on the surface and, on the other hand, to the modification of the surface condition due to encrusting. To better understand this, we need to look at material texture. Indeed, it should be noted that the silty-sandy texture, sandy texture, or sandy-clay texture of soil is a factor that is at the origin of soil crusting. Clearly, a crusted surface will inhibit infiltration [29,32]. The inverse gradient of permeability can be considered the result of an accumulation of alluvia upstream or the consequence of colluvial deposits left by flood damping in major beds. Similarly, the difference in soil permeability between the surface and the lower depth could also be explained by the differences in the layers' structural homogeneity or in the operation between the two infiltration measurement methods. In fact, during the double-ring test, the lateral flow of water is prohibited by the outer ring; this makes the vertical flow more predominant, and the test should not concern many layers. In the case of the Porchet test, the flows are carried out in the vertical and lateral directions, and many layers may be affected by the test. In addition, we can see that the Porchet test provided very spread-out hydraulic conductivity values (upwards) with so many extreme values (332 mm/h and

415 mm/h) (Figure 8). The dispersion of the $K_{sp}$ values could express the effect of the lateral flow during the Porchet experiment.

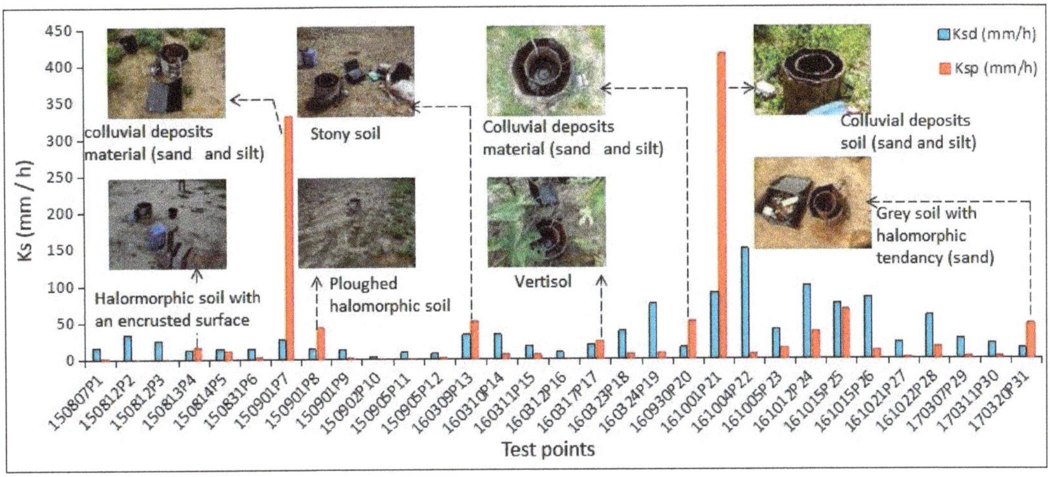


**Figure 7.** Comparison of hydraulic conductivity values between the surface (5 cm) and the soil depth (20 cm).

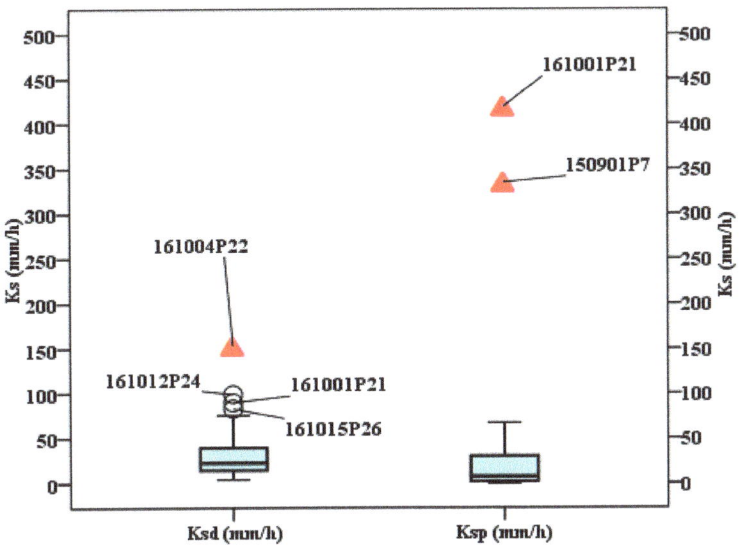


**Figure 8.** Distribution of hydraulic conductivity values obtained using the double-ring test and the Porchet test.

*3.3. Influence of Land Use Type on the Variability of $K_s$ in the Mayo Mizao Watershed*

According to Shukla [33] and Di et al. [34], land use has direct or indirect effects on infiltration capacity. In the case of the Mayo Mizao watershed, four main land use classes were identified and tested: rainfed crops; mixed crops (rainfed crops, off-season crops, irrigated crops)/vegetation (pasture, shrub, forest); mixed vegetation (pasture, shrub, forest)/crops (rainfed crops, off-season crops, market gardening), and finally, the closed to opened landscape (hardwood or softwood; evergreen or deciduous; shrubs). The results of the field experiments show that the cropland landscape is more favorable to infiltration than the covered land (closed to opened landscape, mixt vegetation/crops) (Figure 9).

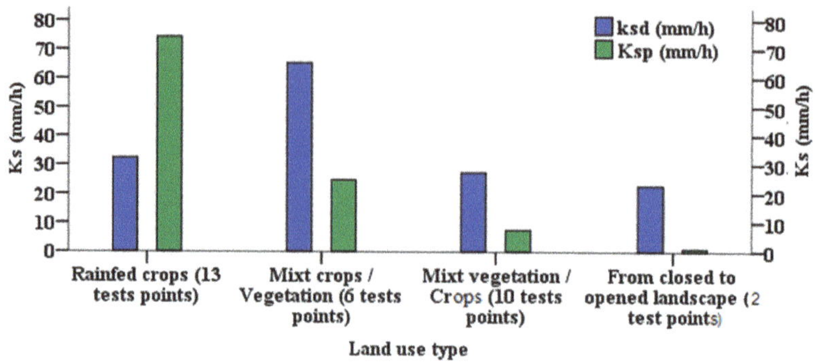


**Figure 9.** Variability in saturated hydraulic conductivity related to land use type in the Mayo Mizao catchment.

The results above could be explained by the fact that tillage, which leads to the creation of clods, aggregates, and pore networks, makes the soil structure looser, less compact, and more permeable to water flow [28–30,35]. The high hydraulic conductivity values of the croplands and plowed lands have also been observed in other West African catchment areas, particularly in the Tougou catchment in northern Burkina Faso [36]. However, it is prudent to express reservations about the involvement of tillage in increasing soil hydraulic conductivity values. According to previous studies that have been carried out in both the Sahelian area [29,30,37,38] and in other bioclimatic environments, in particular in the Mediterranean area [35,39,40], the increase in soil infiltration capacity following plowing constitutes a dynamic state during that growing season. Thus, under the influence of the intensity and cumulative rainfall, the conditions of cropland surface are enhanced over time toward an unfavorable state for infiltration, as under the impact of the energy of successive raindrops, the aggregates and porosity initially generated by tillage fade after a certain time. From then on, encrusting is reconstituted at the surface, and subsequent rains, even of low intensity, will be sufficient to generate runoff. According to Andrieux [35], cumulation and rainfall intensities are not the only factors responsible for the decrease in infiltration capacity in the croplands. In fact, soils with high clay content in the first 10 cm of depth are also responsible for the decrease in soil infiltration capacity. When comparing the values of $K_{sd}$ and $K_{sp}$ obtained in the same given land use class, we can see that $K_{sd} > K_{sp}$, except in the case of plots occupied by rainfed crops, where $K_{sd} < K_{sp}$. This could once again reflect the short-term consequences of tillage activity in association with rainfall intensity and accumulation. However, it seems that the combination of the effects resulting from soil type and land use type could influence hydraulic conductivity in one direction or in the other, depending on the case.

### *3.4. Combined Effects of Soil Type and Land Use Type on the Variability of $K_s$*

By taking into account the combined effects of soil type and land use type (Figure 10), we realized that for the same soil type, $K_s$ vary according to the land use type. For example, in the case of vertisols, $K_s$ is found to be higher on mixed vegetation/crop plots than on rainfed plots. Similarly, it can be seen that for the same land use type, $K_s$ varies according to the soil type. For example, it can be noted that for a plot occupied by mixed vegetation/crops, $K_s$ is higher in gray soils with a halomorphic tendency (at the surface) than in vertisols. Such observations, therefore, allow us to confirm that the preponderance of each soil type over the variability of $K_s$ depends on the land use type, and the opposite is also true. However, the part of the influence of each landscape feature was not determined.

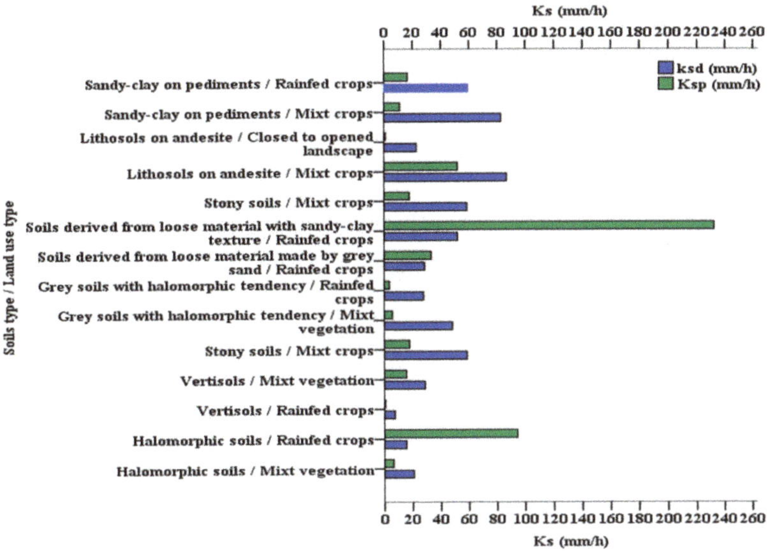


**Figure 10.** Combined effects of soil type—land use type on the variability of the saturated hydraulic conductivity in the Mayo Mizao watershed.

### 3.5. *Impact of the Cultivation Practices on the Variability of $K_s$ in the Mayo Mizao Catchment*

The influence of farming practices was tested on two of the most cultivated soil types in the Mayo Mizao catchment. These are vertisols and undifferentiated halomorphic soils. These two soil classes cover nearly half of the surface area of this catchment.

#### 3.5.1. Farming Practices Conducted on the Vertisols

The spatial variability of saturated hydraulic conductivity as a function of farming practices on vertisols shows that plowed plots have the lowest $K_s$ values (A in Figure 11). Therefore, we can consider that the practice of plowing on vertisols leads to a decrease in infiltration and an increase in runoff. This result is in line with those obtained by Seyni-Boukar [37] and Peugeot [29]. Thus, after studying the influence of tillage on the infiltration capacity of Sahelian soils, the above authors came to the conclusion that tillage is responsible for increasing runoff by reducing the infiltration capacity in vertisols. According to them, plowing helps to bring back to the surface the elementary particles buried in the soil. From then on, and under the effect of the kinetic energy of the raindrops, these elementary particles will cause the intergranular pores to clog. The result is the formation of an impermeable shell that prevents water infiltration. Similarly, Fiès and Castelao-Guegunde [39] also experimented with the effect of tillage on runoff. Thus, by applying kinetic energy rains ranging from 70 to 340 $J \cdot m^{-2}$ to a plowed silty-clayey soil, they noticed that the volume of structural pores decreases by 60% when the kinetic energy of the rain increases. They then estimated a reduction in soil infiltrability by a factor of 10. Such a mechanism of structural soil reorganization induced by tillage is particularly characteristic of vertisols due to their high fine content [37].

Contrary to the effect of plowing on the hydraulic conductivity of vertisols, the results show that slash-and-burn farming and fallowing seem to increase infiltration rather than runoff. This result is supported by [41]. Indeed, slash-and-burn farming has a reducing effect on the infiltration capacity of soils because of its structural modification. The magnitude of the effects of slash-and-burn farming on the hydraulic conductivity of soils would be dependent on the intensity of the calcination or the length of time between the pre-burnt and post-burnt periods. However, due to the high spatial and temporal variability of rainfall, there are many studies conducted under natural rainfall (rather than simulated rainfall) that have shown a more stimulating role for bushfire on soil infiltration capacity. Thus, to talk about the Mayo Mizao watershed, where this study was carried out under simulated rain, one would rather have expected a reducing effect from the practice of slash-and-burn farming on the infiltration capacity of the soil, which is quite the opposite. As such, one could attribute this difference in the hydrodynamic behaviors of burned land to the nature of the soil rather than to an agricultural practice itself. It would, therefore, be interesting to note that in vertisols, there are networks of shrinkage slits that constitute a significant source of permeability and whose densification could come from the desiccation process generated by the practice of burning [28]. In the case of fallowing, it should be noted that the increase in infiltration observed in this study could be explained by the influence of vegetation cover in the infiltration–runoff relationship. Vegetation cover is involved at several levels in the fight against runoff [42]. Thus, thanks to the interception of raindrops by the aerial parts of the plant (leaves, branches, and trunk), vegetation cover contributes to reducing the erosive energy of rainfall by preventing the splash effect. Thanks to the litter and bushes, the vegetation cover also acts directly on runoff by acting as an obstacle to the water flow. In addition, thanks to the root system, the plant cover ensures that the soil's structural cohesion is maintained. Similarly, the development of organic matter from plant residues improves soil porosity and, therefore, promotes infiltration. To summarize, the action of vegetation in increasing infiltration takes place directly and indirectly. In terms of the vertical variability of hydraulic conductivity, it can be seen that whatever the farming practice, the $K_s$ values are higher on the surface than at depth.

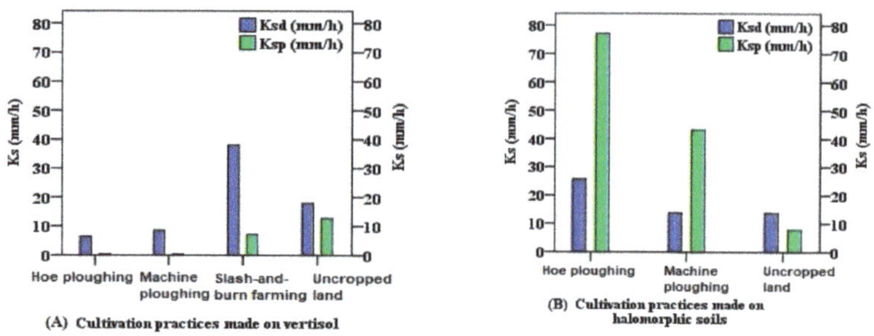


**Figure 11.** Variability of hydraulic conductivity according to the farming practices made on the vertisols and on the halomorphic soils.

### 3.5.2. Farming Practices Conducted on the Halomorphic Soils

The hydraulic conductivity ($K_s$) on halomorphic soils is higher in plowed plots than in fallowed plots (B in Figure 11). This situation is quite the opposite of that observed in vertisols. We can see that the influence of soil type on hydraulic conductivity inhibits farming practices. In fact, halomorphic soils are indurated soils with low organic matter content and a high content of sandy or gravelly materials. Plowing breaks down the compactness of the land, making it looser and more permeable to water. This phenomenon has also been observed by Segalen [31], Casenave and Valentin [28], and Peugeot [29] in the same Sahelian environment. However, the practice of fallowing on halomorphic soils leads to reduced infiltration. Moreover, Fournier et al. [43], who studied the role of fallowing land on runoff and soil erosion in the locality of Bondukury in Burkina Faso, found the opposite results. The results of their study conducted on sandy soils with a low slope (5%) showed that the fallowed plots had the lowest annual runoff coefficients, i.e., 13% compared with 20% for cultivated plots and 50% for bare soil. However, the comparison between the results of the various studies should take into account the stage of evolution of the fallowed land itself [43]. It should, therefore, be noted that the capacity of fallowing to promote infiltration depends on the presence and extent of bedding or the density of the plant cover. In Far North Cameroon and specifically in the Mayo Mizao catchment, vegetation is almost entirely grassy, while the time allocated to fallowing is relatively short. Consequently, the mineral soils of this region are poorly covered with litter [28]. In addition, it should be noted that halomorphic soils in Far North Cameroon have mechanical properties that are not consistent with infiltration in the first 10 cm of depth due to their prismatic and compact structure [31]. During fallowing, the degree of the compactness of this soil should increase, and so runoff should increase as well [30].

To summarize the above paragraphs, we can note that the variability in hydraulic conductivity under different farming practices in the Mayo Mizao watershed suggests two contradictory situations depending on the nature of the soil where the tests were performed. In the case where cultivation practices are carried out on vertisols, plowing seems to be responsible for the decrease in infiltration. The practice of slash-and-burn farming and then fallowing, on the other hand, has less significant impacts, as it is inhibited by the natural properties of the vertisols. In the case of halomorphic soils, plowing seems to increase infiltration. This hydrodynamic behavior of plowed halomorphic soils is linked to their high sand content and to the dismantling of the surface level area, which gives this soil a looser and more permeable structure. On the other hand, halomorphic soils on which fallowing is practiced are against infiltration due to the encrusting that develops at the surface layers.

## 4. Discussion

Many researchers from all over the world have reported that different types of soil and land use have different abilities in passing water into the ground. Bare soil tends to be prone to erosion, reducing the soil's infiltration capacity and increasing surface runoff. Increasing the runoff coefficient will increase the peak discharge in a watershed. The decrease in the river capacity due to sediment can cause a river flood. According to Basga et al. [44], flooded vertisols from northern Cameroon have a high content of clay that displays dispersion and flocculation under rainfall and then inhibits vertical infiltration. Shabtai et al. [45] studied the effects of soil and land use change on the structure and hydraulic conductivity ($K_s$) of vertisols in northern Ethiopia. Their results are similar to ours. Indeed, they observed a difference in saturated hydraulic conductivity between the surface of vertisols and depth. According to them, the low saturated hydraulic conductivity values observed at the surface are associated with the phenomenon of swelling of vertisols in contact with rainwater. These swellings are, therefore, higher at the surface than at depth and in cultivated soils than in savannah and forest soils. They, therefore, concluded that the conversion of natural savannah vegetation to crops and tillage operations must have destabilized the initial soil structure in cultivated plots, which in turn reduced the ability of the soil to infiltrate. Several other studies carried out in the Sahelian zone [37,46–49] have also confirmed the results of this study by concluding that the internal (clay and sandy-clay textures) and external soil factors, such as farming practices (plowing and slash-and-burn farming), lead to the crusting and then to the reduction in soil infiltration capacity. Similarly, numerous studies carried out in other semi-arid regions have produced the same conclusion. It is the case of Chartier et al. [50] in Patagonia (Argentina), Anderson [51] in the lower Mississippi River Valley, and Liu et al. [52] in Shandong (China). To conclude this paragraph, we can say that the involvement of vertisols and halomorphic soils on the one hand and cultivated land and plowing practices on the other hand in the genesis of floods is all the more significant given their considerable extent in the Far North region of Cameroon.

Overall, it should be noted that poor water transfer within the surface formations of a watershed is the cause of runoff and flooding. In the case of the Mayo Mizao watershed, soil types marked by grain seize and its external structural organization have an influence on infiltration and, therefore, on runoff and flooding. In addition, it should be noted that the evolution of landscapes, associated with changes in land use and land cover, and then certain farming techniques, are factors responsible for the genesis of runoff and flooding. Prior identification of the landscape factors behind flooding can help people improve flood hazard management and mitigation strategies. Thus, a number of land management actions can be applied to improve the infiltration capacity of soils across the catchment studied or across a similar environment. Strategies can be agronomic, hydraulic, or ecosystem-based. On the agronomic level, local farmers should limit machine plowing practices or, on the contrary, practice mulching in order to limit the impact of raindrops on soil erosion and crusting [53]. Similarly, practices such as crop rotation, fallowing, humification of cultivated plots [54], and subsoiling should be encouraged to ensure increased soil infiltration capacity. Ecologically, deforestation control and reforestation policies for denuded watersheds should be implemented to limit runoff and soil erosion [55]. Similarly, farmers should practice terrace cropping in steeply sloping areas (hillsides) to minimize runoff and soil water erosion [56]. According to hydraulic measures, farmers should ensure sediment trapping or set up temporary rainwater retention structures (folds, dikes, benches, ponds, hillside reservoirs, embankments, hedges, vegetated cordons) to slow runoff across the watershed [52,57].

To conclude this section, we can say that the involvement of vertisols and halomorphic soils on the one hand and cultivated land and plowing practices on the other in the genesis of floods is all the more significant given their considerable extent in the Far North region of Cameroon. Thus, through watershed redevelopment policies and the practice of agricultural techniques appropriate to each soil type, we could achieve sustainable flood management in the area concerned.

## 5. Conclusions

In a nutshell, this study shows that the genesis of runoff in the Mayo Mizao watershed is dependent on soil type, land use, and farming practices. The influences of these factors are simultaneous. However, with regard to soil type, it should be noted that vertisols, halomorphic soils, and gray soils with a halomorphic tendency are the three main soil types that offer considerable resistance to water infiltration. This is probably due to the high content of fine materials, the compactness that varies, and the advance or earlier stage of pedogenesis. As far as the influence of land use in infiltration, it should be noted that the role of croplands is relevant. With regard to the role of farming practices, it should be noted that, apart from plowing, which favors infiltration during the early stages of the rainy season (but is unfavorable in the advanced stage of the rainy season), the impacts of most agricultural practices on infiltration are strongly inhibited by the soil type.

This study is a contribution to a better understanding of the socio-ecological and hydrological functioning of the Sahelian watersheds, which have recently been dominated by many extreme hydrological events.

**Supplementary Materials:** The following supporting information can be down-loaded at https://www.mdpi.com/article/10.3390/w16121672/s1, Table S1: Methodology of the determination of the field saturated hydraulic conductivity values obtained by the double ring infiltrometry method; Table S2: Methodology of the determination of the field saturated hydraulic conductivity values obtained by the Porchet infiltrometry method.

**Author Contributions:** L.B. designed the project, prepared the figures, and wrote the manuscript. P.-A.A. designed the methodology and supplied the field instruments. S.S., P.-A.A. and L.B. read and improved the manuscript. All authors have read and agreed to the published version of the manuscript.

**Funding:** The authors declare that they have not received any funding.

**Data Availability Statement:** The data supporting the results of this study are available within this paper and in Supplementary Information File.

**Conflicts of Interest:** The authors declare no conflicts of interest.

## References


1. EM-DAT The International Disaster Database. CRED/UCLouvain. 2023. Available online: https://public.emdat.be (accessed on 12 August 2023).
2. Assefa, M.M.; Wossenu, A. *Landscape Dynamics, Soils and Hydrological Processes in Varied Climate*; Springer International Publishing: Cham, Switzerland, 2016; 822p.
3. Rogger, M.; Agnoletti, M.; Alaoui, A.; Bathurst, J.C.; Bodner, G.; Borga, M.; Chaplot, V.; Gallart, F.; Glatzel, G.; Hall, J.; et al. Land use change impacts on floods at the catchment scale: Challenges and opportunities for future research. *Water Resour. Res.* **2017**, *53*, 5209–5219. [CrossRef] [PubMed]
4. Yuan, Y.; Fang, G.; Yan, M.; Sui, C.; Ding, Z.; Lu, C. Flood-Landscape Ecological Risk Assessment under the Background of Urbanization. *Water J.* **2019**, *11*, 1418. [CrossRef]
5. Nwilo, P.C.; Olayinka, C.J.; Okolie, E.I.; Emmanuel, E.I.; Orji, M.J.; Daramola, O.E. Impacts of land cover changes on desertification in northern Nigeria and implications on the Lake Chad Basin. *J. Arid Environ.* **2020**, *181*, 104190. [CrossRef]
6. Cotillon, S.E. *Séries Chronologiques de L'utilisation et de L'occupation des Terres en Afrique de l'Ouest*; Fact Sheet; U.S. Geological Survey: Reston, VA, USA, 2017; 4p. [CrossRef]
7. Amogu, O.; Esteves, M.; Vandervaere, J.P.; Malam, A.M.; Panthou, G.; Rajot, J.-L.; Souley, Y.K.; Boubkraoui, S.; Lapetite, J.-M.; Dessay, N.; et al. Runoff evolution due to land-use change in a small Sahelian catchment. *Hydrol. Sci. J.* **2014**, *60*, 78–95. [CrossRef]
8. Descroix, L.; Mahé, G.; Olivry, J.C.; Albergel, J.; Tanimoun, B.; Amadou, I.; Coulibaly, B.; Bouzou, M.I.; Faran, M.O.; Malam, A.M.; et al. Facteurs anthropiques et environnementaux de la recrudescence des inondations au Sahel. In *Les Sociétés Rurales Face Aux Changements Climatiques et Environnementaux en Afrique de L'ouest*; Sultan, B., Lalou, R., Amadou Sanni, M., Oumarou, A., Soumaré, M.A., Eds.; IRD: Wellington, New Zealand, 2015; pp. 153–170.
9. Hermans, K.; McLeman, R. Climate change, drought, land degradation and migration: Exploring the linkages. *Curr. Opin. Environ. Sustain.* **2021**, *50*, 236–244. [CrossRef]
10. Spinoni, J.; Barbosa, P.; Cherlet, M.; Forzieri, G.; McCormick, N.; Naumann, G.; Vogt, J.V.; Dosio, A. How will the progressive global increase of arid areas affect population and land-use in the 21st century? *Glob. Planet. Chang.* **2021**, *205*, 103597. [CrossRef]
11. Chen, X.; Tian, C.; Meng, X.; Xn, Q.; Cui, G.; Zhang, Q.; Xiang, L. Analyzing the effect of urbanization on flood characteristics at catchment levels. *Proc. IAHS* **2015**, *370*, 33–38. [CrossRef]
12. Guengant, J.P.; Banoin, M.; Quesnel, A. *Dynamique des Populations, Disponibilités en Terres et Adaptation des Régimes Fonciers: Le Cas Du Niger*; FAO et CICRED: Rome, Italy, 2003; 144p.
13. Brandt, M.; Rasmussen, K.; Peñuelas, J.; Tian, F.; Schurgers, G.; Verger, A.; Mertz, O.; Palmer, J.R.B.; Fensholt, R. Human population growth offsets climate-driven increase in woody vegetation in sub-Saharan Africa. *Nat. Ecol. Evol.* **2017**, *1*, 0081. [CrossRef] [PubMed]
14. CILSS. *Les Paysages de L'afrique de L'ouest: Une Fenêtre sur un Monde en Pleine Evolution*; U.S. Geological Survey EROS: Garretson, SD, USA, 2016; 236p.
15. Gbohoui, Y.P.; Paturel, J.-E.; Fowe, T.; Karambiri, H.; Yacouba, H. Impacts des changements climatique et environnemental sur la réponse hydrologique du Nakambé à Wayen (Burkina Faso) à travers le cadre de budyko. *Proc. IAHS* **2021**, *384*, 269–273. [CrossRef]
16. Ndongo, B.; Mbouendeu, L.S.; Hiregued, J.P. Impacts socio-sanitaires et environnementaux de la gestion des eaux pluviales en milieu urbain sahélien: Cas de Maroua, Cameroun. *Afr. Sci.* **2015**, *11*, 237–251.
17. Bouba, L.; Sauvagnargues, S.; Gonné, B.; Ayral, P.-A.; Ombolo, A. Trends in rainfall and flood hazard in the Far North region of Cameroon. *Geo-Eco-Trop* **2017**, *41*, 339–358.
18. Fotsing, E. Small Savannah: An Information System for the Integrated Analysis of Land Use Change in the Far North of Cameroon. Ph.D. Thesis, Universiteit Leiden, Leiden, The Netherlands, 2009; 376p.
19. Wafo, T.G.; Fotsing, J.M. Quantification de l'évolution du couvert végétal dans la réserve forestière de Laf-Madjam au nord du Cameroun par télédétection satellitale. *Sécheresse* **2010**, *21*, 169–178.
20. Leroux, L.; Oszwald, J.; Ngounou, N.B.; Sebag, D.; Penven, M.J.; Servat, E. Le bassin versant du Mayo-Tsanaga (Nord Cameroun): Un bassin versant expérimental pour une compréhension des relations Homme/Milieu. *Rev. Française Photogram. Téléd.* **2013**, *202*, 42–54. [CrossRef]
21. Beidi, E.; Souaré, A.; Adamou, I. Evaluation of anthropization indicators of the floristic landscapes of Kaélé hills in the Sudano-Sahelian zone of Cameroon. *Environ. Chall.* **2021**, *5*, 100393. [CrossRef]
22. Kodji, P.; Tchobsala, A.I. Use of plant resources by refugees from Minawao and their impact on the Sahelian savannah of Cameroon. *Environ. Chall.* **2021**, *5*, 100270. [CrossRef]
23. Olivry, C. Fleuves et Rivières du Cameroun. *Monog. Hydrol. ORSTOM* **1986**, *9*, 781.
24. Morin, S. «Géomorphologie». In *Atlas de la Région de l'Extrême-Nord Cameroun*; Seignobos, C., Iyébi-Mandejeck, O., Eds.; IRD: Paris, France, 2000; pp. 13–17.
25. Chossat, J.-C. *La Mesure de la Conductivité Hydraulique Dans les Sols. Choix des Méthodes*; Lavoisier: Paris, France, 2005; 720p.
26. Collinet, J.; Lafforgue, A. *Mesures de Ruissellement et de L'érosion Sous Pluies Simulées Pour Quelques Types de Sols de Haute Volta*; ORSTOM: Abidjan, Cote d'Ivoire, 1979; 129p.
27. Masse, D. *Amélioration du Régime Hydrique des Sols Dégradés en Vue de Leur Réhabilitation. Cas des Vertisols au Nord Cameroun*; ORSTOM: Paris, France, 1993; 142p.

28. Casenave, A.; Valentin, C. *Les États de Surface de la Zone Sahélienne: Influence sur L'infiltration*; ORSTOM: Paris, France, 1989; 231p.
29. Peugeot, C. Influence de L'encroûtement Superficiel du Sol Sur le Fonctionnement Hydrologique D'un Versant Sahélien (Niger). Expérimentations In-Situ et Modélisation. Ph.D. Thesis, Joseph Fourier University, Grenoble, France, 1995; 355p.
30. Boli, B.Z.; Roose, E.; Bep, A.; Ziem, B.; Kallo, S.; Waechter, F. Effets des techniques culturales sur le ruissellement, l'érosion et la production de coton et maïs sur un sol ferrugineux tropical sableux. Recherche de systèmes de culture intensifs et durables en région soudanienne du Nord-Cameroun (Mbissiri, 1991–1992). *Cah. Orstom Sér. Pédol.* **1993**, *28*, 309–325.
31. Segalen, P. *Carte Pédologique du Nord Cameroun au 1/1000 000. Feuille Maroua*; ORSTOM: Paris, France, 1962; 71p.
32. Tessier, D. Rôle de l'eau sur les propriétés physiques des sols. *Sécheresse* **1994**, *5*, 143–150.
33. Shukla, M.K. *Soil Hydrology, Land Use and Agriculture. Measurement and Modelling*; CAB Intern.: Perai, Malaysia, 2011; 455p.
34. Di, S.; Hong, Y.; Dexin, G.; Ming, Y.; Jiabing, W.; Fenghui, Y.; Changjie, J.; Anzhi, W.; Yushu, Z. The effects of land use change on soil infiltration capacity in China: A metaanalysis. *Sci. Total Environ.* **2018**, *626*, 1394–1401.
35. Andrieux, P. Effets des Pratiques Culturales sur le Ruissellement et L'érosion. Vigne, Sol et Environnement. Une Rencontre Profession-Recherche en Languedoc-Roussillon. Campus Agro-Inra Montpellier. 2006. Available online: http://sol.ensam.inra.fr/lisah/ (accessed on 12 August 2023).
36. Zouré, C.; Queloz, P.; Koïta, M.; Niang, D.; Fowé, T.; Yonaba, R.; Consuegra, D.; Yacouba, H.; Karambiri, H. Modelling the water balance on farming practices at plot scale: Case study of Tougou watershed in Northern Burkina Faso. *Catena* **2019**, *173*, 50–70. [CrossRef]
37. Seiny-Boukar, L.; Floret, C.; Moukouri, K.H.; Pontanier, R. *Dégradation des Vertisols Dans le Nord-Cameroun: Modification du Régime Hydrique des Terres et Tentative de Réhabilitation. Utilisation Rationnelle de L'eau des Petits Bassins Versants en Zone Aride*; AUPERLF-UREF John Libbey Eurotext: Paris, France, 1991; pp. 287–291.
38. Lamachère, J.M. Aptitude au ruissellement et à l'infiltration d'un sol sableux fin après sarclage. Soil Water Balance in the Sudano Sahelian Zone. In Proceedings of the Niamey Workshop, Niamey, Niger, 18–23 February 1991; IAHS Publisher: Wallingford, UK, 1991; 12p.
39. Fiès, J.C.; Castelao-Gegunde, A.M. Modification de l'espace poral des croutes de surface sous l'action des pluies et conséquences sur l'infiltrabilité. *Agronomie* **1996**, *16*, 367–379. [CrossRef]
40. Blavet, D.; De Noni, G.; Roose, E.; Maillo, L.; Laurent, J.-Y.; Asseline, J. Effets des techniques culturales sur les risques de ruissellement et d'érosion en nappe sous vigne en Ardèche (France). *Secheresse* **2004**, *15*, 111–120.
41. Haigh, M.J. Fire effects on soils and restoration strategies. *Land Reconstr. Manag.* **2009**, *5*, 579.
42. Rey, F. *Restaurer les Milieux et Prévenir les Inondations Grâce au Génie Végétal*; Quae: Versailles, France, 2018; 119p.
43. Fournier, J.; Serpantié, G.; Delhoume, J.P.; Gathelier, R. Rôle des jachères sur les écoulements de surface et l'érosion en zone soudanienne du Burkina. Application à la gestion des terres cultivées. *Sud Sci. Technol.* **2000**, *5*, 12. [CrossRef]
44. Basga, D.S.; Tsozué, D.; Temga, J.P.; Balna, J.; Nguetnkam, J.P. Land use impact on clay dispersion/flocculation in irrigated and flooded vertisols from northern Cameroon. *Int. Soil Water Conserv. Res.* **2018**, *6*, 237–244. [CrossRef]
45. Shabtai, I.A.; Shenker, M.; Edeto, W.L.; Warburg, A.; Ben-Hur, M. Effects of land use on structure and hydraulic properties of Vertisols containing a sodic horizon in northern Ethiopia. *Soil Tillage Res.* **2014**, *136*, 19–27. [CrossRef]
46. Collinet, J. Comportements Hydrodynamiques et Érosifs de Sols de L'afrique de L'ouest, Évolution des Matériaux et des Organisations Sous Simulation de Pluies. Ph.D. Thesis, de l'Institut de Géologie Strasbourg, Strasbourg, France, 1988; 630p.
47. Alaoui, A.; Rogger, M.; Peth, S.; Blöschl, G. Does soil compaction increase floods? A review. *J. Hydrol.* **2018**, *557*, 631–642. [CrossRef]
48. Amani, R.; Ibrahimi, K.; Sher, F.; Milham, P.; Ghazouani, H.; Chehaibi, S.; Hussain, Z.; Iqbal, H.M.N. Impacts of different tillage practices on soil water infiltration for sustainable agriculture. *Sustainability* **2021**, *13*, 3155. [CrossRef]
49. Yira, Y.; Bossa, A.Y. Agricultural Expansion-Induced Infiltration Rate Change in a West African Tropical Catchment. *Appl. Environ. Soil Sci.* **2019**, *1*, 2434512. [CrossRef]
50. Chartier, M.P.; Rostagno, C.M.; Pazos, G.E. Effects of soil degradation on infiltration rates in grazed semiarid rangelands of northeastern Patagonia, Argentina. *J. Arid Environ.* **2011**, *75*, 656–661. [CrossRef]
51. Anderson, L.R. Land Use and Soil Property Effects on Infiltration and Soil Aggregate Stability in the Lower Mississippi River Valley. Master's Thesis, University of Arkansas, Fayetteville, AR, USA, 2019. Available online: https://scholarworks.uark.edu/etd/3144 (accessed on 12 August 2023).
52. Liu, Y.; Han, J.; Jiao, J.; Liu, B.; Ge, W.; Pan, Q.; Wang, F. Responses of flood speaks to land use and landscape patterns under extreme rainstorms in small catchments. A case study of the rainstorm of Typhoon Lekima in Shandong, China. *Int. Soil Water Conserv. Res.* **2022**, *10*, 228–239. [CrossRef]
53. Montoroi, J.-P. Rôle des sols sur la genèse des inondations. Problèmes actuels de la protection contre les inondations. In Proceedings of the Symposium Européen, Paris, France, 28–30 March 2012. 6p.
54. Andriambelomanga, E.; Ratsivalaka, S.; Andriamampianina, N.; Randriamboavonjy, J.-C.; Mparany, A. Amélioration de la gestion paysanne de la fertilité des sols des versants cultivés des collines, du bassin versant de Maniandro (Madagascar). In *Lutte Anti-Érosive. Réhabilitation des Sols Tropicaux et Protection Contre les Pluies Exceptionnelles*; Roose, É., Duchaufour, H., De Noni, G., Eds.; IRD: Wellington, New Zealand, 2018; 758p.
55. Duran Zuazo, V.H.; Pleguezuelo, C.R. Soil-erosion and runoff prevention by plant covers. A review. *Agron. Sustain. Dev.* **2008**, *28*, 65–86. [CrossRef]

6. Wakponou, A.; Mainguet, M.; Dumay, F. Les techniques de cultures en terrasse dans les monts Mandara, Extrême Nord Cameroun. In *Lutte Anti-Érosive. Réhabilitation des Sols Tropicaux et Protection Contre les Pluies Exceptionnelles*; Roose, É., Duchaufour, H., De Noni, G., Eds.; IRD: Wellington, New Zealand, 2018; 758p.
7. Poulard, C.; Chastan, B.P.R.; Degoutte, G.; Grelot, F.; Erlendbruch, K.; Nédélec, Y. Prévention des inondations par ralentissement dynamique: Principe et recommandation. *Sci. Eaux Territ.* **2008**, 5–24, Consulté le 26 mai 202 à l'adresse. Available online: https://revue-set.fr/article/view/6263 (accessed on 12 August 2023).

*water*

MDPI

*Article*

# Springs of the Arabian Desert: Hydrogeology and Hydrochemistry of Abu Jir Springs, Central Iraq

**John A. Webb [1,*], Jaafar Jotheri [2,3] and Rod J. Fensham [4,5]**

1 Environmental Geoscience, Discipline of Ecology and Environment, La Trobe University, Bundoora, VIC 3086, Australia
2 Department of Archaeology, University of Al-Qadisiyah, Diwaniyah 88, Iraq; jaafar.jotheri@qu.edu.iq
3 Department of Archaeology, Durham University, Durham DH1 3LE, UK; jaafar.jotheri@durham.ac.uk
4 Queensland Herbarium, Mt Coot-Tha Road, Toowong, QLD 4066, Australia; rod.fensham@des.qld.gov.au
5 School of the Environment, University of Queensland, St. Lucia, QLD 4072, Australia
* Correspondence: john.webb@latrobe.edu.au

**Abstract:** The Arabian Desert is characterised by very low rainfall and high evaporation, yet over 210 springs are on its northeastern edge in central Iraq along the Abu Jir lineament, which represents the western depositional margin of a foreland basin infilled by the floodplain sediments of the Tigris and Euphrates Rivers; there is little evidence of faulting. The springs discharge from gently east-dipping Paleocene–Eocene limestones, either where groundwater flowpaths intersect the ground surface or where groundwater flow is forced to the surface by confining aquitards. Calculated annual recharge to the aquifer system across the Arabian Desert plateau (130–500 million $m^3$) is significant, largely due to rapid infiltration through karst dolines, such that karst porosity is the primary enabler of groundwater recharge. The recharge is enough to maintain flow at the Abu Jir springs, but active management of groundwater extraction for agriculture is required for their long-term sustainability. The hydrochemistry of the springs is determined by evaporation, rainfall composition (high $SO_4$ concentrations are due to the dissolution of wind-blown gypsum in rainfall), and plant uptake of Ca and K (despite the sparse vegetation). Limestone dissolution has relatively little impact; many of the springs are undersaturated with respect to calcite and lack tufa/travertine deposits. The springs at Hit-Kubaysa contain tar and high levels of $H_2S$ that probably seeped upwards along subvertical faults from underlying oil reservoirs; this is the only location along the Abu Jir lineament where deep-seated faults penetrate to the surface. The presence of hydrocarbons reduces the Hit-Kubaysa spring water and converts the dissolved $SO_4$ to $H_2S$.

**Keywords:** Iraq; springs; hydrochemistry; hydrogeology; recharge

## 1. Introduction

The Arabian Desert extends across much of Saudi Arabia and Iraq and is characterised by very low rainfall, high evaporation, and sparse vegetation. Nevertheless, along its northeastern edge in central Iraq, there is an extensive line of over 210 springs along the northwest–southeast geographical divide between the desert and the fertile Mesopotamia plains of the Euphrates and Tigris Rivers (Figure 1). These springs also lie along a cultural divide between the desert, the realm of nomadic people, and the floodplain, where a reliable supply of water has allowed for the development of the flourishing permanent human settlements [1].

For this study, we carried out the first comprehensive mapping of the Abu-Jir Springs and integrated this with a reassessment of the significant amount of available data on the springs in order to determine the reasons for their existence in such an arid region.

The linear feature associated with the springs is generally identified as a major fault (the Abu Jir Fault) [2]. However, geological mapping has failed to identify a near-surface rupture that coincides with the springs, e.g., [3], so the interpretation of this fault seems to be largely conjectural. To understand the nature of the geomorphological/geological structure associated with the springs, we reassessed the regional geological mapping and seismic data for the area.

The springs are supplied by groundwater from the Umm er Radhuma-Dammam aquifer underlying the Arabian Desert to their west [4]. This aquifer has been exploited by recent extraction for irrigated agriculture in some areas [5,6], although much of the groundwater from the aquifer is saline [7]. The exploitation has impacted many of the springs and has been regarded as unsustainable. However, the potential recharge area of the aquifer feeding the springs is enormous, even though it coincides with a landscape of very low rainfall. Here, we assess the sustainability of the springs under current rainfall using two different methods to calculate the recharge potential.

The chemical composition of the springs is characterised by high levels of $SO_4$ and often high salinity; the Hit-Kubaysa springs also contain tar and $H_2S$. The interpretation of these characteristics has invoked many different processes: evaporation [7,8],the dissolution of gypsum [7,9], evaporites [7], and carbonates [5,7,8], the input of connate sea water [10,11], and, for the Hit-Kubaysa springs, the input of petroleum and brines from reservoirs deep beneath the springs [12]. Here, we reassess the hydrochemical evidence using new methods of interpretation, including a comparison with rainfall composition, as this is essential for determining which species in solution were derived from rainfall and which from rock–water interactions in the aquifers.

## 2. Materials and Methods

For the mapping of the springs, we used the CORONA satellite photography derived from the United States intelligence program of satellite reconnaissance from 1959 to 1972 [13,14], which was verified by fieldwork for many springs during this study. The CORONA images dating from August 1968 are essential for locating the springs on the Abu Jir lineament because they predate the groundwater extraction for agricultural development that has deactivated many springs. The mapping was also informed by the detailed map and journal compiled by Alois Musil, a Czech theologian and explorer who made multiple journeys through the Middle East. He travelled southeast along the Abu Jir lineament in 1912, providing locations and names for the springs [15].

Data on the hydrogeology of the region and the spring hydrochemistry were obtained from existing published sources and then integrated and reassessed. For the hydrochemical study, the spring compositions were plotted on a Piper plot. In addition, the median composition of spring water from four different sets of springs along the Abu Jir lineament (Hit-Kubaysa, Haqlaniyah, Shinafiyah, and Najaf) was compared with rainfall (obtained at Riyadh) using a standardised Schoeller plot. A Schoeller plot is a semi-logarithmic diagram of species concentrations from multiple samples, with the advantage that, unlike trilinear plots, the actual concentrations are displayed [16]. The data in the present study were standardised to the $Cl^-$ value of rainfall. This removes any influence of evaporation [17], i.e., if the spring waters represented just evaporated rainwater, with no other influence on their chemistry, all lines on the Schoeller plot for spring water would plot over the top of the rainfall line. This procedure assumes that all $Cl^-$ arrives in rainfall and that processes in the soil or aquifer (apart from evapotranspiration) do not affect the $Cl^-$ concentration.

The saturation indices (*SIs*) with respect to calcite were calculated for the spring water compositions [18]:

$$SI_{calcite} = \frac{Activity\ Product}{K_{calcite}} \quad (1)$$

where

$$Activity\ Product = a_{Ca^{2+}} \cdot a_{CO_3^{2-}} \quad (2)$$

The recharge for the springs was calculated using two methods. The first, the chloride ($Cl^-$) mass balance (CMB) method [19,20], is based on the relationship between $Cl^-$ concentrations in groundwater and in precipitation (rainfall) (Equation (3)). As for the Schoeller plot, this calculation assumes that all $Cl^-$ in the groundwater is derived from rainfall and remains in solution within the groundwater system and that there is no significant loss of chloride in runoff:

$$recharge\ (\text{mm}) = rainfall\ (\text{mm}) \cdot \frac{rainfall\ Cl\ \left(\frac{mg}{L}\right)}{groundwater\ Cl\ \left(\frac{mg}{L}\right)} \tag{3}$$

The second method for calculating recharge used the empirical relationship between rainfall and recharge from MacDonald et al. [21]:

$$\ln recharge\ (\text{mm}) = -5 + 1.388 \cdot \ln rainfall\ (\text{mm}) \tag{4}$$

## 3. Abu Jir Springs

The Abu Jir springs total 214 in number and lie along the Abu Jir lineament, which extends in a slightly arced line for ~520 km (Figure 1). The maximum distance between the springs along the lineament is 42 km northwest of Razazza Lake, and the maximum width across the complex of springs is 26 km in the Hit-Kubaysa area. The springs are closely aligned along the western edge of the Mesopotamian floodplain (Figure 2), which is here called the Abu Jir lineament (discussed further below).

In general, the springs have low flows, with an average discharge of <1 L/s, and provide little inflow to the Euphrates River, which is mostly located to the east of the spring line (Figure 1). At Haqlaniyah and Hit, the springs provide a minor inflow to the main channel of the river, and to the northwest of Qaryat al Gharab (Figure 1), the springs occur on either side of the Al-Atshan River, a channel of the main Euphrates stream crossing an area of marshland. The springs do not discharge into most of the large ephemeral salt lakes to the east of the lineament, although a spring that provides a groundwater source for Sawa Lake (Figure 1) is occasionally exposed when the lake levels recede (Figure 3). Some springs have associated travertine deposits, especially near Hit and Abu Jir (Figure 1).

### 3.1. Geological and Geomorphological Setting

The geology and geomorphology of Iraq are the result of its tectonic history, in particular, the collision between the Arabian and Eurasian plates during the Alpine Orogeny, which began in the Early Paleogene, with a significant phase in the Pliocene–Early Pleistocene [22,23]. Tectonism continues today; the Arabian Plate is still moving northeastwards at ~1.5 cm/year. The thrusting of the Eurasian Plate over the Arabian Plate formed, in increasing distance from the collision zone (i.e., northeast to southwest), the Zagros fold and thrust belt, the Mesopotamian foredeep, and the inner Arabian platform [23,24] (Figure 4). These three tectonic units correspond to the three major geomorphic subdivisions of Iraq (Figure 2). In the northeast are the rugged peaks and linear ridges of the Zagros Mountains, uplifted by the collision and rising to over 3000 m. To the southwest is the flat Mesopotamian Plain, almost entirely < 100 m in elevation, which slopes very gently towards the Persian Gulf; it is crossed by the Tigris and Euphrates Rivers [22]. Beneath the plain are gently folded Cretaceous limestones and sandstones, which are important oil reservoirs in central and southern Iraq [25]. The Mesopotamian Plain represents a foreland basin that subsided in front of the rising Zagros Mountains and was infilled by the floodplain sediments of the rivers. To the southwest is the Arabian desert plateau, rising slowly westwards to over 800 m at the border with Saudi Arabia, and composed largely of Paleogene and Neogene carbonate sediments that dip very gently eastwards.

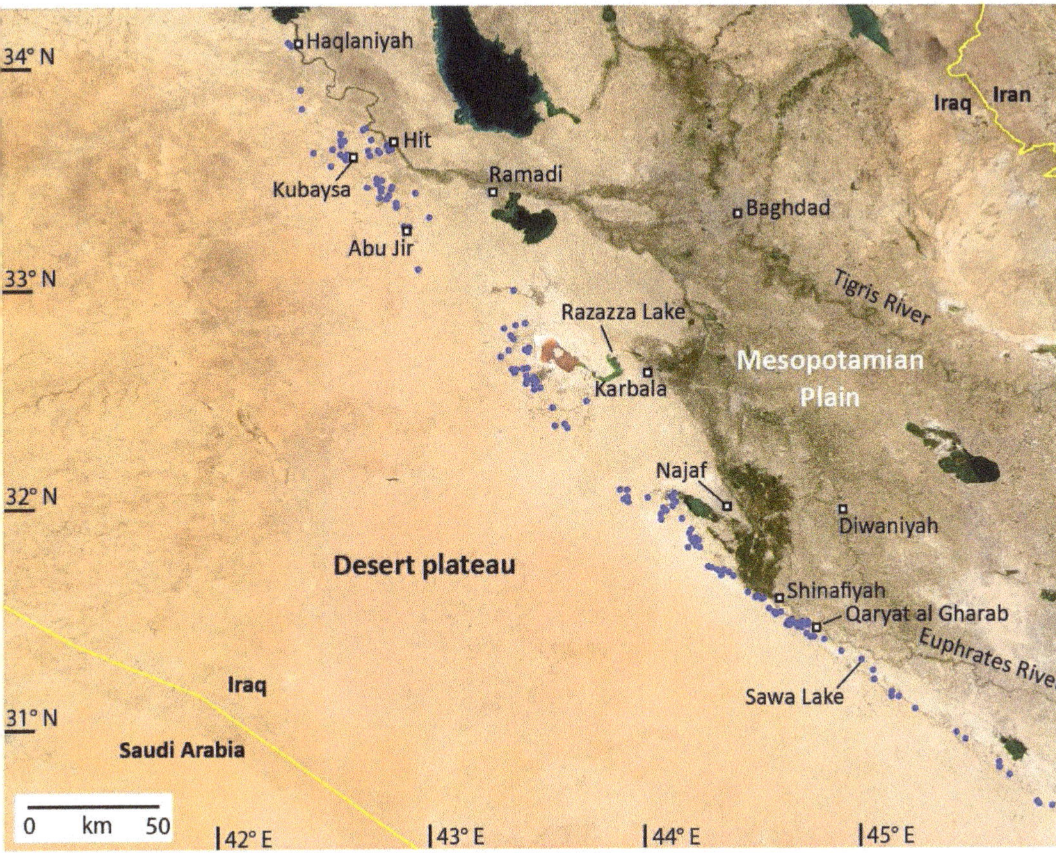


**Figure 1.** Location of springs (blue dots) along the Abu Jir lineament, and modern cities (squares). For location, see Figure 4.

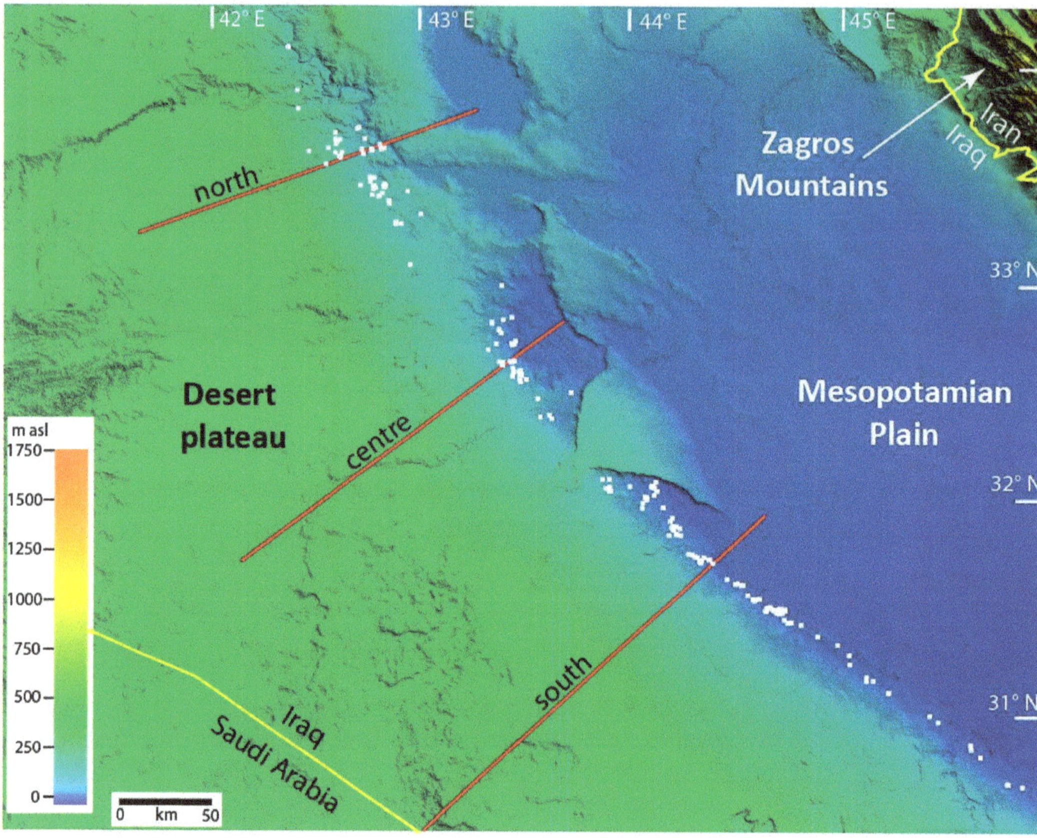


**Figure 2.** Topography of central Iraq showing the three major geomorphic sub-divisions, springs (white dots) located along the Abu Jir lineament, and the locations of the cross-sections in Figure 5. For location, see Figure 4.

**Figure 3.** Spring exposed in the bed of Sawa Lake when the lake receded (see Figure 1 for location).

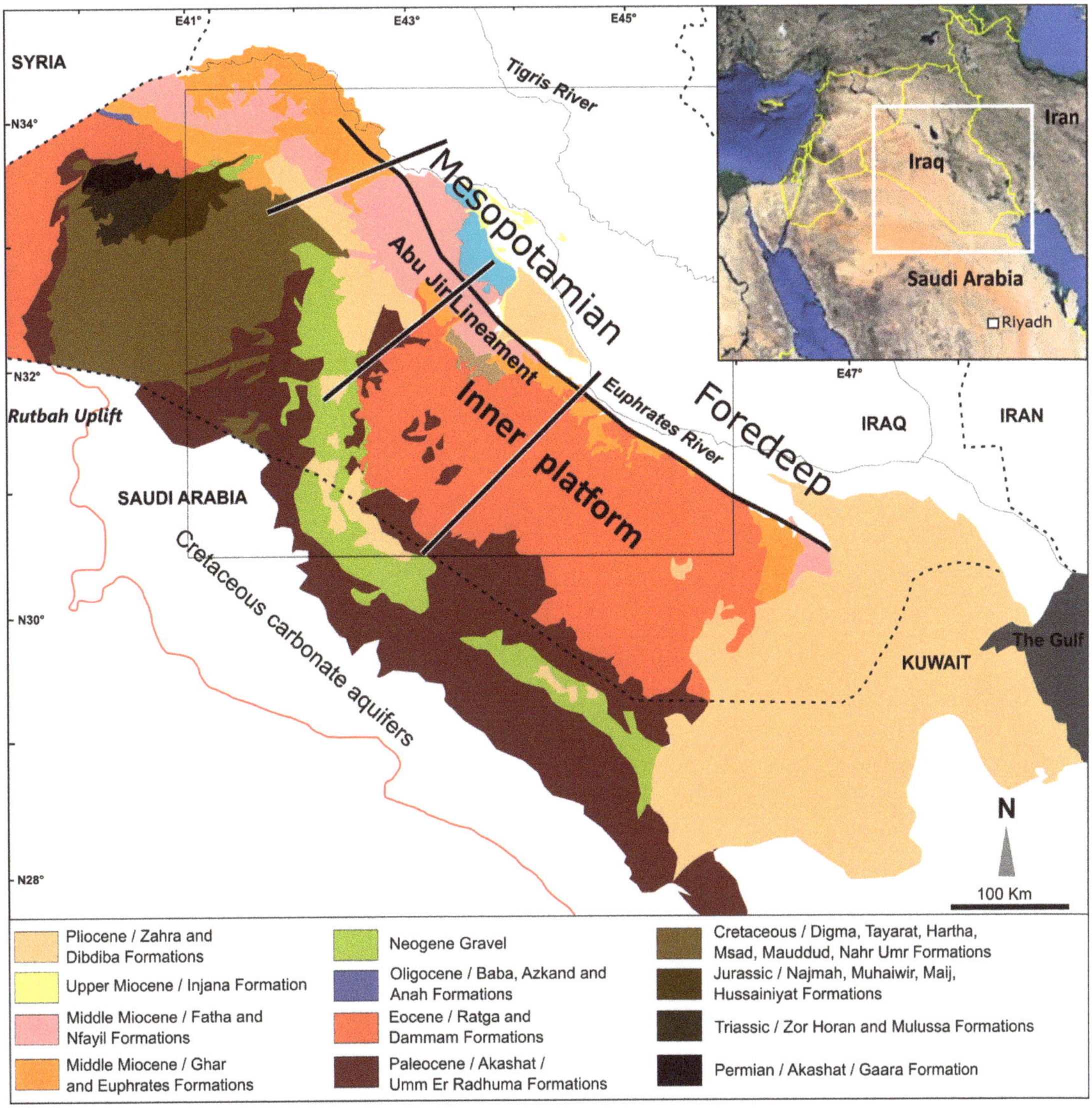


**Figure 4.** Geology of the desert plateau west of Abu Jir lineament, after refs. [23,26,27], showing location of Figures 1 and 2 (box on main figure) and the stratigraphic profiles in Figure 5.

The well-defined topographic boundary between the Mesopotamian Plain and the desert plateau, which trends northwest–southeast for ~520 km (Figure 2), is referred to here as the Abu Jir lineament, because it is a well-marked linear feature. It has also been called the Abu Jir Fault, e.g., [24], but there is little evidence of surface displacement along most of the lineament [28], and cross-sections on the 1:250,000 geological maps covering the lineament show uninterrupted east-dipping strata, e.g., [3]. Even in the Hit-Kubaysa area, where there is evidence of hydrocarbon leakage up steeply dipping faults along or close to the lineament, no displacement has been documented at the surface [8,29]. Seismic sections show that right-lateral strike-slip faults (flower structures) exist at depths beneath the Abu Jir lineament [28], but these have minimal vertical displacement and also occur beneath the Mesopotamian Plain [30].

To confirm this interpretation, new cross-sections were constructed along the lineament using all available geological and geomorphological information (Figure 5); these show clearly that the Abu Jir lineament does not coincide with a fault. It is instead a topographic feature that forms the western edge of the floodplains of the Tigris and Euphrates Rivers. It, therefore, represents the western depositional margin of the foreland basin (Mesopotamian Plain) that subsided in front of the rising Zagros Mountains as the Eurasian Plate was thrust over the Arabian Plate and was infilled by the floodplain sediments of the rivers.

To the west of the Abu Jir lineament is the Arabian Desert plateau (Figure 2). This has very low relief, increasing gradually westwards in elevation from ~100 m at the lineament to over 800 m at the border with Saudi Arabia. The exposed strata in the eastern part of the plateau are predominantly carbonates of Paleocene to Miocene age (Figure 4); older sediments of Permian to Cretaceous age (mostly carbonates) outcrop to the west. The northeastern part of the plateau is dissected by shallow valleys running northeastwards; elsewhere it is characterised by a rocky surface with numerous solution features, such as karst depressions (dolines) (Figure 2), some of which lead to caves. On the desert plateau in eastern Saudi Arabia, horizontally developed shallow caves and vertical shafts have been reported; these are believed to have formed predominantly during wetter climate phases in the Pleistocene [31,32].

### *3.2. Hydrogeological Setting*

The hydrogeology of the springs along the Abu Jir lineament is determined by the geology of the Arabian Desert plateau, which represents the recharge area for the springs. Carbonate aquifers of Cretaceous–Miocene age are exposed across most of the plateau (Figure 4) and extend continuously eastwards beneath the clay-rich aquitard of the Quaternary sediments of the Mesopotamian Plain (Figure 5). Miocene–Pliocene aquitards overlie the carbonate aquifers in places along the easternmost margin of the desert plateau, close to the Abu Jir lineament (Figure 5) [4].

The oldest carbonate units that form significant aquifers are Cretaceous in age; these outcrop in the central and western parts of the desert plateau. In stratigraphic order, these are the Mauddud Formation (horizons up to 50 m thick of limestone and marl), the Rutbah Formation (sandstone and some limestone, 20–30 m thick), and the Ms'ad Formation (about 65 m of limestone with thin sandstone tongues) [2]. These are overlain by the Late Cretaceous Hartha, Tayarat, and Digma Formations, with a total thickness of >100 m; all of these latter units contain marl horizons, particularly the Digma Formation, which may act as aquicludes separating the Cretaceous aquifers from younger Paleogene aquifers.

The two main aquifers of the desert plateau are the Paleocene Umm Er Radhuma Formation and the Eocene Dammam Formation [4] (Figures 4 and 5). Over most of the plateau, the Dammam Formation directly overlies the Umm Er Radhuma Formation [2], and the two form a single unconfined aquifer system. The Umm Er Radhuma Formation consists of microcrystalline, porous, anhydritic, and dolomitic limestones, mostly dull white or buff, with a thickness of 120–180 m. The Dammam Formation comprises whitish grey, porous, dolomitised limestone, sometimes chalky, and is up to 225 m thick. Two members of the Dammam Formation are exposed in the eastern and southern parts of the desert plateau, a Lower Member of whitish grey fossiliferous (nummulites) limestone, and a Middle Member of white, massive shelly limestone; shells are mainly oysters with few small nummulites [2].

Overlying the Dammam Formation is the Early Miocene Euphrates Formation, which consists of white and grey fossiliferous limestone and dolomite. Although it can be over 100 m thick, it is generally much thinner. This formation is also an aquifer, so it represents an additional part of the Umm Er Radhuma Formation/Dammam Formation aquifer system.

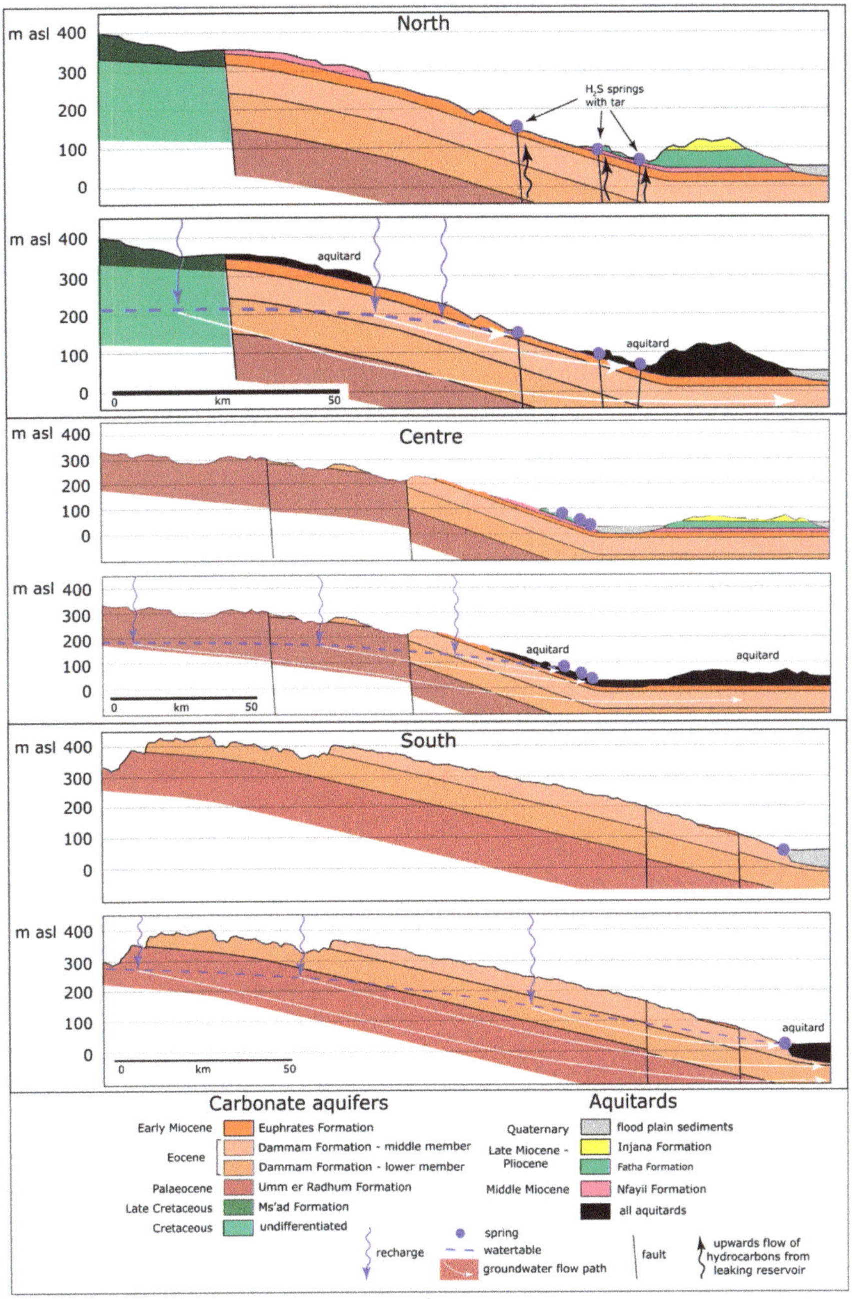


**Figure 5.** Geological cross-sections along the Abu Jir lineament (for locations, see Figures 2 and 4), showing the hydrogeology of the springs. Note the vertical exaggeration (x82); the actual westwards dip of the strata is <1°. Stratigraphy derived from the outcrop distribution and bore logs on the following 1:250,000 geological maps: Karbala [33]; Al Najaf [34]; Baghdad [35]; Al Birreet [36]; Al Ramadi [37]; Shithatha [38].

The porosity and permeability of the carbonate aquifers are due to both dissolution cavities and tectonic fractures [39]; as a result, the hydraulic conductivity of the Umm Er Radhuma and Dammam Formations can reach 20 and 100 m/day, respectively [4]. The Middle Member of the Dammam Formation in the vicinity of the Abu Jir lineament has an average porosity of 22% and hydraulic conductivity of 6 m/day [3]; to the south, in Kuwait, the porosity and permeability of the Dammam Formation can be as much as 53% and 5 m/day, respectively [40].

Groundwater flow within the carbonate aquifers beneath the desert plateau follows the topographic gradient, flowing from southwest to northeast in the northern part of the plateau and west to east in the south [4]. The watertable within the unconfined carbonate aquifers lies up to 300 m below the ground surface along the Saudi Arabian border, and approaches the surface towards the east, as it slopes gently towards the Abu Jir lineament (Figure 5); the slope of the watertable is gentler than that of the topography. East of the lineament, the Umm Er Radhuma and the Dammam Formations extend continuously beneath the sediments of the Mesopotamian Plain as a confined aquifer [4] (Figure 5), and bores within these units close to the lineament are often artesian [3,41].

Along the eastern edge of the desert plateau, the Umm Er Radhuma/Dammam/Euphrates Formation aquifer system is overlain by several thin, Middle and Late Miocene clay-rich units that, together, form an aquitard (Figure 5): the Nfayil Formation (green marl, grey limestone, and red-brown mudstone), the Fat'ha Formation (green marl and bedded limestone), and the Injana Formation (red brown mudstone and sandstone). Each of these units is typically only a few meters thick.

East of the Abu Jir lineament, beneath the Mesopotamian Plain, the Paleogene–Neogene carbonate aquifers are unconformably overlain by Quaternary floodplain muds and sands (Figure 5), up to 100 m thick, forming an extensive aquitard.

### 3.3. *Hydrogeology of Abu Jir Springs*

Along the southern part of the Abu Jir lineament, the springs are mostly sited at the break in slope between the desert plateau and the Mesopotamian Plain. Here, the carbonate aquifers outcrop right up to the lineament (Figure 5), so the boundary between the aquifers and the aquitard of the flood plain sediments lies directly along the lineament. The effect of the clay-rich aquitard sediments of the floodplain is to impede the eastward groundwater flow and force the groundwater to the surface as springs, which are, therefore, located where the watertable intersects the ground surface (Figure 5). Many of the springs are probably fed by conduits dissolved in the limestone, accounting for their occurrence as discrete vents rather than strike-parallel linear seeps.

In contrast, further north along the Abu Jir lineament, the springs are located both at the break in slope and 10 km or more to the west (upslope) due to a different geological setting. In the north, the carbonate aquifers are overlain by Miocene aquitards along the base of the slope immediately west of the Abu Jir lineament (Figure 5); these aquitard outcrops raise the watertable and force the springs higher up the slope. As a result, the springs occur at the upslope extent of the aquitard outcrops but also lower down the slope, where they have broken through the thin aquitard beds. As in the southern area, the springs are most likely conduit-fed.

Among the northern springs are the Hit-Kubaysa springs, which are notable for their tar and $H_2S$ content [12]. These springs may lie along faults that allow for an upwards leakage of hydrocarbons from reservoirs beneath (Figure 5) (discussed further below). However, there is no evident surface displacement of the carbonate aquifers in this area [8,29], so the role of the faults in determining the spring locations is uncertain.

### 3.4. *Recharge to the Abu Jir Springs*

Although the Umm er Radhuma-Dammam aquifer system has recently been categorised as non-renewable [42], recharge is demonstrably occurring at present, as shown by the presence of measurable tritium in the groundwater, indicating recharge in the last

50 years [43,44]. To estimate this recharge, the chloride ($Cl^-$) mass balance method (Equation (3)) was applied, using the precipitation-weighted mean Cl concentration of rainfall in Iraq and Saudi Arabia, (10–20 mg/L) [45–47], the average chloride concentration of the springs at Najaf (Table 1) and the groundwater in the Dammam Formation, (588 and 980 mg/L, respectively) [10], and the average annual rainfall of the desert plateau (100 mm; ranging from 64 mm at Ar'ar in Saudi Arabia to 142 mm on the eastern edge of the desert; much of the rain falls during erratic events of >50 mm). This gave an annual recharge of 1–3.4 mm.

Employing the MacDonald et al. [21] empirical relationship between rainfall and recharge (Equation (4)) using the average annual rainfall of the desert plateau gave an annual recharge of 4 mm. The results of both calculations are close to the measured average recharge of 2.2 mm/yr on the Arabian Peninsula to the southeast [44].

The overall annual recharge to the Neogene carbonate aquifer can be calculated from these recharge estimates and the catchment (outcrop) area in Iraq and immediately across the border to the southwest in Saudi Arabia, which is ~123,000 $km^2$, giving an overall annual recharge of 126–420 million $m^3$ (344–1150 ML/day). If the outcrop of the Cretaceous carbonates to the west in Saudi Arabia is included, the catchment area increases to 208,000 $km^2$, and the annual recharge rises to 210–700 million $m^3$ (580–1900 ML/day).

Recharge over the desert plateau occurs despite the low rainfall (50–150 mm) and the very high evaporation (>2000 mm). This is probably because much infiltration occurs rapidly through the outcrop, assisted by the numerous karst dolines scattered over the plateau, which funnel rainfall underground, so that the amount of recharge after rainfall events can be substantial [43]. Some of these dolines are large enough to be visible on the low-resolution DEM of the region (south-central part of Figure 2). Thus, the karst porosity is the primary enabler of groundwater recharge due to its recognisable effect on the nature of runoff dynamics. In the southwestern part of the plateau, within Saudi Arabia, recharge also occurs directly through overlying sand dunes [43].

If all the recharge to the Neogene carbonates discharged at the 214 springs along the Abu Jir lineament, it would give an average spring flow of 20–60 L/sec (1.6–5.5 ML/day). This is much greater than the average spring discharge of <1 L/sec (0.04–20 ML/day, average < 0.1 ML/day) [12,48,49]. The total volume of discharge through the springs has been estimated as only 137 ML/day [42]. Spring discharge may have been substantially greater prior to extraction for agriculture in some areas, but the excess of recharge over spring discharge indicates that the bulk of groundwater flow through the Umm er Radhuma-Dammam aquifer system bypasses the springs and probably flows through the carbonates beneath them (Figure 5).

The aquifer system is exploited for irrigated agriculture in some areas, and this has been regarded as unsustainable, but the calculated annual recharge for the Neogene carbonates (>344 ML/day) should be sufficient to maintain spring flow along the Abu Jir lineament (~137 ML/day). However, excessive extraction in localised areas will cause the watertable to fall, and many of the springs have already been impacted. Therefore, the long-term sustainability of the springs relies on the active management of groundwater extraction rates.

On the Arabian Peninsula, the groundwater residence times within the Umm er Radhuma and Dammam aquifers are up to 20,000–30,000 years, indicating that recharge may have been greater in the past, such as during the relatively wet 'Pluvial Period' from 9500 to 5000 years ago [44,50].

### 3.5. Hydrochemistry of the Abu Jir Springs

There is high variability in the salinity of the springs along the Abu Jir lineament and even between nearby locations, e.g., Hit-Kubaysa (Table 1), from high (>10,000 μS/cm at Hit), unsuitable for both drinking and irrigation, to moderate (>1800 μS/cm at Najaf). Even the latter levels are marginal for sustained irrigated agriculture, exceeding the recommended levels for drinking water in Europe and noticeably mineralised for unaccustomed users. The levels of fluoride and arsenic are low and safe for human consumption [8,29]. As would be expected, the springs' salinity reflects that of the groundwater in the carbonate aquifers, which increases progressively from, on average, ~1000 mg/L in the west to over 5000 mg/L in the east towards the Euphrates River [4,43].

The spring water composition is variable (Figure 6, Table 1). Anions are dominated by high levels of either Cl or $SO_4$, with much lower amounts of $HCO_3$; for the cations, there are significant concentrations of Ca, Na, and Mg, and relatively minor K. The average composition of the springs broadly reflects that of the average Damman Formation groundwater (Figure 6), as might be expected. However, the Hit-Kubaysa springs show a separate linear trend of increasing Na and Cl; this probably reflects increasing contributions of deep-seated oilfield brine (discussed further below).

**Table 1.** Water chemistry data for the Abu Jir springs (medians: bold, ranges; number of samples: italics) and rainfall at Riyadh (precipitation-weighted mean). All values are mg/L unless indicated otherwise. See Figure 1 for spring locations. The saturation indices with respect to calcite were calculated using Equations (1) and (2).

| Location | T°C | pH | EC µS/cm | Ca | Mg | Na | K | Cl | $SO_4$ | $HCO_3$ | $NO_3$ | $SI_{calcite}$ |
|---|---|---|---|---|---|---|---|---|---|---|---|---|
| Haqlaniyah [8,49] | **29**<br>29–29<br>*3* | **7.2**<br>7.1–7.3<br>*3* | **5068**<br>5038–5523<br>*3* | **312**<br>288–320<br>*3* | **144**<br>134–146<br>*3* | **251**<br>230–709<br>*7* | **22**<br>21–94<br>*3* | **1620**<br>643–1925<br>*7* | **674**<br>403–1260<br>*7* | **265**<br>223–270<br>*3* | **3**<br>2–3<br>*3* | **0.46**<br>0.29–0.5<br>*3* |
| Hit-Kubaysa [8,10,12,29,49] | **27**<br>16–34<br>*49* | **7.2**<br>6.0–7.8<br>*48* | **7100**<br>1800–35418<br>*23* | **400**<br>225–1783<br>*21* | **210**<br>94–607<br>*21* | **619**<br>200–6876<br>*27* | **85**<br>5–540<br>*21* | **1488**<br>320–16100<br>*45* | **480**<br>91–3120<br>*45* | **197**<br>85–1380<br>*21* | **7**<br>2–10<br>*8* | **0.31**<br>−1.54–2.1<br>*21* |
| Najaf [9,11,49] | **25**<br>23–27<br>*20* | **7.2**<br>6.9–7.8<br>*20* | **2360**<br>1820–3200<br>*20* | **188**<br>112–326<br>*10* | **185**<br>46–342<br>*10* | **271**<br>203–457<br>*13* | **54**<br>41–74<br>*9* | **588**<br>350–2591<br>*13* | **785**<br>538–1765<br>*13* | **116**<br>45–140<br>*9* | | **−0.19**<br>−1.11–0.69<br>*9* |
| Shinafiyah [51] | | **7.7**<br>7.0–8.0<br>*12* | **4815**<br>4010–6080<br>*12* | **391**<br>346–496<br>*12* | **242**<br>160–294<br>*12* | | | **817**<br>690–1094<br>*12* | **2610**<br>1330–2861<br>*12* | **178**<br>121–211<br>*12* | | **0.77**<br>−0.18–1.15<br>*12* |
| Riyadh rainfall [47] | | | | 32 | 2 | 6 | 2 | 10 | 17 | | | |

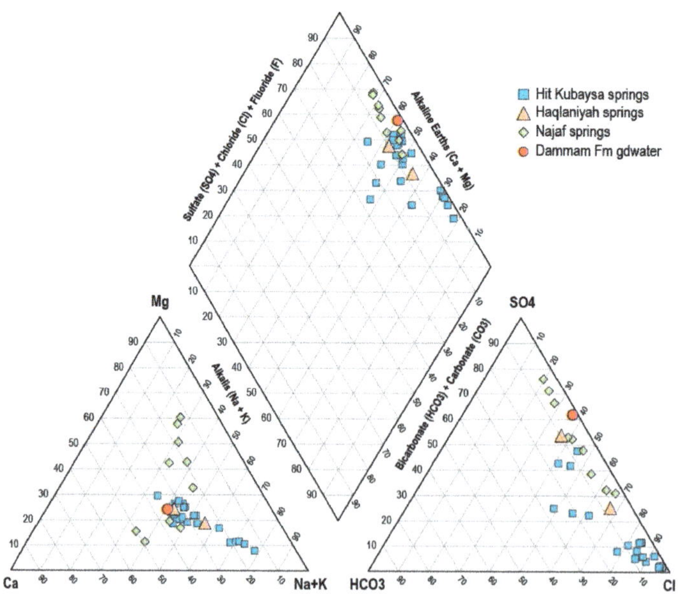


**Figure 6.** Piper plot of compositions of Abu Jir springs (data from) [8,10,11]; note that many available spring compositions, including the Shinafiyah springs, were incomplete as published and could not be plotted.

To better understand the factors responsible for the chemical composition of the spring water, a standardised Schoeller plot was constructed (Figure 7), comparing the median major ion chemistry of the springs with that of the nearest available rainfall data on the desert plateau at Riyadh [34] (Table 1); all data were standardised to the $Cl^-$ value of the rainfall to remove any influence of evaporation. Using median compositions illustrates overall trends, on which processes specific to particular spring groups are superimposed, causing the variability in the composition evident for the Hit-Kubaysa and Najaf springs (Figure 6), as discussed further below.

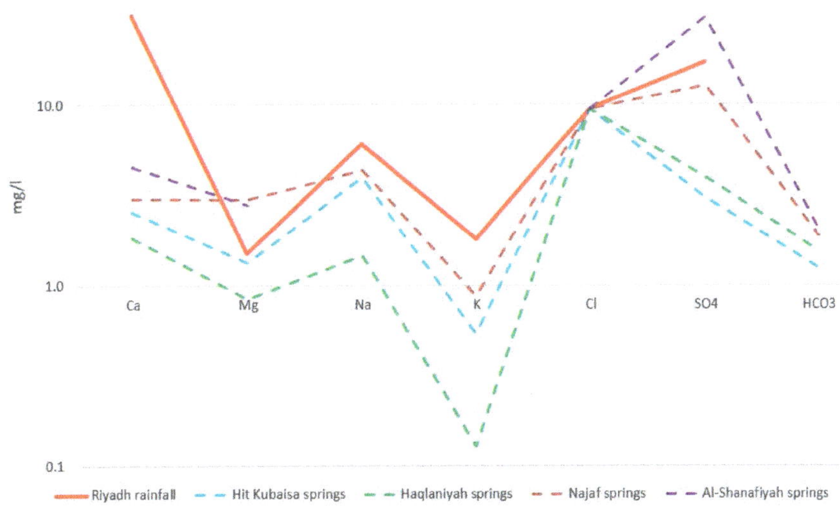


**Figure 7.** Schoeller plot (standardised to Cl) comparing the median compositions of Abu Jir springs with desert plateau rainfall (see Table 1 for data).

The $SO_4$ and Cl contents of the rainfall over the desert plateau (20–50 mg/L and 10–20 mg/L, respectively) reflect the dissolution of wind-blown gypsum and salt (halite), respectively, deflated from the extensive sabhkas in the region [47]. This contrasts with the typical rainfall compositions around the world, which are dominated by Na and Cl among the major ions in places where rainfall derives much of its dissolved content from seaspray, particularly along the coast, e.g., [52], but also inland, often called cyclic salt, e.g., [53]. The higher Ca and $SO_4$ levels in the rainfall relative to Na and Cl (Figure 7) indicate that gypsum dissolution exceeds that of halite.

The standardised Schoeller plot (Figure 7) shows that, apart from Ca and, for the Hit-Kubaysa springs, $SO_4$, the composition of the spring water broadly matches that of the rainfall, so the major source of dissolved Ca, $SO_4$, Na, and Cl in the groundwater within the carbonate aquifers (and therefore, the springs issuing from these aquifers) is likely to be the dissolution of wind-blown evaporites in the rainfall. Therefore, it is not necessary to invoke congruent dissolution of gypsum along the groundwater flow path [7,9] to explain the high $SO_4$ content of the springs. Incongruent dissolution of gypsum is also unlikely; in this process, gypsum dissolves, releasing calcium and sulphate, and some of the calcium precipitates as calcite, resulting in water compositions with relatively lower calcium than sulphate concentrations. Although the spring compositions have, on average, lower Ca than $SO_4$ (Figure 7), many are undersaturated with respect to calcite (negative calcite saturation indices; Table 1) and so cannot precipitate calcite, and the depletion in Ca is more likely due to plant uptake (discussed below). In any case, the carbonate aquifers do not contain gypsum [2,4].

The species concentrations in the springs and groundwater are much greater than those in the rainfall (Table 1), because they were increased by evaporation during recharge [7,8].

The progressive increase in groundwater salinity from west to east within the carbonate aquifers is probably due to the progressive downflow addition of saline soil–water (concentration by evaporation during slow infiltration) to fresher groundwaters that were recharged rapidly through karst dolines; similar increases in salinity down-gradient have been documented elsewhere [52]. Additional evaporation during discharge may have further raised the salinity of the springs.

Interestingly, the standardised Schoeller plot (Figure 7) shows that the spring waters are depleted in Ca compared to the rainfall (but have approximately the same Mg content). There is, therefore, relatively little input to the groundwater of Ca and Mg due to limestone/dolomite dissolution within the carbonate aquifers, contrary to the hypotheses of previous studies [5,7,8]. The relative lack of carbonate dissolution is also evident by the fact that many springs, particularly those at Najaf, are undersaturated with respect to calcite (negative calcite saturation indices) (Table 1). If extensive carbonate dissolution had occurred along the groundwater flow paths, the spring waters would be saturated. In addition, the springs have relatively low $HCO_3$ concentrations (Figure 6) and lack tufa/travertine precipitation around the spring vents. Furthermore, the morphology of limestone caves on the Arabian Peninsula, which formed during wetter climates in the Pleistocene, has been modified only slightly by dissolution under the present-day arid climate [54].

Nevertheless, some carbonate dissolution has contributed to the spring water composition; the variability in the relative Ca and Mg contents of the Najaf springs (Figure 6) is probably due to differences in the amount of limestone and dolomite dissolution along the flow path.

The spring waters show a notable depletion in K as well as Ca compared to rainfall (Figure 7). This is most likely due to the preferential uptake of Ca and K by plants as rainfall infiltrates through the soil. These ions are plant macronutrients; plants take up relatively large quantities of these species directly from the soil solution and/or from the cation exchange sites of clay minerals [55,56]. In contrast, plant uptake of Na and Mg is much lower, so the spring waters are not significantly depleted in these species relative to the rainfall (Figure 7). The mechanism of plant uptake has been identified as significant in determining groundwater composition elsewhere in the world, e.g., [17,57–59]. In Saudi Arabia and Iraq, the depletion in Ca and K due to plant uptake is surprising given the very arid climate and sparse vegetation cover.

There is no evidence that the depletion in Ca and K relative to rainfall is due to cation exchange because this would cause the spring waters to be enriched in Na and/or Mg, and this is not the case (Figure 7). Furthermore, the carbonate aquifers have a low clay content [2,4].

The springs in the Hit-Kubaysa area (Figure 1) are the best studied springs along the Abu Jir lineament because they are characterised by high $H_2S$ contents (up to 305 ppm) [24] and floating spongy tar (Figure 9), as well as elevated nitrate concentrations [8,29]. The hydrocarbon content of the springs makes the waters reducing, and as a result, the dissolved sulphate within the spring water is reduced to $H_2S$. The removal of $SO_4$ from solution means that the Hit-Kubaysa springs are depleted in this species compared to the other Abu Jir springs (Figure 7), also shown by the much higher $Cl/SO_4$ weight ratio of the Hit-Kubaysa springs (2–10) [8,10,49] compared to the regional rainfall, the groundwater within the carbonate aquifers, and the spring water unaffected by $SO_4$ reduction (0.4–0.5) [9–11,45–47,49]. The high nitrate levels in the springs were most likely released by degradation of the organic matter, as crude oils generally contain up to 1 weight % N in compounds, such as pyridines and amines [60], and elevated nitrate levels are typical of oil field brines [29]. The hydrocarbons within the Hit-Kubaysa springs have probably seeped upwards from underlying Jurassic–Cretaceous or Oligocene oil and gas reservoirs [61], driven by the lower density of the oil than the surrounding groundwater (0.85–0.95 g/cc compared to 1 g/cc for pure water). The oil in the springs is the same type as that from the East Baghdad field [49]. The seepage at Hit-Kubaysa is presumably occurring beneath the Abu Jir lineament along subvertical faults (Figure 5), which may be flower structures that have reached the surface [28]; this is the only location along the Abu Jir lineament where deep-seated faults penetrate to the surface. Hussien and Gharbie [12] proposed that deep groundwater (oilfield brine) is seeping upwards along with hydrocarbons at the Hit-Kubaysa springs, and there is good evidence for this. The salinity of the spring water increases to the east towards the Abu Jir lineament [12,29], resulting in a linear trend on the Piper plot (Figure 6), and probably reflecting an increasing contribution of oilfield brine. The maximum temperature of the springs around Hit (34 °C; Table 1), is ~10 °C warmer than the average air temperatures (21–24 °C), so this suggests that the springs are rising from at least 500 m depth; the geothermal gradient in the area is ~20°C/km [62]. The underlying oil reservoirs are at much greater depths [11], indicating that the brines are rising sufficiently slowly that their temperature has partially equilibrated with the aquifer temperature at shallower depths. The hydrocarbons and sulphurous content of the Hit-Kubaysa springs (Figure 5) are considered useful for bathing therapy for the treatment of inflammatory joint disease and psoriatic disease [29].

The above discussion shows that the species in the spring waters were derived either from rainfall, rock–water interaction within the aquifers, or, in the case of some Hit-Kubaysa springs, oilfield brines. Al Dahaan [10] and al Dahan et al. [11] hypothesised that there was also a contribution of connate seawater originally deposited with the marine carbonates of the aquifers. However, the aquifer carbonates are strongly cemented and lack any significant original granular porosity where the connate groundwater could have been stored. In any case, the maximum groundwater ages on the Arabian Peninsula are only 20,000–30,000 years [44,50], indicating that any connate seawater was flushed from the aquifers long ago. Furthermore, the stable isotope composition of the springs (Figure 8) is consistent with an origin entirely from rainfall. The stable isotope composition of seawater is around 0 for both $\delta^{18}O$ and $\delta^{2}H$, and although there is one spring sample with stable isotope values close to this, it has undergone considerable evaporation [7]. The spring samples plot on the evaporation trends (Figure 8), because the spring flow rates are mostly slow enough to allow for significant evaporation soon after the spring water has reached the surface. Groundwater stable isotope data from nearby wells within the carbonate aquifers are mostly located to the right of the Local Meteoric Water Line (LMWL), indicating some evaporation during recharge. A few data lie to the left of the LMWL; this groundwater could have been recharged under a previous wetter climate, e.g., the relatively wet 'Pluvial Period' from 9500 to 5000 years ago [44,50].

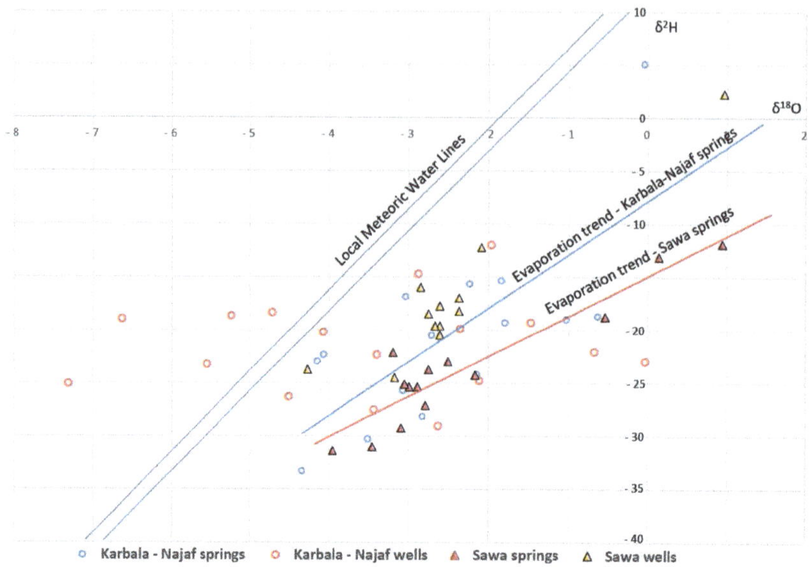


**Figure 8.** Stable isotope data for some Abu Jir springs; groundwater data from nearby wells shown for comparison; data from refs. [7,48].

**Figure 9.** Groundwater spring at Hit showing floating spongy bitumen (see Figure 1 for location).

## 4. Conclusions

A detailed study of the Abu Jir springs, using existing data and satellite mapping, has allowed for a reassessment of the hydrogeology and hydrochemistry of these springs, including the reasons for their existence in such an arid area. They lie along the 520 km-long NW-SE Abu Jir lineament that coincides with the boundary between the Mesopotamian Plain and the Arabian Desert plateau. This lineament is not a fault but the western depositional margin of a foreland basin infilled by the floodplain sediments of the Tigris and Euphrates Rivers. The springs discharge from the gently east-dipping Neogene Umm er Radhuma-Dammam aquifer system, which is composed of carbonates and has karstic permeability. Spring locations along the Abu Jir lineament occur where conduit flow in the carbonate aquifer intersects the ground surface at the break in slope between the Mesopotamian Plain and Desert plateau, or where it is forced to the surface by overlying aquitards, either Neogene marls or Quaternary floodplain sediments. Recharge to the aquifer system occurs across the Arabian Desert plateau and is facilitated by rapid infiltration through karst depressions, such that karst porosity is the primary enabler of groundwater recharge. Annual recharge for the Neogene carbonates is significant (estimated 130–500 million $m^3$), despite the very low rainfall and high evaporation, and is sufficient to maintain spring flow along the Abu Jir lineament. However, excessive extraction for agriculture will negatively impact spring flow, and active management of groundwater extraction rates is needed for the long-term sustainability of the springs. The hydrochemistry of the springs shows high $SO_4$ concentrations (originating from the dissolution of wind-blown gypsum in the rainfall) and depletion in Ca and K relative to the rainfall, probably due to plant uptake as the rainfall infiltrates through the soils, despite the sparse vegetation on the desert plateau. There is little evidence of limestone dissolution in the spring chemistry, even though the groundwater feeding the springs travels through a carbonate aquifer. The springs at Hit-Kubaysa contain tar, high levels of $H_2S$, and a component of oilfield brine; the presence of hydrocarbons makesthe water reducingand converts the dissolved $SO_4$ to $H_2S$. The tar and brine have seeped upwards along faults from underlying oil and gas reservoirs; this is the only location along the Abu Jir lineament where deep-seated faults penetrate to the surface.

**Author Contributions:** Conceptualisation, J.J. and R.J.F.; methodology, J.A.W.; investigation, J.A.W. and R.J.F.; writing—original draft preparation, J.A.W. and R.J.F.; writing—review and editing, J.A.W., R.J.F. and J.J. All authors have read and agreed to the published version of the manuscript.

**Funding:** This research received no external funding.

**Data Availability Statement:** Data supporting the reported results can be obtained upon request from the authors.

**Acknowledgments:** Comments from two anonymous reviewers substantially improved this paper.

**Conflicts of Interest:** The authors declare no conflicts of interest.

## References


1. Altaweel, M.; Marsh, A.; Jotheri, J.; Hritz, C.; Fleitmann, D.; Rost, S.; Lintner, S.F.; Gibson, M.; Bosomworth, M.; Jacobson, M.; et al. New Insights on the Role of Environmental Dynamics Shaping Southern Mesopotamia: From the Pre-Ubaid To the Early Islamic Period. *Iraq* **2019**, *81*, 23–46. [CrossRef]
2. Sissakian, V.K.; Mohammad, B.S. Geology of Iraqi Western Desert: Stratigraphy. *Iraqi Bull. Geol. Min. Spec. Issue* **2007**, 51–124.
3. Thabit, J.M.; Al-yasi, A.I.; Al-shemmari, A.N. Estimation of Hydraulic Parameters and Porosity from Geoelectrical Properties for Fractured Rock Quifer in Middle Dammam Formation at Bahr. *Iraqi Bull. Geol. Min.* **2014**, *10*, 41–57.
4. Saleh, S.A.; Al-Ansari, N.; Abdullah, T. Groundwater Hydrology in Iraq. *J. Earth Sci. Geotech. Eng.* **2020**, *10*, 155–197.
5. Mukhlif, H.N.; Rabeea, R.; Hussien, B.M. Characterization of the Groundwater within Regional Aquifers and Suitability Assessment for Various Uses and Purposes-Western Iraq. *Baghdad Sci. J.* **2021**, *18*, 0670. [CrossRef]
6. Al-Dulaimi, A.M.S.; Al-Kubaisi, Q.Y. Hydrochemistry and Water Quality Index of Groundwater in Abu-Jir Village in Al-Anbar, Western Iraq. *Iraqi Geol. J.* **2022**, *55*, 73–84. [CrossRef]
7. Al Maliki, A.; Kumar, U.S.; Falih, A.H.; Sultan, M.A.; Al-Naemi, A.; Alshamsi, D.; Arman, H.; Ahmed, A.; Sabarathinam, C. Geochemical Processes, Salinity Sources and Utility Characterization of Groundwater in a Semi-Arid Region of Iraq through Geostatistical and Isotopic Techniques. *Environ. Monit. Assess.* **2024**, *196*, 365. [CrossRef] [PubMed]
8. Awadh, S.M.; Al-Ghani, S.A. Assessment of Sulfurous Springs in the West of Iraq for Balneotherapy, Drinking, Irrigation and Aquaculture Purposes. *Environ. Geochem. Health* **2014**, *36*, 359–373. [CrossRef]
9. Al-Dahaan, S.A.J.M.; Hussain, H.M.; Al-Ansari, N.; Knutson, S. Hydrochemistry of Springs, Najaf Area, Iraq. *J. Environ. Hydrol.* **2015**, *23*, 1–12.
10. Al-Dahaan, S.A.J.M. Origin and source of springs, west Iraq. *J. Kufa-Phys.* **2014**, *6*, 1–12.
11. Al-Dahaan, S.A.M.; Alabidi, A.J.; Al-Ansari, N.; Knutsson, S. Relationship between Selected Hydrochemical Parameters in Springs of Najaf Province, Iraq. *Engineering* **2015**, *7*, 337–346. [CrossRef]
12. Hussien, B.M.; Gharbie, M.A. Hydrogeochemical Evaluation of the Groundwater Within Abu Jir Fault Zone, Hit–Kubaisa Region, Central Iraq. *Iraqi Bull. Geol. Min.* **2010**, *6*, 121–138.

13. Casana, J.; Jackson, C. The CORONA Atlas Project: Orthorectification of CORONA Satellite Imagery and Regional-Scale Archaeological Exploration. In *Mapping Archaeological Landscapes from Space, 5*; Comer, D.C., Harrower, M.J., Eds.; Springer Science Business and Media: Berlin/Heidelberg, Germany, 2013.
14. Philip, G.; Donoghue, D.; Beck, A.; Galiatsatos, N. CORONA Satellite Photography: An Archaeological Application from the Middle East. *Antiquity* **2002**, *76*, 109–118. [CrossRef]
15. Musil, A. *Arabia Deserta*; The American Geographical Society: New York, NY, USA, 1927.
16. Schoeller, H. Qualitative Evaluation of Groundwater Resources. In *Methods and Techniques of Groundwater Investigations and Development*; UNESCO: Paris, France, 1965.
17. Dean, J.F.; Webb, J.A.; Jacobsen, G.E.; Chisari, R.; Dresel, P.E. Biomass Uptake and Fire as Controls on Groundwater Solute Evolution on a Southeast Australian Granite: Aboriginal Land Management Hypothesis. *Biogeosciences* **2014**, *11*, 4099–4114. [CrossRef]
18. Drever, J.I. *The Geochemistry of Natural Waters*, 3rd ed.; Prentice Hall: Hoboken, NJ, USA, 1997.
19. Allison, G.; Hughes, M. The Use of Environmental Chloride and Tritium to Estimate Total Recharge to an Unconfined Aquifer. *Soil Res.* **1978**, *16*, 181–195. [CrossRef]
20. Dean, J.F.; Webb, J.A.; Jacobsen, G.; Chisari, R.; Dresel, P.E. A groundwater recharge perspective on locating tree plantations within low-rainfall catchments to limit water resource losses. *Hydrol. Earth Syst. Sci.* **2015**, *19*, 1107–1123. [CrossRef]
21. MacDonald, A.M.; Lark, R.M.; Taylor, R.G.; Abiye, T.; Fallas, H.C.; Favreau, G.; Goni, I.B.; Kebede, S.; Scanlon, B.; Sorensen, J.P.R. et al. Mapping Groundwater Recharge in Africa from Ground Observations and Implications for Water Security. *Environ. Res. Lett.* **2021**, *16*, 034012. [CrossRef]
22. Yacoub, S.Y. Stratigraphy of the Mesopotamia Plain. *Iraqi Bull. Geol. Min.* **2011**, *4*, 47–82.
23. Abdulnaby, W. Structural Geology and Neotectonics of Iraq, Northwest Zagros. In *Tectonic and Structural Framework of the Zagros Fold-Thrust Belt*; Saein, A.F., Ed.; Elsevier: Amsterdam, The Netherlands, 2018; pp. 53–73, ISBN 978-0-12-815048-1.
24. Sissakian, V.K. Geological Evolution of the Iraqi Mesopotamia Foredeep, Inner Platform and near Surroundings of the Arabian Plate. *J. Asian Earth Sci.* **2013**, *72*, 152–163. [CrossRef]
25. Aladwani, N.S.; Alenezi, A.; Diab, A. Investigation of the Cretaceous Total Petroleum System Using Wireline Logs, Core, and Geochemical Data in Bahrah Field, Northern Basin, Kuwait. *J. Pet. Explor. Prod. Technol.* **2023**, *13*, 381–406. [CrossRef]
26. Sissakian, V.K.; Fouad, S.F.A. Geological Map of Iraq, Scale 1: 1,000,000. *Iraqi Bull. Geol. Min.* **2015**, *11*, 1–7.
27. U.S. Geological Survey and Arabian American Oil Company. Geologic Map of the Arabian Peninsula. Washington, D.C., U.S.A. 1963. [CrossRef]
28. Alhadithi, A.A.; Salih, E.M. Behavior of Abu-Jir Fault Zone in Al-Thirthar Valley and near Habbaniya Lake Areas–Comparative Study Using Seismic Reflection Sections. *J. Univ. Anbar Pure Sci.* **2017**, *11*, 47–55. [CrossRef]
29. Al Dulaymie, A.S.; Hussien, B.M.; Gharbi, M.A.; Mekhlif, H.N. Balneological Study Based on the Hydrogeochemical Aspects of the Sulfate Springs Water (Hit-Kubaiysa Region), Iraq. *Arab. J. Geosci.* **2013**, *6*, 801–816. [CrossRef]
30. Fouad, S. Contribution to the Structure of the Abu-Jir Fault Zone, West Iraq. *Iraqi Geol. J.* **1999**, *32*, 63–73.
31. Benischke, R.; Fuchs, G.; Weissensteiner, V. Speleological Investigations in Saudi Arabia. In Proceedings of the 12th International Congress of Speleology, La Chaux-de-Fonds, Geneva, Switzerland, 10–17 August 1997; International Union of Speleology: Geneva, Switzerland, 1997; Volume 1, pp. 425–428.
32. Waltham, T. Asia, Southwest. In *Encyclopedia of Caves and Karst Science*; Gunn, J., Ed.; Fitzroy Dearborn: New York, NY, USA, 2004; pp. 114–116.
33. Barwary, A.M.; Slewa, N.A. *Geological Map of Karbala Quadrangle, Sheet NI-38-14*; State Establishment of Geological Survey and Mining: Baghdad, Iraq, 1995.
34. Barwary, A.M.; Slewa, N.A. *Geological Map of Al-Najaf Quadrangle, Sheet NH-38-2*; State Establishment of Geological Survey and Mining: Baghdad, Iraq, 1996.
35. Deikran, D.B.; Mahammad, S.M. *Geological Map of Baghdad Quadrangle, Sheet NI-38-10*; State Establishment of Geological Survey and Mining: Baghdad, Iraq, 1994.
36. Sissakian, V.K.; Youkhanna, R.Y. *Geological Map of Al-Birreet Quadrangle, Sheet NH-38-1*; State Establishment of Geological Survey and Mining: Baghdad, Iraq, 1995.
37. Sissakian, V.K.; Zwaid, Q.A.; Mohammad, S.M. *Geological Map of Al-Ramadi Quadrangle, Sheet NI-38-9*; State Establishment of Geological Survey and Mining: Baghdad, Iraq, 1995.
38. Mahdi, A.H.; Youkhanna, R.Y. *Geological Map of Shithatha Quadrangle, Sheet NI-38-13*; State Establishment of Geological Survey and Mining: Baghdad, Iraq, 1996.
39. Al-Zubedi, A.S.; Thabit, J.M. Use of 2D Azimuthal Resistivity Imaging in Delineation of the Fracture Characteristics in Dammam Aquifer within and out of Abu-Jir Fault Zone, Central Iraq. *Arab. J. Geosci.* **2016**, *9*, 1–9. [CrossRef]
40. Abdullah, F.H. Porosity and Permeability of Karst Carbonate Rocks along an Unconformity Outcrop: A Case Study from the Upper Dammam Formation Exposure in Kuwait, Arabian Gulf. *Heliyon* **2021**, *7*, e07444. [CrossRef]
41. Al-Abadi, A.M.; Shahid, S. Spatial Mapping of Artesian Zone at Iraqi Southern Desert Using a GIS-Based Random Forest Machine Learning Model. *Model. Earth Syst. Environ.* **2016**, *2*, 96. [CrossRef]
42. UN-ESCWA and BGR. *Inventory of Shared Water Resources in Western Asia*; United Nations Economic and Social Commission for Western Asia; Bundesanstalt für Geowissenschaften und Rohstoffe: Beirut, Lebanon, 2013.

43. Faulkner, R.D. Fossil Water or Renewable Resource: The case for one Arabian aquifer. *Proc. Inst. Civ. Eng. Water Marit. Energy* **1994**, *106*, 325–331. [CrossRef]
44. Rausch, R.; Dirks, H.A. Hydrogeological Overview of the Upper Mega Aquifer System on the Arabian Platform. *Hydrogeol. J.* **2024**, *32*, 621–634. [CrossRef]
45. Awadh, S.A. The Atmospheric Pollution of Baghdad City, Iraq. In Proceedings of the 3rd Scientific Conference College of Science, University of Baghdad, Baghdad, Iraq, 24–26 March 2009; pp. 1727–1740.
46. Handy, A.H.; Tucker, R.A. *Rainfall Quality at Selected Sites in Saudi Arabia. Water Research and Study Division*; Publication No. 2; Ministry of Agriculture and Water, Water Resources Development Department: Riyadh, Saudi Arabia, 1984; Volume 8.
47. Michelsen, N.; Reshida, M.; Siebert, C.; Knoller, K.; Wiese, S.M.; Rausch, R.; Al-Saud, M.; Schuth, C. Isotopic and Chemical Composition of Precipitation in Riyadh, Saudi Arabia. *Chem. Geol.* **2015**, *413*, 51–62. [CrossRef]
48. Ali, K.K.; Ajeena, A.R. Assessment of Interconnection between Surface Water and Groundwater in Sawa Lake Area, Southern Iraq, Using Stable Isotope Technique. *Arab. J. Geosci.* **2016**, *9*, 648. [CrossRef]
49. Awadh, S.M.; Ali, K.K.; Alazzawi, A.T. Geochemical Exploration Using Surveys of Spring Water, Hydrocarbon and Gas Seepage, and Geobotany for Determining the Surface Extension of Abu-Jir Fault Zone in Iraq: A New Way for Determining Geometrical Shapes of Computational Simulation Models. *J. Geochem. Explor.* **2013**, *124*, 218–229. [CrossRef]
50. Engelhardt, I.; Rausch, R.; Keim, B.; Al-Saud, M.; Schüth, C. Surface and Subsurface Conceptual Model of an Arid Environment with Respect to Mid- and Late Holocene Climate Changes. *Environ. Earth Sci.* **2013**, *69*, 537–555. [CrossRef]
51. Matrood, M.J.; Hussein, H.M. A Preliminary Ecological Analysis of Spring Water in Al-Shanafiyah District, Al-Qadisiyah Province, Southern Iraq. *IOP Conf. Ser. Earth Environ. Sci.* **2021**, *790*, 012004. [CrossRef]
52. Bennetts, D.A.; Webb, J.A.; Stone, D.J.M.; Hill, D.M. Understanding the Salinisation Process for Groundwater in an Area of South-Eastern Australia, Using Hydrochemical and Isotopic Evidence. *J. Hydrol.* **2006**, *323*, 178–192. [CrossRef]
53. Imo, T.; Amosa, P.; Latu, F.; Vaurasi, V.; Ieremia, R. Chemical Composition of Rainwater at Selected Sites on Upolu Island, Samoa. *Atmos. Clim. Sci.* **2021**, *11*, 458–468. [CrossRef]
54. Webb, J.A.; White, S. Karst in Deserts. In *Treatise on Geomorphology*; Shroder, J.F., Ed.; Academic Press: San Diego, CA, USA, 2013; Volume 6, p. 397406.
55. Sutcliffe, J.F. *Mineral Salts Absorption in Plants*; Pergamon: London, UK, 1962.
56. Mengel, K.; Kirkby, E.A. *Principles of Plant Nutrition*; Kluwer Academic Publishers: Dordrecht, The Netherlands, 2001.
57. Hudson, R.O.; Golding, D.L. Controls on Groundwater Chemistry in Subalpine Catchments in the Southern Interior of British Columbia. *J. Hydrol.* **1997**, *201*, 1–20. [CrossRef]
58. Moulton, K.L.; West, J.; Berner, R.A. Solute Flux and Mineral Mass Balance Approaches to the Quantification of Plant Effects on Silicate Weathering. *Am. J. Sci.* **2000**, *300*, 539–570. [CrossRef]
59. Edwards, M.; Webb, J. The Importance of Unsaturated Zone Biogeochemical Processes in Determining Groundwater Composition, Southeastern Australia. *Hydrogeol. J.* **2009**, *17*, 1359–1374. [CrossRef]
60. Prado, G.H.C.; Rao, Y.; De Klerk, A. Nitrogen Removal from Oil: A Review. *Energy Fuels* **2017**, *31*, 14–36. [CrossRef]
61. Hussien, B.M.; Rabeea, M.A.; Farhan, M.M. Characterization and Behavior of Hydrogen Sulfide Plumes Released from Active Sulfide-Tar Springs, Hit, Iraq. *Atmos. Pollut. Res.* **2020**, *11*, 894–902. [CrossRef]
62. Pitman, J.K.; Steinshouer, D.; Lewan, M.D. Petroleum Generation and Migration in the Mesopotamian Basin and Zagros Fold Belt of Iraq: Results from a Basin-Modeling Study. *GeoArabia* **2004**, *9*, 41–72. [CrossRef]

hydrology

MDPI

Article

# A Long-Term Evaluation of the Ecohydrological Regime in a Semiarid Basin: A Case Study of the Huangshui River in the Yellow River Basin, China

Lijuan Fan [1,2], Lanxin Liu [3], Jing Hu [3], Fen Zhao [4], Chunhui Li [3,*] and Yujun Yi [3]

1 Qinghai Provincial Key Laboratory of Physical Geography and Environmental Process, College of Geographical Science, Qinghai Normal University, Xining 810008, China; fanlj87@hotmail.com
2 Haidong Municipal Water Resources Bureau of Qinghai Province, Haidong 810600, China
3 Key Laboratory for Water and Sediment Sciences of Ministry of Education, School of Environment, Beijing Normal University, Beijing 100086, China; 202111180043@mail.bnu.edu.cn (L.L.); 202321180028@mail.bnu.edu.cn (J.H.); yiyujun@bnu.edu.cn (Y.Y.)
4 School of Resources and Environmental Engineering, Ludong University, Yantai 264025, China; zhaof@ldu.edu.cn
* Correspondence: chunhuili@bnu.edu.cn

**Abstract:** This study aimed to evaluate the ecohydrological regime and ecological water demand of the Huangshui River Basin under changing environmental conditions, seeking to safeguard its ecosystem. Based on monthly data spanning from 1956 to 2016, the ecohydrological regimes of the Huangshui River and the Datong River were evaluated using methods such as the Pettitt mutation test, the Tennant method, and ecological deficit and surplus analyses. The data were mainly obtained from Xiangtang Station of the Datong River and Minhe Station of the Huangshui River. The results showed the following. (1) The most abrupt increase in measured runoff at Xiangtang Station occurred in 1993, while the point of abrupt change in measured runoff at Minhe Station occurred in 1990. (2) Following an increase in human activities, changes in the ecological surplus at Xiangtang Station were negative in January, April to May, July, and from September to November, while the changes in the ecological deficit were positive from January to April, July to August, and October to December. Changes in the ecological surplus at Minhe Station were negative from March to July and from September to December, while changes in the ecological deficit were positive from January to April and from July to December. (3) The annual average ecological flow of the Datong River, Xiangtang section, was 28.42 $m^3/s$, and the annual average ecological water demand was 896 million $m^3$. The annual average ecological flow of the Minhe section was 19.98 $m^3/s$, and the annual average ecological water demand was 631 million $m^3$. According to a calculation of the degree of ecological water demand and ecological flow satisfaction, prior to the implementation of the Water Diversion Project from the Datong River to Huangshui River, the water volumes in both rivers were generally sufficient to meet the ecological water demand. However, high water consumption during the irrigation period led to an ecological deficit. To address these issues, it is crucial to evaluate the potential impacts of human activities, such as water diversion projects, on river ecological flow. Recommendations include expediting the Water Diversion Project from the Yellow River to Xining to secure sufficient water flow in the Huangshui River and enhancing water conservation efforts in agricultural irrigation.

**Keywords:** ecohydrological regime; ecological surplus; ecological deficit; ecological water demand; Huangshui River Basin

## 1. Introduction

Climate change and increased human activities have led to a decline in river runoff worldwide, posing significant challenges for water resource management and ecological development [1]. Activities such as agricultural irrigation, the construction of water conservancy projects, and changes in land use can cause considerable hydrological variations [2,3].

Human activities significantly impact rivers, leading to considerable changes in hydrological conditions. Traditional methods for evaluating hydrologic alteration, such as the indicators of hydrologic alteration (IHA) and range of variability approach (RVA), mainly rely on the number of IHA index values within a target range obtained before interference [4]. Vogel et al. [5,6] introduced a dimensionally normalized ecological runoff index based on the flow–duration curve (FDC) using two dimensionless indicators: ecological surplus (ES) and ecological deficit (ED). These indicators assess hydrological regime changes in river ecosystems by representing the excess or deficiency in river runoff relative to the protected area. Gao et al. [7] further refined this method by defining the ecological surplus and deficit thresholds using the 75th and 25th percentile FDCs, respectively, rather than relying on the median FDC before mutation. This approach offers a clear conceptual framework and requires fewer indicators. It can be applied at various time scales (monthly, quarterly, or annually) and effectively reflects the overall impact of flow regulation during the specified period. This method provides a better description of changes in hydrological conditions and has been widely adopted by researchers [8–12].

Ensuring ecological flow is crucial for maintaining a healthy aquatic ecosystem amidst changes in river hydrological conditions. Often, the actual runoff of rivers falls short of meeting theoretical ecological water demands. To determine the degree to which the ecological water demand is met, it is essential to analyze the ecological water demand guarantee rate [13] and explore the trends and causes of river hydrological changes [8]. Ecological water demand is influenced by the natural attributes of the ecosystem, including its internal structure, external environment, and resource conditions. This demand exhibits characteristics such as dynamic orientation, target orientation, threshold orientation, and spatiotemporal variability. The numerical value of ecological water demand fluctuates within a specific threshold range. If the flow rate remains within this threshold, the ecosystem can sustain its health. However, if it falls outside this range, ecosystem health may be compromised [14–16]. The ecological water demand of rivers can be assessed based on factors such as river ecosystem characteristics, protected species, and natural hydrological conditions [17]. Common methods for determining ecological water demand include habitat simulation techniques such as the instant flow incremental methodology (IFIM) [18], hydraulic methods such as the wet perimeter method [19], and hydrological methods such as the Tennant method [20], the 7Q10 method [21], and the holistic approach framework [22]. Among these, hydrological methods are particularly favored by water resource management departments due to their efficiency in terms of time, labor, and cost. Consequently, they are the most widely used for assessing ecological water demand [17].

Fish habitats rely on continuous changes in water flow over many years, and fish encounter corresponding habitat conditions while adapting to the natural fluctuations in water flow in their habitats [23]. *Triplophysa siluroides*, *Gymnodiptychus pachycheilus*, *Gymnocypris eckloni*, and *Schizopygopsis pylzovi* are key protected species in the Huangshui River and are listed on the *Red List of Chinese Species*. As a semiarid area, ensuring the ecological water demand is met for key species in the Huangshui River is crucial. Due to high population density and industrial development in the Huangshui River Basin, previous water resource allocations have often failed to fully meet ecological water needs. This has led to the encroachment of ecological water into some rivers and issues of ecological water shortage. Currently, the main agricultural areas in certain tributaries of the Huangshui River Basin require substantial amounts of water for irrigation from April to June, but the inflow is insufficient to meet this demand. This has resulted in a significant mismatch between water supply and demand, with agricultural water use and other factors further reducing the river's ecological water availability. The Water Diversion Project from the Datong River to the Huangshui River undertook a successful trial operation of the main canal at the end of 2015. By September 2023, the main canal projects in northwestern China had also been successfully inaugurated. In recent years, there has been a gradual increase in water diversion projects, such as the planned construction of the Water Diversion Project from the Yellow River to Xining in the Huangshui River Basin. To ensure the stability of river ecosystems, it is essential to closely monitor the hydrological changes and ecological flow of the Datong and Huangshui rivers under the influence of these engineering projects. Several scholars have investigated the ecological water demand of the Huangshui River. Zhang et al. [24] analyzed current runoff, cross sections, and pollution sources employing three methods: the average value of the driest month at different frequencies, an improved same frequency distribution method within the year, and the ecological hydraulic radius method, used to calculate ecological flow in the Xining section of the Huangshui River. Liu et al. [25] examined the ecological water demand of 20 tributaries of the Huangshui River and proposed a method for calculating ecological water demand in river channels. Sha et al. [26] used the GAMS system to allocate and optimize water resources in the area receiving water from the Diversion Project, considering both supply–demand balance and ecological water needs. While these studies provide valuable insights, they tend to approach the issue from a single perspective. This study integrates analyses of hydrological changes, ecological surplus/deficit, and ecological water demand, which is mainly based on the analysis of measured runoff and natural runoff data sequences. Measured runoff refers to the amount of water that passes through a certain cross section of a river during a certain period of time. Natural runoff refers to the amount of water that has been reverted from the measured river runoff, which generally refers to the measured runoff plus the utilization of water above the measured cross section. By comparing and analyzing the measured runoff and natural runoff, we can provide a comprehensive evaluation of the impacts of human activities and climate change on ecological flow. This approach aims to offer scientific support for watershed ecological protection in northwest China.

Therefore, this study aimed to analyze both natural and measured hydrological data for the Datong and Huangshui rivers from 1956 to 2016 and to evaluate the ecohydrological regime using two methods: the Tennant method and the ecological surplus and deficit method. The specific objectives were as follows: (1) examine the characteristics of runoff changes at major stations along the Huangshui River; (2) calculate the ecological surplus/deficit, ecological water demand, and ecological flow guarantee rate; and (3) propose appropriate measures for ensuring ecological flow to provide a theoretical basis for the ecological protection of the Huangshui River. Additionally, recommendations are provided to support decision-makers in basin ecological protection efforts.

## 2. Study Area

The main tributary of the upper stream of the Yellow River is the Huangshui River, while the largest tributary of the Huangshui River is the Datong River. It has gradually shifted from the west to the east, exiting Qinghai Province and entering Gansu Province at Minhe station. The watershed area under the jurisdiction of the Minhe Hydrological Control Station is 15,342 km$^2$. The Datong River empties into the Huangshui River. The watershed area under the jurisdiction of the Xiangtang Hydrological Control Station is 15,126 km$^2$. The Huangshui River Basin is the main agricultural area in the northeast of Qinghai Province, consisting of the Huangshui's main stream area and the Datong River Basin. The Huangshui Valley in the central and southeastern parts of the basin has relatively high temperatures, low-lying terrain, and fertile land, which make it suitable for the development of agriculture and animal husbandry. The Huangshui River Basin features a semiarid plateau continental climate. This place experiences low temperatures and little rainfall throughout the year, with strong solar radiation. This study mainly focused on Minhe Station on the Huangshui River and Xiangtang Station on the Datong River (Figure 1). Minhe Station is located at 102°48′ E longitude and 36°20′ N latitude. It is the main control station for the Huangshui River. Xiangtang Station is located at 102°50′ E longitude and 36°21′ N latitude. It is the main control station for the Datong River. They are both national principal hydrometric stations that measure and report hydrological data such as runoff and sediment, as well as meteorological data such as rainfall and temperature according to national standard methods. This study mainly analyzed runoff and rainfall data measured by the two hydrological stations from 1956 to 2016.

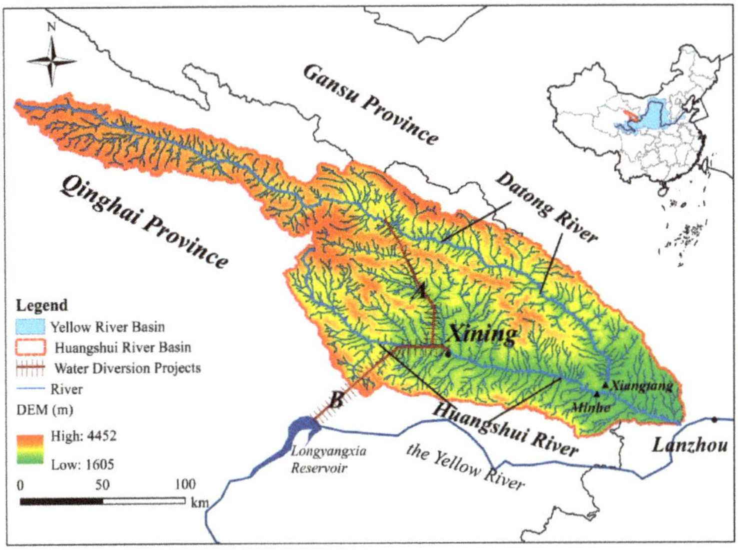


**Figure 1.** Site of the Huangshui River Basin. Note: A. Water Diversion Project from the Datong River to the Huangshui River; B. Water Diversion Project from the Yellow River to Xining City.

According to the Qinghai Province Planning Report on Water Diversion Project from Datong River to Huangshui River, the project is expected to divert 750 million m$^3$ of water overall by 2030. The Water Resources Planning Institute has evaluated and approved the Comprehensive Plan for the Huangshui River Basin (General Water Regulations [2014] no. 1182), which states that the Water Diversion Project from the Datong River to the Huangshui River will transfer 600 million m$^3$ of water by 2030. The diversion of water from the Datong River affects the ecological water requirements of both the Datong and Huangshui rivers' main streams. The Water Diversion Project from the Yellow River to Xining diverts water from the main stream of the Yellow River to the Xining and Haidong areas of the Huangshui River Basin. The preliminary determination of the project is that the water intake will be about 560 million m$^3$ in 2030 and about 960 million m$^3$ in 2040. The implementation of this project will greatly improve the water shortage problem in the Huangshui River Basin.

## 3. Methods

In this research, we analyzed the characteristics of runoff changes at major stations along the Huangshui River using data from 1956 to 2016. The Pettitt test was employed to detect abrupt changes in runoff. Subsequently, we evaluated the ecohydrological regime utilizing two approaches: calculating the ecohydrological condition through the ecological surplus/deficit method and assessing the ecological water demand and ecological flow guarantee rate via the Tennant method. Finally, we propose appropriate measures to ensure ecological flow, providing a theoretical basis for the ecological protection of the Huangshui River.

The analytical framework for this study is shown in Figure 2.

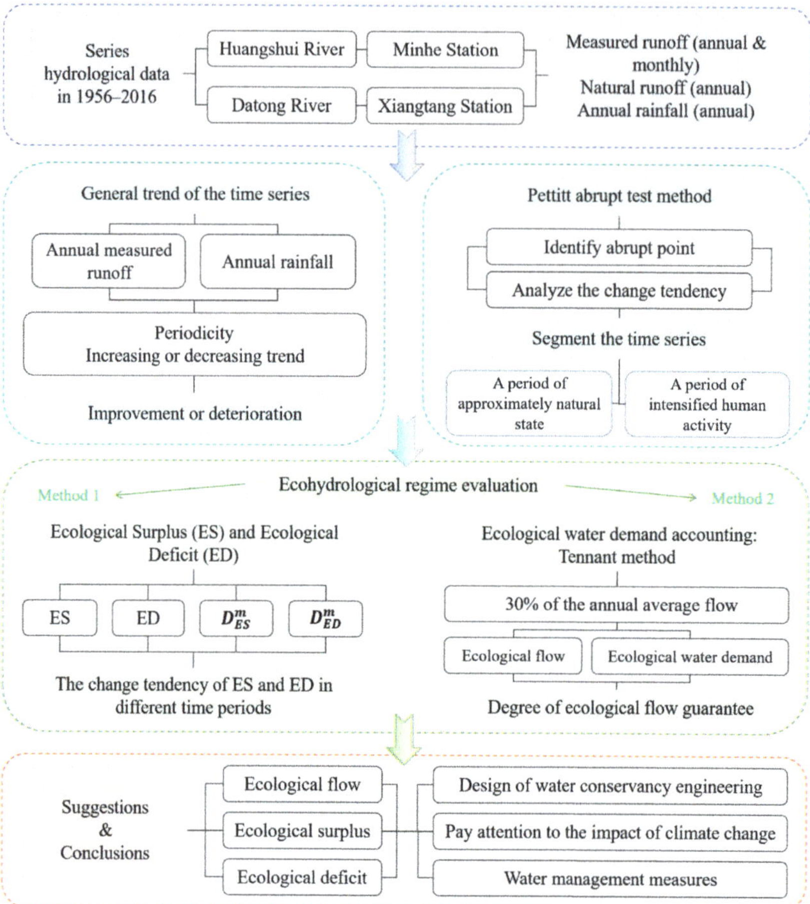


**Figure 2.** The analytical framework. Note: $D^m_{ES}$ is the degree of variation in the *ES* in month *m*. $D^m_{ED}$ is the degree of variation in the *ED* in month *m*.

### 3.1. Pettitt Abrupt Test Method

The trend of and abrupt changes in runoff in the Huangshui River were identified using the Pettitt abrupt test. The Pettitt test is a commonly used non-parametric method for detecting abrupt changes in time series data that is particularly useful when the data distribution is unknown.

To identify abrupt points in long-term time series data, researchers frequently employ the Pettitt method, a reliable and effective non-parametric testing technique [27]. Because it does not rely on presumptions regarding data distribution, including variance stability or a normal distribution, it is very flexible when handling different types of time series data. It is popular because it can efficiently identify abrupt spots and is not dependent on the duration of the time series [28,29].

The null hypothesis $H_0$ is that there is no abrupt at $t$ in the long-term sequence. The alternative hypothesis $H_1$ is that there is an abrupt change at $t$ in the long-term sequence. Assuming that the initial time series featured an abrupt point $x_t$, the time series $X = (x_1, \ldots, x_n)$ with n samples is split into two parts: $x_1, x_2, \ldots, x_t$ and $x_{t+1}, x_{t+2}, \ldots, x_n$. We determine its $U_{t,n}$ statistic via the following:

$$U_{t,n} = U_{t-1,n} + \sum_{j=1}^{n} sgn(x_t - x_j), t \in [2, n] \tag{1}$$

where $sgn(\cdot)$ is a sign function, specifically defined as follows:

$$sgn(x_j - x_k) = \begin{pmatrix} 1 & (x_j - x_k) > 0 \\ 0 & (x_j - x_k) = 0 \\ -1 & (x_j - x_k) < 0 \end{pmatrix} \tag{2}$$

We define the statistic $K_t$ for the time $t$, at which the abrupt point is most likely to occur:

$$K_t = max|U_{t,n}| \; 1 \leq t \leq n \tag{3}$$

We then calculate its significance level $P_t$:

$$P_t = 2 * \exp\left[-6 * K_t^2 / (n^3 + n^2)\right] \tag{4}$$

For a given confidence level $\alpha$ ($\alpha$ = 0.05), if $P_t > \alpha$, we accept the null hypothesis and assume that there is no significant abrupt change at time $t$. If $P_t < \alpha$, we reject the null hypothesis and assume that there is a significant abrupt change at time $t$.

*3.2. Ecological Hydrological Regime: Ecological Flow Surplus and Deficit Method*

Monthly time scales were used to calculate the ecological flow surplus and deficit. We calculated the flow values of the 75% and 25% quantiles of approximate natural state periods based on the division of the period of intensified human activity and the approximate natural state period. The range between the two flow values is here used to represent the changes in ecosystem adaptation. Ecological surplus (ES) is the proportion over the 75% percentile, and ecological deficit (ED) is the proportion below the 25% percentile [30]. The formula is as follows:

$$ES_m = \frac{(Q_i - Q_{i,75\%})}{Q_{i,75\%}} \quad Q_i > Q_{i,75\%} \tag{5}$$

$$ED_m = \frac{(Q_i - Q_{i,25\%})}{Q_{i,25\%}} \quad Q_i < Q_{i,25\%} \tag{6}$$

where $m$ is the month, $m$ = *1,2,. . .,12*; $Q_i$ is the average flow rate of month $i$ in a certain year ($m^3/s$); $Q_{i,75\%}$ is the 75th percentile flow rate for the month ($m^3/s$); and $Q_{i,25\%}$ is the 25th percentile flow rate for the month ($m^3/s$). The ecological surplus for month $m$ is non-negative, and the ecological deficit for month $m$ is non-positive.

Based on the above indicator system, the degree of variation in runoff was quantitatively analyzed, and the specific calculation formula is as follows:

$$TES_{exp}^{m} = \frac{N_{post}}{N_{pre}} \sum_{i=1}^{N_{pre}} ES_{obs,pre,i}^{m} \tag{7}$$

$$D_{ES}^{m} = \frac{\sum_{i=1}^{N_{post}} ES_{obs,post,i}^{m} - TES_{exp}^{m}}{TES_{exp}^{m}} \tag{8}$$

$$TED_{exp}^{m} = \frac{N_{post}}{N_{pre}} \sum_{i=1}^{N_{pre}} ED_{obs,pre,i}^{m} \tag{9}$$

$$D_{ED}^{m} = \frac{\sum_{i=1}^{N_{post}} ED_{obs,post,i}^{m} - TED_{exp}^{m}}{TED_{exp}^{m}} \tag{10}$$

where $TES_{exp}^{m}$ is the sum of the expected monthly $ES$ values for all years in the approximate natural state period based on the average $ES$ value of month $m$ in the approximate natural state period; $N_{post}$ is the total number of years during which human activity intensified; $N_{pre}$ is the total number of years in the approximate natural state time period; $ES^{m}_{obs,pre,i}$ and $ES^{m}_{obs,post,i}$, respectively, represent the *ES* value of month *m* in the *i* year calculated according to Equation (5) in the approximate natural state period and intensified human activity period; $\sum_{i=1}^{N_{pre}} ES^{m}_{obs,pre,i}$ is the sum of the *m*-month's *ES* of all years in the approximate natural state period; $\sum_{i=1}^{N_{post}} ES^{m}_{obs,post,i}$ is the sum of the *m*-month's ES of all years during the period of intensified human activity; and $D^{m}_{ES}$ is the degree of variation in the *ES* in month *m*. The indicators related to ecological deficit (ED) are similar to the aforementioned.

If the degree of variation in the ecological surplus is greater than 0, it indicates that the mean value of the ecological surplus at various time scales is greater than the approximate natural state period during the period of intensified human activity. The larger the degree of change, the longer the approximate natural state period. If the degree of variation in the ecological surplus is lower than 0, it indicates that the mean value of the ecological surplus at various time scales is lower than the approximate natural state period during the period of intensified human activity. The smaller the degree of change, the smaller the approximate natural state period.

### 3.3. River Ecological Water Demand: Tennant Method

The recommended levels of water demand to maintain the ecological environment in the river are split into two categories: a general water use period that spans from October to March of the following year, and a fish-spawning and juvenile period from April to September. The Tennant approach comprises eight levels, and the recommended value is based on the percentage of runoff.

The recommended ecological flows using the Tennant method are listed in Table 1.

**Table 1.** The ecological flow recommended by the Tennant method.

| Qualitative Description of Habitats | Recommended Base Flow Standard (Percentage of Annual Average Flow) | |
|---|---|---|
| | General Water Use Period (from October to March of the Following Year) | Fish Spawning and Juvenile Period (from April to September) |
| Maximum | 200 | 200 |
| Optimum flow | 60~100 | 60~100 |
| Excellent | 40 | 60 |
| Very good | 30 | 50 |
| Good | 20 | 40 |
| Becomes vestigial | 10 | 30 |
| Poor or minimum | 10 | 10 |
| Extremely poor | <10 | <10 |

This method considers 10%, 30%, and 60% to 100% of the annual average flow as the minimum ecological water demand in the river, the optimal flow required to ensure the survival of aquatic organisms, and ecological flow required to maintain the original natural river ecosystem, respectively. The formula is:

$$Q_T = \sum_{i}^{12} Q_i \times Z_i \quad (11)$$

where $Q_T$ is the ecological water demand of the river channel ($m^3$); $Q_i$ is the average annual flow rate in *i* month of a given year ($m^3$); and the recommended runoff percentage for the *i* month corresponds to $Z_i$ (%).

### 3.4. Ecological Flow Guarantee Rate and Evaluation Standard

In this study, we evaluated the ecological flow guarantee status of rivers based on the monthly average flow guarantee degree. The degree of ecological flow guarantee is here defined as the ratio of the number of months in which the monthly flow value was greater than the monthly ecological flow value to the total number of months corresponding to the long-term runoff year. The formula is:

$$D_i = (T_{bi}/T_i) \times 100\%, \ 1 \leq i \leq 12 \quad (12)$$

where $D_i$ is the ecological flow guarantee rate for the $i$ month; $T_{bi}$ is the number of months in the $i$ month of the calculation year that meet the ecological flow for that month; and $T_i$ is the total number of months in the $i$ month of the calculation year.

The evaluation criteria for the ecological flow guarantee rate are listed in Table 2.

**Table 2.** Evaluation criteria for ecological flow guarantee rate.

| Index | Evaluation Criterion/% | | | | |
|---|---|---|---|---|---|
| | Excellent | Good | Medium | Poor | Inferior |
| Ecological flow guarantee rate | 100 | 95~100 | 90~95 | 80~90 | <80 |

## 4. Results

### *4.1. Analysis of Runoff Trend in Huangshui River*

#### 4.1.1. Runoff Change Trends

According to the analysis of hydrological and meteorological data from 1956 to 2015 for the Huangshui River Basin, it can be seen that the annual precipitation and runoff generally show a decreasing trend. The decrease in runoff was 10 million $m^3/10a$, and the precipitation fluctuated at a rate of $-2.0$ mm/10a (Figure 3). The annual runoff went through a stage of increase before 1989 and a stage of decrease after 1989. In the increasing stage, the years of high flow and low flow were basically the same, that is, 11a and 13a, respectively. In the decreasing stage, the years of high flow and low flow were 15a and 6a, respectively. This indicates that the degree of runoff reduction was severe and that the sensitivity of runoff to precipitation changes was strong. The decreasing trend in annual runoff showed quasi-periodic changes in 4a, 9a, and 20a.

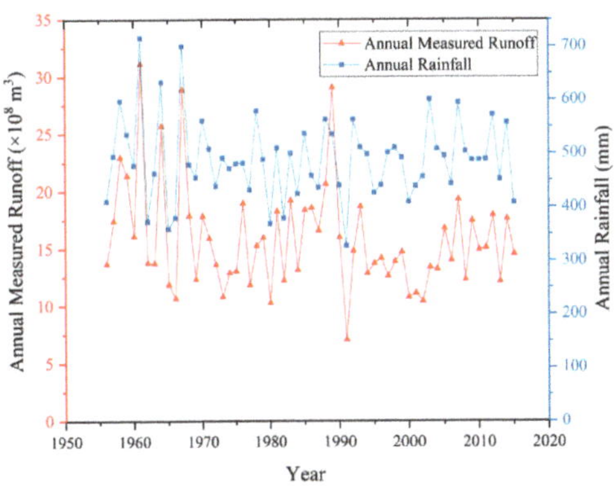


**Figure 3.** The annual rainfall and measured annual runoff characteristics observed at Minhe Station on the Huangshui River from 1956 to 2015.

#### 4.1.2. Analysis of Abrupt Changes of Runoff

The processes of variation in natural runoff and measured runoff at Minhe Station and Xiangtang Station are shown in Figure 4. It is believed that the difference between natural and measured runoff is due to the combined effects of human activities and climate change, such as agricultural irrigation and reservoir regulation, which can affect river flow. The Pettitt abrupt test was performed using the long-term runoff data from the Xiangtang and Minhe stations from 1956 to 2016, and the results are shown in Figure 5. From the position of the red dashed line in the figure, it can be seen that the point of abrupt change in the measured runoff at Xiangtang Station occurred in 1993 and the abrupt change in the measured runoff at Minhe Station occurred in 1990. For Xiangtang Station, the measured runoff showed an increasing trend before 1993 and a decreasing trend after 1993. For Minhe Station, the measured runoff increased first and then gradually stabilized before 1970. From 1970 to 1980, it showed a fluctuating decrease, and then gradually increased from 1980 to 1990. Since 1990, the measured runoff at Minhe Station has sharply decreased, indicating that human activities began to intensify at this time. It was not until 2004 that the measured runoff began to slowly increase again. Similar results were obtained using the M-K test method.

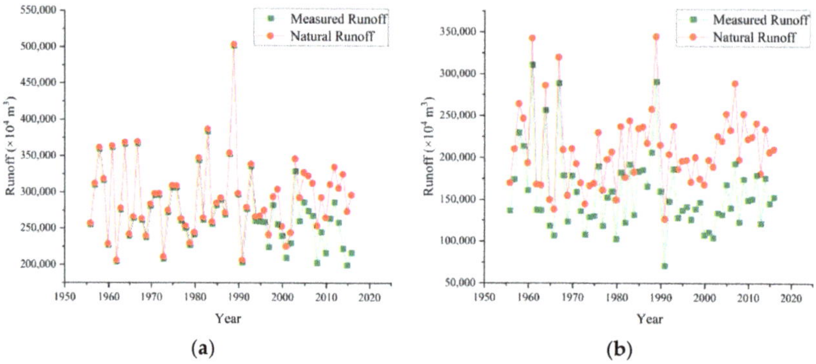


**Figure 4.** Changes in natural and measured runoff at Xiangtang (**a**) and Minhe (**b**) stations from 1956 to 2016.

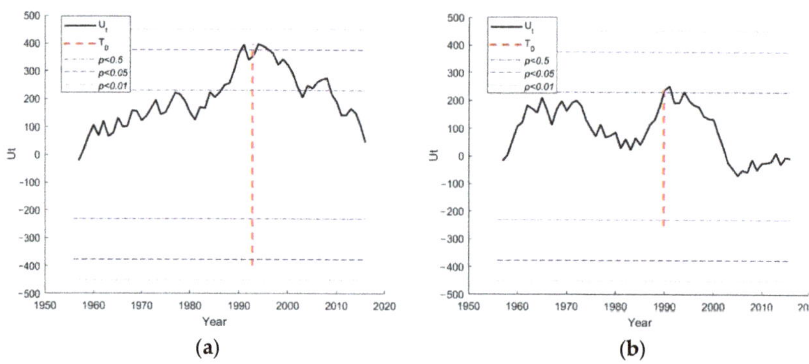


**Figure 5.** Pettitt abrupt test for annual measured runoff observed by Xiangtang (**a**) and Minhe (**b**) stations from 1956 to 2016.

Therefore, the period of research performed at the two stations can be divided into two sections. For Xiangtang Station, 1956–1993 represents an approximate period with the natural state, and 1994–2016 represents a period of intensified human activity. For Minhe Station, 1956–1990 represents an approximate period with the natural state, and 1991–2016 is a period of intensified human activity.

### *4.2. Analysis of Ecological Flow Surplus and Deficit Results*

The results obtained from the ecological flow surplus and ecological flow deficit methods are listed in Tables 3 and 4, respectively. The ecological surplus changes during the period of intensified human activity at the Xiangtang Hydrological Station (1994–2016) were negative in January, April to May, July, and September to November compared to the natural state period (1956–1993), while the ecological deficit changes were positive from January to April, July to August, and October to December. The ecological and hydrological conditions of the rivers deteriorated during these months. The ecological surplus decreased the most in September, reaching −71.4%, and in February, it increased the most, reaching 475%. The ecological deficit increased the most in April, reaching 285%, and the most significant decrease was in June, reaching −9.5%. The degree of change in the ecological deficit exceeded 100%, most notably from March to April and November to December. Ecological runoff has adverse effects on river ecosystems.

**Table 3.** Ecological surplus and deficit results of Xiangtang Station.

| Month | Ecological Surplus | | $D_{ES}^{m}$/% | Ecological Deficit | | $D_{ED}^{m}$/% |
|---|---|---|---|---|---|---|
| | 1956–1993 | 1994–2016 | | 1956–1993 | 1994–2016 | |
| January | 0.026 | 0.022 | −15.4 | −0.025 | −0.039 | 56.0 |
| February | 0.012 | 0.069 | 475.0 | −0.027 | −0.034 | 25.9 |
| March | 0.034 | 0.042 | 23.5 | −0.013 | −0.032 | 146.1 |
| April | 0.034 | 0.020 | −41.2 | −0.020 | −0.077 | 285.0 |
| May | 0.058 | 0.030 | −48.3 | −0.042 | −0.042 | 0 |
| June | 0.048 | 0.052 | 8.3 | −0.041 | −0.037 | −9.5 |
| July | 0.071 | 0.035 | −50.7 | −0.056 | −0.061 | 8.9 |
| August | 0.047 | 0.054 | 14.9 | −0.049 | −0.053 | 8.2 |
| September | 0.084 | 0.024 | −71.4 | −0.036 | −0.035 | −2.8 |
| October | 0.056 | 0.034 | −39.3 | −0.033 | −0.051 | 54.5 |
| November | 0.035 | 0.031 | −11.4 | −0.014 | −0.036 | 157.1 |
| December | 0.032 | 0.051 | 59.4 | −0.012 | −0.035 | 191.7 |

**Table 4.** Ecological surplus and deficit results of Minhe Station.

| Month | Ecological Surplus | | $D_{ES}^{m}$/% | Ecological Deficit | | $D_{ED}^{m}$/% |
|---|---|---|---|---|---|---|
| | 1956–1990 | 1991–2016 | | 1956–1990 | 1991–2016 | |
| January | 0.038 | 0.047 | 23.7 | −0.022 | −0.031 | 40.9 |
| February | 0.024 | 0.044 | 83.3 | −0.021 | −0.026 | 23.8 |
| March | 0.064 | 0.029 | −54.7 | −0.035 | −0.038 | 8.6 |
| April | 0.081 | 0.047 | −42.0 | −0.049 | −0.055 | 12.2 |
| May | 0.197 | 0.042 | −78.7 | −0.117 | −0.114 | −2.6 |
| June | 0.090 | 0.065 | −27.8 | −0.110 | −0.077 | −30.0 |
| July | 0.065 | 0.043 | −33.8 | −0.050 | −0.071 | 42.0 |
| August | 0.054 | 0.055 | 1.9 | −0.037 | −0.057 | 54.1 |
| September | 0.092 | 0.060 | −34.8 | −0.024 | −0.060 | 150.0 |
| October | 0.075 | 0.040 | −46.7 | −0.029 | −0.056 | 93.1 |
| November | 0.093 | 0.039 | −58.0 | −0.028 | −0.040 | 42.9 |
| December | 0.065 | 0.018 | −72.3 | −0.023 | −0.029 | 26.1 |

The ecological surplus changes during the period of intensified human activity at the Minhe Hydrological Station (1991–2016) were negative from March to July and September to December compared to the natural state period (1956–1990), while the ecological deficit changes were positive from January to April and July to December. The ecological and hydrological conditions of rivers deteriorated in these months. The ecological surplus decreased the most in May, reaching −78.7%, and in February, it increased the most, reaching 83.3%. The ecological deficit increased the most in September, reaching 150%, and the most significant decrease was in June, reaching −30%.

Overall, the months when the ecological flow surplus increased were mainly concentrated in the dry seasons for the rivers, which were primarily the discharge period of the reservoirs and not the agricultural irrigation period. These changes in human activity led to an increase in river flow. The ecological flow deficit in most months increased significantly after human activity intensified, with a slight decrease from May to June.

Figure 6 shows the monthly ecological surplus and deficit levels at Xiangtang and Minhe stations from 1956 to 2016. From the chart, it can be seen that the distribution of ecological surplus and deficit at Xiangtang Station over the past 61 years is relatively uniform. Before 1980, the ecological deficit during the rainy season was severe. From 1980 to 2005, there was a significant increase in the ecological surplus, but after 2014, the ecological deficit worsened. For Minhe Station, the degree of ecological surplus was relatively light, while there were two periods with obvious ecological deficits: from April to July around 1969–1982 and from 1991 to 2004. Both the Datong and Huangshui rivers showed significant ecological deficits before 1980, which may have been due to the impact of climate conditions at that time. After 1990, due to human activities such as water diversion, the ecological deficit of the Huangshui River improved, while the reduction in water volume in the Datong River exacerbated the ecological deficit of the Datong River.

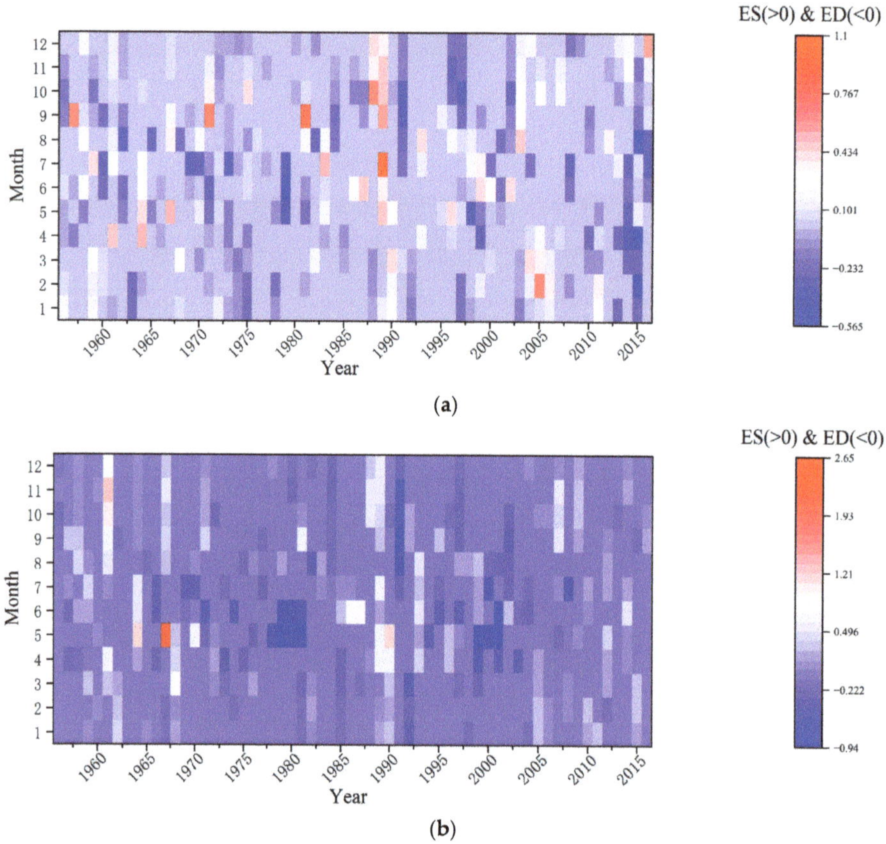


**Figure 6.** Monthly flow surplus and deficit chart of Xiangtang (**a**) and Minhe (**b**).

## 4.3. Ecological Flow Guarantee Rate of Huangshui River and Datong River

Given the paramount ecological significance of the Huangshui River, in this study, we employed the Tennant method and designated 30% of the natural average flow spanning 1956 to 2016 as the ecological flow benchmark for this stretch. Adhering to the Implementation Rules of the Yellow River Water Regulation, the minimum flow required for a 95% guarantee rate at the Datong River Xiangtang Station is here set at 10 $m^3/s$, while the warning flow level at Minhe Station stands at 8 $m^3/s$. Integrating these criteria, we determine the monthly ecological flow thresholds for the Minhe and Xiangtang sections. The results are shown in Table 5.

**Table 5.** Ecological flow (unit: $m^3/s$).

| Month \ Section | Xiangtang | Minhe |
|---|---|---|
| January | 10 | 8.32 |
| February | 10 | 8.65 |
| March | 10 | 10.13 |
| April | 17.8 | 22.16 |
| May | 28.87 | 18.57 |
| June | 37.80 | 20.72 |
| July | 60.87 | 28.43 |
| August | 58.54 | 31.28 |
| September | 52.19 | 31.14 |
| October | 29.95 | 31.13 |
| November | 15 | 18.43 |
| December | 10 | 10.21 |

The annual average ecological flow for the Xiangtang section of the Datong River was calculated as 28.42 $m^3/s$, translating to an annual ecological water demand of 896 million $m^3$. Similarly, the Minhe section's annual average ecological flow was 19.98 $m^3/s$, equating to an annual ecological water demand of 631 million $m^3$.

After comparing and analyzing the monthly measured runoff and ecological flow data from Minhe Station and Xiangtang Station, we derived monthly ecological flow guarantee rates spanning from 1956 to 2016 (Table 6). Our calculations indicate that the ecological flow guarantee rate for the Xiangtang section of the Datong River in January stands at 98%, whereas the guarantee rates for the remaining months are set consistently at 100%. Notably, in March, August, and September, the guarantee rates for the same section also dipped to 98%, yet they remained at 100% for all other months. This fluctuation is primarily attributed to the influence of irrigation water on the guarantee rate of ecological water demand.

**Table 6.** Evaluation of ecological flow guarantee rate.

| Month | Ecological Flow Guarantee Rate/% | | Ecological Flow Guarantee | |
|---|---|---|---|---|
| | Datong River | Huangshui River | Datong River | Huangshui River |
| January | 98 | 100 | Good | Excellent |
| February | 100 | 100 | Excellent | Excellent |
| March | 100 | 98 | Excellent | Good |
| April | 100 | 100 | Excellent | Excellent |
| May | 100 | 100 | Excellent | Excellent |
| June | 100 | 100 | Excellent | Excellent |
| July | 100 | 100 | Excellent | Excellent |
| August | 100 | 98 | Excellent | Good |
| September | 100 | 98 | Excellent | Good |
| October | 100 | 100 | Excellent | Excellent |
| November | 100 | 100 | Excellent | Excellent |
| December | 100 | 100 | Excellent | Excellent |

From these results, it is evident that the ecological flow guarantee of the Datong River in January can be categorized as "good", whereas for the remaining months, it can be deemed "excellent". Similarly, the ecological flow guarantee of the Huangshui River in March, August, and September can also be labeled "good", with the other months achieving "excellent" status. This signifies that the water volumes of both the Datong River and Huangshui River adequately fulfill the requirements for ecological water demand.

## 5. Discussion

### *5.1. Impact of Climate Change and Human Activities on Hydrological Regime*

Hydrological changes in the Huangshui River Basin are influenced by both climate change and human activities. Studies reveal that climate change causes 35.46% of the reduction in runoff within the basin [31]. Temperature plays a pivotal role, as it directly affects evaporation, a critical factor in determining runoff. The interplay between temperature and precipitation can either exacerbate or mitigate changes in runoff. For instance, combinations such as "rising temperature + decreasing precipitation" and "decreasing temperature + increasing precipitation" tend to worsen runoff changes, whereas "increasing temperature + increasing precipitation" and "decreasing temperature + decreasing precipitation" tend to lessen these changes [32]. Increased runoff due to climate change would raise the river's ecological base flow and enhance the ecological flow guarantee rate. Conversely, decreased runoff would lower both the ecological base flow and the guarantee rate. With the projected increase in precipitation and decrease in runoff under climate change, it is crucial to manage water resources effectively in order to sustain the ecological flow.

The analysis also highlights that intensified human activities have led to abrupt changes in river flow, significantly affecting ecological surplus and deficit. For example, in 1994, following the completion of the main canal under the Water Diversion Project from the Datong River to Qinwangchuan, there was a marked change in the flow rate of the Datong River. During the construction of the Datong River to Huangshui River Water Diversion Project [33] from 1996 to 2018, both the ecological surplus and deficit of the Datong and Huangshui rivers underwent significant changes, showing a deteriorating trend. Although the water volumes of these rivers during the study period generally met the ecological water demands, their future trends need continuous monitoring due to their sensitivity to human activities.

### *5.2. Different Ecological Water Demand Accounting Methods*

The ecological water demand in this study was primarily estimated using the Tennant method. Various techniques can be employed to estimate the ecological water demand of rivers, including hydrological, hydraulic, habitat, and holistic analysis methods [34]. Among these, hydrological methods are widely used due to their ability to quickly provide results based on recorded hydrological data. In contrast, the other three methods are more computationally complex and require additional data, such as hydraulic parameters and fish data. The Tennant method, along with the historical flow curve method, forms the basis of the historical flow approach, a research-oriented hydrological technique. The Tennant approach is known for its simplicity and user-friendliness, making it a relatively quick and straightforward method. Hydrological technology or monitoring stations can directly supply the necessary data for this method [35]. However, this method also has its limitations. For example, the reliance on simple flow percentage calculations may not capture the true needs of complex ecosystems. Additionally, the method primarily considers flow, overlooking the impact of other environmental factors such as temperature and seasonal variations. Zhang et al. [24] used three methods to calculate the ecological water demand of the Xining section of the Huangshui River and compared the results with the Tennant method for rationality. The result was that the ecological flow from April to June was 9.3 $m^3/s$, from July to October, it was 12.2 $m^3/s$, and from November to March of the following year, it was 5.6 $m^3/s$. This result is comparatively smaller than those obtained in this study, as the Xining section is located upstream of the investigated river segment, resulting in a correspondingly lower flow rate. According to the Comprehensive Plan for the Huangshui River Basin (2019), the ecological water demand for the Minhe section varied, ranging from 28.3 $m^3/s$ from April to June to 41.7 $m^3/s$ from July to October and then decreasing to 8.6 $m^3/s$ for November to March of the following year. For the Xiangtang section, the minimum ecological flow ranged from 14.8 $m^3/s$ to 24.1 $m^3/s$ for April to June and remained stable at 24.1 $m^3/s$ from July to October. The findings of this study closely align with the projections set out in the Huangshui River Basin Comprehensive Plan.

In this study, due to the lack of specific water requirements for aquatic organisms in the region, only the overall runoff of the river could be used to roughly calculate the ecological water demand. This may affect the accuracy of the calculation results. Therefore, it is recommended to conduct a more comprehensive analysis of the ecological water demand of rivers after obtaining more sufficient water demand data for aquatic organisms in the future. It is also advisable to further explore the potential impacts of climate change and water transfer projects on the ecological water demand guarantee of the Huangshui River and its tributaries. This exploration should utilize daily flow data and detailed engineering water transfer scheduling plans to ensure more accurate assessments.

## 6. Conclusions and Suggestions

Based on the comprehensive analysis of hydrological data from 1956 to 2016, this study evaluated the ecohydrological regimes and ecological water demands of the Huangshui River Basin, focusing on the Minhe and Xiangtang stations. The key findings reveal the significant influence of human activities and climate change on river runoff, ecological surplus, and deficit patterns. These findings emphasize the need for targeted water management strategies. Below are the main conclusions and actionable suggestions.

### 6.1. Conclusions

(1) Abrupt changes in measured runoff were observed at Xiangtang Station in 1993 and Minhe Station in 1990, primarily due to increased human activities such as agricultural irrigation and water diversion projects.
(2) Human activities, particularly during the irrigation season, have exacerbated the ecological deficits at both stations. The most significant ecological deficits were observed in critical months, including April and September, where the flow rates were insufficient to meet the ecological water demands.
(3) The implementation of the Water Diversion Project from the Datong River to the Huangshui River improved water availability. However, high water consumption for irrigation has resulted in ecological deficits during key periods, stressing the need for further intervention to balance water usage.

Our results show that the ecological deficit in most months has increased, indicating that intensified human activities have had a significant negative impact on river ecosystems. This study provides critical scientific support for watershed ecological protection in northwest China, particularly in the context of climate change and increased human activities.

### 6.2. Suggestions

This analysis of the Huangshui River's ecological water demand guarantee rate and deficit indicates that high water consumption for agricultural irrigation adversely affects runoff and the ecological water demand guarantee rate. The impact of water diversion projects on river runoff is also significant. Therefore, the suggestions are as follows.

(1) Optimizing Agricultural Water Use: The significant ecological deficits during the irrigation season suggest a need for enhanced water-saving agricultural practices. Farmers should adopt advanced irrigation technologies such as drip irrigation and soil moisture monitoring systems to reduce water wastage. Incentives for adopting water-efficient crops that require less water should be provided, particularly in arid zones.

(2) Prioritizing Ecological Flow Maintenance in Future Water Diversion Projects: While water diversion projects, like the Yellow River to Xining initiative, are essential for sustaining agricultural and urban needs, ecological flow preservation must be integrated into project planning. A dynamic water allocation model should be implemented, ensuring that minimum ecological flows are maintained year-round, especially during dry months. Monitoring stations along the rivers should be enhanced to track the impacts of these projects on ecological water demands.

(3) Strengthening the Management of Water Resource Allocation: A basin-wide integrated water resource management system should be established to balance the competing demands for agricultural, industrial, and ecological water. This includes the development of a centralized management system that coordinates the operation of reservoirs and water diversion projects, ensuring ecological water demands are consistently met. Public awareness campaigns should be initiated to promote water conservation among all stakeholders, particularly in the agriculture and industrial sectors.

(4) Climate Change Adaptation Strategies: As climate change is expected to further alter precipitation patterns, a long-term water resource planning framework should be developed. This should include adaptive measures to cope with the increased variability in runoff, ensuring that the ecological integrity of the rivers is preserved under future climate scenarios. The framework should incorporate predictive models to simulate potential future scenarios, facilitating proactive decision-making.

(5) Enhancing Scientific Research and Data Collection: Future research should focus on collecting more granular data on aquatic ecosystems to refine the estimation of ecological water demand. This includes detailed studies on the water needs of key protected species and their habitats. Additionally, daily flow data should be collected to improve the accuracy of ecohydrological assessments, allowing for more precise ecological flow management.

By implementing these strategies, the Huangshui River Basin can move towards sustainable water management, ensuring both human and ecological needs are met. Long-term monitoring and adaptive management will be crucial to safeguarding the river ecosystem in the face of increasing anthropogenic pressures and climate change.

**Author Contributions:** L.F.: conceptualization, data curation, formal analysis, and writing—original draft; L.L.: writing—original draft; J.H. and F.Z.: writing—original draft; C.L.: investigation, methodology, and project administration; Y.Y.: funding, resources, software, supervision, and writing—review and editing. All authors have read and agreed to the published version of the manuscript.

**Funding:** This work was supported by the Joint Funds of the National Natural Science Foundation of China (U2243236), the National Natural Science Foundation of China (grant 42201046), National Science Fund for Distinguished Young Scholars (52025092), and the Qinghai Haidong Urban-Rural Eco-Development Project (L3443-PRC-HD-CB-CS4). The authors express their gratitude to the editor and anonymous reviewers for their valuable comments, which have enhanced the quality of the article.

**Data Availability Statement:** Publicly available datasets were analyzed in this study. This data can be found here: doi: 10.6084/m9.figshare.27201213.

**Conflicts of Interest:** The authors declare that the research was conducted in the absence of any commercial or financial relationships that could be construed as a potential conflict of interest.

## References

1. Wang, H.; Liu, J.G.; Klaar, M.; Chen, A.F.; Gudmundsson, L.; Holden, J. Anthropogenic climate change has influenced global rive flow seasonality. *Science* **2024**, *383*, 1009–1014. [CrossRef] [PubMed]
2. Lv, X.Z.; Liu, S.S.; Li, S.P.; Ni, Y.X.; Qin, T.L.; Zhang, Q.F. Quantitative estimation on contribution of climate changes an watershed characteristic changes to decreasing streamflow in a typical basin of yellow river. *Front. Earth Sci.* **2021**, *9*, 75242 [CrossRef]
3. Wang, S.J.; Yan, M.; Yan, Y.X.; Shi, C.X.; He, L. Contributions of climate change and human activities to the changes in runo increment in different sections of the Yellow River. *Quat. Int.* **2012**, *282*, 66–77. [CrossRef]

4. Cui, T.; Tian, F.Q.; Yang, T.; Wen, J.; Khan, M.Y.A. Development of a comprehensive framework for assessing the impacts of climate change and dam construction on flow regimes. *J. Hydrol.* **2020**, *590*, 125358. [CrossRef]
5. Vogel, R.M.; Fennessey, N.M. Flow-Duration Curves. I: New Interpretation and Confidence Intervals. *J. Water Resour. Plan. Manag.* **1994**, *120*, 485–504. [CrossRef]
6. Vogel, R.M.; Sieber, J.; Archfield, S.A.; Smith, M.P.; Apse, C.D.; Huber-Lee, A. Relations among storage, yield, and instream flow. *Water Resour. Res.* **2007**, *43*. [CrossRef]
7. Gao, Y.X.; Vogel, R.M.; Kroll, C.N.; Poff, L.R.; Olden, J.D. Development of representative indicators of hydrologic alteration. *J. Hydrol.* **2009**, *374*, 136–147. [CrossRef]
8. Liu, L.L.; Wei, L.X.; Xu, Y.; Xin, X.G.; Xiao, C. Projection of climate change impacts on ecological flow in the Yellow River basin. *Adv. Water Sci.* **2021**, *32*, 824–833. (In Chinese)
9. Wang, Y.K.; Wang, D.; Lewis, Q.W.; Wu, J.C.; Huang, F. A framework to assess the cumulative impacts of dams on hydrological regime: A case study of the yangtze river. *Hydrol. Process.* **2017**, *31*, 3045–3055. [CrossRef]
10. Deng, L.; Guo, S.; Tian, J.; Wang, H. Assessment of ecological flow alterations induced by hydraulic engineering projects in the Han River, China. *Hydrol. Res.* **2023**, *55*, 33–50. [CrossRef]
11. Zhu, S.N.; Dong, Z.C.; Wu, S.J.; Fu, G.B.; Li, Z.Z.; Meng, J.Y.; Shao, Y.; Zhang, K. Evaluation of reservoir-induced ecological flow with consideration of ecological responses to multiple hydrological alterations: A case study in the Feiyun River Basin, China. *Ecol. Indic.* **2023**, *154*, 110859. [CrossRef]
12. Wang, F.L.; Guo, X.M.; Zhang, S.; Hu, T. A Risk Assessment Method for Ecological Flows under Different Time Scales. *J. Chang. River Sci. Res. Inst.* **2023**, *40*, 29–35. (In Chinese)
13. Yang, Y.; Wang, Z.H.; Wang, X.L.; Zhao, C.S.; Zhang, C.B.; Pan, T.L. Calculation of environment flow and analysis of spatial environment flow satisfaction rate: A case study of Ji'nan City. *Adv. Earth Sci.* **2020**, *35*, 513–522. (In Chinese)
14. Zhang, X.X. Study on Theory and Process of Ecological Water Requirement of River Basin. Master's Thesis, Tianjin University of Technology, Tianjin, China, 2013. (In Chinese).
15. Tang, J.; She, X.Y.; Lin, N.F.; Ma, S.T. Advances in research on the theories and methods of eco-environmental water demand. *Sci. Geogr. Sin.* **2005**, *25*, 3367–3373. (In Chinese)
16. Ma, L.K. Catchment Eco-Environmental Water Requirement through a Case Study in the Yanhe Basin. Ph.D. Thesis, Peking University, Beijing, China, 2008. (In Chinese).
17. Zhao, F.; Pang, A.P.; Li, C.H.; Zheng, X.K.; Wang, X.; Yi, Y.J. Progress in ecological water requirements in the mainstream and estuary ecosystem of the Yellow River Basin. *Acta Ecol. Sin.* **2021**, *41*, 6289–6301. (In Chinese)
18. Bovee, K.D. Stream habitat analysis using the instream flow incremental methodology. *Reports* **1982**, *2*, 19–28.
19. Bartschi, D. A habitat-discharge method of determining instream flows for aquatic habitat. *Proc. Symp. Specility Conf. Instream Flow Needs II* **1976**, 285–294.
20. Tennant, D.L. Instream flow regimens for fish, wildlife, recreation and related environmental resources. *Fisheries* **1976**, *1*, 6–10. [CrossRef]
21. Boner, M.C.; Furland, L.P. Seasonal treatment and variable effluent quality based on assimilative capacity. *J. Water Pollut. Control Field* **1982**, *54*, 1408–1416.
22. Arthington, A.H.; Pusey, B.J. Instream flow management in Australia: Methods, deficiencies and future direction. *Aust. Biol.* **1993**, *6*, 52–60.
23. Zhang, Z.G.; Tan, Q.L.; Zhong, Z.G.; Jin, Y.; Du, J.M. Study on ecological flow regime based on habitat requirements of fish. *Water Power* **2016**, *42*, 13–17. (In Chinese)
24. Zhang, Y.; Hou, Q.Z.; Tian, F.C.; Cheng, P.R. Ecological flow calculation and simulation analysis of Huangshui River in Xining section based on water quality and water quantity coupling model. *Water Power* **2023**, *49*, 6–13+120. (In Chinese)
25. Liu, B.J.; Cui, C.Y.; Wang, L.W.; Hou, B.J. Study on ecological water quantity of typical branch ditch replacement in Huangshui river valley. *Water Resour. Plan. Des.* **2020**, *11*, 58–63. (In Chinese)
26. Sha, J.X.; Xie, X.M.; Zhang, S.P.; Lu, H.Y. Study on optimization and readjustment of water allocation for reception basin of Datonghe River-to-Huangshui River Water Diversion Project. *Water Resour. Hydropower Eng.* **2012**, *43*, 1–3. (In Chinese)
27. Pettitt, A.N. A non-parametric approach to the change-point problem. *J. R. Stat. Soc.* **1979**, *28*, 126–135. [CrossRef]
28. Karim, S.A.; Sedigheh, B.A. Evaluation of Terra Climate gridded data in investigating the changes of reference evapotranspiration in different climates of Iran. *J. Hydrol. Reg. Stud.* **2024**, *52*, 101678.
29. Mendes, M.P.; Rodriguez-Galian, V.; Aragones, D. Evaluating the BFAST method to detect and characterise changing trends in water time series: A case study on the impact of droughts on the Mediterranean climate. *Sci. Total Environ.* **2022**, *846*, 157428. [CrossRef] [PubMed]
30. Zhang, S.; Guo, X.M.; Zhou, M.; Hu, T. New method for evaluating changes in hydrologic regimes based on eco-flows metrics. *J. Hydroelectr. Eng.* **2021**, *40*, 65–76. (In Chinese)
31. Zhang, T.F.; Zhu, X.D.; Wang, Y.J.; Li, H.M.; Liu, C.H. The impact of climate variability and human activity on runoff changes in the Huangshui River Basin. *Resour. Sci.* **2014**, *36*, 2256–2262. (In Chinese)
32. Yang, J.J. Research on Runoff Simulation of Huangshui River Basin Based on SWAT Model. Master's Thesis, Qinghai Normal University, Xining, China, 2009. (In Chinese)

33. Li, H.C.; Qin, T.; Wang, Z.Q.; Qiu, Y.; Lei, X.H. Study on Management System of Inter-Basin Water Diversion Project—A Case Study of the Project of Drawing Water from Datong River into Huangshui River. *IOP Conf. Ser. Earth Environ. Sci.* **2018**, *189* 022047. [CrossRef]
34. King, J.; Louw, D. Instream flow assessments for regulated rivers in South Africa using the Building Block Methodology. *Aquat Ecosyst. Health Manag.* **1998**, *1*, 109–124. [CrossRef]
35. Li, C.W. Key Technologies of River Ecological Flow on Modified Tennant Method and Sensitive Ecological Requirements Ph.D. Thesis, Huazhong University of Science and Technology, Wuhan, China, 2016. (In Chinese)

water

MDPI

Article

# The Surface Water Potentiality in Arid and Semi-Arid Basins Using GIS and HEC-HMS Modeling, Case Study: Gebel El Sibai Watershed, Red Sea

Abdelfattah Elsayed Elsheikh [1], Mahmoud A. El Ammawy [2], Nessrien M. Hamadallah [3], Sedky H. A. Hassan [4,*], Sang-Eun Oh [5], Kotb A. Attia [6,*] and Mahmoud H. Darwish [7]

1 Hydrology Department, Desert Research Center, Cairo 11753, Egypt; elsheikh70@hotmail.com
2 Geology Department, Desert Research Center, Cairo 11753, Egypt; elammawymahmoud@yahoo.com
3 Department of Plant Ecology and Ranges, Desert Research Center, Cairo 11753, Egypt; nessrien2023@gmail.com
4 Biology Department, College of Science, Sultan Qaboos University, Muscat 123, Oman
5 Department of Biological Environment, Kangwon National University, Hyoja-2-dong, Chuncheon-si 24341, Gangwon-do, Republic of Korea; ohsangeun@kangwon.ac.kr
6 Department of Biochemistry, College of Science, King Saud University, P.O. Box 2455, Riyadh 11451, Saudi Arabia
7 Geology Department, Faculty of Science, New Valley University, El Kharga 72511, Egypt; mahmoud.hamed68@sci.nvu.edu.eg
* Correspondence: s.hassan@squ.edu.om (S.H.A.H.); kattia1.c@ksu.edu.sa (K.A.A.)

**Abstract:** The Red Sea region is considered one of the regions that suffer most from water scarcity among the Egyptian areas. This situation reinforces the importance of maximizing the utilization of available water sources. Rainwater and flood harvesting may form a good water source if good harvesting practices are applied. Natural pastures, Bedouin communities, and wild plants may be affected by severe droughts expected due to climate change. Additional water resources are very important to enhance the resilience of the Bedouin communities to probable droughts. Five main hydrographic basins are issued from Gebel El Sibai (+1435 m), including Wadi Esel, Wadi Sharm El Bahari, Wadi Sharm El Qibli, Wadi Wizr, and Wadi Umm Gheig. Detailed investigation of morphometric parameters, runoff/rainfall relationship, and flood volume using GIS and HEC-HMS model of each basin were estimated as well as natural vegetation. This study reveals that rainfall ranges from 84 mm to 0 mm, and a storm of 84 mm (highest event) is expected to occur every 42 years with a probability of 2.4%. Quantitative morphometric analysis implies that the area has good potential for flooding, especially Wadi Sharm El Qibli and Wadi Umm Gheig, where Wadi Sharm El Bahri represents the lowest priority for flooding. The flood volume of Umm Gheig basin is the greatest: 12 million $m^3$ at the basin outlet with a rainfall event of 15 mm. Wadi Esel is expected to collect 8.7 million $m^3$ due to the ratio of the impervious soil and rainfall quantity, Wadi Sharm El Bahari 2.1 million $m^3$, Wadi Sharm El Qibli 1.6 million $m^3$, and Wadi Wizer 1.04 million $m^3$. Seven storage dams (SD1-SD7) were proposed to enhance the utilization of the surface water potentialities of this study area.

**Keywords:** morphometric parameters; flood risk analysis; wild plants; HEC-HMS software

## 1. Introduction

Egypt is indeed facing a significant challenge with water scarcity, with an annual water deficit of approximately seven billion cubic meters. A major challenge in managing the country's water resources is the imbalance between the growing demand for water and the limited available supply, especially in the Red Sea Coast region, where limited surface and groundwater are present. These issues are exacerbated by climate change, overpopulation, and environmental degradation. It is a complex issue that requires careful management and sustainable practices. The surface water resources in Egypt primarily include the Nile River. The Nile River originates outside the country and is Egypt's lifeblood, contributing about 97% of the renewable water resources, restricted to the Nile Valley and Delta [1,2]. Egypt is almost without rain, except on the north coast, where rain falls at an average yearly rate of 50–250 mm. The Egyptian Red Sea coast is periodically exposed to flash floods that cause severe human and economic losses [3]. Flash floods occur occasionally and are destructive, especially in the Red Sea and South Sinia mountainous areas. The occasional rain can be harvested and stored using runoff control systems such as dykes and dams [4]. Unfortunately, these resources vary because of climatic changes and other factors that negatively affect natural vegetation. Significant efforts are directed to utilize these water resources and avoid risks. Flash floods are indeed a significant risk in the Red Sea Mountains, Quseir, and Gebel El Sibai areas. High flood frequencies (F) and densities (D), which are attributable to the existence of impervious rocks of rough, steep surface and low permeable subsoil material, consequently limit infiltration of rainfall to the groundwater [5]. These regions are particularly susceptible to flash floods because of various contributing factors, such as topography, geomorphology, drainage patterns, rainfall intensity and duration, evaporation and infiltration rates, and both environmental and human activities. Most of the previous literature addresses the flood risk assessment due to its geomorphological basis, and less attention is paid to quantifying the flood volume. In addition, the relationship between the availability of water resources and the keeping of natural vegetation is important for Bedouin communities. Effective management of these risks is crucial to reducing the potential impact on human lives and infrastructure [6]. A previous study [6] estimated the rainfall-runoff relationship and hydrographic parameters necessary for managing seasonal floods in certain hydrographic basins south of El Quseir City, employing morphometric parameters and the SCS-CN method, which is widely used to estimate surface runoff volume for specific rainfall events [7]. El Alfy [8] integrates geographic information systems (GIS), remote sensing, and rainfall-runoff modeling to evaluate the effects of urbanization on flash floods in arid regions. In areas lacking hydro-meteorological data, runoff modeling can be a valuable approach to improve the livability of such regions [9]. HEC-HMS, a hydrologic modeling software developed by the US Army Corps of Engineers-Hydrologic Engineering Centre (HEC), is frequently used to estimate a basin's hydrological response to precipitation [10]. Gebel El Sibai is a significant landform in the area south of El Quseir city of +1435 m height, forming an outstanding and distinctive watershed. This situation gives the chance to form important deep, steep slopes and high-density drainage basins. Five basins are encountered from north to south: Wadi Esel, Wadi Sharm El Bahari, Wadi Sharm El Qibli, Wadi Wizer, and Wadi Umm Gheig. Drought is a natural hazard that arises from various parameters, such as a reduction in rainfall amount, high evapotranspiration, and overexploitation of water resources, or a combination of all the above, which results in serious impacts on social, economic, and environmental levels, as well as vegetation [11], which provides human being and other species with food, shelter, medicine and everything else [12]. Drought is also associated with high temperatures, winds, and low relative humidity; its severity is based on the demands that human activities and vegetation impose on the water supply in a given area [13]. Climate change and human-modified landscapes have increased the risk of floods and droughts globally, while biodiversity has declined [14]. These processes will further increase in the future [15]. This paper aims to evaluate the surface water resources of these ungagged basins through detailed estimation of morphometric parameters, runoff/rainfall relationship, hydrograph generation of each basin using GIS and HEC-HMS model, and explore wild plant species of this study area. The results will recommend adopting a robust water management policy in this area. In addition, it aims to search for new water sources to support the Bedouin communities that live in these areas and suffer from water scarcity.

## 2. Study Area

This study area is in the Red Sea area, nearly in the middle part of the Eastern Desert between latitudes 25°20′ and 25°55′ N and longitudes 34°0′ and 34°35′ E (Figure 1). The area is located south of El Quseir city by about 20 km with a total area of about 2000 km$^2$.

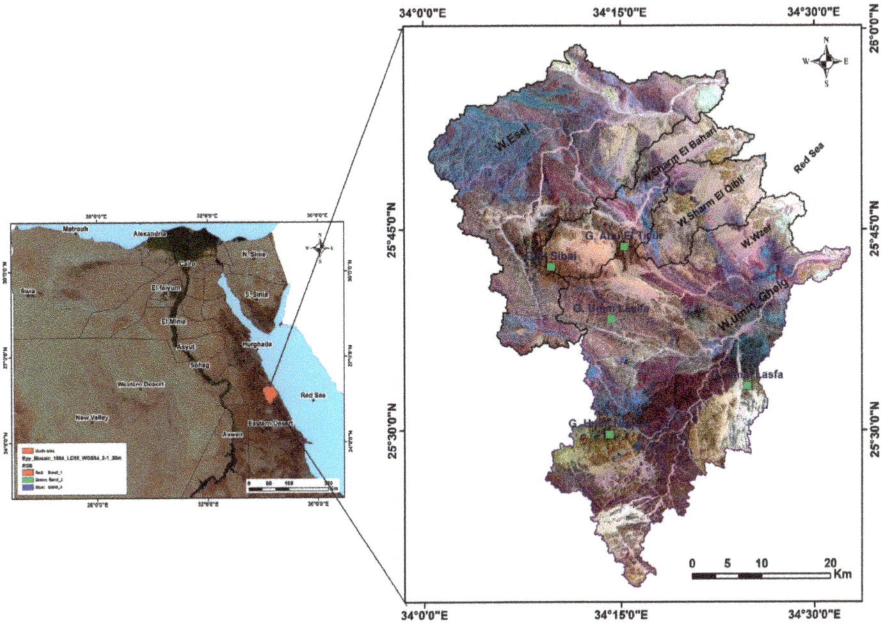


**Figure 1.** Location map of this study area.

The area has a cost front of about 30 km, extending from the outlet of Wadi Esel in the north to Marsa Umm Gheig in the south. The area is bounded by the water divide of Wadi Esel and Wadi El Ambagi to the North and the southern water divide of Wadi Umm Gheig to the south, with an average width of 40 km and average length of 55 km. The Eastern Desert is located in the extremely arid provinces of Egypt. This arid climate influences not only the hydrological properties of the drainage basins in the area but also the forms of developmental opportunities. The maximum and minimum recorded temperatures in the area are 41 °C and 21 °C, respectively, and increase from north to south, while the relative humidity (RH) ranges between 56% and 30%, averaging about 43% in summer and 48% in winter. The area is characterized by a large number of mountainous crests that belong to the basement Red Sea complex, of whom are Gebel El Sibai (+1435), Gebel Abu El Tiyur (+1100 m), Gebel Umm Lasifa (+1165), Gebel Umm Naqqat (+1280), and finally Gebel Umm Lasfa (+570). This study area also includes five main drainage patterns, named from north to south as Wadi Esel, Wadi Sharm El Bahari, Wadi Sharm El Qibli, Wadi Wizer, and Wadi Umm Gheig.

The topographic and geologic maps indicate a complex physiographic setting of this study area where the ground surface suffers from many undulations with various rock types (Figure 2). Geologically, the Eastern Desert is composed of both sedimentary and crystalline rocks. The oldest exposed crystalline rocks belong to the pre-Cambrian age and are overlain by the Nubian sequence and post-Nubian deposits. The crystalline rocks of the pre-Cambrian basement complex form massive formations extending parallel to the Red Sea graben and consisting mainly of metamorphic rocks, as well as acidic and basic igneous rocks [16–18]. The area is geologically complicated where the western parts are characterized by metamorphic and igneous rocks, and the eastern area is characterized by sedimentary rocks of quaternary and tertiary age. The area is also affected by the tectonic movements that occurred in the Eastern Desert, coinciding with the formation of the Red Sea Canyon and the subsequent violent tectonic movements, whose impact was reflected in the density of faults and the shape of water drainage basins in the region. Topographically, the area is characterized by a wavy surface. The ground elevation increases from east to west. Elevations range from zero level at the Red Sea coast at the extreme east to about +1435 m at G. El Sibai (Figure 3).

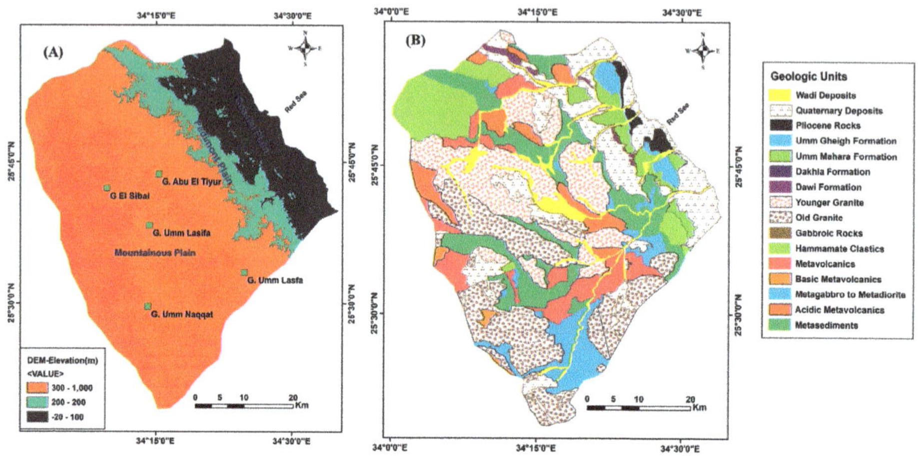


**Figure 2.** Geomorphologic units (**A**) and Geologic units (**B**) prevail over this study area [19].

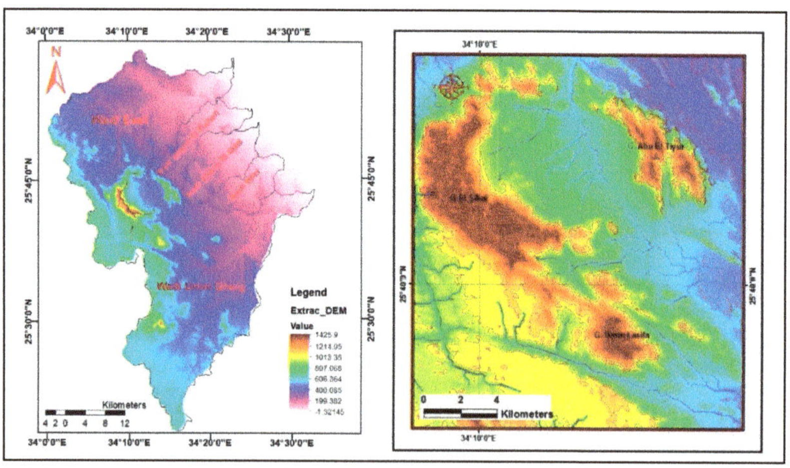


**Figure 3.** Digital Elevation Model (DEM) of this study area (**left**), mountainous area (**right**).

Two profiles were constructed, one from east to west and the other crossing the area from north to south. **Profile A-A′**, crosses the area from east to west through G. El Tiyur and G. El Sibai area with 40 km. The ground surface gradually increases westward. The surface gradient is estimated at 13 m/km. Two mountain crests are present, separated by a wide depression of 7 km. These mountains, G. El Tiyur and G. El Sibai, are characterized by steep slopes. **Profile B-B′** crosses the area from north to west through the mountains G. El Sibai, G. Umm Lasifa, and G. Umm Naqqat with a 55 km length. The area in this direction suffers from high undulations. G. El Sibai has a longitudinal crest of 7 km wide, whereas G. Umm Lasifa and G. Umm Naqqat have point tops (Figure 4).

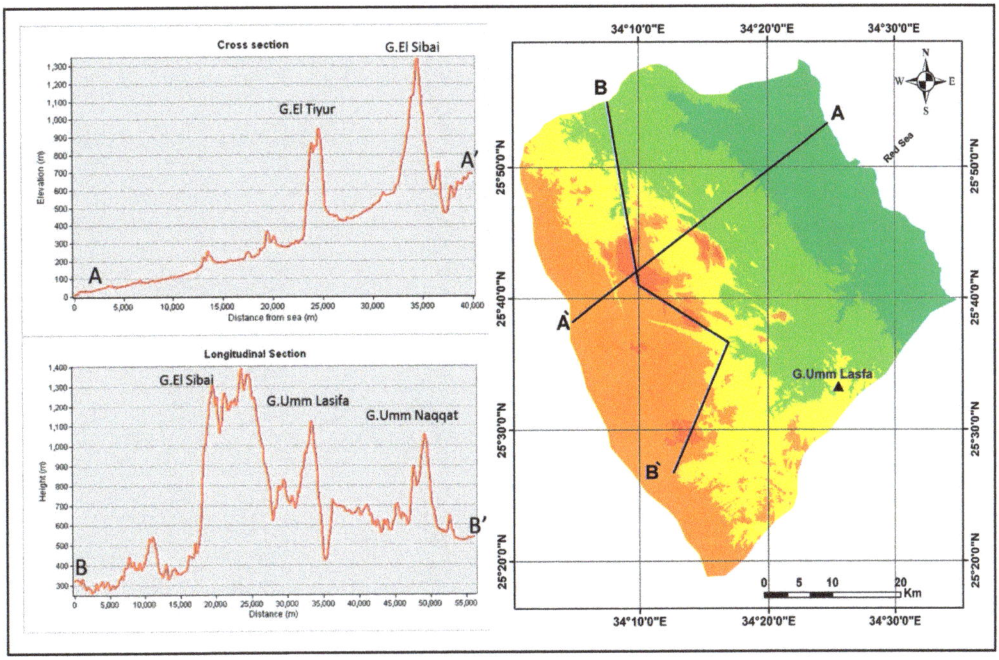


**Figure 4.** Terrain variations of this study area.

**Geomorphologically**, four geomorphological units characterize the area. **1. Coastal Plain** extends parallel to the Red Sea coast between Contour lines (0–100 m), where the width of this plain is about 7 km, the breadth of this plain increases from one region to another. It reaches its maximum width at the main basin's outlets, i.e., Wadi Esel, Wadi Umm Gheig, Wadi Sharm El Bahari, and Wadi Sharm El Qibli. The widening of the coastal plain area at the valley's deltas may be due to the large water discharge resulting from tolerance of surface runoff and the large number of sediments associated with the addition to the receding of the rift edge due to the water erosion activity prevailing in the area because of the thunderstorms that occur from time to time. **2. Pediment Plain** is represented by the sedimentary zone, which is characterized by steep slopes. This plain extends between the contour line 100 m and 200 m with a breadth of about 5 km. This zone is characterized by steep slopes, irregular shapes, and flood fans. This plain reaches its maximum width at Wadi Esel and Wadi Umm Gheig, where the contour line 200 m retreated to the west. This plain is characterized by straightforwardness because of long SE-NW faults parallel to the great Red Sea canyon. **3. Mountainous Plain** covers about 75% of this study area and borders by the 200 m contour line from the east. This zone is highly terrain plain. Many tributaries cross this plain, forming the water collectors of the main wadis. It is composed mainly of fractured Igneous and metamorphic rocks with little sedimentary rocks in the main valley courses. The highest mountains represented by G. El Sibai (+1435), G. Abu El Tiyur (+1100 m), G. Umm Lasifa (+1165 m), and G. Umm Naqqat (+1280 m) are present in this plain. The mountain masses are characterized by their cliff edges and pointed peaks because of the hardness of their rocks and their resistance to erosion factors. **4. Hydrographic Basins** represent five main hydrographic basins. They include Wadi Esel, Wadi Sharm El Bahari, Wadi Sharm El Qibli, Wadi Wizr, and Wadi Umm Gheig (Figure 5). These wadis collect rainwater and sediments from the mountainous area in the west to the outlets at the Red Sea Coast. The main morphometric parameters vary from one basin to another. Wadi Umm Gheig is the largest area and longest perimeter among the Wadis, where Wadi Esel represents the tallest main course. On the other hand,

Wadi Sharm El Bahari represents the steep slope among the whole wadis. These Wadis play an important role in erosion and plant cover in the area under concern.

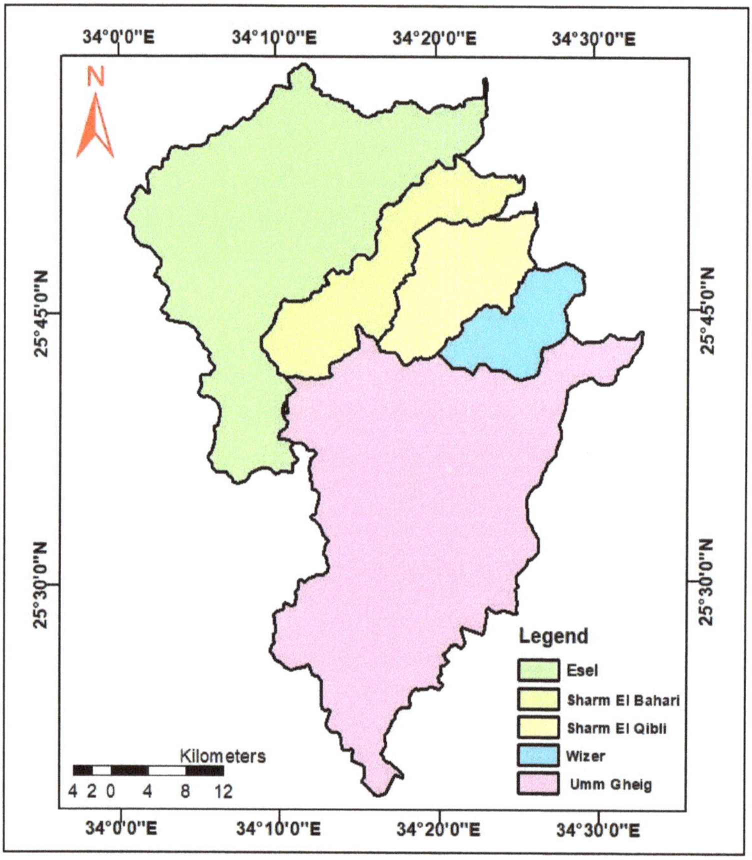


**Figure 5.** Drainage basins of this study area.

Geologically, the Central Eastern Desert of Egypt is primarily composed of a complex mix of tectonically assembled ophiolitic mélange and interoceanic arcs. Over time, these formations were overlain by molasse-type sediments and continental volcanic deposits and further shaped by the intrusion of granitoid suites and dyke swarms [18,20–22]. The area under concern is geologically complicated; it consists of basement rocks and sedimentary rocks that are affected by many folding and faulting structures [19]. The geologic units of the area are subdivided into two main divisions. The first one represents the Precambrian rocks, which form the main erosion unit, and the second one is the sedimentary rocks, which form the main depositional unit [23]. **Basement complex:** These rocks constitute the oldest rocks outcropping in the area. It includes igneous and metamorphic rocks. These rocks form the base where the recent sedimentary rocks were deposited. It includes Metagabbro and Metadiorite, metavolcanics, Hammamat Classics, Old Granite, and Younger Granite Rocks. **Sedimentary Rocks:** These rocks are formed from many formations, ranging in age from Cretaceous to Holocene periods. **Dawi Formation** is composed of three phosphate zones, separated by clay and limestone layers rich in marine fossils. It has limited distribution in this study area except in the area extended from Wadi Sharm El Bahari and Wadi Esel. **Dakhla Formation** belongs to the Pliocene period. It is formed from dark, grey, shallow marine marl and clay with limestone intercalations. It is an outcropping in the middle parts of Wadi Sharm El Qibli and lower parts of Wadi Esel. **Mahara Formation** represents the Middle Miocene rocks unconformably with older rocks. It comprises sandy limestone in the lower and gypsiferous limestone in the upper part. It appears in many parts of the coastal plain in Wadi Umm Gheig and Wadi Esel. **Umm Gheig Formation** is composed of limestone rich in algal remains, organic debris, and fossils of plecepods and gastropods. Its thickness ranges from 8–10 m. It belongs to the Upper Miocene period. It appears in the lower parts of Wadi Umm Gheig and Wadi Esel. **Shagara Formation** belongs to the Pliocene period. It is composed of sandstone and marl rich in fossils of 22 m thickness. This formation is well distributed in the coastal plain area at the outlets of Wadi Esel, Wadi Sharm El Bahari, Wadi Sharm El Qibli, and Wadi Wizr. **Quaternary Deposits** are widespread all over this study area, i.e., fluvial deposits, sabkhas, shore dunes, and marine terraces. **Wadi deposits** are restricted to the water collectors of the drainage basins and are mainly formed of conglomerates and sand deposits resulting from erosion by water during intensive flash floods.

These geological formations and topographic features played the greatest role in forming the valleys and will have the greatest impact on the amount of water produced from these valleys in the event of heavy rainfall, as igneous rocks act as a positive factor in collecting rainwater and the possibility of harvesting, while sedimentary rocks are useful in feeding groundwater reservoirs.

## 3. Materials and Methods

Archival data, such as long-term rainfall records required for estimating the probability of exceedance $G(x)$ and the return period $T$, were collected from the Desert Research Center (DRC) database, Egypt, in addition to recent rainfall records from the World Wide Web through the NASA website (Table 1). The POWER Data Access Viewer (DAV) is a web-based application developed by the NASA Prediction of Worldwide Energy Resources (POWER) project. It provides access to community-based analysis-ready meteorology and solar-related parameters. These parameters are specifically formulated for assessing and designing renewable energy systems. The relationship between the probability of exceedance $G(x)$ and the return period T is given by the following formula [24].

$$G(x) = \frac{1}{T} \tag{1}$$

**Table 1.** Rainfall data of the El Sibai watershed area in the period (1981–2021).POWER | Data Access Viewer (nasa.gov).

| YEAR | JAN | FEB | MAR | APR | MAY | JUN | JUL | AUG | SEP | OCT | NOV | DEC | ANN |
|---|---|---|---|---|---|---|---|---|---|---|---|---|---|
| 1981 | 0 | 0 | 0 | 0 | 0 | 0 | 0 | 0 | 0 | 0 | 0 | 0 | 0 |
| 1982 | 0 | 0 | 0 | 0 | 0 | 0 | 0 | 0 | 0 | 0 | 0 | 0 | 0 |
| 1983 | 0 | 0 | 0 | 0 | 0 | 0 | 0 | 0 | 0 | 0 | 0 | 0 | 0 |
| 1984 | 0 | 0 | 0 | 0 | 0 | 0 | 0 | 0 | 0 | 0 | 0 | 0 | 0 |
| 1985 | 0 | 0 | 5.27 | 0 | 0 | 0 | 0 | 0 | 0 | 0 | 0 | 0 | 5.27 |
| 1986 | 0 | 0 | 0 | 0 | 0 | 0 | 0 | 0 | 0 | 52.73 | 21.09 | 10.55 | 84.38 |
| 1987 | 0 | 0 | 0 | 0 | 0 | 0 | 0 | 0 | 0 | 0 | 0 | 0 | 0 |
| 1988 | 0 | 0 | 0 | 0 | 0 | 0 | 0 | 0 | 0 | 0 | 0 | 0 | 0 |
| 1989 | 0 | 0 | 0 | 0 | 0 | 0 | 0 | 0 | 0 | 0 | 0 | 0 | 0 |
| 1990 | 0 | 0 | 0 | 0 | 0 | 0 | 0 | 0 | 0 | 0 | 0 | 0 | 0 |
| 1991 | 5.27 | 0 | 0 | 0 | 0 | 0 | 0 | 0 | 0 | 0 | 0 | 0 | 5.27 |
| 1992 | 0 | 0 | 0 | 0 | 0 | 0 | 0 | 0 | 0 | 0 | 0 | 0 | 0 |
| 1993 | 0 | 0 | 0 | 0 | 0 | 0 | 0 | 0 | 0 | 26.37 | 0 | 0 | 26.37 |

**Table 1.** *Cont.*

| YEAR | JAN | FEB | MAR | APR | MAY | JUN | JUL | AUG | SEP | OCT | NOV | DEC | ANN |
|---|---|---|---|---|---|---|---|---|---|---|---|---|---|
| 1994 | 0 | 0 | 0 | 0 | 0 | 0 | 0 | 0 | 0 | 0 | 10.55 | 0 | 10.55 |
| 1995 | 0 | 0 | 0 | 0 | 0 | 0 | 0 | 0 | 0 | 0 | 0 | 0 | 0 |
| 1996 | 0 | 0 | 0 | 0 | 0 | 0 | 0 | 0 | 0 | 0 | 5.27 | 0 | 5.27 |
| 1997 | 15.82 | 0 | 0 | 0 | 5.27 | 0 | 0 | 0 | 0 | 0 | 0 | 0 | 21.09 |
| 1998 | 0 | 0 | 0 | 0 | 0 | 0 | 0 | 0 | 0 | 0 | 0 | 0 | 0 |
| 1999 | 0 | 0 | 0 | 0 | 0 | 0 | 0 | 0 | 0 | 0 | 0 | 0 | 0 |
| 2000 | 0 | 0 | 0 | 0 | 0 | 0 | 0 | 0 | 0 | 0 | 0 | 0 | 0 |
| 2001 | 0 | 0 | 0 | 0 | 0 | 0 | 0 | 0 | 0 | 0 | 0 | 0 | 0 |
| 2002 | 0 | 0 | 0 | 0 | 0 | 0 | 0 | 0 | 0 | 0 | 0 | 0 | 0 |
| 2003 | 0 | 0 | 0 | 0 | 0 | 0 | 0 | 0 | 0 | 0 | 0 | 0 | 0 |
| 2004 | 0 | 0 | 0 | 0 | 0 | 0 | 0 | 0 | 0 | 0 | 0 | 0 | 0 |
| 2005 | 0 | 0 | 0 | 0 | 0 | 0 | 0 | 0 | 0 | 0 | 0 | 0 | 0 |
| 2006 | 0 | 0 | 5.27 | 0 | 0 | 0 | 0 | 0 | 0 | 0 | 0 | 0 | 5.27 |
| 2007 | 0 | 0 | 0 | 0 | 0 | 0 | 0 | 0 | 0 | 0 | 0 | 0 | 0 |
| 2008 | 0 | 0 | 0 | 0 | 0 | 0 | 0 | 0 | 0 | 0 | 0 | 0 | 0 |
| 2009 | 0 | 0 | 0 | 0 | 0 | 0 | 0 | 0 | 0 | 0 | 0 | 0 | 0 |
| 2010 | 5.27 | 0 | 0 | 0 | 0 | 0 | 0 | 0 | 0 | 0 | 0 | 0 | 5.27 |
| 2011 | 0 | 0 | 0 | 0 | 0 | 0 | 0 | 0 | 0 | 0 | 0 | 0 | 0 |
| 2012 | 0 | 0 | 0 | 0 | 0 | 0 | 0 | 0 | 0 | 0 | 0 | 0 | 0 |
| 2013 | 10.55 | 0 | 0 | 0 | 0 | 0 | 0 | 0 | 0 | 0 | 0 | 0 | 10.55 |
| 2014 | 0 | 0 | 0 | 0 | 0 | 0 | 0 | 0 | 0 | 0 | 0 | 0 | 0 |
| 2015 | 0 | 0 | 0 | 0 | 0 | 0 | 0 | 0 | 0 | 0 | 0 | 0 | 0 |
| 2016 | 0 | 0 | 0 | 0 | 0 | 0 | 0 | 0 | 0 | 15.97 | 0 | 0 | 15.97 |
| 2017 | 0 | 0 | 0 | 0 | 0 | 0 | 0 | 0 | 0 | 0 | 0 | 0 | 0 |
| 2018 | 0 | 0 | 0 | 0 | 0 | 0 | 0 | 0 | 0 | 0 | 0 | 0 | 0 |
| 2019 | 0 | 0 | 0 | 0 | 0 | 0 | 0 | 0 | 0 | 0 | 0 | 0 | 0 |
| 2020 | 0 | 0 | 0 | 0 | 5.27 | 0 | 0 | 0 | 0 | 0 | 0 | 0 | 5.27 |
| 2021 | 0 | 0 | 0 | 0.09 | 0 | 0 | 0 | 0 | 0 | 0 | 0.04 | 0 | 0.14 |

The most universally used formula for calculating the return period ($T$) for a particular flood peak is that of Weibull [25,26].

$$T = \frac{(N+1)}{m} \tag{2}$$

where $N$ is the number of events in the series (dimensionless), and m is the rank (from largest to smallest) of each event in the series (dimensionless).

The hazard degree approach is applied to estimate the basin's hazard degree, which is dependent on the values of the affected estimated morphometric parameters. Using the maximum and minimum values of the morphometric parameter, [27] proposed the following two equations to calculate the hazard degree of each parameter.

$$\text{Hazard degree} = (4(X - Xmin)/(Xmax - Xmin)) + 1 \tag{3}$$

for direct proportion parameters

$$\text{Hazard degree} = (4(X - X\text{max})/(X\text{min} - X\text{max})) + 1 \tag{4}$$

for inverse proportion parameters
where X is the value of the assessed morphometric parameter. X max is the maximum value of the assessed morphometric parameter of all studied basins. X min is the minimum value of the assessed morphometric parameter of all studied basins. The hazard value of each basin is the sum of the hazard degree of its affected morphometric parameters. A hazard scale starting with 1 (lowest) to 5 (highest) has been assigned, i.e., the higher the value, the greater the hazard.

Additionally, one of the most widely used methods for estimating surface runoff volume from a specific rainfall event is the SCS-CN method, originally developed by the Soil Conservation Service (SCS) in 1956 and now referred to as the NRCS-CN method. This method relies on the principles of water balance and is founded on two key hypotheses. The first hypothesis posits that the ratio of direct runoff to the potential maximum runoff is equivalent to the ratio of infiltration to the possible maximum retention. The second hypothesis suggests that the initial abstraction is directly proportional to the potential maximum retention. The water balance equation and the two hypotheses are expressed mathematically, respectively, as:

$$P = I_\alpha + F + Q \tag{5}$$

$$\frac{Q}{P - I_\alpha} = \frac{F}{S} \tag{6}$$

$$I_\alpha = \lambda S \tag{7}$$

The variables are defined as follows: $P$ represents the total precipitation (in millimeters), $I_\alpha$ is the initial abstraction before runoff (in millimeters), $F$ stands for the cumulative infiltration after runoff begins (in millimeters), $Q$ indicates direct runoff (in millimeters), $S$ denotes the potential maximum retention (in millimeters), and $\lambda$ is the coefficient for the initial abstraction ratio. Storm events with $P \geq 5$ mm were used to determine Curve Number (CN) values during the calibration period [28]. The original SCS-CN model assumes $\lambda = 0.2$.

The general runoff equation combination of Equations (6) and (7) is shown in Equation (8):

$$Q = \frac{(P - I_\alpha)^2}{P - I_\alpha + S} \text{ for } P > I_\alpha \quad = 0 \text{ otherwise} \tag{8}$$

The potential maximum retention $S$ (mm) can vary in the range of $0 \leq S \leq \infty$, and it is directly linked to $CN$. Parameter $S$ is mapped to the $CN$ using Equation (9) as follows:

$$S = \frac{25,400}{CN} - 254 \tag{9}$$

The (CN) is influenced by factors such as land use, hydrologic soil group, hydrologic condition, and antecedent moisture condition (AMC), and its value can range from 0 to 100. Three AMCs are defined as follows: dry (representing the lower moisture limit or upper limit of S), moderate (normal soil moisture condition), and wet (representing the upper moisture limit or lower limit of S), labeled as AMC I, AMC II, and AMC III, respectively. A higher antecedent moisture level and CN value suggest increased runoff, while lower levels indicate less runoff. Consequently, the median CN, calculated from a range of CN values, is often used for a catchment area [29]. The weighted CN for areas with mixed land use can be calculated using the following equation:

$$\text{CN} = \sum \text{Ai CNi} / \sum \text{Ai} \tag{10}$$

where CNi is the CN for a part of the watershed of an area Ai.

The HEC-HMS model is used to estimate the volume runoff of the studied basins. It is a software designed to simulate the complete hydrologic processes of watershed systems. The software features a range of traditional hydrologic analysis techniques, including event infiltration, unit hydrographs, and hydrologic routing. Additionally, HEC-HMS is equipped with tools essential for continuous simulation, such as evapotranspiration, snowmelt, and soil moisture accounting [30]. The time of concentration (Tc) refers to the duration, in minutes, it takes for a drop of water to travel from the most hydraulically distant point in a watershed to a downstream gauging point [31]. Kirpich [32] formula was applied to estimate the time of concentration for each sub-basin. This formula is a popularly used formula relating the time of concentration of the length of travel and slope of the catchment as follows:

$$Tc = 0.0078 \, [L0.77 \, S\text{-}0.385] \quad (11)$$

Time of concentration (Tc) is measured in minutes and represents the time it takes for water to travel from the most distant point in the catchment to the outlet. It is influenced by the maximum length of travel (L, in kilometers) and the slope of the catchment (S, in meters per meter). The lag time (TL), which is the interval between the centroid of the net rainfall and the centroid of the runoff, is typically around 60% of Tc as per the SCS method. The final step in generating the basin hydrograph is routing the watershed. Routing involves the movement of runoff through the various watershed outlets, along streams, and ultimately to the watershed's outlet or sink [31].

The Hydrologic Modeling System (HEC-HMS) is designed to simulate the precipitation-runoff processes of drainage basins. HEC-HMS rainfall-runoff model has many flood modeling and water resource planning and management applications. In most studies, HEC-HMS rainfall-runoff modeling was efficient and dependable in predicting runoff accuracy in various watersheds. As a result, the model can simulate runoff in an ungauged basin for water resource planning, development, management, and decision-making. These needed data include watershed parameters, i.e., CN, area drainage density . . . etc., storm parameters, i.e., duration and intensity, and routing parameters, i.e., time of concentration and time lag. The HEC-HMS model provides several routing options, including the Muskingum, Modified Puls, Kinematic Wave, and Muskingum–Cunge methods. The runoff coefficient quantifies the relationship between the depth of runoff and rainfall, indicating the proportion of rainfall that contributes to runoff. The runoff depth and runoff coefficient are calculated using the following equations [33]:

$$\text{Runoff depth (mm)} = \text{Runoff volume/Drainage area}$$

$$\text{Runoff coefficient (\%)} = \text{Runoff depth/Rainfall depth}$$

## 4. Results and Discussion

### 4.1. Rainfall Analysis

As Quseir meteorological station is located so far from this study area, about 44 km long, NASA archival data were used to download rainfall data of the Gebel El Sibai watershed area at latitude 25.7257 and longitude 34.2038 (Table 1). These data imply that rainfall is generally scarce. January rainfall data records from 1981 to 2021 indicate that October, November, and December are the rainy months. Rainfall ranges from 84 mm to zero. These conditions reflect the aridity of the area, and the flash rainfalls that occur from time to time should be utilized. Bennet et al. [26] were used to calculate the recurrence of events (T) from the downloaded series of observations (Table 2 and Figure 6). So, the recurrence of any event of flash flooding can be predicted. As the annual rainfall increases, the recurrence period becomes longer, and vice versa, i.e., a storm of 84 mm is expected to occur every 42 years with a probability of 2.4%, while 21 mm is of rainfall expected to occur every 14 years with a probability of 7.1% . . . etc. The event of 5.27 mm is expected to repeat many times every 6 years. The relation between the annual rainfall and recurrence period is given by the best-fit relation: P = 1.9314 log T—3.66191.

**Table 2.** Rainfall recurrence period in the Gebel El Sibai watershed area during the period 1981 to 2021.

| Year | Annual pp (mm) | Ranked PP (mm) | Rank | Return Period (Year) | Probability % |
|---|---|---|---|---|---|
| 1981 | 0 | 84.38 | 1 | 42.0 | 2.4 |
| 1982 | 0 | 26.37 | 2 | 21.0 | 4.8 |
| 1983 | 0 | 21.09 | 3 | 14.0 | 7.1 |
| 1984 | 0 | 15.97 | 4 | 10.5 | 9.5 |
| 1985 | 5.27 | 10.55 | 5 | 8.4 | 11.9 |
| 1986 | 84.38 | 10.55 | 6 | 7.0 | 14.3 |
| 1987 | 0 | 5.27 | 7 | 6.0 | 16.7 |
| 1988 | 0 | 5.27 | 8 | 5.3 | 19.0 |
| 1989 | 0 | 5.27 | 9 | 4.7 | 21.4 |
| 1990 | 0 | 5.27 | 10 | 4.2 | 23.8 |
| 1991 | 5.27 | 5.27 | 11 | 3.8 | 26.2 |
| 1992 | 0 | 5.27 | 12 | 3.5 | 28.6 |
| 1993 | 26.37 | 0.14 | 13 | 3.2 | 31.0 |
| 1994 | 10.55 | 0 | 14 | 3.0 | 33.3 |
| 1995 | 0 | 0 | 15 | 2.8 | 35.7 |
| 1996 | 5.27 | 0 | 16 | 2.6 | 38.1 |
| 1997 | 21.09 | 0 | 17 | 2.5 | 40.5 |
| 1998 | 0 | 0 | 18 | 2.3 | 42.9 |
| 1999 | 0 | 0 | 19 | 2.2 | 45.2 |
| 2000 | 0 | 0 | 20 | 2.1 | 47.6 |
| 2001 | 0 | 0 | 21 | 2.0 | 50.0 |
| 2002 | 0 | 0 | 22 | 1.9 | 52.4 |
| 2003 | 0 | 0 | 23 | 1.8 | 54.8 |
| 2004 | 0 | 0 | 24 | 1.8 | 57.1 |
| 2005 | 0 | 0 | 25 | 1.7 | 59.5 |
| 2006 | 5.27 | 0 | 26 | 1.6 | 61.9 |
| 2007 | 0 | 0 | 27 | 1.6 | 64.3 |
| 2008 | 0 | 0 | 28 | 1.5 | 66.7 |
| 2009 | 0 | 0 | 29 | 1.4 | 69.0 |
| 2010 | 5.27 | 0 | 30 | 1.4 | 71.4 |
| 2011 | 0 | 0 | 31 | 1.4 | 73.8 |
| 2012 | 0 | 0 | 32 | 1.3 | 76.2 |
| 2013 | 10.55 | 0 | 33 | 1.3 | 78.6 |
| 2014 | 0 | 0 | 34 | 1.2 | 81.0 |
| 2015 | 0 | 0 | 35 | 1.2 | 83.3 |
| 2016 | 15.97 | 0 | 36 | 1.2 | 85.7 |
| 2017 | 0 | 0 | 37 | 1.1 | 88.1 |

**Table 2.** *Cont.*

| Year | Annual pp (mm) | Ranked PP (mm) | Rank | Return Period (Year) | Probability % |
|---|---|---|---|---|---|
| 2018 | 0 | 0 | 38 | 1.1 | 90.5 |
| 2019 | 0 | 0 | 39 | 1.1 | 92.9 |
| 2020 | 5.27 | 0 | 40 | 1.1 | 95.2 |
| 2021 | 0.14 | 0 | 41 | 1.0 | 97.6 |

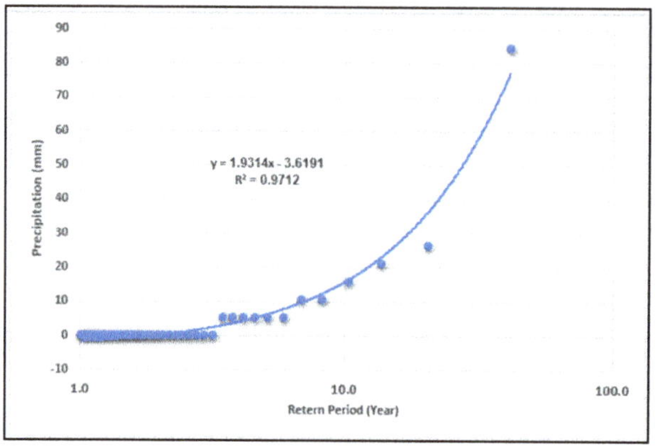


**Figure 6.** The relation between the annual precipitation (mm) and the recurrence period (year) of the Gebel El Sibai watershed area.

## *4.2. Quantitative Morphometric Analysis*

Five drainage basins of five and six ordered basins are issued from the Gebel El Sibai watershed: Esel, Sharm El Bahari, Sharm El Qibli, Wizer, and Umm Gheig, of a total area equal to 1954 km$^2$ (Figure 7). These wadis are variable morphometric parameters controlled by structural elements, rock types, and erosion factors. Detailed quantitative morphometric analysis was estimated using topographic maps of scale 1:50,000, DEM, and GIS techniques (Table 3). The basic morphometric parameters can be summarized as follows:

**Table 3.** Morphometric parameters of the studied basins.

| Parameter | Wadi Esel | Wadi Sharm El Bahari | Wadi Sharm El Qibli | Wadi Wizer | Wadi Umm Gheig |
|---|---|---|---|---|---|
| Stream Order (u) | 6.00 | 5.00 | 5.00 | 5.00 | 6.00 |
| Total Stream Numbers (Nu) | 1229.00 | 322.00 | 246.00 | 172.00 | 1435.00 |
| Total Stream length (Lu) (km) | 1149.61 | 354.37 | 281.26 | 175.91 | 1470.68 |
| Mean bifurcation ratio (Rbm) | 4.44 | 4.08 | 3.89 | 3.64 | 4.12 |
| Weighted Mean bifurcation ratio WMRb | 3.79 | 4.18 | 4.24 | 4.96 | 4.8 |
| Basin perimeter(km) (P) | 170.52 | 95.47 | 62.08 | 48.8 | 191.81 |
| Steam length (Ls)(Km) | 30.93 | 36.86 | 8.72 | 8.31 | 41.96 |
| Basin Length (Lb) (km) | 68.95 | 43.05 | 25.4 | 20.25 | 60.66 |
| sinuosity ratio (Si) | 0.45 | 0.86 | 0.34 | 0.41 | 0.69 |
| Basin Area (km$^2$) (A) | 658.85 | 187.92 | 147.83 | 95.47 | 863.66 |
| Form factor Ratio (Rf) | 0.14 | 0.1 | 0.23 | 0.23 | 0.23 |
| Elongation Ratio (Re) | 0.42 | 0.36 | 0.54 | 0.54 | 0.55 |
| Texture Ratio (T) | 5.11 | 2.56 | 3 | 2.79 | 5.91 |

**Table 3.** *Cont.*

| Parameter | Wadi Esel | Wadi Sharm El Bahari | Wadi Sharm El Qibli | Wadi Wizer | Wadi Umm Gheig |
|---|---|---|---|---|---|
| Circulatory Ratio (Rc) | 0.28 | 0.26 | 0.48 | 0.5 | 0.29 |
| Drainage density (Dd) | 1.74 | 1.89 | 1.9 | 1.84 | 1.7 |
| Fitness Ratio (Fr) | 0.4 | 0.45 | 0.41 | 0.41 | 0.32 |
| Constant of channel maintenance (CCM) (km) | 0.57 | 0.53 | 0.53 | 0.54 | 0.59 |
| Stream Frequency (Fs) | 1.87 | 1.71 | 1.66 | 1.8 | 1.66 |
| Infiltration Number (FN | 3.25 | 3.23 | 3.17 | 3.32 | 2.83 |
| Form Factor (Rf) | 0.14 | 0.1 | 0.23 | 0.23 | 0.23 |
| Basin shape | 7.22 | 9.86 | 4.36 | 4.3 | 4.26 |
| Basin shape Index (Ish) | 0.18 | 0.13 | 0.29 | 0.3 | 0.3 |
| Length of overland flow (Lo) | 0.33 | 0.28 | 0.28 | 0.29 | 0.34 |
| Compactness Constant (Cc) | 1.87 | 1.96 | 1.44 | 1.41 | 1.84 |
| max elevation (R) | 1099.00 | 1477.00 | 1477.00 | 438.00 | 1264.00 |
| min. elevation (r) | 0.00 | 0.00 | 0.00 | 0.00 | 0.00 |
| relative relief (Rr) | 1099.00 | 1477.00 | 1477.00 | 438.00 | 1264.00 |
| Relief ratio (Rr) | 0.02 | 0.03 | 0.06 | 0.02 | 0.02 |
| Dissection index (Di) | 1.00 | 1.00 | 1.00 | 1.00 | 1.00 |
| Ruggedness index (Rn) | 1.92 | 2.79 | 2.81 | 0.81 | 2.15 |
| Melton Ruggedness number (MRn) | 0.04 | 0.11 | 0.12 | 0.04 | 0.04 |
| Lb (m) | 68,953.00 | 43,050.00 | 25,397.00 | 20,250.00 | 60,660.00 |
| H (m) | 1099.00 | 1477.00 | 1477.00 | 438.00 | 1264.00 |
| slope (s) | 0.02 | 0.03 | 0.06 | 0.02 | 0.02 |

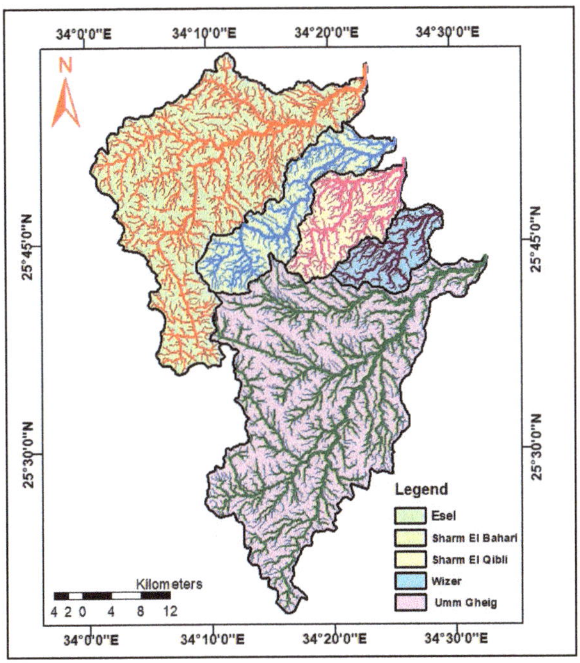


**Figure 7.** Streams distribution of the drainage basins in the Gebel El Sibai watershed area.

### 4.3. Linear Characteristics

Streams numbers and lengths: The streams collect rainfall to the wadi outlet. The greater the number and longer the length of the streams, the higher the potential of the wadis to drain water. Wadi Umm Gheig is the greatest and longest stream's basin, 1435/1470 km, whereas Wadi Wizer is the smallest and shortest at 172/175 km.

The bifurcation ratio (Rb) is the ratio between the number of stream segments of a given order and the number of segments in the next higher order [34]. Horton [35] regarded the bifurcation ratio as an important index for assessing relief and dissection. This dimensionless value represents the relationship between the number of streams in one order and those in the subsequent higher order within a drainage network. A higher bifurcation ratio indicates a greater likelihood of flooding, making it a valuable tool for flood risk assessment. The higher the bifurcation ratio, the greater the probability of flooding. It normally lies in the range of 2–5 [35,36]. In Wadi El Sibai watershed studied basins it ranges between 3.64 and 4.44 with an average of 4.03. On the other hand, the weighted mean bifurcation Ratio (WMRb) ranges between 3.97 and 4.96. The high Rb values of the area reflect the essential effects of the tectonic and erosion elements in forming and deepening the streams and the high potential for flooding.

The Stream length (Ls) refers to the length of the drainage basin. The stream length of the studied basins ranges from 8.31 to 41.96 km (Wadi Wazar and Umm Gheig, respectively), with an average of 25.4 km. On the other hand, the basin length ranges from 20.25 to 68.95 km in Wadi Wizer and Wadi Esel, respectively.

The sinuosity ratio (Si) is the ratio between the stream length and the basin length. This ratio quantifies how much a stream deviates from a straight path. A higher sinuosity ratio indicates a more meandering stream. The sinuosity ratio is a measure to quantify the degree to which stream meanders (i.e., bends and turns). The sinuosity ratio ranges between 0.34 for Wadi Sharm El Qibli and 0.86 for Wadi Sharm El Bahari, reflecting lithological and structural control.

The length of overland flow (Lo) is a key parameter in hydrological modeling. It represents the maximum length of surface flow or runoff. **Horton** [35] recommended using one-half the reciprocal of the drainage density. It is a measure of credibility, which can be described as the length of the flow of water over the surface before it becomes concentrated in definite stream channels [37]. The shorter the Lo value, the quicker the surface runoff will enter the stream. The values of the overland flow range between 0.28 for Wadi Sharm El Qibli and 0. 34 km for Wadi Umm Gheig.

Stream frequency (F) is defined as the ratio of the total number of stream segments across all orders within a watershed to the area of the basin or watershed [35]. This metric represents the number of streams per unit area. It varies between 1.66 $m^{-2}$ (Wadi Sharm El Qibli) to 1.87 $m^{-2}$ (Wadi Esel), with an average value of 1.74 $km^{-2}$. The stream frequency values indicate that Wadi Esel and Wadi Wizer have high flooding potential.

## 5. Areal Characteristics

Drainage area (A) is an area of land where all flowing surface water converges to a single point, such as a lake or sea. The drainage area is measured in a horizontal plane enclosed by the drainage divide, outlining the basin using the GIS tool. The drainage area of Gebel El Sibai watershed basins ranges from 95.47 $km^2$ (Wadi Wizer) to 863.66 $km^2$ (Wadi umm Gheig), with an average value of 390.75 $km^2$. These values reflect the good rainwater harvesting potential of the area under concern.

Perimeter (P) refers to the length of the line enclosing the catchment area of a drainage basin. The basin perimeter was calculated using GIS analysis. The basin's perimeters range from 48.80 km (Wadi Wizer) to 191.81 km (Wadi Umm Gheig).

Drainage density (D) refers to the total length of stream channels within a drainage basin per unit area. It is determined by dividing the combined length of all streams by the basin's total area. It indicates how well stream channels, measured in km/$km^2$, drain a watershed. High drainage densities indicate a high bifurcation ratio and may approximate the average length of overland flow. It ranges between 1.7 $km^{-1}$ (Wadi Esel) and 1.9 $km^{-1}$ (Wadi Sharm El Qibli), with an average value of 1.8 $km^{-1}$. The estimated drainage density values indicate high flood potentiality.

The constant of channel maintenance (Rm) is defined as the inverse of the drainage density. **Schumm** [34] introduced it as the ratio between the area of a drainage basin and the total length of all the channels. The value of the constant of channel maintenance increases with the size of the watershed, and it gives an idea of how much area (in $m^2$) is needed to sustain one meter of a stream channel. The Rm values range from 0.53 km to 0.59 km (Wadi Sharm El Qibli and Wadi Umm Gheig, respectively) with an average value of 0.114 km. The Rm values indicate the risk of flooding and soil erosion in a watershed [38]. A lower Rm value (and hence higher drainage density) indicates a well-dissected watershed with less risk of flooding and soil erosion and vice versa.

## 6. Shape Characteristics

The elongation ratio (Re) measures the shape of a watershed. The elongation ratio varies between 0.36 (Wadi Sharm El Bahari) to 0.55 (Wadi Umm Gheig). According to the elongation ratio values, Wadi umm Gheig, Wadi Sharm El Qibli, and Wadi Wizer are more hazardous than Wadi Esel and Wadi Sharm El Bahari.

The circularity ratio (Rc) is a measure of the shape of a watershed. It is defined as the ratio between the areas of a watershed to the area of a circle having the same circumference as the perimeter of the watershed. It is a significant ratio that indicates the dendritic stage of a watershed. The estimated values of the circulation ratio indicate that Wadi Wizer is the most circular basin.

The form factor ratio (F)

It is a dimensionless shape factor that is an indicator of the shape of the river basin. It is calculated as the ratio of the area of the basin (A) to the square of the basin length ($L^2$). It ranges between 0.1 (Wadi Sharm El Bahari) to 0.23 (Wadi Umm Gheig).

## 7. Relief Characteristics

Relative Relief (H) refers to the maximum altitudinal difference in a basin. It is the difference between the highest and lowest points in a particular basin. This measure can provide insights into the overall variation in the basin's elevation, geomorphology variations, basin potential erosion, and hydrological behavior. The relative relief ranges from 1477 m (Wadi Sharm El Bahari) and 438 m (Wadi Wizer). These values reflect high potential erosion and floods in Wadis Sharm El Bahari, Sharm El Qibli, Umm Gheig, and Esel.

The relief Ratio (Rr) is a measure used to quantify the overall steepness of a basin surface. It is calculated by dividing the total relief by the length of the main channel. A high Relief Ratio suggests a young, steep river, while a low Relief Ratio indicates a more mature, less steep basin [39]. It ranges between 0.02 (Wadi Esel, Umm Gheig) and 0.06 (Wadi Sharm El Qibli) with an average value of 0.004.

The ruggedness number (Rn) is used to represent the actual conditions of sub-basin watersheds. It is calculated as the product of the maximum basin relief and drainage density. Since Rn depends on both slope and drainage density, a lower value suggests slower stream flow velocity, indicating a lower susceptibility to soil erosion. This parameter provides a straightforward and efficient tool for precise planning and more equitable economic use of land based on its suitability. The ruggedness number in the studied area ranges from 0.81 (Wadi Wizer) to 2.79 (Wadi Sharm El Bahari), with an average of 2.096, indicating a high potential for erosion in the Gebel El Sibai watershed basins.

The basin relief (R) is a geomorphological parameter that represents the elevation difference between the highest point on the drainage divide and the outlet point of the subbasin. This parameter is significant as it provides insights into the overall topography and slope of the watershed, which can influence factors such as water flow velocity and potential for soil erosion.

## 8. Hazard Degree Model

The degree of hazard of drainage basins is a measure of the potential risk they pose, particularly in terms of flooding. This is an indicator of the behavior of direct surface runoff. The hazard degree of a drainage basin can be assessed using various parameters that contribute to the behavior of direct surface runoff and alter the risk of direct surface runoff [40]. Nine parameters are used to estimate the hazard degree of the studied basins (Table 4). These parameters are directly proportional to the basin's hazard except the weighted mean bifurcation ratio. Equations (3) and (4) were used to calculate the hazard degree of the different basins. Accordingly, Wadi Sharm El Qibli is the highest hazard degree, and Wadi Sharm El Bahari is the lowest.

**Table 4.** The estimated hazard degree values of different values.

| Effective Parameter | Wadi Esel | Wadi Sharm El Bahari | Wadi Sharm El Qibli | Wadi Wizer | Wadi Umm Gheig |
|---|---|---|---|---|---|
| Basin Area ($km^2$) (A) | 3.934 | 1.481 | 1.273 | 1.000 | 5.000 |
| Drainage density (Dd) | 1.800 | 4.800 | 5.000 | 3.800 | 1.000 |
| Stream Frequency (Fs) | 5.000 | 1.952 | 1.000 | 3.667 | 1.000 |
| Circulatory Ratio (Rc) | 1.333 | 1.000 | 4.667 | 5.000 | 1.500 |
| Length of overland flow (Lo) | 4.333 | 1.000 | 1.000 | 1.667 | 5.000 |
| Basin shape Index (Ish) | 2.176 | 1.000 | 4.765 | 5.000 | 5.000 |
| Ruggedness index (Rn) | 3.220 | 4.960 | 5.000 | 1.000 | 3.680 |
| Relief ratio (Rr) | 1.000 | 2.000 | 5.000 | 1.000 | 1.000 |
| Weighted Mean bifurcation ratio (WMRb) | 1.000 | 2.333 | 2.538 | 5.000 | 4.453 |
| Total Hazard Degree | 23.797 | 20.527 | 30.242 | 27.133 | 27.633 |
| Hazard Degree | Low | V. low | V.high | high | high |

## 9. Rainfall/Runoff Relationship

The relationship between rainfall and runoff is a key element in hydrology, influenced by the dynamic interplay of rainfall intensity, soil infiltration, and surface storage. This relationship is vital for effective catchment management, including sustainable water resource development and flood hazard mitigation [41]. The curve number (CN) value, as identified by Ponce and Hawkins [33], is a primary source of uncertainty impacting the outcomes of this process. The (CN) is estimated for each basin depending on the land use type, hydrologic soil group, hydrologic condition, and the antecedent moisture condition (AMC) (Table 5). Due to the scarce rainfall, the area is almost dry and poor with vegetation. The affected character of the CN value is the soil type group. Four main soil types are well recognized: Wadi fill deposits, quaternary deep sand soil, limestone rocks, and fractured basement. The value of the CN and the area of each soil type are estimated for each basin. The weighted CN attends 89.34 (Wadi Esel), 89.44 (Wadi Sharm El Bahari), 83.60 (Wadi Sharm El Qibli), 76.81 (Wadi Wizer) and 91.39 (Wadi Umm Gheig). Wadi Umm Gheig has the highest priority to flood, where initial abstraction (Ia) is 4.79 mm. This may be due to the prevalence of the basement rock surface of the basin (83.99%). On the other hand, Wadi Sharm El Bahri represents the lowest priority for flooding, where Ia attains 15.34 mm. These conditions reflect the prevalence of the quaternary sand deposits of high infiltration rate of the basin. The Ia value records 9.97 mm in (Wadi Sharm El Qibli, 6 mm in (Wadi Wizer, and 6.06 mm in Wadi Umm Gheig (Figure 8).

**Table 5.** Run-off equation of Gebel El Sibai watershed basins using the SCS_CN method (**SCS, 1972**).

| Basin Name | Area km² | AMC | Land Use | Soil Type | Area km² | Soil Area % | CN | Weighted CN | S mm | Ia mm | Basin Equation Q (mm) |
|---|---|---|---|---|---|---|---|---|---|---|---|
| Esel | 658.85 | dry | poor | Wadi fill deposits | 58.804 | 8.93 | 63 | 89.34 | 30.32 | 6.06 | $(P-6.06)^2/(p+24.26)$ |
| | | | | Sand deposits | 73.357 | 11.13 | 71 | | | | |
| | | | | Limestone rocks | 12.808 | 1.94 | 88 | | | | |
| | | | | Basement rocks | 513.881 | 78.00 | 95 | | | | |
| Sharm El Bahari | 187.92 | dry | poor | Wadi fill deposits | 4.984 | 2.65 | 63 | 89.44 | 30.00 | 6.00 | $(P-6.0)^2/(p+24.0)$ |
| | | | | Sand deposits | 27.599 | 14.69 | 71 | | | | |
| | | | | Limestone rocks | 31.912 | 16.98 | 88 | | | | |
| | | | | Basement rocks | 123.425 | 65.68 | 95 | | | | |
| Sharm El Qibli | 147.83 | dry | poor | Wadi fill deposits | 8.381 | 5.67 | 63 | 83.60 | 49.83 | 9.97 | $(P-9.97)^2/(p+39.87)$ |
| | | | | Sand deposits | 51.033 | 34.52 | 71 | | | | |
| | | | | Limestone rocks | 27.509 | 18.61 | 88 | | | | |
| | | | | Basement rocks | 60.907 | 41.20 | 95 | | | | |
| Wizer | 95.47 | dry | poor | Wadi fill deposits | 17.5 | 18.33 | 63 | 76.81 | 76.70 | 15.34 | $(P-15.34)^2/(p+61.36)$ |
| | | | | Sand deposits | 41.029 | 42.98 | 71 | | | | |
| | | | | Limestone rocks | 27.451 | 28.75 | 88 | | | | |
| | | | | Basement rocks | 9.49 | 9.94 | 95 | | | | |
| Umm Gheig | 863.66 | dry | poor | Wadi fill deposits | 34.194 | 3.96 | 63 | 91.39 | 23.94 | 4.79 | $(P-4.79)^2/(p+19.15)$ |
| | | | | Sand deposits | 76.348 | 8.84 | 71 | | | | |
| | | | | Limestone rocks | 27.701 | 3.21 | 88 | | | | |
| | | | | Basement rocks | 725.417 | 83.99 | 95 | | | | |

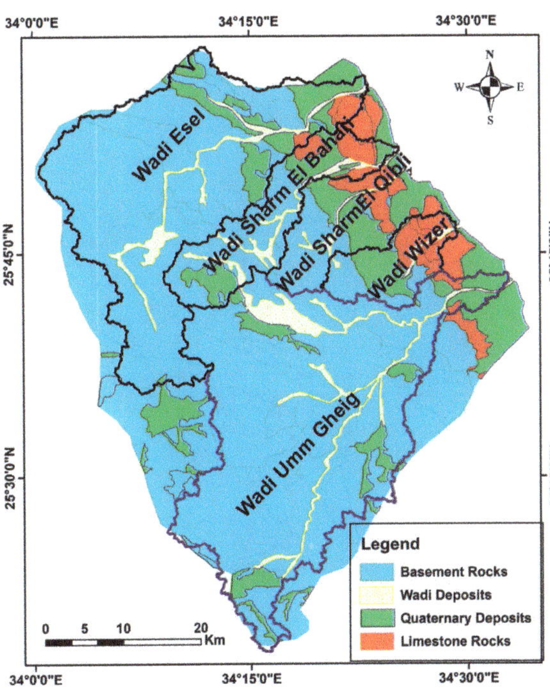


**Figure 8.** Map showing kinds of soil covering this study area.

## 10. Run-Off Volume

The runoff volume of the un-gauged basins of the Gebel El Sibai watershed was estimated through the hydrograph generation using the HEC-HMS modeling software (V. 3.5), which was designed to simulate the hydrologic response of the watersheds. The HEC-HMS was used to estimate the flood volume in many areas all over the world. Hessein et al. [42] investigated recharge mechanisms in structurally controlled terrain under arid environments using HEC-HMS in Wadi Morra, Sinai, Egypt. A total flood volume was estimated as $29.47 \times 10^6$ m$^3$. Goodarzi et al. [43]) evaluate food forecasting using the weather research and forecasting (WRF) model and the Hydrologic Engineering Center-Hydrologic Modeling System (HEC-HMS) model in the Talesh catchment. The model was calibrated and validated using previous runoff events and by comparing observed and simulated streak flow and peak discharge against those reported in previous studies. It was found that the model is efficient and can be used in similar regions. Khan et al. [44] stated that the Hydrologic Engineering Center-Hydrologic Modeling System (HEC-HMS) model was fed with information about precipitation, slope, soil type, as well as land use and land cover. An event that occurred in October 2016 was used to examine the response of the studied basins to this event. It has a 22 h duration and a total rainfall quantity of 15.98 mm (Figure 9). The hydrologic elements of every basin were represented, i.e., the Esel basin has four sub-basins. Run-off sub-basins 1 and 2 are collected at Junction 1. Floodwater routines to Junction 2 (outlet point) through reach 1. The input parameters of The HEC-HMS model include the CN of each subbasin, the potential maximum retention (s), the initial abstraction (Ia), the ratio of the impervious soil, time of concentration (Tc), and lag time (Lg) (Table 6). The time of concentration of the different subbasins is calculated using the Kirpich formula. The values range from 64.2 min (sub-basin 2, Wadi Sharn El Qibli) to 123.3 min (sub-basin 1, Wadi Umm Gheig) with an average value of 85.2 min (Table 6). On the other hand, the lag time values range from 38.5 to 74 min, with an average value of 51.1 min. The results indicated that the different wadis received about 26.421 million m$^3$ of rainwater due to that event, where subbasin-1 of Wadi Umm Gheig represents the largest one collecting rainwater (6.281 million m$^3$) and subbasin-2 of Wadi Sharm El Qibli is the smallest (742,900 m$^3$) (Table 7). The subbasin's total losses range from 1.04 million m$^3$ (Subbasin-4, Wadi Esel), where basement, impervious rocks prevail, to about 20,900 m$^3$ (Subbasin-2, Wadi Esel), where quaternary and wadi deposits prevail. The discharge volume of the studied basins at outlets varies from 8.718 million m$^3$ at Wadi Esel, 2.099 million m$^3$ at Wadi Sharm El Bahari, 1.595 million m$^3$ at the outlet of Wadi Sharm El Qibli, 1.072 million m$^3$ at Wadi Wizer and 12.024 million m$^3$ at the outlet of Wadi Umm Gheig (Figures 10 and 11) (Table 7). Sensitivity and uncertainty analysis was addressed here by carefully using rainfall data and relying on rainfall values from influential points in the valleys, as there are no rainfall stations in this study area. Rain data values were used for a period of 41 years, from 1981 to 2001. Comparing the results with previous studies, it was found that there is a slight difference in the degrees of risk, which is mostly due to the applied method, i.e., Gad et al. [6] showed that the Essel, Sharm El-Bahri, and El-Qebly basins are the riskiest. To determine the flood volume, the HEC-HMS model was used, which is a widely used and reliable method, as the inputs were calculated using GIS techniques, and the mathematical equations of Kirpich [32] formula were applied. It is worth mentioning that the drainage basins lack the presence of a weir in the outlet areas to calibrate the results of the mathematical model. This point will be recommended.

**Table 6.** Time of concentration (Tc) and Lag time (Lg) calculations of the studied sub-basins.

| Main Wadis | Sub-Basins | Area km$^2$ | Pervious % | Impervious % | CN | S (mm) | Ia (mm) | Longest Stream (m) | Slope m/m | Tc Min. | Lg Min. |
|---|---|---|---|---|---|---|---|---|---|---|---|
| Esel | Sub1 | 253.0 | 6.7 | 93.3 | 93 | 19.1 | 3.8 | 41,032 | 0.029 | 108.0 | 64.8 |
| | Sub2 | 158.4 | 2.0 | 98.0 | 97 | 7.9 | 1.6 | 24,213 | 0.026 | 75.8 | 45.5 |
| | Sub3 | 77.3 | 44.2 | 55.8 | 78 | 71.6 | 14.3 | 19,886 | 0.012 | 87.0 | 52.2 |

Table 6. *Cont.*

| Main Wadis | Sub-Basins | Area km$^2$ | Pervious % | Impervious % | CN | S (mm) | Ia (mm) | Longest Stream (m) | Slope m/m | Tc Min. | Lg Min. |
|---|---|---|---|---|---|---|---|---|---|---|---|
| Esel | Sub4 | 168.5 | 39.9 | 60.1 | 83 | 52.0 | 10.4 | 29,188 | 0.016 | 106.3 | 63.8 |
| Sharm El Bahari | Sub1 | 110.7 | 25.0 | 75.0 | 84 | 48.4 | 9.7 | 25,790 | 0.047 | 63.1 | 37.9 |
| | Sub2 | 78.5 | 40.9 | 59.1 | 82 | 55.8 | 11.2 | 19,873 | 0.019 | 72.5 | 43.5 |
| Sharm El Qibli | Sub1 | 100.5 | 23.8 | 76.2 | 90 | 28.2 | 5.6 | 26,884 | 0.038 | 71.0 | 42.6 |
| | Sub2 | 46.5 | 59.2 | 40.8 | 71 | 103.7 | 20.7 | 17,708 | 0.021 | 64.2 | 38.5 |
| Wizer | | 96.0 | 32.6 | 67.4 | 76 | 80.2 | 16.0 | 22,977 | 0.013 | 95.4 | 57.2 |
| Umm Gheig | Sub1 | 392.8 | 8.2 | 91.8 | 92 | 22.1 | 4.4 | 43,868 | 0.024 | 123.3 | 74.0 |
| | Sub2 | 170.1 | 5.3 | 94.7 | 93 | 19.1 | 3.8 | 29,008 | 0.032 | 80.2 | 48.1 |
| | Sub3 | 140.1 | 26.3 | 73.7 | 83 | 52.0 | 10.4 | 32,052 | 0.035 | 83.6 | 50.2 |
| | Sub4 | 157.7 | 27.6 | 72.4 | 82 | 55.8 | 11.2 | 22,268 | 0.021 | 76.6 | 45.9 |

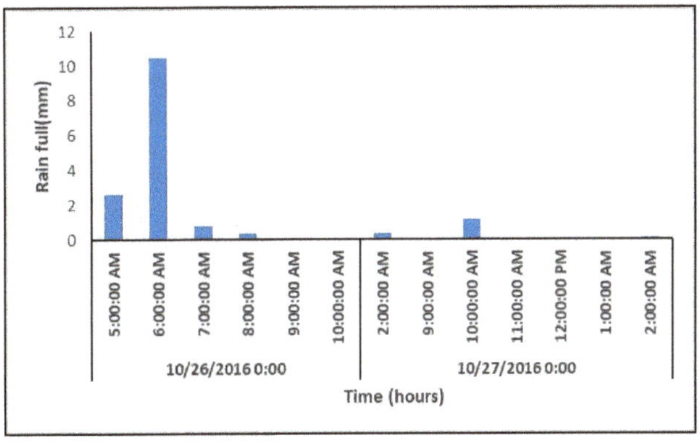


**Figure 9.** A rainfall event occurred in the Gebel El Sibai watershed area (October 2016).

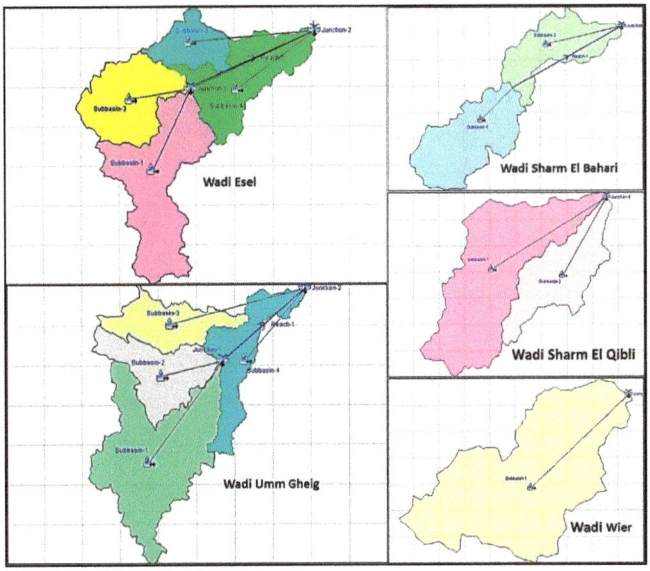


**Figure 10.** The hydrological model parameters of the Gebel El Sibai Watershed sub-basins.

**Table 7.** Flood volume of the studied sub-basins using HEC-HMS program software.

| Basins | Hydrologic Element | Basin Area ($km^2$) | Total Precipitation (1000 $m^3$) | Total Loss (1000 $m^3$) | Peak Discharge ($m^3$/s) | Time of Peak | Volume (1000 $m^3$) |
|---|---|---|---|---|---|---|---|
| Esel | Subbasin-1 | 253.01 | 4045.6 | 190.6 | 436 | 26 Oct. 2016, 18:30 | 3855 |
| | Subbasin-2 | 158.429 | 2533.3 | 20.9 | 344.5 | 26 Oct. 2016, 18:15 | 2512.4 |
| | Junction-1 | - | - | - | 762.8 | 26 Oct. 2016, 18:30 | 6367.4 |
| | Subbasin-4 | 168.541 | 2695.0 | 1039.9 | 187.1 | 26 Oct. 2016, 18:30 | 1655.1 |
| | Subbasin-3 | 77.328 | 1236.5 | 546.4 | 87.9 | 26 Oct. 2016, 18:30 | 690.1 |
| | Reach-1 | - | - | - | - | 26 Oct. 2016, 16:00 | - |
| | Junction-2 | 657.308 | - | - | 1037.7 | 26 Oct. 2016, 18:30 | 8712.6 |
| Sharm El Bahari | Subbasin-1 | 110.627 | 1768.9 | 422.6 | 196.8 | 26 Oct. 2016, 18:15 | 1346.4 |
| | Reach-1 | - | - | - | 194.5 | 26 Oct. 2016, 18:30 | 1346.4 |
| | Subbasin-2 | 78.425 | 1254.0 | 501.0 | 104.2 | 26 Oct. 2016, 18:15 | 753 |
| | Junction-1 | - | - | - | 293.9 | 26 Oct. 2016, 18:15 | 2099.4 |
| Sharm El Qibli | Subbasin-1 | 100.542 | 1607.7 | 315.7 | 181.1 | 26 Oct. 2016, 18:15 | 1291.9 |
| | Subbasin-2 | 46.462 | 742.9 | 439.8 | 44.5 | 26 Oct. 2016, 18:15 | 303.1 |
| | Junction-1 | - | - | - | 225.5 | 26 Oct. 2016, 18:15 | 1595.1 |
| Wizer | Subbasin-1 | 95.997 | 1535.0 | 499.8 | 127.4 | 26 Oct. 2016, 18:30 | 1035.2 |
| | Junction-1 | - | - | - | 127.4 | 26 Oct. 2016, 18:30 | 1035.2 |
| Umm Gheig | Subbasin-1 | 392.804 | 6280.9 | 388.0 | 621.3 | 26 Oct. 2016, 18:45 | 5892.9 |
| | Subbasin-2 | 170.184 | 2721.2 | 101.9 | 345.9 | 26 Oct. 2016, 18:15 | 2619.3 |
| | Junction-1 | - | - | - | 931.1 | 26 Oct. 2016, 18:30 | 8512.2 |
| | Subbasin-4 | 157.652 | 2520.9 | 679.7 | 249.3 | 26 Oct. 2016, 18:15 | 1841.1 |
| | Subbasin-3 | 140.137 | 2240.8 | 570.0 | 213.5 | 26 Oct. 2016, 18:30 | 1670.8 |
| | Reach-1 | - | - | - | - | 26 Oct. 2016, 16:00 | - |
| | Junction-2 | | - | - | 1382.3 | 26 Oct. 2016, 18:30 | 12,024.2 |

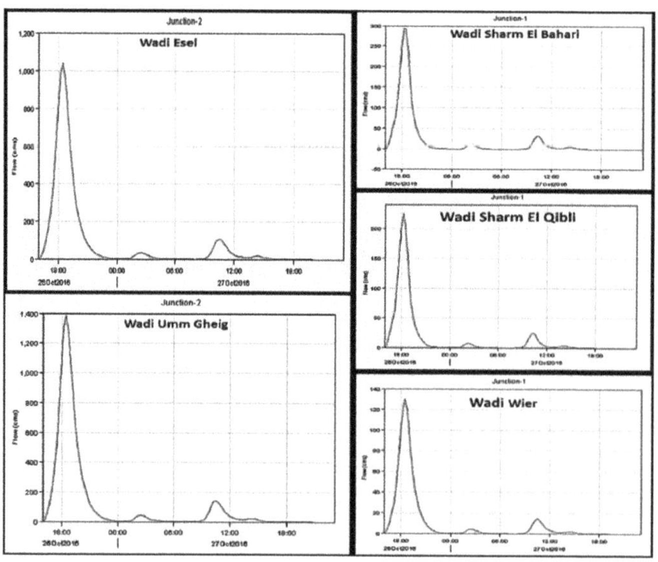


**Figure 11.** Run-off hydrograph of Gebel El Sibai Watershed basins applying HEC-HMS hydrological model.

## 11. Ecological Studies

Based on ecological studies published in previous research in various places in this study area, including (Gebel Abu El Tiyur, Wadi Umm Gheig, Wadi Sharm El Bahri, Wadi Esel, and Wadi Sharm El Qibli), as sites representative of the vegetation. The results showed that it is among the important recorded plant taxa were *Tamarix nilotica, Juncus rigidus, Aerva javanica, Leptadenia pyrotechnica, Zygophyllum album, Moringa peregrina, Solenostemma arghel, Taverniera aegyptiaca, Nitraria retusa, Cleome droserifolia, Acacia species, Panicum turgidum, Achillea fragrantissima*, and *Zilla spinosa* are being important for the environment, and local communities because of their ecological (e.g., prevents soil erosion and serves as a habitat for wildlife), economical (fuelwood, construction, fiber producing, animal fodder, food, and edible fruits), and medicinal benefits. Biodiversity is threatened by intensified human pressures that lead to habitat degradation, the removal of vegetation from large areas (over-collection of medicinal plants), in particular those with low abundance such as *Solenostemma arghel* (needs high priority for conservation), also *Nitraria retusa, Moringa peregrina*, and *Tamarix nilotica* (providing shade, support large number of insects, birds and fruit-eating mammals, firewood, soil stabilization, etc.) the conservation of these plant species will protect those animals, and reduce land degradation [45]. *Moringa peregrina* has become one of the most endangered trees due to overgrazing and excessive collection, which has led to a sharp decline in its population size and numbers [46], resulting in the loss of an essential natural resource for local Bedouin communities and their domestic animals [47]. Vegetation cover is also vulnerable to catastrophic drought phenomena, which can significantly affect its density in any region, especially arid and semi-arid regions [48–50]. If plant species populations are destroyed by drought or other unforeseen events, recolonization should occur more quickly than the remaining populations of species in surrounding habitats and sites, as well as the cultivation of economically important plants and other medicinal species in ecologically suitable environments, with large water resources [47,50]. So, providing a permanent source of water for plant species will contribute to overcoming the effects of climate change (especially lack of rain) that negatively affect the vegetation cover as well as the environmental balance and to ensure their sustainability and protect them from extinction. On the other hand, the Bedouin communities will be encouraged to adapt to the harsh nomadic life and continue animal husbandry to ensure food security, improving Bedouin livelihoods through the cultivation of medicinal plants, rehabilitation and restoration programs to conserve vegetation and increase population size, and rehabilitation of rain-fed pastures and some rain-fed cultivation of cereal crops. This can only be achieved by maximizing the use of water sources in the region. Harvesting rainwater and floods is the priority in this context.

## 12. Conclusions and Recommendations

A storm of 84 mm is expected to occur every 42 years with a probability of 2.38%. The basins have a total area of about 1953.93 km$^2$, whereas Wadi Um Gheig is the largest at 863.66 km$^2$. The basin has five to six stream orders, reflecting the high stream density and high effect of the structure and erosional elements. The morphometric parameters reflect the essential effects of the tectonic and erosion elements in creating and excavating the streams and the high potential for flooding. Although Wadi Esel and Wadi umm Gheig are the largest, regarding the area, Wadi SharmEl Qibli is the highest-risk basin among wadis. This may be attributable to the steep slope, high circularity ratio, and Ruggedness index. On the other hand, the basin rainfall/runoff relationship reflects the dynamic interaction between rainfall intensity, soil infiltration, and surface storage. The calculations of the CN reflect the impervious soil ratio that may prevail over the wadi surface. Wadi Umm Gheig has the highest priority for flooding, where it attends the highest CN value (91.39) and lowest initial abstraction (Ia) 4.79 mm. This may be due to the prevalence of the basement rock of the basin surface (83.99%). The flood volume calculated by applying the HEC-HMS model reveals that Umm Gheig basin has the greatest ability to collect water with a flood volume of about 12 million m$^3$ at the basin outlet of rainfall event of 15 mm. Wadi Esel is expected to collect 8.7 million $m^3$ depending on the ratio of the impervious soil and rainfall quantity.

According to the results of estimating the runoff volume of each sub-basin, seven storage dams (SD1 to SD7) are proposed to enhance the utilization of the surface potentialities of this study area (Figure 12). SD1 is imposed to collect the runoff water of Sub-basin 1 and 2 of Wadi Esel with an expected water volume of 6.367 million $m^3$, whereas SD2 is imposed at the outlet of sub-basin 3 with a storage capacity of 0.69 million $m^3$. SD3, SD4, and SD5 are suggested at the outlets of Wadi Sharm El Bahari, Sharm El Qibli, and Wizer. SD6 is suggested at the location of Junction 1 and SD7 at the outlet of sub-basin 3 of the Umm Gheig basin. A quantity of water is 10.182 million m3 is expected to be harvested using these storage dams at Wadi Umm Gheig. Current threats and the prevailing harsh habitat conditions are among the most important reasons that emphasize the need to search for additional water sources to preserve and resettle these species. Floodwater harvesting is the most important policy used in arid regions worldwide. In addition, implementing of controlled structure is very important for good surface water management of the studied basins.

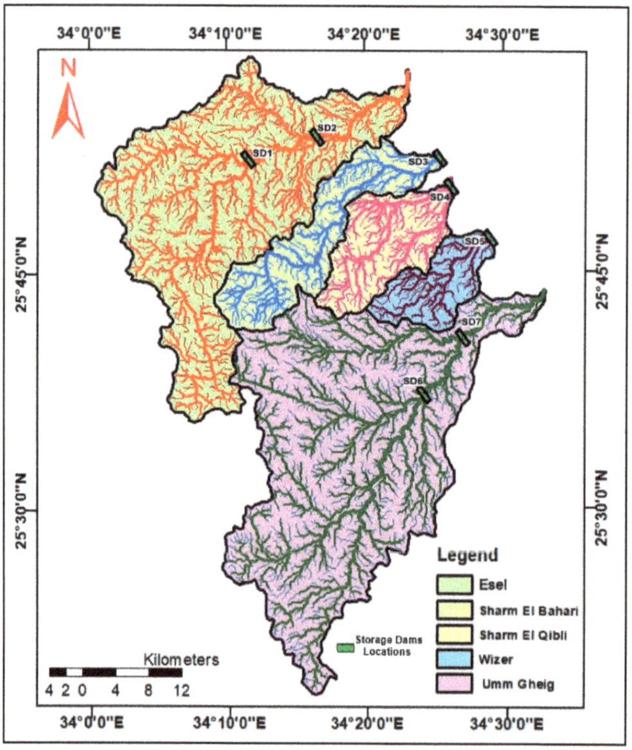


Figure 12. Recommended storage dam locations of the Gebel El Sibai watershed area.

**Author Contributions:** Formal analysis, M.A.E.A.; Data curation, N.M.H.; Writing—original draft, A.E.E.; Writing—review & editing, S.H.A.H., S.-E.O. and M.H.D.; Funding acquisition, K.A.A. All authors have read and agreed to the published version of the manuscript.

**Funding:** The authors extend their appreciation to the Researchers Supporting Project number (RSP-2024R369), King Saud University, Riyadh, Saudi Arabia.

**Data Availability Statement:** Data are contained within the article.

**Acknowledgments:** The authors extend their appreciation to the Researchers Supporting Project number (RSP-2024R369), King Saud University, Riyadh, Saudi Arabia. High appreciation for the Desert Research Center and its president for supporting and funding this research.

**Conflicts of Interest:** The authors declare no conflict of interest.

## References

1. El-Rawy, M.; Ismail, E.; Abdalla, O. Assessment of groundwater quality using GIS, hydrogeochemistry, and factor statistical analysis in Qena governorate, Egypt. *Desalination Water Treat.* **2019**, *162*, 14–29. [CrossRef]
2. El-Rawy, M.; Abdalla, F.; El Alfy, M. Water Resources in Egypt. In *The Geology of Egypt. Regional Geology Reviews*; Hamimi, Z., El-Barkooky, A., Martínez Frías, J., Fritz, H., Abd El-Rahman, Y., Eds.; Springer: Berlin/Heidelberg, Germany, 2020.
3. Rashwan, M.; Mohamed, A.K.; Alshehri, F.; Almadani, S.; Khattab, M.; Mohamed, L.; Mohamed, L. Flash Flood Hazard Assessment along the Red Sea Coast Using Remote Sensing and GIS Techniques Remote Sensing and GIS Techniques. *Int. J. Geo-Inf.* **2023**, *12*, 465. [CrossRef]
4. Abdel Moneim, A.A. Overview of the geomorphological and hydrogeological characteristics of the Eastern Desert of Egypt. *Hydrogeol. J.* **2005**, *13*, 416–425. [CrossRef]
5. Zaid, S.M. Potential Flash flooding of drainage basins of Quseir area and risk evaluation. In Proceedings of the Fourth Environmental Conference, Zagazig, Egypt, 2009; pp. 1–16.
6. Gad, M.I.; El-Shiekh, A.E.; Khalifa, R.A.; Ahmed, K.A. Flash flood risk assessment applying multicriteria analysis for some northwestern coastal basins, Egypt. *Eur. Law Rev.* **2016**, *8*, 39–62.
7. SCS. *National Engineering Handbook, Section 4, Hydrology*; US Department of Agriculture, US Government Printing Office: Washington, DC, USA, 1972.
8. El Alfy, M. Assessing the impact of arid area urbanization on flash floods using GIS, remote sensing, and HEC-HMS rainfall-runoff modeling. *Hydrol. Res.* **2016**, *47*, 1142–1160. [CrossRef]
9. Gheith, H.; Sultan, M. Construction of a hydrologic model for estimating Wadi runoff and groundwater recharge in the Eastern Desert. *Egypt. J. Hydrol.* **2002**, *263*, 36–55. [CrossRef]
10. Jyothi, M.; Vishnu, B. Modeling the Hydrology of Watershed by using HEC-HMS. *Int. J. Eng. Sci. Comput.* **2017**, *7*, 13129–13132.
11. Bhuiyan, C. Various Drought Indices For Monitoring Drought Condition in Aravalli Terrain of India. In Proceedings of the XXth ISPRS Congress, Istanbul, Turkey, 12–23 July 2004; pp. 12–23.
12. Shehata, H.S.; Galal, T.M. Factors affecting the distribution and associated species of *Malva parviflora* in the Nile Delta, Egypt. *Weed Biol. Manag.* **2014**, *15*, 42–52. [CrossRef]
13. Azarakhshi, M.; Mahdavi, M.; Arzani, H.; Ahmadi, H. Assessment of the Palmer drought severity index in arid and semi-arid rangeland: (Case study: Qom province, Iran). *Desert* **2011**, *16*, 77–85.
14. Penning, E.; Burgos, R.P.; Mens, M.; Dahm, R.; de Bruijn, K. Nature-based solutions for floods and droughts and biodiversity: Do we have sufficient proof of their functioning? *Camb. Prism. Water* **2023**, *1*, e11. [CrossRef]
15. IPCC, Climate Change. *The Physical Science Basis*; Contribution of Working Group I to the Sixth Assessment Report of the Intergovernmental Panel on Climate Change; Cambridge University Press: Cambridge, UK; New York, NY, USA, 2021.
16. Said, R. *The Geology of Egypt*; Elsevier Publication: Amsterdam, The Netherlands, 1962.
17. Said, R. *The Geology of Egypt*; A.A. Balkema: Rotterdam, The Netherlands; Brookfield: Toronto, ON, Canada, 1990; 734p.
18. El Ramly, M.F. A new geological map for the basement rocks in the eastern and south Western Deserts of Egypt. *Ann. Geol. Surv. Egypt* **1972**, *2*, 1–18.
19. CONCO. Geologic Map of Egypt. Egyptian General Authority for Petroleum (UNESCO Joint Map Project), 20 Sheets, Scale, 1987, 1:500,000. Cairo.
20. EI Gaby, S.; List, F.K.; Tehrani, R. Geology, evolution and metallogenies of the Pan African belt in Egypt. In *The Pan-African Belt of Northeast Africa and Adjacent Areas*; EI Gaby, S., Greiling, R.O., Eds.; Vieweg & Sohn: Braunschweig/Wiesbaden, Germany, 1988; pp. 17–68.
21. EI Gaby, S.; List, F.K.; Tehrani, R. The basement complex of the Eastern Desert and Sinai. In *The Geology of Egypt*; Said, R., Ed.; Balkema: Rotterdam, The Netherlands, 1990; pp. 175–184.
22. Greiling, R.O.; Abdeen, M.M.; Dardir, A.A.; El Akhal, H.; El Ramly, M.F.; Kamal El Din, G.M.; Osman, A.A.; Rashwan, A.A.; Rice, A.H.N.; Sadek, M.F. A structural synthesis of the Proterozoic Arabian-Nubian Shield in Egypt. *Geol Rundsch* **1994**, *83*, 484–501. [CrossRef]
23. Abu Raya Ahmed, M. The Area Extending Between USEI and UMM GHEIG: Geomorphological Study. Ph.D. Thesis, Faculty of Arts, Geographic and GIS Department, Alexandria University, Alexandria, Egypt, 2007; 393p.
24. Ponce, V.M. *Engineering Hydrology, Principles and Practices*; Prentice-Hall: Upper Saddle River, NJ, USA, 1989.
25. Raghunath, H.M. *Groundwater*; Wiley Eastern Limited: Hoboken, NJ, USA, 1990; 563p.
26. Bennett, M.R.; Doyle, P. *Geology and the Human Environment Chichester*; John Wiley & Sons: New York, NY, USA, 1997.
27. Davis, J.C. *Statics and Data Analysis in Geology*; Wiley: New York, NY, USA, 2002.
28. Gundalia, M.; Dholakia, M. Impact of Monthly Curve Number on Daily Runoff Estimation for Ozat Catchment in India. *Open J. Mod. Hydrol.* **2014**, *4*, 144–155. [CrossRef]

29. Misra, D.; Kadhane, U.; Singh, Y.P.; Tribedi, L.C.; Fainstein, P.D.; Richard, P. Interference Effect in Electron Emission in Heav Ion Collisions with H2 by Comparison with the Measured Electron Spectrum from Atomic Hydrogen. *Phys. Rev. Lett.* **200** *92*, 153201. [CrossRef] [PubMed]
30. *HEC-HMS Tutorials & Guides*; US Army Corps of Engineers, Hydrologic Engineering Center: Washington, DC, USA, 2023.
31. Chow, V.T.; Maidment, D.R.; Mays, L.W. *Applied Hydrology*, International ed.; McGraw-Hill Book Company: New York, N USA, 1988.
32. Kirpich, Z.P. Time of concentration of small agricultural watersheds. *J. Civ. Eng.* **1940**, *10*, 362.
33. Ponce, V.; Hawkins, R. Runoff Curve Number: Has It Reached Maturity? *J. Hydrol. Eng.* **1996**, *1*, 11–19. [CrossRef]
34. Schumm, S.A. Evolution of Drainage Systems & Slopes in Badlands at Perth Amboy, New Jersey. *Bull. Geol. Soc. Am.* **1956**, *6* 597–646.
35. Horton, R.E. Erosional Development of Streams and their Drainage Basins. *Bull. Geol. Soc. Am.* **1945**, *56*, 275–370. [CrossRef]
36. White, R.S.; McKenzie, D.; O'Nions, K. Oceanic crustal thickness from seismic measurements and rare earth element inversior *J. Geophys. Res.* **1992**, *97*, 19683–19715. [CrossRef]
37. Krishnamurthy, J.; Venkatesa, K.N.; Jayaraman, V.; Manivel, M. An approach to demarcate ground water potential zones throug remote sensing and a geographical information system. *Int. J. Remote Sens.* **1996**, *17*, 1867–1884. [CrossRef]
38. Sakthivel, R.; Jawahar Raj, N.; Sivasankar, V.; Akhila, P.; Omine, K. Geo-spatial technique-based approach on drainage morph metric analysis at Kalrayan Hills, Tamil Nadu, India. *Appl. Water Sci.* **2019**, *9*, 24. [CrossRef]
39. Mahala, A. The significance of morphometric analysis to understand the hydrological and morphological characteristics in tw different morpho-climatic settings. *Appl. Water Sci.* **2019**, *10*, 33. [CrossRef]
40. Alqreai, F.N.; Altuwaijri, H.A. Assessing the Hazard Degree of Wadi Malham Basin in Saudi Arabia and Its Impact on Nort Train Railway Infrastructure. *ISPRS Int. J. Geo-Inf.* **2023**, *12*, 380. [CrossRef]
41. Elsheikh, A. Management of water resources in Wadi Dara area using mathematical modeling for sustainable developmer Eastern Desert, Egypt. *Egypt J. Geol.* **2016**, *60*, 37–59.
42. Hussiena, M.; Yousif, M.; El Sheikhb, A. Investigation of groundwater occurrences in structurally controlled terrain, based o geological studies and remote sensing data: Wadi El Morra, South Sinai, Egypt. *NRIAG J. Astron. Geophys.* **2020**, *9*, 512–53 [CrossRef]
43. Goodarzi, M.R.; Poorattar, M.J.; Vazirian, M.; Talebi, A. Evaluation of a weather forecasting model and HEC-HMS for foo forecasting: Case study of Talesh catchment. *Appl. Water Sci.* **2024**, *14*, 34. [CrossRef]
44. Khan, I.R.; Elmahdy, S.I.; Rustum, R.; Khan, Q.; Mohamed, M.M. Floods modeling and analysis for Dubai using HEC-HMS mod and remote sensing using GIS. *Sci. Rep.* **2024**, *14*, 25039. [CrossRef]
45. Hegazy, A.K.; Lovett-Doust, J.; Hammouda, O.; Gomaa, N. Vegetation distribution along the altitudinal gradient in the nort western Red Sea region. *J. Community Ecol.* **2007**, *8*, 151–162. [CrossRef]
46. Abd El-Wahab, R.H.; Zaghloul, M.S.; Moustafa, A.A. Conservation of Medicinal Plants in St. Catherine Protectorate, South Sina I. Evaluation of ecological status and human impact. In Proceedings of the First International Conference on Strategy of Egyptia Herbaria, Giza, Egypt, 9–11 March 2004; pp. 231–251.
47. Zaghloul, M.S.; Abd El-Wahab, R.H.; Moustafa, A.A. Ecological assessment and phenotypic and fitness variation of Sinai' remnant populations of *Moringa peregrina*. *Appl. Ecol. Environ. Res.* **2010**, *8*, 351–366. [CrossRef]
48. Ahmadi, S.; Azarnivand, H.; Khosravia, H.; Dehghan, P.; Behrang Manesh, M. Assessment the effect of drought and land us change on vegetation using Landsat data. *Desert* **2019**, *24*, 23–31.
49. Alamdarloo, E.H.; Manesh, M.B.; Khosravi, H. Probability assessment of vegetation vulnerability to drought based on remot sensing data. *Environ. Monit. Assess.* **2018**, *190*, 702–710. [CrossRef] [PubMed]
50. Shawky, R.; El-Khouly, A. Plant species diversity of some wadis at Red Sea Coast, Egypt. *Res. Rev. J. Bot. Sci.* **2017**, *6*, 25–35.

atmosphere

Article

# Topographic Precipitation Diagnosis: Model Design and Validation in a Two-Dimensional Context

Xiangqian Wei, Yi Liu *, Cong Cheng, Xinyu Chang and Jun Guo

School of Civil and Hydraulic Engineering, Huazhong University of Science and Technology, Wuhan 430074, China; d202080495@hust.edu.cn (X.W.); m202371761@hust.edu.cn (C.C.); xinyu_chang12@hust.edu.cn (X.C.); prof_guojun@hust.edu.cn (J.G.)
* Correspondence: prof_liuyi@hust.edu.cn

**Abstract:** This study develops a two-dimensional (*z*-*x* direction) atmospheric dynamics model based on a set of simplified atmospheric motion equations, designed for rapid simulation of atmospheric flow characteristics over complex terrains. The model effectively captures the influence of topography on the atmospheric flow field, offering a new research tool for the in-depth investigation of atmospheric dynamic phenomena under complex terrain conditions. Furthermore, the model takes into account water vapor transport and condensation processes, and employs a simplified algorithm for the conversion of cloud droplets to raindrops to estimate the intensity and spatial distribution of precipitation. The innovative use of the *z*-coordinate system allows for a focused simulation of dynamic processes in complex terrains, capable of real-time computation of the temporal variations in precipitation processes. The model exhibits high simulation precision and has a wide range of potential practical applications.

**Keywords:** complex terrains; vapor transport and condensation processes; precipitation

## 1. Introduction

In recent years, significant attention has been devoted to the study of convective precipitation phenomena in complex terrain regions. Particularly in mountainous and valley areas, the distinctive topographical features readily trigger extreme weather events such as thunderstorms and short-duration heavy rainfall [1]. These weather phenomena often lead to geological disasters such as flash floods and debris flows [2]. The underlying cause is the substantial impact of complex terrain on wind flow patterns in these regions [3]. When moisture-laden warm and moist air encounters topographical barriers, it is forced to ascend due to the pushing of air behind and the blocking by the mountainous terrain ahead [4,5]. This ascent results in intense updrafts [6]. Such updrafts facilitate the condensation of water vapor into clouds [7], thereby significantly increasing the probability and intensity of precipitation events [8,9].

Studying weather processes under complex terrain conditions presents significant challenges from model development to validation [10], which are primarily encapsulated in four key aspects. Firstly, the complexity of model mechanisms, which involves the intricate interplay of multiple physical processes [11]. Secondly, the scarcity of empirical data [12], as precipitation events in complex terrains tend to be abrupt and localized [13], making the collection of high-quality empirical data exceptionally difficult [14]. Thirdly, in mountainous and valley regions, small-scale precipitation events occur with high frequency; thus, research often concentrates on these microscale phenomena [15]. Lastly, the triggering and formation mechanisms of precipitation are not yet fully elucidated [16,17], lacking sufficient experimental research and data support [18,19], which poses a considerable challenge in the current research endeavors [20].

Built upon atmospheric dynamics models and incorporating algorithms for water vapor phase transitions and condensation [21,22], a precipitation diagnostic model is constructed. The model takes into account phenomena such as the horizontal movement of air masses, vertical ascent, and convergence of air flows. In terms of the precipitation mechanism, the model comprehensively considers processes including the transport of water vapor, condensation of water vapor, formation of water droplets, and the descent of precipitation particles, aiming to accurately diagnose the precipitation process [23].

Simulating the precipitation process involves modeling the relevant physical processes and effectively integrating these models [24]. This necessitates the development of precise algorithmic models to comprehensively simulate these complex physical processes as fully as possible.

To diagnose precipitation over complex terrain areas, it is necessary to establish a small-scale complex terrain precipitation diagnostic model based on the characteristics of the terrain flow field. The model employs a $z$-coordinate system in the vertical direction and establishes an Eulerian grid system, where information such as water vapor and liquid water in the air is stored within the grid cells and undergoes material exchange between grid points through fluid transport mechanisms.

Although real atmospheric motion is a three-dimensional process, the substantial workload associated with constructing a three-dimensional model led to the selection of a two-dimensional structure in the experimental design, considering the vertical ($z$-direction) and horizontal ($x$-direction) dimensions. Within this framework, a precipitation diagnostic model is introduced to conduct the calculation and analysis of precipitation.

## 2. Data Description

ERA5 is the fifth-generation global atmospheric reanalysis dataset released by the European Centre for Medium-Range Weather Forecasts (ECMWF). ERA5 provides hourly estimates of a wide range of atmospheric, terrestrial, and oceanic climate variables, with a maximum resolution of $0.25^\circ \times 0.25^\circ$. The data cover the Earth's surface on a 30 km grid and utilize 137 vertical levels, ranging from the surface to 80 km, to represent the atmosphere, including uncertainty information for all variables when spatial and temporal resolutions are reduced. ERA5 integrates model data with observational data from across the globe to form a comprehensive and consistent global dataset. The dataset incorporates various sources of observational data, including remote sensing data, ground station measurements, and other data sources. This data assimilation approach enables ERA5 to provide more comprehensive and accurate meteorological information, earning it the designation of 'reanalysis' data. The ERA5 dataset includes variables such as temperature, pressure, wind, water vapor content, humidity, and precipitation. ERA5 data accurately reflect the spatiotemporal distribution patterns of local meteorological variables, and its accessibility and global applicability make it widely used in various fields.

## 3. Dynamic Process Framework

This diagnostic model is built with a fundamental structure and an integrated precipitation diagnostic unit. The fundamental structure performs calculations for atmospheric dynamics, pressure, temperature, and water vapor flux. The precipitation diagnostic unit specifically deals with the calculations of water vapor condensation, the formation of raindrops, and their falling from the clouds.

### 3.1. Fundamental Formulas

Considering the model's focus on atmospheric motion over small-scale complex terrain, it is necessary to simplify the equations of motion as much as possible. In this scenario, the effect of the Coriolis force can be disregarded, as it generally becomes significant only over ranges exceeding hundreds of kilometers. In conventional meteorological models, pressure is divided into two components: base pressure and perturbation pressure. The base pressure is used to calculate the mesoscale pressure gradient force, which represents a weak interacting force, while the perturbation pressure is used to simulate the intercompression processes between air masses. In the two-dimensional ($z$-$x$) simulation framework, the calculation of the pressure gradient force only needs to consider the variations in perturbation pressure.

**Atmospheric Motion:** Within the scope of small-scale meteorological models, the movement of air masses is primarily influenced by perturbation pressure. In the vertical direction, the atmosphere maintains a state of hydrostatic equilibrium, where gravity and the pressure gradient force are in balance. Consequently, perturbation pressure becomes the decisive factor affecting vertical motion [25,26].

$$\frac{du}{dt} = -\lambda \cdot \frac{1}{\rho} \cdot \frac{\partial \delta p}{\partial x} + F_x \tag{1}$$

$$\frac{dw}{dt} = -\lambda \cdot \frac{1}{\rho} \cdot \frac{\partial \delta p}{\partial z} + g \cdot \left( \frac{\theta'}{\theta} + 0.61 \cdot \delta q_v - \delta q_c - \delta q_r \right) + F_z \tag{2}$$

In this context, $\lambda$ represents the adaptive coefficient of the grid (in practical calculations, a parameter is required to adjust the proportional relationship between the perturbation pressure $\delta p$ and the grid spacing), $F_x$ and $F_z$ quantify the frictional forces on air masses in the horizontal and vertical directions, respectively, $\rho$ denotes the density of air mass, $g$ is the acceleration due to gravity, $\theta$ is the potential temperature, and $\theta'$ refers to the perturbation potential temperature. ($u$, $w$) is the wind speed vector, $\delta p$ is the perturbation pressure, which is involved in calculating the expansion and compression effects between air masses. $q_v$ denotes the water vapor mixing ratio (units: kg/kg), and its variation $\delta q_v$ reflects the increase or decrease in water vapor content; an increase in water vapor leads to a decrease in air density, thereby triggering upward air currents. $q_c$ is the cloud droplet mixing ratio, with a positive $\delta q_c$ indicating the condensation of water vapor into cloud droplets; $q_r$ represents the raindrop mixing ratio, and a positive $\delta q_r$ indicates the transformation of water vapor and cloud droplets into raindrops. These physical phenomena all influence the vertical velocity of the air.

**Perturbation Pressure ($\delta p$):** Minor fluctuations in atmospheric pressure caused by the movement of air masses, which typically reflect the interactions and dynamic motion states between air masses. The perturbation pressure can also be approximated as follows [27]:

$$\delta p = -\gamma \cdot p \cdot \left( \frac{\partial u}{\partial x} + \frac{\partial w}{\partial z} \right) \cdot \delta t \tag{3}$$

In the above equation, $\gamma$ denotes the grid adaptive coefficient (in practical calculations, a parameter is required to adjust the proportional relationship between the pressure $p$ and the grid spacing, so that $\delta p$ can be adapted to the current grid), and $\delta t$ represents the time step. The equation estimates the degree of compression or expansion experienced by the air mass by analyzing the velocity changes at the grid points surrounding the air mass.

**Potential temperature ($\theta$):** The temperature reached by an air mass when it rises or descends vertically along its dry adiabatic process [28,29] (i.e., without exchanging heat with the surroundings) to a standard pressure level (usually $p_0$ = 1000 hPa).

$$\theta = T \cdot \left( \frac{p_0}{p} \right)^{\frac{R}{c_p}} \tag{4}$$

where $c_p$ represents the specific heat capacity of air, with a value of $c_p$ = 1005 J/(kg·K), and $R$ typically denotes the gas constant, which for dry air is 287.15 J/(kg·K).

The form of the perturbation potential temperature variation over time is expressed as follows [30]:

$$\frac{d\theta'}{dt} = -\rho \cdot w \cdot \frac{\partial \theta'}{\partial z} + \kappa \cdot \nabla^2 \theta' \tag{5}$$

where ρ denotes air density, and κ represents the turbulent diffusion coefficient.

**Pressure Equation:** This represents the time-dependent changes in pressure across the background grid. The formula for its computation is presented below [31,32]:

$$\frac{dp}{dt} = -\frac{p}{RT} \cdot \left( g + \frac{dw}{dt} \right) \cdot w \tag{6}$$

In this context, $g$ denotes gravitational acceleration. This formula is applicable for calculating the pressure changes experienced by an air mass during its displacement process.

Utilizing the aforementioned equations, an atmospheric dynamics model was constructed, with its dynamic framework based on the computational strategy outlined in "*Applicability Study of Euler–Lagrange Integration Scheme in Constructing Small-Scale Atmospheric Dynamics Models*" [33]. The transport processes of water vapor and liquid water were calculated using the semi-Lagrange method.

### *3.2. Time Step Constraints*

To ensure numerical stability, the time step must adhere to the Courant–Friedrichs–Lewy (CFL) condition [34], which states that the product of the time step ($\delta t$) and the characteristic velocity should not exceed the spatial step ($\delta x$). Therefore, the maximum permissible time step can be expressed as follows:

$$\Delta t_{\max} = \frac{l_{\min}}{v_{\max}} \tag{7}$$

Herein, $l_{min}$ represents the minimum value of the grid points, and $v_{max}$ denotes the maximum speed.

Usually, the grid spacing in atmospheric models is not less than 400 m. In this experiment, the maximum value of the wind vector is restricted to within 10 m/s. Accordingly, the maximum allowable time step can be set to 40 s.

## 4. The Fundamental Framework of the Precipitation Process

(I) Scheme for Detecting Water Vapor Condensation

Generally, when the water vapor content in the air reaches the saturation vapor pressure, the excess water vapor condenses onto aerosol particles to form minute droplets. The saturation vapor pressure in the air is commonly calculated using the Clausius–Clapeyron equation. The following is an initial description of the computational process [35,36]:

$$es = 611.2 \cdot e^{\frac{17.67 \cdot T_C}{T_C + 243.5}} \tag{8}$$

Herein, $es$ represents the saturation vapor pressure of water vapor in the air, with units of pascals (Pa), and $T_c$ denotes the air temperature in degrees Celsius.

Consequently, the saturation specific humidity corresponding to the temperature $T_c$ is given as follows [37]:

$$q_{sat} = 0.622 \cdot \frac{es}{p - es} \tag{9}$$

Here, $q_{sat}$ denotes the threshold of gaseous water that 1 kg of air can hold, with units of kg/kg. When the water vapor content in the air exceeds the threshold $q_{sat}$, the excess water vapor will condense into liquid water. In the actual modeling process, the relative humidity threshold is set to $RH$ = 1.05 (this value is chosen as 1.05 instead of 1 to allow for some numerical perturbations during the simulation, providing a certain degree of tolerance in the model). It is assumed that when the water vapor content exceeds $1.05 \cdot q_{sat}$, the program considers the excess water vapor to condense into liquid water.

Liquid water droplets in the air are categorically divided into two primary types based on their size: cloud droplets and raindrops. Generally, cloud droplets span a diameter range from a few nanometers to about 0.25 mm, while raindrops are defined as having a diameter greater than 0.25 millimeters [38]. Assuming that the condensation of water vapor results in cloud droplets, in the model design, cloud droplets are converted into raindrops according to a certain conversion rate [39].

(II) Conversion of cloud droplets into raindrops

The conversion rate is calculated using a simplified formula [40]:

$$\tau = \frac{q_c}{q_c + q_r} \tag{10}$$

In the equation, $q_c$ and $q_r$ represent the water content of cloud droplets and raindrops, respectively, with units in kg/kg.

$$P_{au} = \begin{cases} 0, & q_v \leq q_{sat} \\ k \cdot \tau \cdot (q_c - q_{c_0}), & q_v > q_{sat} \end{cases} \tag{11}$$

In the equation, $k$ represents the conversion coefficient, and $q_{c_0} = 10^{-3}$ g/kg denotes a constant.

The raindrop increment $\Delta q_r$ is calculated based on the conversion rate, expressed as follows [41]:

$$\Delta q_r = q_c \cdot P_{au} \tag{12}$$

$$q_{r\_new} = q_r + \Delta q_r \tag{13}$$

$$q_{c\_new} = q_c - \Delta q_r \tag{14}$$

It is assumed that precipitation is triggered when the raindrop content $q_r$ in a grid point reaches a certain threshold.

(III) Precipitation determination process

In the model, the gaseous and liquid water content in each grid point is calculated at each time step.

As illustrated in Figure 1, the iterative process for precipitation diagnosis is as follows. Initially, at time step $t_{n-1}$, water vapor and liquid water (inclusive of cloud droplets and raindrops) are transported within the grid cells along the wind direction. The saturation ratio at each grid point is calculated, and based on this ratio, the contents of water vapor and liquid water within the grid are updated, marking this intermediate time as $t_n'$ (indicating the time between $t_{n-1}$ and $t_n$). Subsequently, the liquid water content in the grid at $t_n'$ is partitioned into cloud droplets and raindrops based on the ratio of cloud droplet (small water droplets) to raindrop content at $t_{n-1}$. The contents of cloud droplets and raindrops are then updated according to the conversion rate formula for the aggregation of cloud droplets into raindrops. When the mass of larger raindrops exceeds a certain threshold, it is deemed as precipitation. After accounting for the raindrops that have fallen, the remaining raindrop content at time step $t_n$ is obtained.

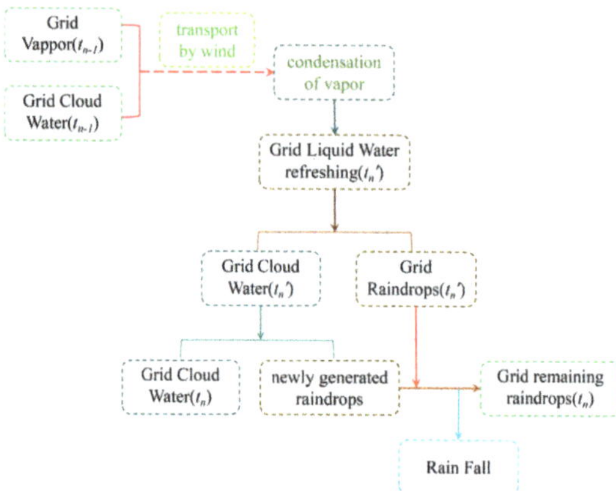


**Figure 1.** Iterative computation of the contents of water vapor, liquid water, cloud droplets, and raindrops. Here, $t_n$ denotes the time at the nth time step, while $t_n{}'$ represents the time between $t_{n\text{-}1}$ and $t_n$.

## 5. Result

### 5.1. Topographic Wind Field

The case study is situated in the Enshi area of Hubei Province, China, with the precipitation event on 25 September 2018 selected as the research case for analysis.

As shown in Figure 2, Figure 2a illustrates the wind field characteristics at 06:00 on 25 September. Influenced by an outer typhoon, the atmospheric flow field at 850 hPa undergoes dramatic changes, with the region exhibiting an easterly wind pattern. Due to the model's 2D framework, as depicted in Figure 2b, a z-directional cross-section is selected as the research target within this region. The cross-section corresponds to the red dashed line in Figure 2b. As shown in the figure, the cross-section is aligned parallel to the prevailing wind direction, minimizing disruption from winds perpendicular to the cross-section. Figure 2c illustrates the relative humidity at the 850 hPa height level, revealing that the relative humidity near the target region approaches 100%. Figure 2d displays the water vapor content at the 850 hPa height level (unit: g/kg), indicating that the water vapor content around the target region fluctuates around 10 g/kg.

A 2D atmospheric flow field terrain model was established based on the cross-section corresponding to the red dashed line in Figure 2b. In terms of grid setup, the horizontal grid spacing is 500 m (dx = 500 m), the vertical grid spacing is 100 m (dz = 100 m), and the time integration step is 30 s (dt = 30 s). To observe the model's sensitivity to actual terrain, a virtual wind field was imported prior to introducing the ERA wind field data.

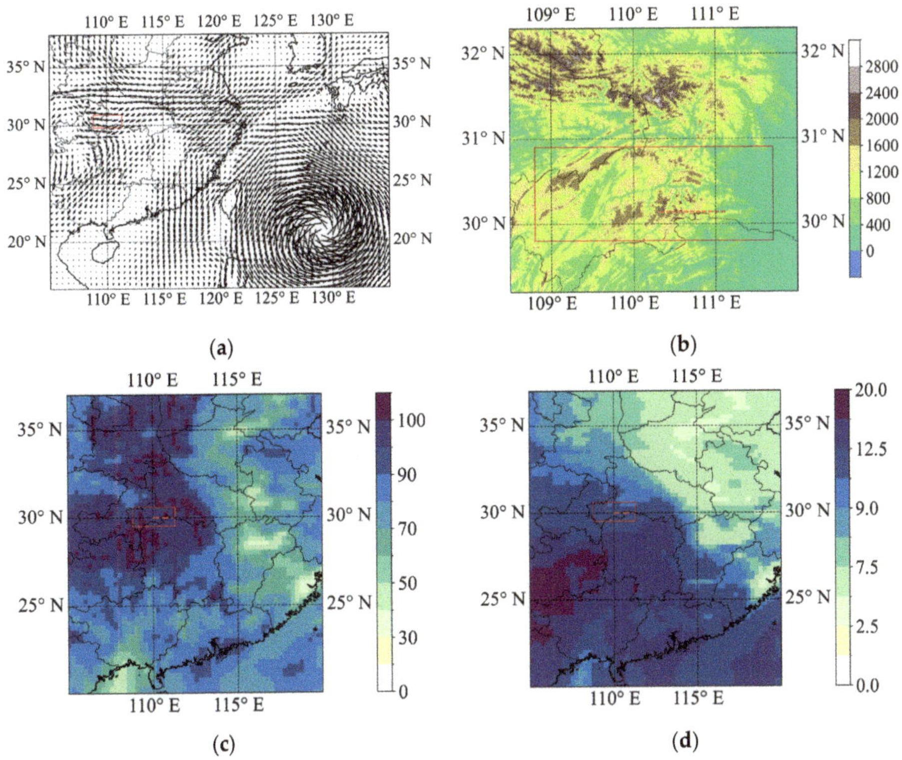


**Figure 2.** Weather characteristics of the study area (the red box indicates the study area, and the red curve denotes the selected simulation cross-section). (**a**) Atmospheric circulation characteristics at 850 hPa; (**b**) topographical features of the study area; (**c**) relative humidity of the atmosphere at 850 hPa; (**d**) moisture content of the atmosphere at 850 hPa.

As shown in Figure 3, the model incorporates real terrain and a simulated wind field, with the simulated wind field set to a horizontal wind speed of −3 m/s. Figure 3a depicts the wind field at the initial moment, while Figure 3b shows the wind field after 10 min of simulation. The filled contour plot represents the horizontal wind speed. Wind vector arrows are displayed at intervals of eight data points along the horizontal direction. As indicated in Figure 3b, the absolute value of the wind speed is greater than 3 m/s over the mountain ridge, whereas in the valley area, the wind speed is significantly reduced. This simulation result aligns with actual observations.

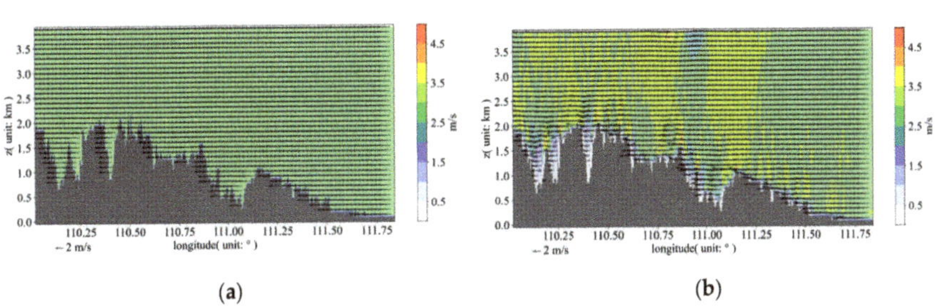


**Figure 3.** Simulated terrain and wind field, with filled contours representing the initial horizontal wind speed values. (**a**) Initial virtual wind field; (**b**) wind field simulated by the model after 10 min.

As shown in Figure 4, it can be observed from Figure 4a,b that the model exhibits good stability over time. Furthermore, the figures illustrate that on the windward slope, the airflow ascends along the slope, while on the leeward slope, the airflow descends. These results indicate that the wind field simulated by the model conforms to the laws of real atmospheric motion.

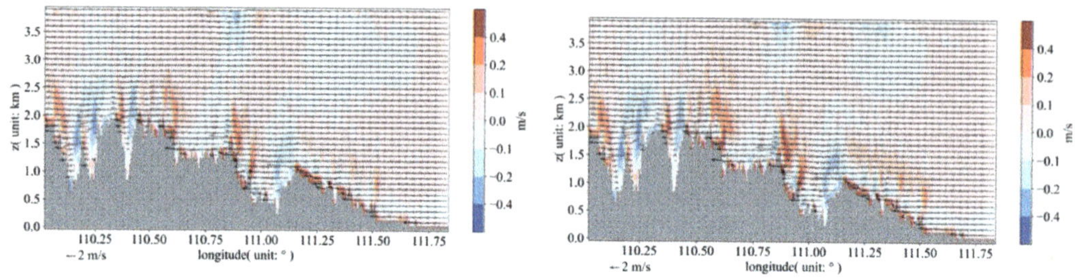

(a) (b)

**Figure 4.** Simulated topographic wind field by the model based on the virtual wind field, with filled contours representing vertical wind speed values. (**a**) Wind field simulated by the model after 10 min; (**b**) wind field simulated by the model after 20 min.

Key meteorological data such as wind vectors, air temperature, pressure, and humidity were extracted from the ERA (European Centre for Medium-Range Weather Forecasts reanalysis data). After undergoing filtering processes, these data were used to establish corresponding weight relationships and were interpolated onto the grid system of the diagnostic model.

Simulate the changes in weather conditions from 0 to 59 min. Initially, we will analyze and present the wind field characteristics under the influence of terrain.

As depicted in Figure 5, the filled contour plots in Figure 5a represent the magnitude of the horizontal wind speed, while those in Figure 5b represent the magnitude of vertical wind speed. It can be observed from the figures that the background wind field is quite strong, with significant wind speeds. The background data obtained through interpolation from ERA data take into account the influence of terrain to some extent; however, due to the spatial resolution limitations of the ERA data itself, the estimation of the terrain is relatively coarse.

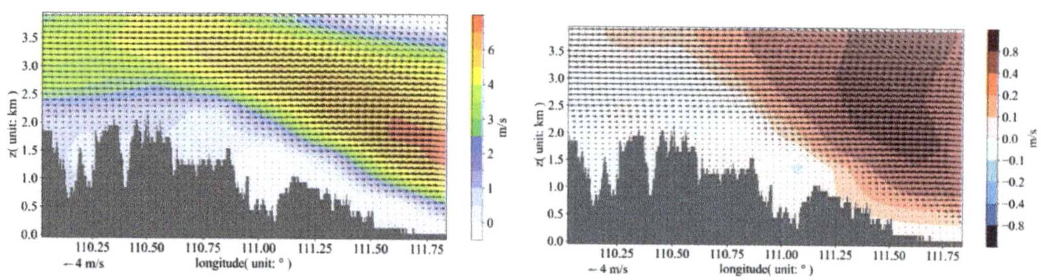

(a) (b)

**Figure 5.** Wind field imported from ERA data. (**a**) Initial wind vectors and horizontal wind speed (shaded contour plot); (**b**) initial wind vectors and vertical wind speed (shaded contour plot).

Figure 6 illustrates the temporal evolution of the wind field after the initial background wind field has been processed through the diagnostic model. As seen in Figure 6a, there is no significant change in the horizontal wind speed after the model's correction. Figure 6b reveals that, influenced by the strong winds at high altitudes, a distinct terrain-induced updraft occurs near the windward slopes. This phenomenon is attributable to a continuous inflow of air from the horizontal direction, which, upon encountering topographical obstacles, is forced to ascend, thereby initiating vertical movement in the air current.

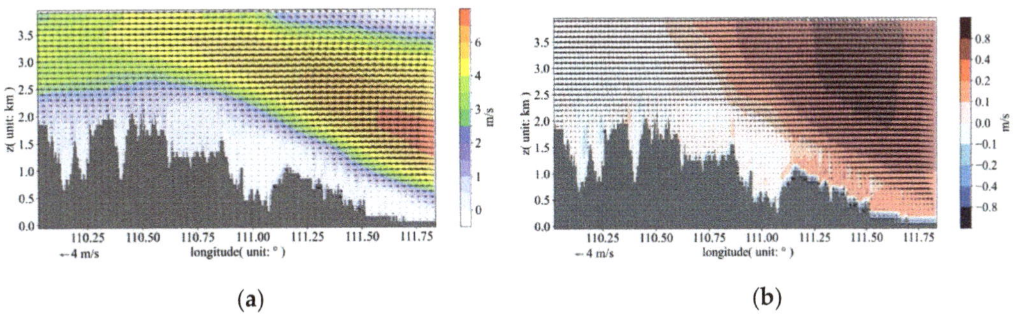


**Figure 6.** Wind field after model correction. (**a**) Wind vectors and horizontal wind speed at the 10 min mark (contour plot); (**b**) wind vectors and vertical wind speed at the 10 min mark (contour plot).

*5.2. Precipitation Process*

The transport of water vapor in the atmosphere is influenced by variations in wind speed and direction. Its migration process is a result of the combined effects of the atmospheric flow field, terrain characteristics, and various meteorological factors.

As depicted in Figure 7, water vapor in the air is subject to the driving force of the wind field, which continuously supplies external input. On the other hand, it is impeded by terrain features. The transport of water vapor is restricted by high mountains, forcing it to ascend along the mountain slopes.

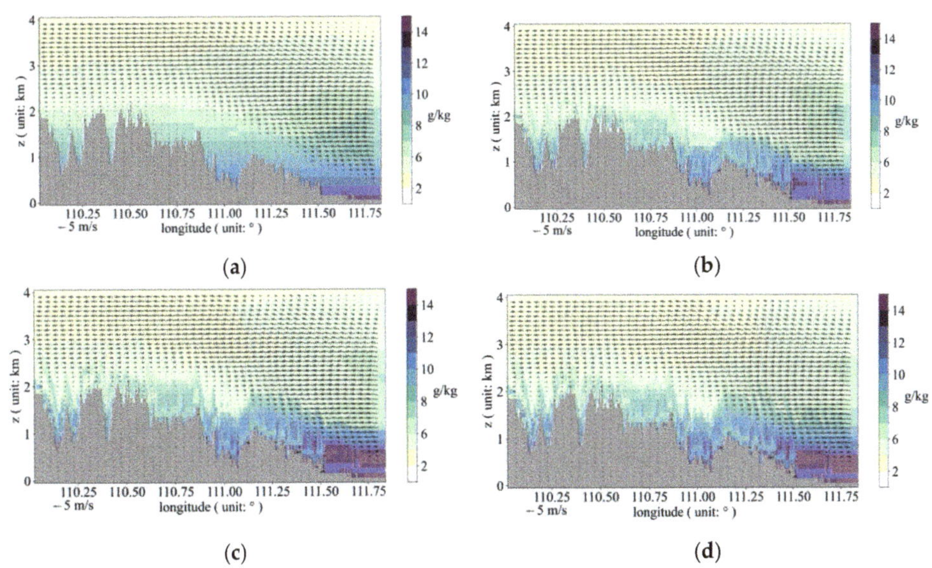


**Figure 7.** Temporal variation in the wind field and specific humidity $q_v$ (water vapor mixing ratio, unit: g/kg). (**a**) State at time 0; (**b**) state at 10 min; (**c**) state at 20 min; (**d**) state at 30 min.

Following the attainment of saturation, water vapor condenses into small raindrops. Figure 8 illustrates the variation in cloud water mixing ratio ($q_c$, where $q_c$ denotes the mass of cloud droplets per unit mass of air, unit: g/kg) over time. Panel (a) shows that at 15 min, a cloud mass forms over the mountain, which then rapidly intensifies with time; panel (b) reveals that by 20 min, the water content within the cloud mass has significantly increased; panel (c) indicates that as the transport of water vapor continues, a cloud mass forms over a canyon near 111° longitude at 30 min; and panel (d) demonstrates that at 50 min, the water content of the cloud mass over the canyon near 111° longitude has increased substantially.

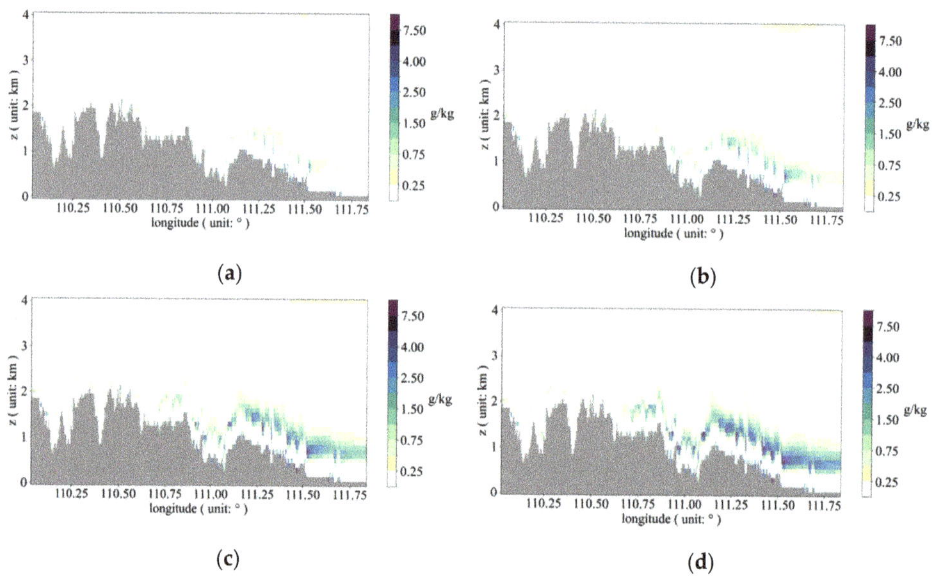


**Figure 8.** Temporal variation in cloud droplet water content $q_c$ (cloud water mixing ratio, unit: g/kg). (**a**) State at 15 min; (**b**) state at 20 min; (**c**) state at 30 min; (**d**) state at 50 min.

Figure 9 depicts the temporal evolution of the rainwater mixing ratio ($q_r$, unit: g/kg) within a rain cloud cluster. Figure 9a indicates that at the 20 min time point, the rain cloud cluster emerges above the mountain, after which the mass of raindrops increases progressively and rapidly over time. Figure 9b shows an increase in raindrop mass at the 20 min mark. Figure 9c,d demonstrate that the mass of raindrops reaches a stable condition.

In the precipitation cloud cluster, raindrops fall to the ground and produce rainfall once they reach a certain mass. The experiment involves measuring the mass of the falling raindrops and converting this measurement into the equivalent precipitation depth upon impact, expressed in millimeters (mm).

Figure 10 illustrates raindrops falling from high altitudes, which contribute to surface precipitation. Figure 10a shows that sporadic precipitation occurs at high altitudes. Figure 10b indicates an increase in the intensity and extent of the precipitation. Figure 10c reveals that the precipitation extends to the vicinity of the canyon above longitude 111°. Figure 10d demonstrates that the intensity of the precipitation remains constant.

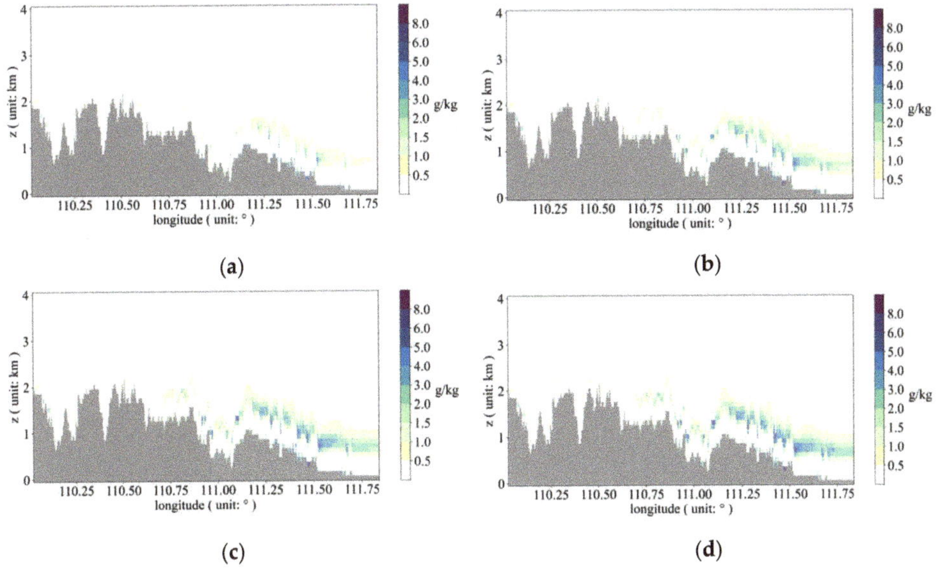


**Figure 9.** Variation in rainwater content (rainwater mixing ratio, unit: g/kg) in the rain cloud cluster with time. Panel (**a**) Status at 20 min; panel (**b**) status at 30 min; panel (**c**) status at 40 min; panel (**d**) status at 50 min.

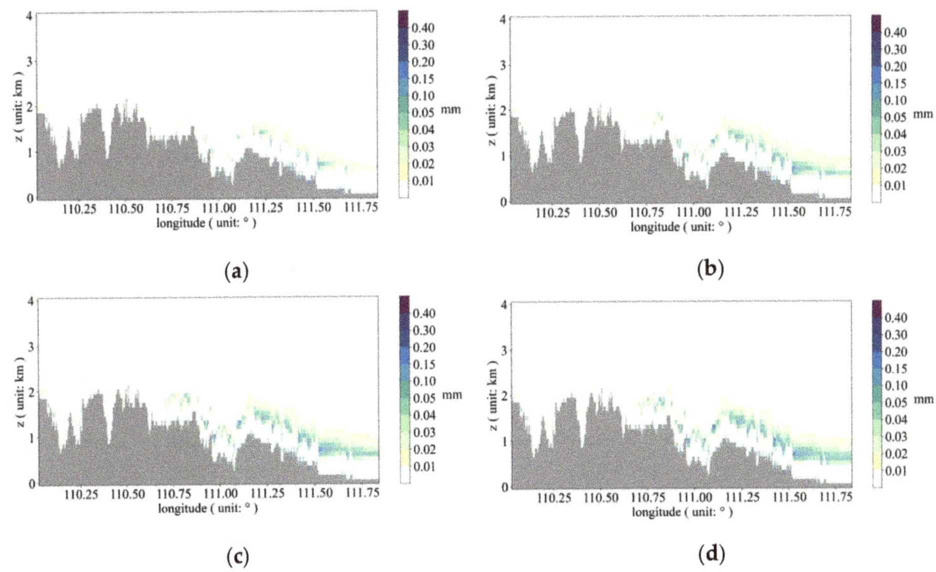


**Figure 10.** Precipitation amounts triggered by the model. Panel (**a**) status at 20 min; panel (**b**) status at 30 min; panel (**c**) status at 40 min; panel (**d**) status at 50 min.

Figure 11 presents the spatial total precipitation amount triggered within one hour, as shown in Figure 11a. Figure 11b displays the cumulative precipitation curve (model result) as output by the model. Due to the high volatility of the curve, a smoothing process is applied to obtain the smoothed precipitation curve (model smooth). Figure 11c employs the results from the smoothed precipitation curve (model smooth) and compares them with ERA precipitation data. It can be observed from the figure that both datasets exhibit a similar trend in precipitation, although there are some discrepancies in the details of the rainfall events.

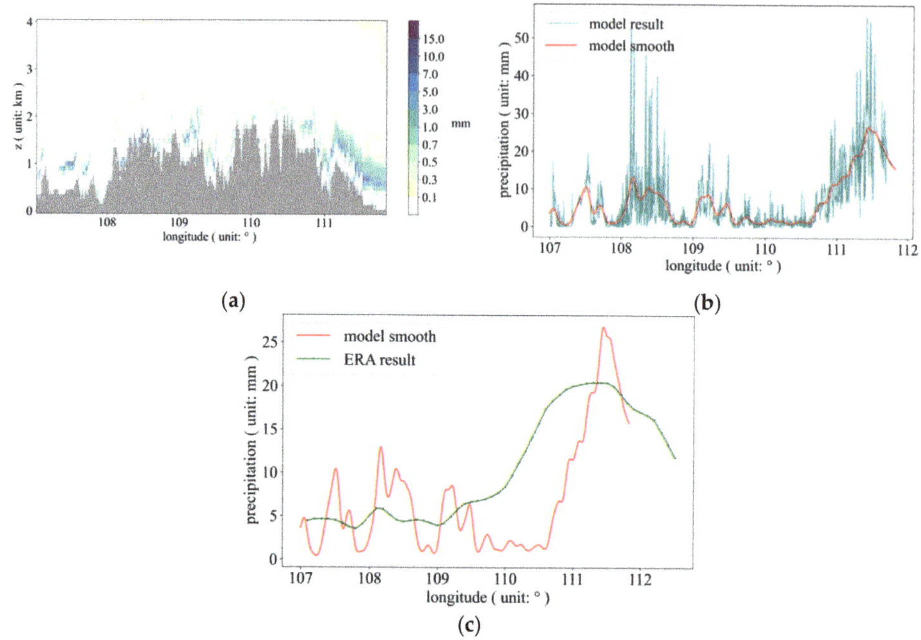


**Figure 11.** Panel (**a**) spatial distribution of the total precipitation triggered by the model over a 1 h period; panel (**b**) accumulated precipitation over 1 h from the model (model result) and the smoothed model output (model smooth); panel (**c**) smoothed curve of the 1 h accumulated precipitation from the model (model smooth) compared to the ERA precipitation results.

## 6. Discussion

The existing meteorological models are generally adequate for routine weather forecasting tasks but have limitations when applied to specific scenarios, such as river basin flood forecasting and short-term predictions in complex terrain regions. Typically, mainstream mesoscale meteorological models (e.g., WRF) require a certain "spin-up time" before they can provide reliable forecast results. This time is usually at least 3 to 6 h, which restricts the real-time forecasting capability of these models in the event of sudden meteorological phenomena. Moreover, river basin flood forecasting heavily relies on high-resolution precipitation predictions. However, due to architectural constraints, current mesoscale models show diminishing improvements when increasing grid resolution below 3 km. Specifically, in the case of the WRF model, terrain data at mesoscale resolution are typically interpolated onto the model grid for simulations and forecasts, rather than using high-resolution terrain data directly. When WRF directly incorporates high-resolution terrain data (e.g., grid sizes below 3 km), it may result in numerical instability, leading to potential model interruptions during operation.

Radar extrapolation models offer unique advantages in short-term forecasting, particularly in flat areas. However, their effectiveness significantly diminishes in complex terrain regions, such as mountains and canyons. Therefore, it is essential to develop rapid precipitation forecasting models tailored to complex terrain to supplement radar-based predictions. Additionally, the development of small-scale meteorological models that are both high-resolution and capable of rapid response is crucial to enhancing short-term forecasting accuracy.

The precipitation diagnostic model used in the experiment calculates a precipitation amount that is generally in agreement with the ERA precipitation trends, albeit with some deviation. The discrepancies between the two are attributed to several factors. On one hand, there is a significant difference in spatial resolution between the two, with the model having a resolution of 500 m grid spacing, whereas the ERA resolution is 0.25° (approximately 30 km). On the other hand, the model is a 2D cross-sectional model, whereas weather processes are three-dimensional, leading to inadequate consideration of certain precipitation factors in the model.

The experimental results presented above suggest that the modeling approach proposed in this paper for complex terrain conditions is feasible. Subsequent efforts should focus on developing the model into a 3D version based on the existing theoretical framework, to be used for precipitation diagnostics in complex terrain settings.

In the context of application, subsequent research should integrate established cloud microphysics schemes into the model. These schemes are models based on physical processes that can accurately depict the formation, growth, coalescence, and dissipation of microphysical particles such as cloud droplets and ice crystals. The incorporation of these mature schemes will offer a more precise foundation for precipitation forecasting.

Additionally, integrating radar data for precipitation diagnosis is an indispensable component in subsequent research. Radar data are characterized by their high spatiotemporal resolution, enabling real-time monitoring of the dynamic changes in precipitation systems. By assimilating radar data into the model and conducting comparative analysis of the simulation results, we can effectively diagnose the strengths and weaknesses of the model, thereby facilitating optimization and refinement of the model.

## 7. Conclusions

This study integrates fundamental atmospheric principles within an atmospheric dynamics model framework to develop a water vapor transport and condensation model. It simulates the physical processes of water vapor transport and condensation in complex terrains. Furthermore, it introduces a simplified model for cloud droplet to raindrop conversion. By categorizing liquid water in the atmosphere into cloud droplets and raindrops, the model framework efficiently estimates their distribution and enables rapid computation of surface precipitation amounts. The following conclusions were drawn from the study:

(1) The $z$-coordinate-based motion model possesses unique advantages in simulating atmospheric motion over complex terrain, effectively capturing the movement characteristics that align with real atmospheric conditions.

(2) In this study, a two-dimensional structural design was adopted for the model, with the objective of assessing the applicability of this technology in the construction of weather system models. To improve the model's ability to simulate complex atmospheric flow fields, future efforts will concentrate on developing a three-dimensional dynamic framework. In addressing small-scale weather processes, the model involved a suitable simplification of the dynamic equations. In subsequent research, the model will incorporate key factors such as pressure gradient force and Coriolis force to more effectively simulate and interpret complex weather phenomena.

(3) The atmospheric model based on the $z$-coordinate system offers distinct advantages in precipitation simulation for specific application scenarios, particularly in the lower atmosphere. It is especially effective in addressing challenges posed by complex terrain and urban topographic conditions in meteorological simulations.

**Author Contributions:** Methodology, X.W.; Software, X.W.; Formal analysis, X.W.; Resources, Y.L.; Visualization, C.C.; Data curation, X.C.; Writing—original draft, X.W.; Writing—review and editing,

C.C. and J.G.; Visualization, X.C.; Project administration, Y.L. All authors have read and agreed to the published version of the manuscript.

**Funding:** This research was funded by Project U2340211 supported by National Natural Science Foundation of China (Grant NO.: U2340211), The National Key R&D Program of China (Grant NO.: 2022YFC3002704), The National Key R&D Program of China (Grant NO.: 2023YFC3209105), and the Strategic Consulting Project supported by the Chinese Academy of Engineering (CAE) (Grant NO.: HB2024C18). Special thanks are extended to the editors and anonymous reviewers for their constructive comments.

**Institutional Review Board Statement:** Not applicable.

**Informed Consent Statement:** Not applicable.

**Data Availability Statement:** Additional data associated with this article can be found at https://figshare.com/s/e750c2829d36e96e53c6, accessed on 1 March 2020.

**Conflicts of Interest:** The authors declare no conflict of interest.

## References


1. Loginov, V.F.; Volchek, A.A.; Shelest, T.A. Analysis and Simulation of Rain Flood Hydrographs in Belarus Rivers. *Water Resour* **2015**, *42*, 292–301. [CrossRef]
2. Morell, D.A.; Pereira Filho, A.J.; Beltrán, R.P. Quantitative Rainfall Estimation with a Mobile XPOL Weather Radar. *Pure Appl Geophys.* **2022**, *179*, 2957–2968. [CrossRef]
3. Dong, W.; Fritts, D.C.; Hickey, M.P.; Liu, A.Z.; Lund, T.S.; Zhang, S.; Yan, Y.; Yang, F. Modeling Studies of Gravity Wave Dynamics in Highly Structured Environments: Reflection, Trapping, Instability, Momentum Transport, Secondary Gravity Waves, and Induced Flow Responses. *J. Geophys. Res. Atmos.* **2022**, *127*, e2021JD035894. [CrossRef]
4. Ikeda, T.; Onaga, H.; Sakurai, Y. Influence of surrounsding weather on the prevention of vapssor condensation and vapo condensation damage by using hygroscopic materials in dwellings: Understanding on properties of moisture content in porous materials and its estimation by condition of cons. *J. Archit. Plan.* **2000**, *65*, 9–16. [CrossRef]
5. Takayabu, Y.N. Large-Scale Cloud Disturbances Associated with Equatorial Waves. *J. Meteorol. Soc. Jpn. Ser. II* **2009**, *72*, 433–449 [CrossRef]
6. Wei, X.; Liu, Y.; Chang, X.; Guo, J.; Li, H. Numerical Simulation of Terrain-Adaptive Wind Field Model Under Complex Terrain Conditions. *Water* **2024**, *16*, 2138. [CrossRef]
7. Merritt, G.E.; Weatherston, R.C. Condensation of mercury vapor and drop growth processes in a nitrogen flow. *Aiaa J.* **2015**, *5* 721–728. [CrossRef]
8. Zhou, Y.; Wang, S.; Fang, J. Diurnal Cycle Dipolar Pattern of Precipitation Over Borneo During the MJO: Linear Theory Nonlinear Sensitivity Experiments. *J. Geophys. Res. Atmos.* **2023**, *128*, 2022JD037616. [CrossRef]
9. Miao, J.E.; Yang, M.J. Amodeling study of the severe afternoon thunderstorm event at Taipei on 14 June 2015: The roles of Sea Breeze microphysics terrain. *J. Meteorol. Soc. Jpn.* **2020**, *98*, 129–152. [CrossRef]
10. Turner, R.; Mcconney, P.; Monnereau, I. Climate Change Adaptation and Extreme Weather in the Small-Scale Fisheries of Dominica. *Coast. Manag.* **2020**, *48*, 436–455. [CrossRef]
11. Hiraga, Y.; Meza, J. Extreme precipitation modeling Probable Maximum Precipitation (PMP) estimation in Chile. *J. Hydrol. Reg. Stud.* **2025**, *58*, 102274. [CrossRef]
12. Wang, H.; Li, Z.; Zhang, T.; Chen, Q.; Guo, X.; Zeng, Q.; Xiang, J. Downscaling of GPM satellite precipitation products based on machine learning method in complex terrain and limited observation area. *Adv. Space Res.* **2023**, *72*, 2226–2244. [CrossRef]
13. Gai, Y.; Long, X.; Wei, Z.; Sui, X.; Jiang, B.; Chen, X.; Yu, D.; Qin, S.; Cui, Y. Toward a better understanding of uncertainty for satellite precipitation products over complex terrain with sparse rain gauge data. *Atmos. Res.* **2023**, *288*, 106742. [CrossRef]
14. Zhang, S.; Solari, G.; De Gaetano, P.; Burlando, M.; Repetto, M.P. A refined analysis of thunderstorm outflow characteristics relevant to the wind loading of structures. *Probabil. Eng. Mech.* **2017**, *54*, 9–24. [CrossRef]
15. Zhang, Y.; Deng, C.; Xu, W.; Zhuang, Y.; Jiang, L.; Jiang, C.; Guan, X.; Wei, J.; Ma, M.; Chen, Y.; et al. Long-term variability of extreme precipitation with WRF model at a complex terrain River Basin. *Sci. Rep.* **2025**, *15*, 156. [CrossRef] [PubMed]
16. Casellas, E.; Bech, J.; Veciana, R.; Pineda, N.; Rigo, T.; Miró, J.R.; Sairouni, A. Surface precipitation phase discrimination in complex terrain. *J. Hydrol.* **2021**, *592*, 125780. [CrossRef]
17. Crochet, P. Enhancing radar estimates of precipitation over complex terrain using information derived from an orographic precipitation model. *J. Hydrol.* **2009**, *377*, 417–433. [CrossRef]

18. Henn, B.; Newman, A.J.; Livneh, B.; Daly, C.; Lundquist, J.D. An assessment of differences in gridded precipitation datasets in complex terrain. *J. Hydrol.* **2018**, *556*, 1205–1219. [CrossRef]
19. de la Fraga, P.; Del-Toro-Guerrero, F.J.; Vivoni, E.R.; Cavazos, T.; Kretzschmar, T. Evaluation of gridded precipitation datasets in mountainous terrains of Northwestern Mexico. *J. Hydrol. Reg. Stud.* **2024**, *56*, 102019. [CrossRef]
20. Zhao, R.; Chen, B.; Zhang, W.; Yang, S.; Xu, X. Formation mechanisms of persistent extreme precipitation events over the eastern periphery of the Tibetan Plateau: Synoptic conditions, moisture transport and the effect of steep terrain. *Atmos. Res.* **2024**, *304*, 107341. [CrossRef]
21. Dheebakaran, G.; Geethalakshmi, V.; Ramanathan, S.P.; Ragunath, K.P.; Kokilavani, S. WRF's microphysics options on the temporal variation in the accuracy of cluster of village level medium range rainfall forecast in Tamil Nadu. *J. Agrometeorol.* **2022**, *24*, 133–137. [CrossRef]
22. Tiwari, S.; Kar, S.C.; Bhatla, R. Dynamic downscaling over western Himalayas: Impact of cloud microphysics schemes. *Atmos. Res.* **2018**, *201*, 1–16. [CrossRef]
23. Xue, L.; Fan, J.; Lebo, Z.J.; Wu, W.; Morrison, H.; Grabowski, W.W.; Chu, X.; Geresdi, I.; North, K.; Stenz, R.; et al. Idealized Simulations of a Squall Line from the MC3E Field Campaign Applying Three Bin Microphysics Schemes: Dynamic and Thermodynamic Structure. *Mon. Weather Rev.* **2017**, *145*, 4789–4812. [CrossRef]
24. Pflaum, J.C.; Johnson, H.L.; Poellot, M.R. Preliminary Investigations of a Dynamic Seeding Strategy for Oklahoma Convective Clouds. *J. Weather. Modif.* **1989**, *21*, 54–61.
25. Dobashi, Y.; Kusumoto, K.; Nishita, T.; Yamamoto, T. Feedback control of cumuliform cloud formation based on computational fluid dynamics. *Acm Trans. Graph.* **2008**, *27*, 1–8. [CrossRef]
26. Dale, R.D. Improving the anelastic approximation. *J. Atmos. Sci.* **1989**, *46*, 1453–1461.
27. Grell, G.A.; Dudhia, J.; Stauffer, D.R. A description of the fifth-generation Penn State/NCAR Mesoscale Model (MM5). *NCAR Tech. Note* **1994**, *10*, 1–121. [CrossRef]
28. Heymsfield, A.J.; Sabin, R.M. Cirrus Crystal Nucleation by Homogeneous Freezing of Solution Droplets. *J. Atmos. Sci.* **1989**, *46*, 2252–2264. [CrossRef]
29. Rothenberg, D.; Wang, C. Metamodeling of Droplet Activation for Global Climate Models. *J. Atmos. Sci.* **2016**, *73*, 1255–1272. [CrossRef]
30. Garcia-Dorado, I.; Aliaga, D.G.; Bhalachandran, S.; Schmid, P.; Niyogi, D. Fast Weather Simulation for Inverse Procedural Design of 3D Urban Models. *Acm Trans. Graph.* **2017**, *36*, 21. [CrossRef]
31. Çelik, F.; Marwitz, J.D. Droplet Spectra Broadening by Ripening Process. Part I: Roles of Curvature and Salinity of Cloud Droplets. *J. Atmos. Sci.* **2010**, *56*, 3091–3105. [CrossRef]
32. Yang, F.; Kollias, P.; Shaw, R.A.; Vogelmann, A.M. Cloud droplet size distribution broadening during diffusional growth: Ripening amplified by deactivation and reactivation. *Atmos. Chem. Phys.* **2018**, *18*, 7313–7328. [CrossRef]
33. Wei, X.; Liu, Y.; Guo, J.; Chang, X.; Li, H. Applicability Study of Euler–Lagrange Integration Scheme in Constructing Small-Scale Atmospheric Dynamics Models. *Atmosphere* **2024**, *15*, 644. [CrossRef]
34. Abdi, D.S.; Giraldo, F.X.; Constantinescu, E.M.; Carr, L.E.; Wilcox, L.C.; Warburton, T.C. Acceleration of the implicit-explicit non-hydrostatic unified model of the atmosphere (NUMA) on manycore processors. *Int. J. High Perform. Comput. Appl.* **2017**, *33*, 242–267. [CrossRef]
35. Bolton, D. The Computation of Equivalent Potential Temperature. *Mon. Weather Rev.* **1980**, *108*, 1046–1053. [CrossRef]
36. Wei, X.; Liu, Y.; Chang, X.; Guo, J.; Li, H. Mechanism Study of Two-Dimensional Precipitation Diagnostic Models Within a Dynamic Framework. *Atmosphere* **2025**, *16*, 380. [CrossRef]
37. Schneider, T.; O'Gorman, P.A.; Levine, X.J. Water vapor and the dynamics of climate change. *Rev. Geophys.* **2010**, *48*, RG3001. [CrossRef]
38. Seifert, A.; Beheng, K.D. A double-moment parameterization for simulating autoconversion, accretion and selfcollection. *Atmos. Res.* **2001**, *59*, 265–281. [CrossRef]
39. Nober, F.J.; Graf, H.F. A new convective cloud field model based on principles of self-organisation. *Atmos. Chem. Phys.* **2004**, *4*, 3669–3698. [CrossRef]
40. Berry, E.X. Modification of the warm rain process. In Proceedings of the National Conference on Weather Modification Amer, Albany, NY, USA, 28 April–1 May 1968.
41. Lebo, Z.J.; Seinfeld, J.H. A continuous spectral aerosol-droplet microphysics model. *Atmos. Chem. Phys. Discuss.* **2011**, *11*, 23655–23705. [CrossRef]

*agriculture*

MDPI

*Article*

# Sensitivity of Soil Moisture Simulations to Noah-MP Parameterization Schemes in a Semi-Arid Inland River Basin, China

**Yuanhong You [1,2,3], Yanyu Lu [1,*], Yu Wang [3], Houfu Zhou [1], Ying Hao [4], Weijing Chen [5] and Zuo Wang [2]**

1 Anhui Province Key Laboratory of Atmospheric Science and Satellite Remote Sensing, Anhui Institute of Meteorological Sciences, Hefei 230031, China
2 School of Geography and Tourism, Anhui Normal University, Wuhu 241002, China
3 School of Earth and Space Sciences, University of Science and Technology of China, Hefei 230031, China
4 Huaihe River Basin Meteorological Center, Anhui Meteorological Bureau, Hefei 230031, China
5 School of Aeronautic Engineering, Changsha University of Science and Technology, Changsha 410000, China
* Correspondence: ahqxlyy@163.com

**Abstract:** Soil moisture simulations in semi-arid inland river basins remain highly uncertain due to complex land–atmosphere interactions and multiple parameterization schemes in land surface models. This study evaluated the ability of the Noah-Multiparameterization Land Surface Model (Noah-MP) to simulate soil moisture at meteorological sites representing the upstream, midstream and downstream regions of a semi-arid inland river basin with contrasting climates. A large physics-ensemble experiment (17,280 simulations per site) combining different parameterization schemes for 10 main physical processes was conducted. Natural selection, Tukey's test and uncertainty contribution analysis were applied to identify sensitive processes and quantify their contributions to simulation uncertainty. Results indicate that Noah-MP captures soil moisture variability across the basin but with notable biases. Three physical processes—frozen soil permeability, supercooled liquid water in frozen soil and ground resistance to sublimation—were sensitive at all sites, whereas radiation transfer and surface albedo were consistently insensitive. At the upstream and midstream sites, supercooled liquid water contributed about half of the ensemble uncertainty, and at the downstream site ground resistance to sublimation contributed roughly 51%. These findings reveal which physical processes most strongly affect Noah-MP soil moisture simulations in semi-arid basins and provide guidance for improving parameterization schemes to reduce uncertainty.

**Keywords:** multi-parameterization scheme ensemble simulation; sensitivity analysis; uncertainty contribution rate; soil moisture

## 1. Introduction

Soil moisture is a crucial component of the terrestrial water cycle and a significant influencing factor in the distribution of surface radiation [1]. Accurately estimating soil moisture is of paramount importance for understanding surface conditions, studying land-atmosphere interactions, and managing agricultural production [2–4]. It serves as a key variable in studies related to land-atmosphere interactions and hydrological cycles, exerting important impacts on surface energy fluxes, runoff, radiation balance, and material transport [5,6]. As a critical parameter in land surface processes, soil moisture plays a vital role in climate and regional environmental changes. It affects these changes by altering surface reflectance, soil heat capacity, surface evaporation, transpiration, and vegetation


https://doi.org/10.3390/agriculture1521228[6]

growth conditions, leading to the redistribution of surface energy and moisture. This, in turn, modifies the sensible, latent, and longwave radiation fluxes from the surface to the atmosphere. Furthermore, the thermal properties and water transport processes within the soil were influenced by the changes in soil moisture, causing variations in various surface parameters and subsequently impacting climate [7]. Against the backdrop of ongoing global warming, accurately simulating soil moisture and understanding its spatial-temporal distribution characteristics and intrinsic properties are critical for comprehending the Earth's surface response to climate change. This knowledge holds significant scientific importance for sustainable water resource utilization, regional climate change predictions, runoff estimation, and agricultural production irrigation.

Numerical simulation is a crucial approach for studying the spatiotemporal distribution characteristics of soil moisture. Internationally, numerous land surface models focusing on various surface elements have emerged, and those that incorporate multiple parameterization schemes for diverse surface physical processes can accurately simulate soil moisture. The Noah-MP model represents a new generation of land surface models developed based on the Noah model framework [8,9], incorporating various parameterization schemes for different physical processes and plays a reliable candidate for simulating soil moisture [10,11]. In recent years, researchers have extensively evaluated the model at both single-point and regional scales [11–13]. For instance, the runoff simulation performance of the Noah-MP model was assessed in the Mississippi River Basin, revealing its high sensitivity to surface dryness coefficients, hydraulic conductivity, and saturated soil moisture [14]. To further clarify the simulation performance of the Noah-MP model, a study evaluated its ability to simulate water and energy fluxes in the continental United States and results demonstrated that the Noah-MP model effectively captures the spatiotemporal distribution characteristics of net radiation, snow cover area, and runoff [15]. Compared with the original Noah model, significant improvements were found in simulating surface fluxes, dry season surface temperature, snow water equivalent, snow depth, and runoff [10,16,17]. Due to the high applicability and excellent land surface process modeling performance of the Noah-MP model, it has gradually become a strong candidate for soil moisture simulation. The water budget components including soil moisture simulated by Noah-MP model were evaluated in main river basin, China and the results showed that the simulations effectively reproduce the observed spatial pattern of soil moisture during the warm seasons across most regions, however, the results also demonstrated that there is significant room for improvement in the model's simulation performance [18]. In response to the issue of Noah-MP does not consider surface water ponding and overland flow, lateral terrestrial water flow schemes have been developed for Noah-MP to improve the simulation of soil moisture [19]. The bias of the Noah-MP model in simulating soil water-heat transport in freeze-thaw process were investigated and results exhibited that the biases in simulating soil moisture in freeze-thaw process are difficult to be eliminated [20]. Indeed, assimilating satellite observation data, like leaf area index and SMAP soil retrievals [21–23], Sentinel-1 backscatter observations [24], remotely sensed MODIS evapotranspiration [25], vegetation optical depth retrievals [26], etc. into the Noah-MP model can effectively improve soil moisture simulation results. However, the soil moisture data assimilation technique aims to obtain the optimal posterior estimation of the model state by fully considering model errors and observation errors, in order to improve the accuracy of soil moisture simulations, without adjusting the model's physical parameterization schemes. Efforts are still needed to reduce errors caused by model structural uncertainties by enhancing the model's ability to describe actual physical processes. The most distinctive feature of the Noah-MP model is its integration of multiple parameterization schemes for the same physical process, greatly enhancing its applicability [11,27]. However, this also makes it difficult to determine which combination of schemes results in the least structural uncertainty and simulation error. To address this issue, the sensitivity of sensible and latent heat simulations to parameterization schemes was analyzed by conducting a total number of 4608 Noah-MP physics-ensemble simulations [13]. Similar studies have been conducted to explore the sensitivity of snow simulation results to parameterization schemes and the uncertainties in the results under different snow climate conditions worldwide through ensemble simulations with multiple parameterization schemes [28]. Additionally, the sensitivity of surface heat flux to parameterization schemes has been investigated [29]. The sensitivities of the simulated runoff to parameterizations were quantified by varying the optional parameterization schemes of six land surface processes [30].

Although researchers have explored the sensitivity of simulation results to parameterization schemes for various surface elements, there are currently few studies investigating the sensitivity of soil moisture simulation results to the Noah-MP model parameterization schemes under different surface types and climate conditions. Considering this, based on previous research efforts, this study selects the Heihe River Basin (HRB) in China, which encompasses various underlying surfaces and climate types, as the study area. Soil moisture simulation experiments are conducted in the upper, middle, and lower reaches of the Heihe River to evaluate the soil moisture simulation performance of the Noah-MP model. Building upon this, given that the Noah-MP model incorporates multiple parameterization schemes for its key physical processes, resulting in a diverse set of parameterization scheme combinations, the study designs experiments using multiple parameterization schemes to explore the sensitivity of soil moisture simulation results to different parameterization schemes under various underlying surface conditions. This study is an attempt to provide a scientific basis for improving the soil moisture simulation performance of the Noah-MP model. All simulations in this study were based on meteorological and soil moisture observations from sites in the upper, middle, and lower reaches of the HRB. The main goal of this research is to address the following issues: (1) How does the Noah-MP model perform in simulating soil moisture in the HRB? (2) What is the sensitivity of soil moisture simulations to parameterization schemes?

Section 2 contains the study sites, input meteorological dataset, the Noah-MP model and the sensitivity analysis methods. Section 3 provides the results and discussions of the experiment. Section 4 summarizes the findings and presents main conclusion of this study.

## 2. Data and Methods

### 2.1. Study Sites and Data

The Heihe river is the second-largest inland river in China, located in the Qilian Mountains and the middle section of the Hexi Corridor. It spans approximately 819 KM, with coordinates ranging from 97°24′ to 102°10′ E longitude and 37°41′ to 42°42′ N latitude. As shown in Figure 1, the southern part of HRB is the Qilian Mountain area, has an average elevation of over 3600 m and is the source of the Heihe river's main stream. The central part is characterized by a corridor plain with fault basins, at an elevation of around 1200 m. The northwest part of HRB consists of mid-low mountains, with elevations ranging from 1400 to 2800 m, and the whole basin covers a total area of over 1300 square kilometers [31]. The upper mountainous region serves as the runoff-producing area of the Heihe river, with an average annual precipitation of 330 mm, good vegetation, and a humid, cold climate. The middle reaches rely on the Heihe river for water supply and have developed into an artificial oasis, making this area has abundant light and heat resources. The lower reaches are typical arid regions characterized by dry climate, sparse vegetation, and low annual precipitation. The complex surface features of the HRB provide favorable conditions for studying land surface processes. Li et al. (2009) [32] conducted the Heihe River Basin

Upper Reaches Cold Region Hydrological Remote Sensing and Ground Synchronous Observation Experiment, establishing a series of hydrological and ecological observation stations and collecting a substantial amount of observational data, laying a foundation for further in-depth research on land surface processes in the HRB.

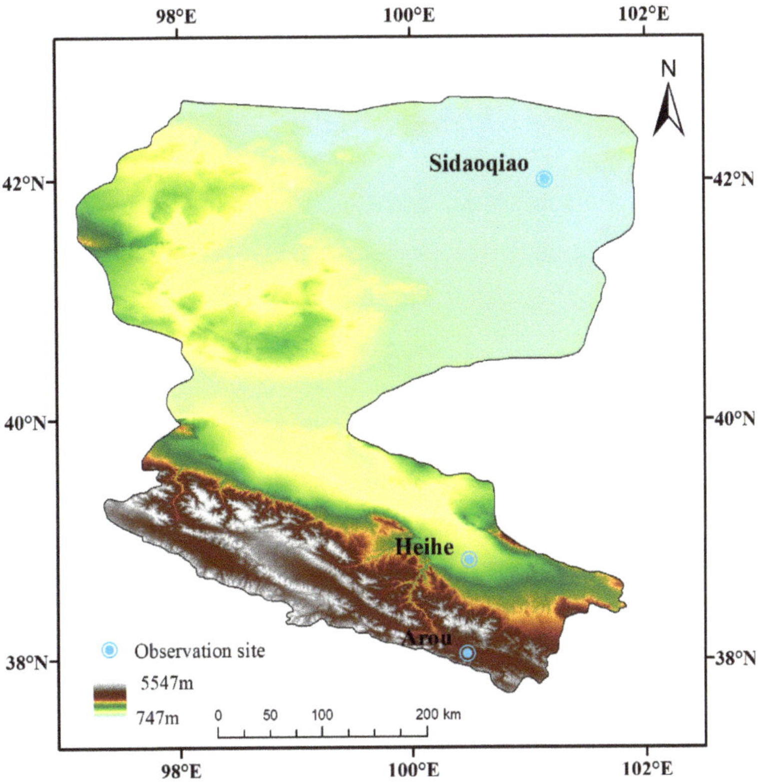


**Figure 1.** Terrain height of the HRB and the location of experiment sites.

Considering the high spatial heterogeneity of land cover types in the HRB and the significant differences in climate types across different sections of the basin, this study selected a typical meteorological station in the upper, middle, and lower reaches of the HRB to conduct soil moisture simulation experiment, respectively. This was done to compare the performance of the Noah-MP model in simulating soil moisture under different climate and land cover conditions. The three selected sites are the Arou, Heihe, and Sidaoqiao site. Arou site is located in the upper reaches of the HRB, with coordinates 100.46° E, 38.03° N, and an elevation of 3032.8 m. The underlying surface type at Arou site is subalpine mountain meadow. According to soil texture information for the HRB provided by the National Cryosphere Desert Data Center, the soil type at this site is dark cold calcareous soil. The Heihe site is located in the middle reaches of the HRB, with coordinates 100.48° E, 38.83° N, and an elevation of 1560 m. The underlying surface at Heihe site is crop land, and the soil type is sandy loam. The surrounding area is mainly farmland planted with corn. In addition to observing conventional meteorological elements, this site also monitored soil moisture at different depths. Sidaoqiao site is located in the lower reaches of the HRB, with coordinates 101.14° E, 42.00° N, and an elevation of 873 m. The underlying surface is dominated by sparse vegetation, and the soil type is saline-alkali soil. This area has a typical extremely arid continental climate characterized by dryness, low rainfall, high evaporation, and significant diurnal temperature variation. Unlike the upper and middle reaches, the lower reaches have sparse vegetation and suffer from serious desertification and salinization due to wind erosion. The locations of the three sites are signed in Figure 1, and the detail information of the three sites are listed in Table 1.

**Table 1.** Geographic features of the three sites in HRB.

| Site Name | Arou | Heihe | Sidaoqiao |
|---|---|---|---|
| Latitude (N) | 38.03 | 38.83 | 42.00 |
| Longitude (E) | 100.46 | 100.48 | 101.14 |
| Elevation (m) | 3032.8 | 1560 | 873 |
| Study Period | 2016 | 2017 | 2016 |
| Climate Type | Highland Continental Climate | Temperate Continental Arid Climate | Temperate Monsoon Climate |
| Vegetation Type | Grassland | Crop | Desert |
| Soil Type | Dark Cold Calcareous Soil | Coral Sandy Soil | Mountain Shrub Meadow Soil |
| Evaluation Data | Soil Moisture | Soil Moisture | Soil Moisture |

### 2.2. Noah-MP Default Parameterization and Physics Ensemble Numerical Experiment

The Noah-MP model has undergone significant improvements based on the original Noah model framework. Compared to the original Noah model, Noah-MP has redesigned the internal structure of the model. The most notable improvement is the design of an independent vegetation canopy to separately calculate canopy temperature and surface temperature, and to separate the vegetation canopy energy balance from the surface energy balance, making the calculation of canopy temperature more reasonable [10,11]. The improvements to the Noah-MP model on the basis of the Noah model mainly include a 2-m-deep soil column, a snowpack stratified according to snow depth, and an unconfined aquifer. The model divides the soil column into four layers, with soil thicknesses of 0.1 m, 0.3 m, 0.6 m, and 1.0 m respectively [11]. Additionally, the Noah-MP model has incorporated several new physical processes into the original Noah model framework, such as dynamic water table [10,33], interactive vegetation canopy [34], and an unconfined aquifer for groundwater storage [11].The most notable feature of the Noah-MP model is that it provides 2–4 different parameterization schemes for the major physical processes in the model, greatly enhancing the applicability of the Noah-MP model under different environmental conditions. The newest version of the Noah-MP 5.0 includes multiple parameterization schemes for physic options, The main physic options which may affect the soil moisture simulations are shown in Table 2. Considering that the Noah-MP model is currently the most widely used in different land surface simulation experiment, this study examines the performance of the Noah-MP model in simulating soil moisture at the selected sites.

The meteorological driving elements required by the Noah-MP model include near-surface air temperature, air pressure, relative humidity, precipitation, downward longwave and shortwave radiation, as well as near-surface wind speed and wind direction. Model initialization requires surface datasets such as soil type, land cover type, and elevation. All three sites selected in this study have high-precision meteorological and soil moisture observation data. The time scale of the meteorological observation data used as model driving data is 1 h, meeting the model's requirements for driving data in the experiment. A spin-up simulation was performed with meteorological forcing data from the year prior to the study period to attain soil-state equilibrium. The equilibrium criterion followed Gao et al. (2015) [12] and Cai et al. (2014) [14], requiring that the difference between consecutive annual mean values from two one-year simulations be less than 0.1% of the mean. The selected time period data in the study have all undergone quality control, ensuring data quality. All the site data can be downloaded from the National Tibetan Plateau Data Center (https://data.tpdc.ac.cn).

**Table 2.** The physical process options employed for ensemble simulation in Noah-MP model.

| Physical Process | Parameterization Schemes |
|---|---|
| Soil moisture factor controlling stomatal resistance (BTR) | * 1. Noah 2. CLM 3. SSiB |
| Surface layer drag or exchange coefficient (SFC) | * 1. M-O 2. Original Noah (Chen97) |
| Frozen soil permeability (INF) | * 1. Linear effects, more permeable 2. Nonlinear effects, less permeable |
| Soil supercooled liquid water (FRZ) | * 1. No iteration 2. Koren's iteration |
| Canopy radiation transfer (RAD) | 1. Modified two-stream 2. Two-stream applied to grid-cell (gap = 0) * 3. Two-stream applied to vegetated fraction (gap = 1-VegFrac) |
| Snow surface albedo (ALB) | * 1. BATS snow albedo 2. CLASS snow albedo |
| Partitioning precipitation into rainfall and snowfall (PCP) | * 1. Jordan (1991) 2. BATS 3. Noah 4. Use WRF microphysics output 5. Wet-bulb temperature-based |
| Lower boundary condition of soil temperature (TBOT) | 1. Zero-flux scheme * 2. Noah scheme |
| Snow or soil temperature time scheme (only layer 1) (TEMP) | * 1. Semi-implicit; flux top boundary condition 2. Full-implicit (original Noah); temperature top boundary condition 3. Same as 1, but snow cover for skin temperature calculation |
| Ground resistant to evaporation or sublimation (SRE) | * 1. Sakaguchi and Zeng, 2009 2. Sellers (1992) 3. Adjusted Sellers to decrease RSURF for wet soil 4. Option 1 for non-snow; rsurf = rsurf_snow for snow |

* Represents the default parameterization scheme.

### 2.3. Analysis and Evaluation Methods

Soil moisture was recorded as a simulation variable, and the simulation performance of Noah-MP model was evaluated by comparing the soil moisture simulation results with the observed soil moisture data. The Root Mean Square Error (RMSE) between the simulations and observations is used as the evaluation metric for simulation accuracy,

$$RMSE = \sqrt{\frac{1}{N}\sum_{i=1}^{n}\left[sim(i) - obs(i)\right]^2} \quad (1)$$

where $N$ is the number of samples, $sim$ is the simulated value, and $obs$ is the observed value. The smaller the RMSE value, the closer the simulated value is to the observed value, indicating higher simulation accuracy. In the ensemble simulation experiment, two sensitivity analysis methods, Natural Selection and Tukey's Test, were used to analyze the sensitivity of the parameterization schemes based on the simulation results. Both methods evaluate the model's simulation performance by calculating the RMSE between the simulated and observed values. They analyze different physical processes one by one to identify the soil moisture simulations sensitive to parameterization schemes. The two sensitivity analysis methods are introduced as follows.

Natural Selection: First, the RMSE of simulated soil moisture for each combination scheme was calculated based on the observed values, and all the RMSE values was sorted in ascending order. The top 5% (about 864 ensemble members) and the bottom 5% (about 864 ensemble members) were extracted from the ensemble. Obviously, the combination schemes with RMSE values in the top 5% perform better than those in the bottom 5%. Therefore, the set of all RMSE values in the top 5% was labeled as "Best members", and the set of all RMSE values in the bottom 5% was labeled as "worst members". Based on the newly formed sets, the frequency of a given scheme for each physical process occurring in the "best members" and "worst members" sets were counted. Obviously, the higher the frequency of a given scheme in the "best members", the greater its contribution to improving simulation accuracy. Conversely, the higher the frequency of a given scheme in the "worst members", the greater its contribution to reducing simulation accuracy. This method allows for a preliminary macro-level assessment of the relative merits of different parameterization schemes within the same physical process and helps determine the sensitivity of the simulated soil moisture to the parameterizations based on frequency differences.

Tukey's Test: as a method of hypothesis testing, is used to examine the sensitivity of simulated soil moisture to parameterizations. The concept of this method is to use hypothesis testing to compare whether there are significant differences between parameterizations within a given physical process. If significant differences exist, the simulated soil moisture is sensitive to the given parameterization scheme. First, the mean RMSE values for each parameterization scheme was calculated, $\overline{Y_1}, \overline{Y_2}, \cdots, \overline{Y_a}$. Then, the method of hypothesis testing was used to compare whether the differences between each pair of means, $u_i$ and $u_j (i \neq j)$, are significant. The null hypothesis and alternative hypothesis are shown in Equation (2).

$$\begin{aligned} H_0 &: u_i = u_j \\ H_1 &: u_i \neq u_j \end{aligned} \tag{2}$$

The indicator of significant differences between parameterizations is

$$q = \frac{\overline{Y}_{\max} - \overline{Y}_{\min}}{\sqrt{\frac{MS_E}{2}\left(\frac{1}{n_i} + \frac{1}{n_i}\right)}} = \frac{\overline{Y}_{\max} - \overline{Y}_{\min}}{\sqrt{\frac{SS_E}{2(N-a)}\left(\frac{1}{n_i} + \frac{1}{n_i}\right)}} \tag{3}$$

where $\overline{Y}_{\max}$ represents the larger mean RMSE value of the two parameterization schemes being compared, and $\overline{Y}_{\min}$ is the smaller one. $n_i$ and $n_j$ are the sample sizes of the $i-th$ and $j-th$ parameterizations, respectively. $MS_E$ represents the mean square error, calculated using Equation (4).

$$MS_E = \frac{1}{N-a}\sum_{k=1}^{a}\sum_{l=1}^{n_k}\left(Y_{kl} - \overline{Y}_k\right)^2 \tag{4}$$

$SS_E$ is the sum of squares, which can be calculated using Equation (5),

$$SS_E = \sum_{k=1}^{a}\sum_{l=1}^{n_k}\left(Y_{kl} - \overline{Y}_k\right)^2 \tag{5}$$

where $N$ is the total sample size including all schemes, and $N-a$ is the degrees of freedom. If the mean RMSE values of the two parameterization schemes satisfy Equation (6), then the difference between them is deemed statistically significant.

$$\left|\overline{Y}_i - \overline{Y}_j\right| \Big/ \sqrt{\frac{MS_E}{2}\left(\frac{1}{n_i} + \frac{1}{n_i}\right)} > q_\alpha(a, N-a) \tag{6}$$

where $\alpha$ is the significance level of the hypothesis test, set to 0.01 in this study. The value of $q_\alpha(a, N-a)$ can be obtained by looking up the hypothesis test distribution table. For example, the physical process of radiation transfer has three parameterizations (i.e., $a = 3$), the hypothesis test method is used to examine whether there is a significant difference between the first and second parameterization scheme. The null hypothesis and alternative hypothesis are set as $H_0 : u_2 = u_3$ and $H_1 : u_2 \neq u_3$, respectively. The total sample size $N$ is 17,280, and the sample size for each parameterization scheme is 5760 (i.e., $n_1 = n_2 = n_3 = 5760$). The degrees of freedom $N-a$ is 17,277, and the mean square error is $MS_E = \sum_{k=1}^{3}\sum_{l=1}^{n_k}\left(Y_{kl} - \overline{Y}_k\right)^2 \Big/ 17277$. The significance level is $\alpha = 0.01$. The value of $q_{0.01}(3, 17277)$ can be obtained by looking up the hypothesis test distribution table. If the values of $\overline{Y}_i$ and $\overline{Y}_j$ satisfy Equation (6), the null hypothesis will be rejected at the significance level of 0.01. Conversely, the null hypothesis will be accepted at this significance level.

Natural Selection and Tukey's Test are two sensitivity analysis methods that determine the sensitivity of parameterization schemes from a macro and micro perspective, respectively. In this study, if either method identifies sensitivity in a parameterization scheme, the simulated soil moisture is considered sensitive to that scheme.

Uncertainty Contribution Analysis: To elucidate the impact of sensitive physical processes to the uncertainty in ensemble simulation results, this study employed a comprehensive approach to quantify the contribution of individual physical processes to the overall uncertainty in soil moisture simulations. Ten main physical processes within Noah-MP model were selected to configure the multi-parameterization scheme ensemble simulation experiment, with each process represented by multiple parameterization schemes, resulting in a total of 17,280 soil moisture simulations. For instance, the BTR physical process has three different parameterization schemes. To quantify BTR's contribution to overall uncertainty, a conditional uncertainty analysis was performed. The simulation outputs were grouped according to the BTR parameterization scheme used: BTR1, BTR2, and BTR3. For each group, the mean simulation result was computed,

$$\overline{R}_k = \frac{1}{n_k}\sum_{i=1}^{n_k} R_i^k \tag{7}$$

where $\overline{R}_k$ represents the mean of the simulation results for the $k$th BTR scheme, $n_k$ is the number of simulations using the $k$th BTR scheme, and $R_i^k$ is the individual simulation result for the $i$th simulation under the $k$th BTR scheme. The between-group variance, representing the variability in simulation results due to differences in the BTR schemes, was calculated as,

$$BV = \frac{1}{K}\sum_{k=1}^{K} \left(\overline{R}_k - \overline{R}\right)^2 \tag{8}$$

where $K$ is the total number of BTR schemes, and $\overline{R}$ is the overall mean of all simulation results across all BTR schemes,

$$\overline{R} = \frac{1}{N}\sum_{i=1}^{N} R_i \tag{9}$$

with $N$ being the total number of simulations. The total variance, representing the overall model uncertainty, was calculated as,

$$TV = \frac{1}{N}\sum_{i=1}^{N} \left(R_i - \overline{R}\right)^2 \tag{10}$$

Finally, the contribution of the BTR process to the total model uncertainty was quantified as the ratio of the between-group variance to the total variance,

$$Con = BV/TV \tag{11}$$

This ratio reflects the proportion of total variability in soil moisture simulations attributable to differences in the BTR parameterization schemes. By applying this methodology, the contribution of sensitive processes to the uncertainty of ensemble simulations can be quantified, offering valuable insights into their role in driving model uncertainty.

# 3. Results

## 3.1. Soil Moisture Simulated by Default Parameterization Combination

This section evaluates the performance of the Noah-MP model in simulating soil moisture across different layers at three sites-Arou, Heihe, and Sidaoqiao, which come from the upper, middle and lower reaches of the HRB. The comparison between simulated and observed soil moisture data covers the period from January to December 2016 in Arou and Sidaoqiao site, with an extension into 2017 in Heihe site. The analysis is performed across four distinct soil layers, ranging from the surface layer (0–10 cm) to deeper layers, progressively down to the fourth layer. Figure 2 shows the comparison of soil moisture at the three sites simulated by Def with the corresponding observations.

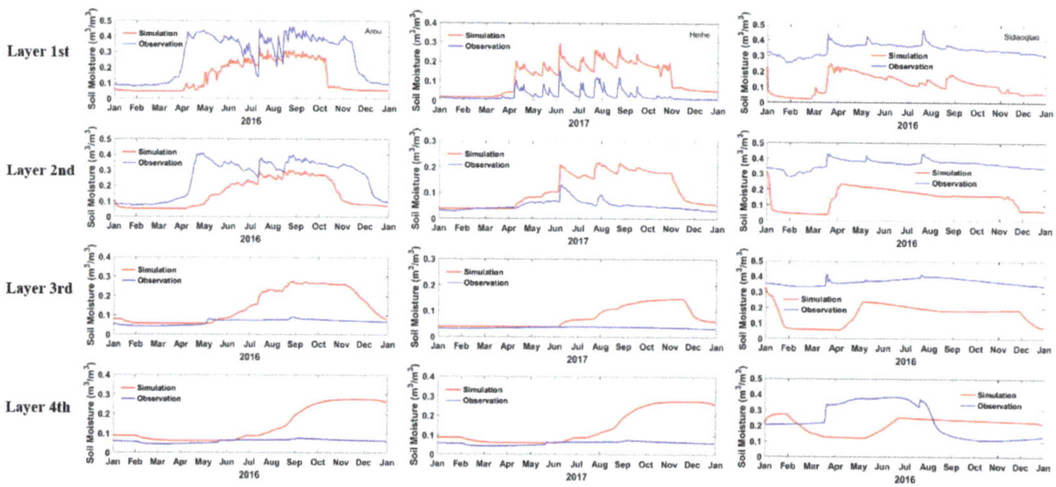


**Figure 2.** Variation of soil moisture across three sites in the HRB, observed and simulated, by default parameterization combination, from top to bottom are the first to fourth soil layer, respectively.

At the Arou site, the simulation of soil moisture exhibits varying degrees of accuracy across the different layers. The first layer, representing the surface soil moisture, shows significant variability throughout the year, particularly during the growing season (May to September). While the model captures the overall trend, there are substantial deviations between the simulated and observed values, especially during the wetter months (July and August). The model tends to underestimate soil moisture during these periods, as indicated by the consistently higher simulated values compared to the observations. Moving to the second layer, the model's performance slightly improves, though discrepancies remain evident. The underestimation trend persists, particularly during the peak moisture periods. The third layer shows a similar pattern, with the simulation capturing the seasonal fluctuations but demonstrating an overestimation tendency, particularly during the latter half of the year. The fourth layer presents the most significant challenge for the model. Here, the simulation not only overestimates soil moisture but also fails to capture the observed drying trends seen in the latter part of the year. Since the Arou site is located in the cold mountainous region of the upper reaches of the HRB, the freeze-thaw process of the soil is very intense, and the liquid water content in the soil rapidly increases when the temperature rises. The soil moisture simulation results for the first and second layers at the Arou site indicate that the Noah-MP model can adequately simulate the soil freeze-thaw process. However, it should be noted that the soil moisture during this process is consistently underestimated. Furthermore, the model's inability to accurately simulate the deeper soil moisture dynamics suggests potential issues in the parameterization of subsurface processes within the Noah-MP model, at least this is the case in the cold mountainous region.

The Heihe site, located in the midstream oasis of HRB, presents a more complex scenario. Since this area is a desert oasis, the soil moisture content is relatively low and is greatly influenced by precipitation. Unlike the Arou site, the Noah-MP model tends to overestimate soil moisture at the Heihe site. As shown in the Figure 2, in the first layer, the model exhibits a mixed performance. While it captures the seasonal variation, including the sharp increase in soil moisture during the monsoon season (July and August), it again tends to overestimate the moisture content during the peak season. For the second layer, the model shows a decline in performance. The simulated soil moisture during the wet season is notably higher than the observed values, with a noticeable lag in capturing the drying trend post-monsoon. Soil moisture in the third and fourth layers at the Heihe site is very low, with deep soil moisture showing almost no variation throughout the year and no occurrence of freeze-thaw processes. As shown in Figure 2, the overestimation in the third and fourth layers are significant and the simulated values remain high throughout the latter part of the year, contrary to the observed data. Moreover, the fourth layer exhibits the most considerable deviation from the observations. The model not only overestimates soil moisture but also shows a delayed response to the drying trends observed from September onwards. This suggests that the default parameterization scheme combination of Noah-MP model for deeper soil layers may not be adequately capturing the subsurface hydrological processes at Heihe site.

At the Sidaoqiao site, located in the downstream desert area, the model's performance is somewhat consistent with the patterns observed at the other two sites. The model is able to accurately simulate the trend of the first layer of soil moisture, but the simulation results show a significant underestimation. This underestimation continues into the post-monsoon period, although the trend matches the observed data more closely than at the other two sites. In the second layer, the model begins to deviate more noticeably from the observations, and the model fails to capture the rapid decrease in soil moisture observed towards the end of the year. Compared to the other two sites, soil moisture in the third and fourth layers at the Sidaoqiao site is noticeably higher and more stable. The third layer follows a similar trend, the underestimation here is substantial, and the model does not adequately reflect the observed soil moisture dynamics. The fourth layer presents the most significant challenge, similar to the Arou and Heihe sites. The model almost underestimates soil moisture throughout the year, and the drying trend observed in the data is almost entirely absent from the simulation. Previous studies have shown that the annual mean soil moisture (SM) in the driest and wettest regions was underestimated, while SM was generally overestimated in other regions [30]. Here, we found that the soil moisture simulations at the Arou and Sidaoqiao stations were underestimated, whereas at the Heihe station the soil moisture simulation was overestimated, even though the HRB is located in an arid and semi-arid region.

### 3.2. Sensitivities of Physical Parameterization Schemes

The simulation results of the default parameterization scheme combination at different sites indicate that the accuracy of soil moisture simulations is relatively low across all sites, whether upstream, midstream, or downstream. This is especially true for the third and fourth soil layers, where the model almost fails to accurately capture the soil moisture trends. There are many methods to improve the accuracy of soil moisture simulations, such as optimizing model parameters, adjusting the model structure, or using data assimilation techniques to obtain the optimal posterior estimates of soil moisture. However, the goal of this study is not to improve the accuracy of the Noah-MP model's soil moisture simulations in the HRB. Instead, we are more interested in the sensitivity of soil moisture simulation results to the parameterization schemes. Given the lower accuracy of the Noah-MP model in simulating soil moisture in the third and fourth layers, this study focuses only on the sensitivity of shallow soil moisture (first layer) simulation results to parameterization schemes and compares the sensitivity under different climate types in the HRB. To this end, two classic sensitivity analysis methods were employed to analysis the sensitivity of soil moisture simulations to parameterization schemes at different sites.

#### 3.2.1. Natural Selection Results

First, the shallow soil moisture observation data from three sites were used to calculate the RMSE values of soil moisture simulations for all parameterization scheme combinations. Then, all the RMSE values were sorted in ascending order, the members concentrated below the fifth percentile of RMSE for soil moisture were considered as "best members", and the "worst members" contains the members which are those above the 95th percentile. Subsequently, the frequency of different parameterization schemes being selected were determined for the two groups. The selected frequency of different parameterization schemes for soil moisture was shown in Figure 3, and above the horizontal axis in the subfigure is the frequency of each parameterization scheme for the best members, and below is the frequency of the worst members. The frequency with which different parameterization schemes within the same physical process are selected in the ensemble simulation results at the same site varies significantly, as seen in Figure 3. For example, at the Arou site, the selection frequency of parameterization schemes for the INF, FRZ, PCP, and TEMP physical processes shows considerable differences. Conversely, some physical processes exhibit minimal differences in the selection frequency of their parameterization schemes. For instance, the two parameterization schemes of the ALB physical process show almost no difference in their selection frequency at the Heihe and Sidaoqiao sites; the frequency with which these two schemes are selected in the "best members" and "worst members" is nearly identical. Additionally, it can be observed that the selection frequency of parameterization schemes for partial physical processes shows little difference at some sites but significant differences at others, like the parameterization schemes for the PCP physical process display large differences in selection frequency at the Arou site, but not at the Heihe and Sidaoqiao sites. Clearly, the significant differences in the selection frequencies of parameterization schemes within the same physical process between the "best members" and "worst members" indicate that altering the parameterization scheme for this process can lead to substantial variations in soil moisture simulation results. And this suggests a high sensitivity of soil moisture simulations to the choice of parameterization schemes for this physical process.

There are three parameterization schemes in the BTR physical process, the default is the Noah type scheme which uses soil moisture, the second is CLM type which uses matric potential and the third is the SSIB type which also uses matric potential. The three schemes result in significant differences in the calculation of the soil moisture controlling stomatal resistance, $\beta$ factor [11]. At the Arou site, the selection frequencies of the three parameterization schemes for the BTR physical process in the "best members" and "worst members" groups are nearly identical. Changing the parameterization scheme for this process does not affect the accuracy of soil moisture simulations, indicating that soil moisture simulation results at the Arou site are not sensitive to the BTR physical process. However, at the Heihe site, we found that the selection frequencies of schemes two and three for the BTR process in the "best members" group are identical and higher than that of scheme one. In the corresponding "worst members" group, the selection frequencies of schemes two and three are zero, while that of scheme one is 1.0. This suggests that at the

Heihe site, changing the parameterization scheme for the BTR process leads to significant differences in soil moisture simulation results, with scheme one degrading the accuracy, indicating sensitivity to the parameterization scheme at this site. Although scheme one is not favored at the Heihe site, Figure 3 shows that in the "worst members" group at the Sidaoqiao site, the selection frequencies of the three schemes are nearly identical. However, in the "best members" group, the selection frequency of scheme three is zero, scheme one is selected with a frequency of 0.9, and scheme two with a frequency of 0.1, indicating that scheme one enhances the accuracy of soil moisture simulations at this site.

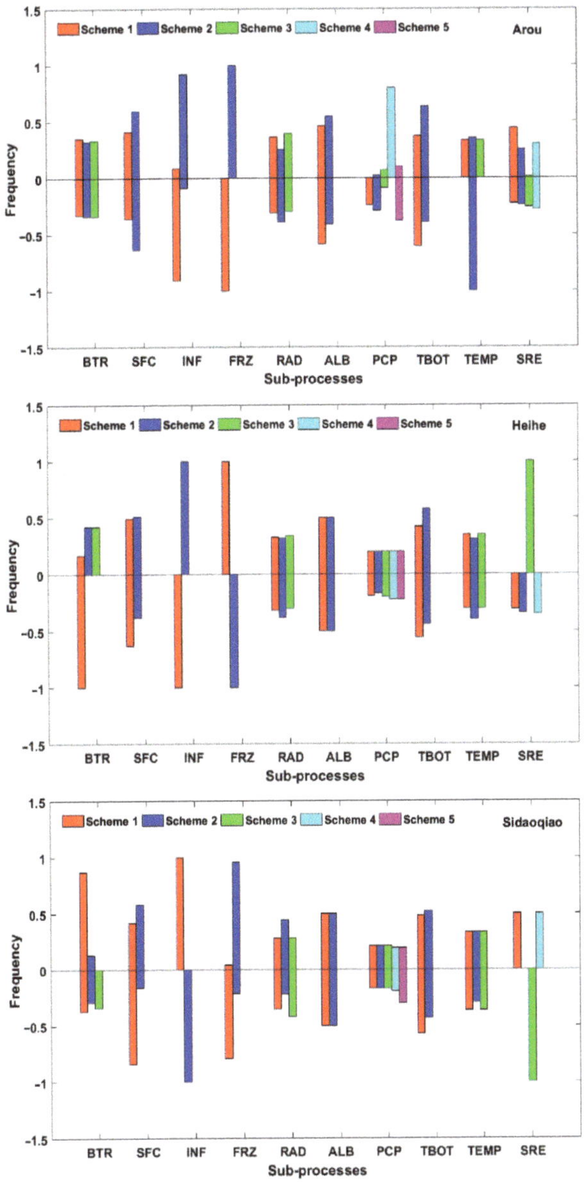


**Figure 3.** The selected frequency of different parameterization schemes for soil moisture of the three sites, by best members (0–1) or worst members (−1–0), in ensemble experiments.

Physical process INF is the frozen soil permeability and two schemes were integrated, the first scheme considers that a model grid cell includes both permeable and impermeable regions, thereby calculating the soil's hydraulic properties based on the total soil moisture content. In contrast, the second option determines hydraulic properties using only the volume of liquid water. The difference between the two approaches results in more permeable permafrost being generated by the first scheme compared to the second. It can be observed that the two parameterization schemes of the INF physical process showed significant differences in selection frequency across all three sites. At the Arou site, in the "best members" group, scheme INF(1) was selected with a frequency of 0.08, while INF(2) was selected with a frequency of 0.92. However, in the "worst members" group, INF(1) was selected 91% of the time, and INF(2) only 9% of the time. This indicates that at the Arou site, scheme INF(2) outperforms INF(1), yielding soil moisture simulation results with smaller errors. A similar pattern was observed at the Heihe site, where INF(2) was chosen 100% of the time in the "best members" group and 0% in the "worst members" group. Conversely, INF(1) was never selected in the "best members" ensemble but was selected 100% of the time in the "worst members" group. Clearly, soil moisture simulation results at both the Arou and Heihe sites are highly sensitive to the INF physical process. For the Sidaoqiao site, although it is evident from Figure 3 that soil moisture simulation results are also highly sensitive to the INF process, the key difference from the previous two sites is that scheme INF(1) was chosen with a frequency of 1.0 in the "best members" group and 0 in the "worst members" group. In contrast, INF(2) was selected with a frequency of 0 in the "best members" group and 1.0 in the "worst members" group. Clearly, for the desert region where the Sidaoqiao site is located, scheme INF(1) is more suitable. However, for the Arou site in the alpine region and the Heihe site in the artificial oasis area, INF(2) is better suited for simulating soil moisture. To be honest, there could be many reasons for this phenomenon, such as differences in climate type, vegetation type, and soil type, etc. However, it is undeniable that in the upper and middle reaches of the HRB, water permeability in permafrost is significantly lower than in the lower reaches area where the Sidaoqiao site is located.

Based on the results presented in Figure 3, it is evident that the soil moisture simulation outcomes at all three sites are highly sensitive to the parameterization schemes of the FRZ physical process. At the Arou site, scheme FRZ(1) was selected with a frequency of 0 in the "best members" group, but with a frequency of 1.0 in the "worst members" group. Conversely, scheme FRZ(2) was chosen with a frequency of 1.0 in the "best members" group and 0 in the "worst members" group. This suggests that at Arou, FRZ(2) outperforms FRZ(1), resulting in more accurate soil moisture simulations. A similar pattern was observed at the Sidaoqiao site, where FRZ(2) was selected with a frequency of 0.96 in the "best members" group, while FRZ(1) was selected with a frequency of 0.79 in the "worst members" group, indicating that FRZ(2) is also more favorable at Sidaoqiao. At the Heihe site, although soil moisture simulations are highly sensitive to the FRZ process, Figure 3 shows that FRZ(1) was selected with a frequency of 1.0 in the "best members" group, whereas FRZ(2) was selected with a frequency of 1.0 in the "worst members" group. This indicates that FRZ(1) is the more suitable option at this site, producing more accurate soil moisture simulation results compared to FRZ(2). Why is there a difference in sensitivity when all three sites are located in the HRB? Further analysis reveals that scheme FRZ(1) uses the more general form of the freezing-point depression equation, while scheme FRZ(2) employs a variant of this equation with an added extra term. This modification in FRZ(2) results in the generation of more liquid water compared to FRZ(1). Arou and Sidaoqiao are both located in field environments, while the Heihe site is situated in farmland within a midstream artificial oasis. During winter, when the soil freezes, the land remains fallow due to low temperatures and the inability to cultivate. Additionally, limited precipitation and the absence of artificial irrigation lead to low liquid water content within the frozen soil. As a result, FRZ(1) is more suitable for simulating soil moisture in this region.

Precipitation is the primary source of soil moisture, as illustrated in Figure 3, the rain-snow partitioning process (PCP) may not have a direct impact on the accuracy of soil moisture simulations, at least at the Heihe and Sidaoqiao sites. This is evidenced by the fact that the five parameterization schemes for the PCP process were selected with similar frequencies in both the "best members" and "worst members" groups at these sites, indicating that variations in the PCP parameterization schemes do not affect the soil moisture simulation accuracy at these locations. In contrast, at the Arou site, scheme PCP(4) was chosen with a frequency of 0.8 in the "best members" group, which is significantly higher than the other four schemes, while PCP(4) was not selected in the "worst members" group. This demonstrates that, at the Arou site, PCP(4) exhibits a clear advantage over the other parameterization schemes. A similar pattern is observed with the physical process TEMP across the three sites. At both Heihe and Sidaoqiao sites, the frequencies of selecting the three TEMP parameterization schemes are similar in both the "best members" and "worst members" groups. This suggests that changing the parameterization schemes for this physical process does not affect the accuracy of soil moisture simulations at these sites. However, at the Arou site, TEMP(2) was selected with a frequency of 1.0 in the "worst members" group, indicating that TEMP(2) tends to produce poorer soil moisture simulation results at this site. The sensitivity testing results of the physical process of surface resistance to ground evaporation/sublimation (SRE) vary significantly across the three sites. At the Arou site, the frequencies of the four different SRE schemes being chosen are quite similar, and SRE(3) was not selected in the "best members" group, making it difficult to determine the sensitivity of soil moisture simulation results to the parameterization scheme of this physical process. At the Heihe site, SRE(3) was selected with a frequency of 1.0 in the "best members" group and 0 in the "worst members" group. At the Sidaoqiao site, SRE(3) was selected with a frequency of 1.0 in the "worst members" group but was not chosen in the "best members" group. The frequency statistics of SRE(3) at these two sites indicate that soil moisture simulation results at these sites are sensitive to the parameterization scheme of this physical process, with SRE(3) improving the simulation accuracy of soil moisture at the Heihe site, whereas at the Sidaoqiao site, SRE(3) decreases the accuracy of soil moisture simulation.

Some physical processes, such as RAD, ALB, and TBOT, show similar frequencies for parameterization schemes in both the "best members" and "worst members" groups. This suggests that soil moisture simulations are not particularly sensitive to the parameterization schemes for these processes. However, discrepancies in frequency statistics are observed for other processes, such as the two parameterization schemes for the SFC process across the three sites. In such cases, it remains uncertain whether the soil moisture simulation results are sensitive to the parameterization schemes for these processes.

#### 3.2.2. Tukey Test Results

In this section, the Tukey's test was employed to examine the difference of parameterization schemes for a certain physical process. First, a total number of 17,280 RMSEs of all scheme combinations for soil moisture were calculated, and all of the RMSE values are independent of each other. Moreover, before applying Tukey's test, the assumptions of normality and equality of variances were examined. Taking the physical process SFC as an example, SFC has two parameterization schemes. For each scheme, 8640 out of 17,280 combination schemes selected the corresponding parameterization, resulting in 8640 RMSE values for each scheme. Figure 4 shows the kernel density distribution of the

RMSE samples for each parameterization scheme across the ten physical processes, along with the results of Tukey's test.

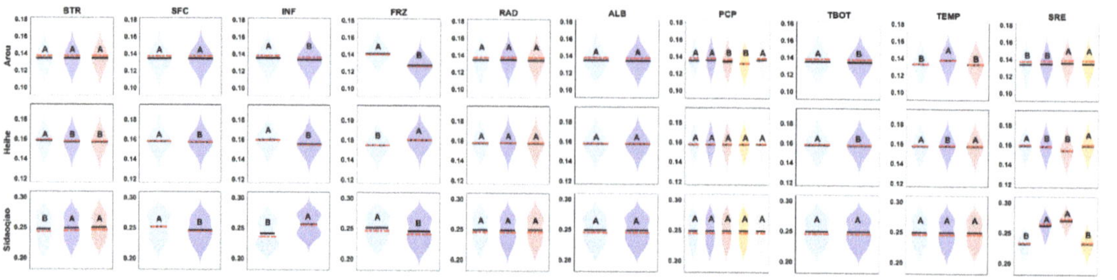


**Figure 4.** Tukey's test for RMSE. For each subfigure, letters A and B represent the categories of corresponding schemes, and schemes of the same physical process that do not share a letter are significantly different; the black line indicates the mean, and the red line indicates the median; the kernel bandwidth was 0.01.

As shown in Figure 4, if different parameterization schemes for the same physical process are marked with the same letter, such as "A", it indicates that Tukey's test results show no significant difference between these schemes. Conversely, if parameterization schemes for the same physical process are marked with different letters, such as "A" and "B", it signifies a significant difference between the schemes, with the scheme marked "B" being slightly better due to its lower corresponding RMSE values. Clearly, if there is a significant difference between parameterization schemes for the same physical process, it indicates that soil moisture simulation results are sensitive to the parameterization scheme of that physical process. If all parameterization schemes for a specific physical process in the figure are marked with the same letter, such as "A", it implies that soil moisture simulation results are not sensitive to the parameterization scheme of that physical process. For example, for the SFC physical process at the Arou site, there is no significant difference between SFC(1) and SFC(2). Meanwhile, we noticed that at the Heihe and Sidaoqiao sites, the two parameterization schemes for SFC were marked with different letters, with the SFC(2) scheme marked with the letter "B". This indicates that the soil moisture simulation results at these two stations are sensitive to the parameterization scheme of this physical process, and the performance of the SFC(2) scheme is slightly better. Further analysis and comparison reveal that at the Arou site, there is no significant difference between the parameterization schemes of the BTR and SFC physical processes, indicating that the soil moisture simulation results at this station are not sensitive to the parameterization schemes of these two physical processes. At the Heihe site, the parameterization schemes for the RAD, ALB, and PCP physical processes do not show significant differences, meaning that changing the parameterization schemes for these three physical processes does not result in substantial changes in the soil moisture simulation results at this station. At the Sidaoqiao site, the parameterization schemes for the physical processes FRZ, RAD, ALB, PCP, TBOT, and TEMP do not show significant differences, indicating that the soil moisture simulation results at this site are not sensitive to the parameterization schemes of these physical processes.

Figure 4 shows that the sensitivity detection results from Tukey's test are largely consistent with those from the Natural Selection method. However, there are differences in the sensitivity of some physical process parameterization schemes. For example, it is difficult to determine the sensitivity of the SFC physical process parameterization scheme using the Natural Selection method, but Tukey's test results indicate that the SFC process is sensitive at both the Heihe and Sidaoqiao sites. There is no contradiction between the results of the two sensitivity analysis methods. The Natural Selection method identifies sensitivity from a macro statistical perspective, based on the differences in the frequency with which different parameterization schemes are selected in the best and worst members groups. Tukey's test, on the other hand, takes a more micro-level approach, using hypothesis testing to detect subtle differences between different parameterization schemes within the same physical process, thereby determining sensitivity. Therefore, in this study, if either method detects sensitivity in a physical process, it is concluded that soil moisture simulation results are sensitive to that parameterization scheme of that physical process. We integrated the results of the two sensitivity analysis methods, and Figure 5 illustrates the sensitivity of major physical processes to soil moisture.

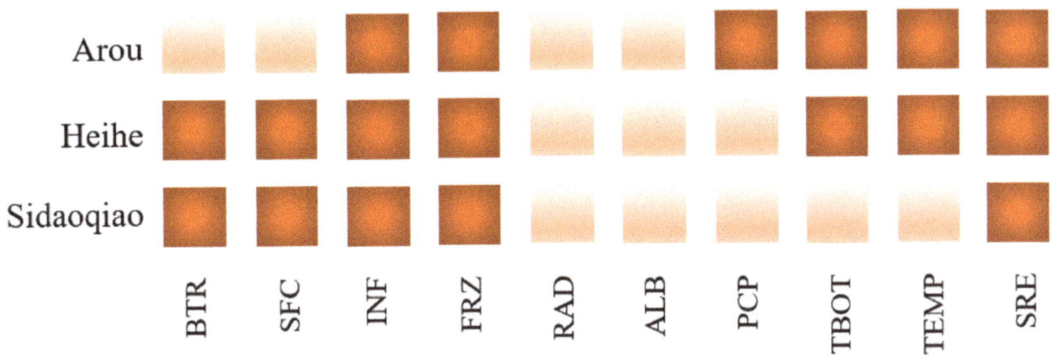


**Figure 5.** Sensitivity diagram of soil moisture. Light color indicates that soil moisture is insensitive to the physical process; dark color indicates that soil moisture is sensitive to the physical process.

### *3.3. Uncertainty Contribution Analysis of Physical Options*

Our previous research results indicated that the uncertainty in ensemble simulation results with multiple parameterization schemes primarily stems from sensitive physical processes, and the greater the sensitivity of the simulation results to a specific physical process, the higher the uncertainty in the ensemble simulation results [28,35]. Although this conclusion was derived from snow depth simulation results, it is evidently applicable to soil moisture simulations as well. In this study, we use 'Uncertainty Contribution Analysis' to assess the contribution of each physical process to the uncertainty in soil moisture ensemble simulation results, as shown in Figure 6. Clearly, the greater a physical process's contribution to the uncertainty in ensemble simulation results, the more varied its parameterization scheme will be, leading to significant differences in simulated soil moisture, thereby generating greater uncertainty. As shown in Figure 6, at the Arou site, most physical processes contribute less than 1% to the overall uncertainty. However, certain processes contribute more significantly, such as INF (14.03%), FRZ (48.70%), PCP (6.31%), and TEMP (17.86%). At the Heihe site, INF contributes 30.86%, FRZ contributes 52.96%, TEMP contributes 4.56%, and SRE contributes 6.05%. At the Sidaoqiao site, SFC contributes 4.46%, INF contributes 31.07%, FRZ contributes 4.80%, and SRE contributes 50.92%. Contributions from other physical processes at these sites are all below 1%.

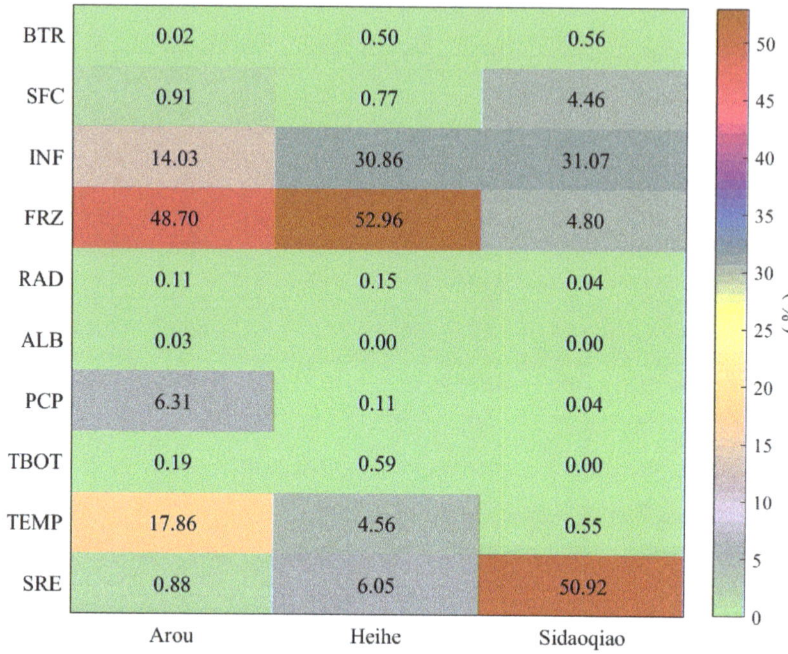


**Figure 6.** Uncertainty contribution diagram of physical options at three sites, the results are all expressed in percentage format.

## 4. Discussion

Across all three sites, the Noah-MP model, with its default parameterization, consistently overestimates soil moisture at Heihe site which is located in an oasis region, particularly during the wet season. However, the soil moisture simulations show significant underestimation at Arou and Sidaoqiao site. Among the three sites, the default parameterization scheme combination of the Noah-MP model is generally able to accurately simulate the soil moisture trends in the first and second layers, although overestimation or underestimation occurs to varying degrees at each site. However, for the third and fourth layers, the Noah-MP model is generally unable to simulate the trends in soil moisture. The overestimation of soil moisture by the Noah-MP model may be attributed to several factors. One potential reason is the model's representation of soil hydraulic properties, which may not be adequately parameterized for the specific conditions in the HRB. The model's default parameterization scheme combination may not account for the unique soil texture and structure of the region, leading to an overestimation of water retention capacity, especially in the deeper layers. Additionally, the model's inability to capture the drying trends in the deeper layers suggests that the subsurface drainage processes might be inadequately represented, possibly due to oversimplified assumptions in the soil moisture parameterization schemes. The variations in model performance across the different sites and layers also highlight the spatial heterogeneity of the HRB. The differences in soil texture, vegetation cover, and topography between the sites likely contribute to the varying levels of model accuracy. The Arou site, for example, with its more complex terrain and vegetation, poses a greater challenge for the model, especially the freeze-thaw process of the soil, resulting in larger discrepancies. In contrast, the Heihe site, with its more uniform conditions, shows somewhat better model performance, although significant issues remain. In conclusion, while the Noah-MP model, with its default parameterization, can capture the general seasonal trends of soil moisture across the HRB, significant biases and inaccuracies remain, particularly in the deeper soil layers. The consistent underestimation and overestimation of soil moisture suggest that the default parameterization scheme combination may not be adequately capturing the subsurface hydrological processes in the region. Future work should focus on optimizing the model's parameterization for the specific conditions of the HRB, potentially through site-specific calibration or the integration of more detailed soil and vegetation data. Additionally, the implementation of advanced data assimilation techniques could help reduce the model's biases and improve the accuracy of soil moisture simulations, particularly in the deeper layers.

From the Figure 5, it can be observed that the INF, FRZ, and SRE processes exhibit sensitivity at all three sites, with the parameterization schemes for INF and FRZ being particularly sensitive. The parameterization schemes for the BTR, SFC, TBOT, and TEMP processes show sensitivity at two of the sites, while the PCP process exhibits sensitivity only at the Arou site. Furthermore, the parameterization schemes for the RAD and ALB processes did not exhibit sensitivity at any of the three sites. The RAD process is primarily designed for areas with tall shrub vegetation, but the land surface types at the three sites in this study are grassland, crop, and desert, respectively. Therefore, altering the parameterization scheme for RAD does not lead to changes in the simulation results. The parameterization scheme for the ALB process is mainly used to calculate the albedo of the snow surface and significantly impacts snow depth simulations [28]; however, in this study, it clearly did not affect the soil moisture simulation results. For the PCP process, the parameterization scheme PCP(4) calculates the proportion of solid precipitation using the total amount of snow, graupel, and hail from the forcing data. The Arou site is located in the upstream alpine region, where the average temperature is lower, and the average precipitation is higher compared to the midstream and downstream sites. This gives the PCP(4) scheme a distinct advantage at the Arou site. Our findings regarding differences in Noah-MP parameterization sensitivity align with earlier work in similar contexts. For example, Hu et al. (2023) [36] demonstrated spatial and seasonal variations of sensitivity across the Tibetan Plateau, notably in parameterizations such as SFC and ALB.

An analysis combining Figures 5 and 6 shows that the contributions of sensitive physical processes to the uncertainty in ensemble simulation results at all three sites are non-negligible, indicating that these processes are primarily responsible for the overall uncertainty. However, does this mean that all sensitive physical processes contribute significantly to the uncertainty in the simulation results? The answer is no. For example, at the Arou site, although TBOT and SRE both exhibit sensitivity, their contributions to the uncertainty are below 1%. Similarly, at the Heihe site, the BTR, SFC, and TBOT processes, as well as the BTR process at the Sidaoqiao site, show sensitivity but contribute less than 1% to the overall uncertainty. The results demonstrate that the contributions of the RAD and ALB physical processes to the uncertainty in ensemble simulation results are minimal at all three stations. In particular, the ALB physical process contributes virtually nothing to the uncertainty at any of the stations. Although the FRZ physical process shows significant sensitivity across all three stations, its contribution to the uncertainty in ensemble soil moisture simulation results varies considerably. It is relatively high at the upstream Arou site and the midstream Heihe site, but lower at the downstream Sidaoqiao site. At Sidaoqiao, the SRE physical process contributes 50.92% to the uncertainty in ensemble soil moisture simulation results, accounting for half of the total uncertainty.

## 5. Conclusions

In this study, a representative research site from the upstream, midstream, and downstream regions of the HRB were chosen to evaluate the performance of the Noah-MP model in simulating soil moisture, respectively. For the first soil layer, we conducted ensemble simulations using multiple parameterization schemes and employed two sensitivity analysis methods to examine the sensitivity of soil moisture to parameterization schemes from both macro and micro perspectives. Building on this, the contribution of specific physical processes to the uncertainty in ensemble soil moisture simulation results was further quantified and the impact of sensitive physical processes on this uncertainty was assessed. The main findings are as follows:

The Noah-MP model is capable of simulating soil moisture variations in the first and second layers at the three sites located in the upstream, midstream, and downstream areas of the HRB. However, the simulations exhibit varying degrees of overestimation and underestimation, indicating that there is room for improvement in the model's accuracy. In contrast, the model fails to capture the variations in soil moisture in the third and fourth layers at these sites, with significant discrepancies compared to the observed data.

Sensitivity analysis reveals that the physical processes of infiltration (INF), freeze-thaw cycles (FRZ), and surface resistance to evaporation (SRE) demonstrate sensitivity across all three sites. However, processes such as bare soil evaporation (BTR) and surface exchange coefficients (SFC) show sensitivity only at the midstream and downstream sites. The bottom temperature (TBOT) and canopy air temperature (TEMP) processes do not exhibit sensitivity at the downstream site. Precipitation (PCP) sensitivity is observed only in the upstream alpine region, while radiation (RAD) and albedo (ALB) processes do not show sensitivity in this study's soil moisture simulation experiments.

The contribution of specific physical processes to the uncertainty in ensemble simulation results for soil moisture highlights that this uncertainty mainly stems from the sensitive processes. Variations in parameterization schemes within these sensitive processes are the primary drivers of uncertainty in the ensemble simulations. The FRZ process, for instance, accounts for nearly 50% of the uncertainty in soil moisture simulations at both the upstream Arou site and the midstream Heihe site, making it the dominant factor for uncertainty at these locations. At the downstream Sidaoqiao site, the SRE process contributes 50.92% of the uncertainty, making it the primary source of uncertainty in soil moisture simulations at that site.

In this study, we only considered grassland, farmland, and desert, and did not include shrub vegetation. This may limit the representativeness of our soil moisture simulations. Future work should consider more land cover types to better capture soil moisture variability. In forthcoming research, we plan to incorporate data assimilation techniques to integrate high-precision soil moisture observations within the model framework, aiming to further enhance the accuracy of soil moisture simulations. We hope that the findings of this study can serve as a reference for selecting parameterization schemes for soil moisture simulation, provide insights into the applicability of parameterization schemes in specific regions, and offer a scientific basis for developing more accurate land surface process parameterization schemes.

**Author Contributions:** Conceptualization, Y.Y.; methodology, Y.Y. and H.Z.; software, Y.Y. and Y.W.; validation, Y.H.; data curation, Z.W. and W.C.; writing—original draft preparation, Y.Y.; writing—review and editing, Y.W., Z.W. and W.C.; supervision, Y.L. and Y.W. All authors have read and agreed to the published version of the manuscript.

**Funding:** This study was supported by several sources, including the National Natural Science Foundation of China (U2342206, 42201425, 42101361), Jianghuai Meteorological Joint Project of Anhui Natural Science Foundation (2408055UQ006), the Scientific Research Project of Higher Education Institutions in Anhui province (2023AH050143), and the China Postdoctoral Science Foundation (2024M753092).

**Data Availability Statement:** The original data presented in the study are openly available on the website of the National Tibetan Plateau Data Center (TPDC) (https://data.tpdc.ac.cn).

**Conflicts of Interest:** The authors declare no conflicts of interest.

## References

1. Sepulcre-Canto, G.; Horion, S.; Singleton, A.; Carrao, H.; Vogt, J. Development of a Combined Drought Indicator to detect agricultural drought in Europe. *Nat. Hazards Earth Syst. Sci.* **2012**, *12*, 3519–3531. [CrossRef]
2. Kumar, S.V.; Peters-Lidard, C.D.; Santanello, J.A.; Reichle, R.H.; Draper, C.S.; Koster, R.D.; Nearing, G.; Jasinski, M.F. Evaluating the utility of satellite soil moisture retrievals over irrigated areas and the ability of land data assimilation methods to correct for unmodeled processes. *Hydrol. Earth Syst. Sci.* **2015**, *19*, 4463–4478. [CrossRef]
3. Massari, C.; Brocca, L.; Tarpanelli, A.; Moramarco, T. Data assimilation of satellite soil moisture into rainfall-runoff modelling: A complex recipe? *Remote Sens.* **2015**, *7*, 11403–11433.
4. Azimi, S.; Dariane, A.B.; Modanesi, S.; Bauer-Marschallinger, B.; Bindlish, R.; Wagner, W.; Massari, C. Assimilation of sentinel 1 and SMAP-based satellite soil moisture retrievals into SWAT hydrological model: The impact of satellite revisit time and product spatial resolution on flood simulations in small basins. *J. Hydrol.* **2020**, *581*, 124367.
5. Vereecken, H.; Huisman, J.A.; Bogena, H.; Vanderborght, J.; Vrugt, J.A.; Hopmans, J.W. On the value of soil moisture measurements in vadose zone hydrology: A review. *Water Resour. Res.* **2008**, *44*, W00D06. [CrossRef]
6. Heathman, G.C.; Cosh, M.H.; Han, E.; Jackson, T.J.; McKee, L.; McAfee, S. Field scale spatiotemporal analysis of surface soil moisture for evaluating point-scale in situ networks. *Geoderma* **2012**, *170*, 195–205. [CrossRef]
7. Seneviratne, S.I.; Corti, T.; Davin, E.L.; Hirschi, M.; Jaeger, E.B.; Lehner, I.; Orlowsky, B.; Teuling, A.J. Investigating soil moisture-climate interactions in a changing climate: A review. *Earth-Sci. Rev.* **2010**, *99*, 125–161.
8. Chen, F.; Dudhia, J. Coupling an advanced land surface hydrology model with the Penn State-NCAR MM5 modeling system. Part I: Model implementation and sensitivity. *Mon. Weather. Rev.* **2001**, *129*, 569–585.
9. Ek, M.B.; Mitchell, K.E.; Lin, Y.; Rogers, E.; Grunmann, P.; Koren, V.; Gayno, G.; Tarpley, J.D. Implementation of Noah land surface model advances in the National Centers for Environmental Prediction operational mesoscale Eta model. *J. Geophys. Res.-Atmos.* **2003**, *108*, D22.
10. Niu, G.Y.; Yang, Z.L.; Mitchell, K.E.; Chen, F.; Ek, M.B.; Barlage, M.; Kumar, A.; Manning, K.; Niyogi, D.; Rosero, E.; et al. The community Noah land surface model with 9 multi-parameterization options (Noah-MP): 1. Model description and evaluation with local-scale measurements. *J. Geophys. Res.-Atmos.* **2011**, *116*, D12109.
11. Yang, Z.L.; Niu, G.Y.; Mitchell, K.E.; Chen, F.; Ek, M.B.; Barlage, M.; Longuevergne, L.; Manning, K.; Niyogi, D.; Tewari, M.; et al. The community Noah land surface model with multiparameterization options (Noah-MP): 2. Evaluation over global river basins. *J. Geophys. Res.-Atmos.* **2011**, *116*, D12110. [CrossRef]
12. Gao, Y.H.; Li, K.; Chen, F.; Jiang, Y.; Lu, C. Assessing and improving Noah-MP land model simulations for the central Tibetan Plateau. *J. Geophys. Res.-Atmos.* **2015**, *120*, 9258–9278. [CrossRef]
13. Zhang, G.; Chen, F.; Gan, Y.J. Assessing uncertainties in the Noah-MP ensemble simulations of a cropland site during the Tibet Joint International Cooperation program field campaign. *J. Geophys. Res.-Atmos.* **2016**, *121*, 9576–9596. [CrossRef]
14. Cai, X.T.; Yang, Z.L.; David, C.H.; Niu, G.-Y.; Rodell, M. Hydrological evaluation of the Noah-MP land surface model for the Mississippi River Basin. *J. Geophys. Res.-Atmos.* **2014**, *119*, 23–38. [CrossRef]
15. Ma, N.; Niu, G.Y.; Xia, Y.L.; Cai, X.; Zhang, Y.; Ma, Y.; Fang, Y. A systematic evaluation of Noah-MP in simulating land-atmosphere energy, water, and carbon exchanges over the continental United States. *J. Geophys. Res.-Atmos.* **2017**, *122*, 12245–12268. [CrossRef]
16. Kuribayashi, M.; Noh, N.J.; Saitoh, T.M.; Tamagawa, I.; Wakazuki, Y.; Muraoka, H. Comparison of Snow Water Equivalent Estimated in Central Japan by High-Resolution Simulations Using Different Land-Surface Models. *Meteorol. Soc. JPN.* **2013**, *9*, 148–152. [CrossRef]
17. Fang, Y.H.; Zhang, X.N.; Niu, G.Y.; Zeng, W.; Zhu, J.; Zhang, T. Study of the Spatiotemporal Characteristics of meltwater contribution to the total runoff in the upper Changjiang river basin. *Water* **2017**, *9*, 165. [CrossRef]
18. Liang, J.J.; Yang, Z.L.; Lin, P.R. Systematic hydrological evaluation of the Noah-MP land surface model over China. *Adv. Atmos. Sci.* **2019**, *36*, 1171–1187. [CrossRef]
19. Meng, C.L.; Jin, H.D.; Zhang, W.L. Lateral terrestrial water flow schemes for the Noah-MP land surface model on both natural and urban land surfaces. *J. Hydrol.* **2023**, *620*, 129410. [CrossRef]
20. Yang, K.; Yang, J.T.; Zhaoye, P.H.; Zhang, F.; Zhang, G.; Wang, C. Optimization and validation of soil Frozen-Thawing parameterizations in Noah-MP. *J. Geophys. Res.-Atmos.* **2023**, *128*, e2022JD038217. [CrossRef]
21. Hassani, F.; Zhang, Y.; Kumar, S.V. Improved representation of vegetation soil moisture coupling enhances soil moisture data assimilation in water limited regimes: A Case Study over Texas. *Water Resour. Res.* **2024**, *60*, e2023WR035558. [CrossRef]

22. Ahmad, J.A.; Forman, B.A.; Kumar, S.V. Soil moisture estimation in South Asia via assimilation of SMAP retrievals. *Hydrol. Earth Syst. Sci.* **2022**, *26*, 2221–2243. [CrossRef]
23. Zhao, L.; Yang, K.; He, J.; Zheng, H.; Zheng, D. Potential of mapping global soil texture type from SMAP soil moisture product: A pilot study. *IEEE Trans. Geosci. Remote Sens.* **2022**, *60*, 4406410. [CrossRef]
24. Bechtold, M.; Modanesi, S.; Lievens, H.; Baguis, P.; Brangers, I.; Carrassi, A.; Getirana, A.; Gruber, A.; Heyvaert, Z.; Massari, C.; et al. Assimilation of Sentinel-1 Backscatter into a Land Surface Model with River Routing and Its Impact on Streamflow Simulations in Two Belgian Catchments. *J. Hydrometeorol.* **2023**, *24*, 2389–2408. [CrossRef]
25. Gavahi, K.; Abbaszadeh, P.; Moradkhani, H. How does precipitation data influence the land surface data assimilation for drought monitoring? *Sci. Total Environ.* **2022**, *831*, 154916. [CrossRef]
26. Kumar, S.V.; Holmes, T.R.; Bindlish, R.; de Jeu, R.; Peters-Lidard, C. Assimilation of vegetation optical depth retrievals from passive microwave radiometry. *Hydrol. Earth Syst. Sci.* **2020**, *24*, 3431–3450. [CrossRef]
27. You, Y.H.; Huang, C.L.; Hou, J.L.; Zhang, Y.; Wang, Z.; Zhu, G. Improving the estimation of snow depth in the Noah-MP model by combining particle filter and Bayesian model averaging. *J. Hydrol.* **2023**, *617*, 128877. [CrossRef]
28. You, Y.H.; Huang, C.L.; Yang, Z.L.; Zhang, Y.; Bai, Y.; Gu, J. Assessing Noah-MP parameterization sensitivity and uncertainty interval across snow climates. *J. Geophys. Res.-Atmos.* **2020**, *125*, e2019JD030417. [CrossRef]
29. Chang, M.; Liao, W.H.; Wang, X.M.; Zhang, Q.; Chen, W.; Wu, Z.; Hu, Z. An optimal ensemble of the Noah-MP land surface model for simulating surface heat fluxes over a typical subtropical forest in South China. *Agric. For. Meteorol.* **2020**, *281*, 107815. [CrossRef]
30. Li, J.; Miao, C.Y.; Zhang, G.; Fang, Y.-H.; Shangguan, W.; Niu, G.-Y. Global Evaluation of the Noah-MP Land Surface Model andSuggestions for Selecting Parameterization Schemes. *J. Geophys. Res.-Atmos.* **2022**, *127*, e2021JD035753. [CrossRef]
31. Pan, X.D.; Li, X.; Shi, X.K.; Han, X.; Luo, L.; Wang, L. Dynamic downscaling of near-surface air temperature at the basin scale using WRF-a case study in the Heihe River Basin, China. *Front. Earth Sci.* **2012**, *6*, 314–323. [CrossRef]
32. Li, X.; Li, X.W.; Li, Z.Y.; Ma, M.; Wang, J.; Xiao, Q.; Liu, Q.; Che, T.; Chen, E.; Yan, G.; et al. Watershed Allied Telemetry Experimental Research. *J. Geophys. Res.-Atmos.* **2009**, *114*, D22103. [CrossRef]
33. Niu, G.Y.; Yang, Z.L.; Dickinson, R.E.; Gulden, L.E.; Su, H. Development of a simple groundwater model for use in climate models and evaluation with Gravity Recovery and Climate Experiment data. *J. Geophys. Res.-Atmos.* **2007**, *112*, D7. [CrossRef]
34. Dickinson, R.E.; Shaikh, M.; Bryant, R.; Graumlich, L. Interactive canopies for a climate model. *J. Clim.* **1998**, *11*, 2823–2836. [CrossRef]
35. You, Y.; Huang, C.; Gu, J.; Li, H.; Hao, X.; Hou, J. Assessing snow simulation performance of typical combination schemes within Noah-MP in northern Xinjiang, China. *J. Hydrol.* **2020**, *581*, 124380. [CrossRef]
36. Hu, W.; Ma, W.Q.; Yang, Z.L.; Ma, Y.; Xie, Z. Sensitivity analysis of the Noah-MP land surface model for soil hydrothermal simulations over the Tibetan Plateau. *J. Adv. Model. Earth Syst.* **2023**, *15*, e2022MS003136. [CrossRef]

MDPI AG
Grosspeteranlage 5
4052 Basel
Switzerland
Tel.: +41 61 683 77 34

MDPI Books Editorial Office
E-mail: books@mdpi.com
www.mdpi.com/books

www.ingramcontent.com/pod-product-compliance
Lightning Source LLC
Chambersburg PA
CBHW061814210926
51850CB00006B/704